DECIDUOUS FORESTS
OF
EASTERN NORTH AMERICA

DECIDUOUS FORESTS OF EASTERN NORTH AMERICA

By E. Lucy Braun, Ph.D.

(Facsimile of the Edition of 1950)

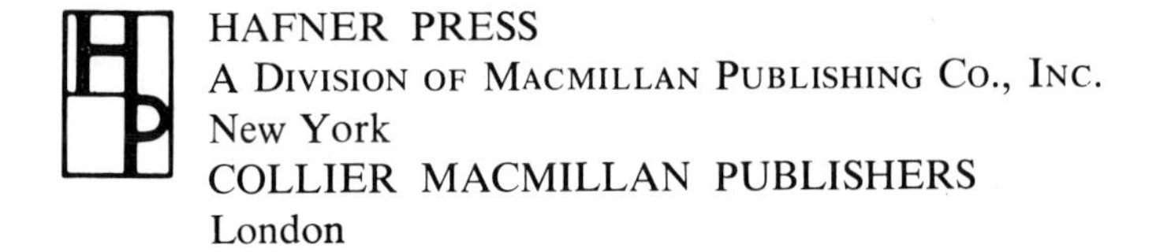
HAFNER PRESS
A DIVISION OF MACMILLAN PUBLISHING CO., INC.
New York
COLLIER MACMILLAN PUBLISHERS
London

Reprinted by arrangement 1974

HAFNER PRESS
A Division of Macmillan Publishing Co., Inc.
866 Third Avenue, New York, N.Y. 10022
Collier-Macmillan Canada Ltd.

Library of Congress Catalog Card Number: 64-20220
ISBN: 02-841910-3

Printed in U.S.A. by
NOBLE OFFSET PRINTERS, INC.
New York, N.Y. 10003

To
My sister Annette
Who has shared
The joys and the hardships
In the field

Preface

This book was conceived as a result of many years of study throughout the deciduous forest. Its purpose is to give a broad and coordinated account of the entire deciduous forest—a comprehensive view which may make possible the better understanding of fragments of forest and give a setting for past and future local and detailed studies. Toward this end, I have attempted, *first,* to portray what is (or was) present in any geographic area and to reconstruct the pattern of original forest insofar as the fragments remaining permit; *second,* to give data on composition and aspect of forest communities in all parts of the deciduous forest; and *third,* to trace through geologic time the development of the present pattern of forest distribution.

While primarily a reference book, this volume will serve as a textbook for advanced students. It is hoped that it will be of interest to the general reader as well as to the forester and ecologist. It is suggested that in the study of plant and animal distribution, ranges of species be given in terms of forest regions, rather than by states. Some species may be characteristic of or reach their maximum abundance in certain regions, and be absent or rare in others.

As the years go by, it becomes increasingly difficult to form any concept of the original forest cover. The virgin forests have been cut, the land is either cleared and farmed or is clothed with second-growth forest which may in no way suggest the original forest. In many sections no single tract of virgin forest remains today.

The book is based on some 25 years of field study throughout the deciduous forest involving in the last 12 or 15 years over 65,000 miles of travel, on many years of familiarity with the deciduous forest as a whole, and on intimate association with and study of parts of this forest. The limitations of personal knowledge of so vast an area are obvious. The literature of forestry and plant ecology in all the regions has been freely drawn upon to make the picture more complete. However, I have made my own interpretations, and have maintained a comparative viewpoint which is not always possible when studies are confined to any one region.

I am fully aware that all regions are not treated with equal detail. This is in part due to differences in the amount of time I have spent in the several regions; it is in part due to the very unequal amounts of published information available concerning the different regions, to the almost complete lack of material in some states; it is in part due to the poor representation of the original forest cover in certain areas, which makes reconstruction of the picture of forest cover almost impossible. Any weakness or paucity of material on any region should be a challenge to other workers.

In the course of the field work, multitudinous problems presented themselves—problems which await intensive research. These are the province of the local investigator, who can make detailed studies of the factors involved.

When used in accord with the concept of the association-abstract, the names of the several major associations are capitalized. So, also, are the names of formations and of forest regions.

During the progress of the field studies upon which much of this book is based, grants-in-aid were received from the American Association for the Advancement of Science (1932), the National Research Council (1934–35), and the American Association for the Advancement of Science through the Ohio Academy of Science (1938). A grant for typing of manuscript was received from the Research Fund of the Ohio Academy of Science (1947). In 1943 and 1944, I was the recipient of John Simon Guggenheim Fellowships. Without this opportunity for uninterrupted investigation, the work could not have been consummated, and I wish here to express my grateful appreciation.

To the many individuals and organizations whose hospitality, helpful suggestions or encouragement stimulated the work, I extend sincere thanks. Of these but a few can be mentioned. Through the cooperation of the Stearns Coal and Lumber Company, it was possible to visit isolated virgin tracts in a section where little virgin forest remained. To Dr. W. H. Camp, who read all of Part III and contributed critical notes, to Dr. Ralph W. Chaney, who read the chapter on the paleobotanic record, to Dr. Margaret Fulford, for critical reading of much of the manuscript, and to Mrs. Anneliese S. Caster for suggestions in regard to the making of the maps, grateful acknowledgment is made. Finally, thanks are due my sister, Dr. Annette F. Braun, for her aid in the final revision of the manuscript, in the construction of the maps, and in the reading of proof.

Illustrations without credit lines were made by the author and are here published for the first time; the source of all other illustrations is acknowledged in the credit lines of figure legends. However, special acknowledgment should be made to the United States Forest Service for the many forest photographs used in this work. In the preparation of the large map of forest regions and sections, I have derived data from many sources; the general method of delineation of topographic features is that employed by A. K. Lobeck and Erwin Raisz.

E. LUCY BRAUN

Contents

List of Maps for Reference

Tables Illustrating Forest Composition

Distributed in the text, are Tables giving percentage composition of the canopy layer of representative forest tracts and, in some instances, of second layer also. In a few of the Tables, species present in lower layers are noted by x; X and XX are used for subdominant species. Unless otherwise stated, the forest stands for which data are given are old-growth primary forests. Virgin forests or forests in as nearly undisturbed condition as possible have been used. The Tables, except where noted, are based upon the author's data.

Some of the communities for which composition is given are readily referable to "forest cover types," as defined by the Society of American Foresters. However, an attempt to classify all communities as to "cover type" would be artificial and, in fact, impossible for a large proportion of mixed mesophytic communities.

The Tables are distributed in the text as follows:

Part I

INTRODUCTION

CHAPTER 1

The Forest

Let us envision the forest as an aggregate of plant and animal life: an aggregate of trees of the canopy and lower layers, shrubs and vines, both large and small, herbaceous plants of the forest floor; mosses and lichens on the ground, on rocks, and on tree trunks, fungi, both parasitic and saprophytic, and also of the animals—grazing, burrowing, leaf-eating, seed-eating —the mammals, the birds, the insects, and the micro-organisms (plant and animal) above and below ground; an aggregate of interdependent and interacting organisms. From among this aggregate of multitudinous forms, the trees stand out pre-eminent, because of their superior size and because of their pronounced influence upon all of the forest inhabitants. Yet each member of the forest community plays its part, in competition, in contribution to organic detritus, in giving to the forest its many aspects, both local and seasonal.

SEASONAL RHYTHM

Deciduous forests, or "summer green" forests, are familiar in one or all of their seasonal aspects to anyone who has traversed or lived within any one of the great areas of the world where this vegetation type prevails. In eastern North America, eastern Asia, the Caucasus region, middle Europe, and small areas of southern South America, there are forests in which the seasonal cycle is such that either appellation, "deciduous" or "summer green," is fitting. The northern hemisphere deciduous forests possess a certain unity in physiognomy and seasonal rhythm which is correlated with similarity in generic composition. The similarity of European and eastern American forests—or rather of their constituent species—is emphasized by the popular names applied to American trees (and lesser plants) by the earlier settlers from Europe. On the new continent they found the American counterpart of the European beech, of ash, of maple, of oak—indeed, of many trees. They found, as well, many unfamiliar things, unknown in the forests of Europe. Outstanding among these are such common American trees as the hickories, tuliptree, and magnolias. They found, as seasonal rhythm was observed, other differences between the forests of Europe and those of America.

The annual march of the seasons, with their changing temperature, light, and moisture conditions, is accentuated by the changing aspects of our deciduous forest.

A cycle has no end and no beginning. The cold of winter induces dormancy in many of our plants; life processes are slowed down; the deciduous forest "sleeps." The trees are leafless. The branch and twig arrangement and form and the bark contour and color dominate the forest aspect. Here and there are trees on which a few brown and lifeless leaves cling, but mostly the leaves of summer are now transferred to the forest floor, where the leaf litter protects from sudden temperature changes the root environment of all the forest plants and the living quarters of thousands of unseen forest inhabitants. Days are short. The sun is low in the southern skies and shadows are long. Sun reaches much of the forest floor, except on northerly slopes, accentuating the rapidity of diurnal temperature changes, of thawing after the cold of night. Snow lies on north slopes long after it has melted away from warmer slopes. The leaves of even the hardiest herbaceous plants suffer from too rapid thawing in sun after a freezing night; hence the winter woods displays, in its ground-layer, some differing slope aspects. Herbaceous plants which retain their leaves in winter, as *Hepatica, Hydrophyllum, Tiarella,* are often localized on northerly slopes where diurnal temperature changes are slow. The green leaves of these plants and the semi-evergreen leaves of such plants as *Smilax hispida* stand out conspicuously in the brown and gray background of the winter woods.

The length of day increases; temperatures moderate; new growth starts. From beneath the protecting carpet of leaves on the forest floor, the first green shoots of the pre-vernal flora appear. From buds on subterranean parts, buds well formed during the previous warm season, growth is rapid. Earliest spring may find the ground carpeted with flowers while yet there is no sign of green on the trees. But the buds are swelling; flowers appear on elm, on poplar, on soft maple, changing the color of the forest canopy to richer though subdued tones of brown and red or dotting it with spots of crimson where red maple is a forest inhabitant. Buckeye leaves unfold; soft greens appear on many trees, or delicate reds where young leaves (as of white oak) are strongly colored.

The vernal aspect prevails. Renewed growth is everywhere apparent. The trees are leafing; the forest floor is lightly shaded. The riot of brightly colored spring flowers, the vernal flora of our deciduous woods, appears. But not equally in all types. Many of the amentiferous trees are flowering; oak pollen dusts the dry leaves of the forest floor. In rapid sequence, the growth of buds and the flowering of different trees take place. The soft multicolor tones of early spring give way to various shades of green of later spring, with here and there the blossoms of some of the showy and conspicuous of the later blooming trees—the tuliptree, magnolias, chestnut, and sourwood. The woods are shady now; the sun-loving species of earliest spring have almost completed their growth period and are ripening their seeds. Bright colored spring flowers are still present, for leaf growth is not yet complete and its shade not so dense as later.

Soon the uniform green of summer prevails, the "summer green" which has given this forest one of its names. In all the denser forests, light is insufficient for most showy flowers; the delicate leaves of ferns, which did not begin to uncurl until shade was developing, have reached maturity. Green is the prevailing color, in canopy and in undergrowth. Buds are forming, on tree, on shrub, on herb—buds which will rest during the long dormant season, but which will be ready for the rapid growth of the next season. Flowering and, in many species, fruiting are completed; buds are full grown; life processes slow down. The absciss-layer is slowly forming, cutting down the water and nutrient supply to the leaves. Chemical changes set in; the green of summer changes and yellows and reds begin to appear.

The autumnal aspect of the American deciduous forest is a glorious sight, not shared by her European counterpart, although it is by the forest of eastern Asia. The innumerable brilliantly colored species of the American deciduous forest give to that forest one of its outstanding characteristics.

Leaf fall is in progress. Some species early, some later, take on the leafless aspect of winter. The cycle of the seasons is completed.

INTERRELATIONS

Deciduous forest originally covered much of the land surface of the United States east of the 98th meridian and of adjacent Canada in the lower Great Lakes region and upper St. Lawrence Valley. The western margin is strongly indented by a great prairie lobe, which in the latitude of Illinois and Indiana extends far eastward in what is now generally known as the Prairie Peninsula.

This vast area, although sufficiently similar throughout to permit development of one great vegetation type, is nevertheless far from uniform in climate, in soil, in topography, in stage of development of the erosion cycle, and, particularly, in age of land surface. All of these are interrelated and, while it is generally conceded that climate controls climaxes in vegetation, it does not stand alone in determining vegetational boundaries within the deciduous forest formation. Climatic differences do exist between the North and the South, the East and the West, and the major vegetational differences between widely separated parts are due largely to climate; however, other factors are influential.

Soils undergo development during which they are affected by occupying vegetation and by climate. They may lag behind vegetation in their development and by this lag may then modify the influence of climate, particularly in tension zones.

Topography, although apparently exerting its principal control in seral development, nevertheless, in its larger features, affects not only the ground water relation, but also regional as well as local climate. Its control cannot be divorced from that of climate, for the region of high relief is climatically different from the hypothetical base-leveled region of the same geographic location. Hence, the stage in the development of the erosion cycle affects, and has affected in the past, the distribution of major vegetational units. Erosion cycles of the past and their subsequent partial or extensive peneplains, together with slow diastrophic movements, resulted in configurations of the North American continent far different from that of today. The past cannot be ignored in interpreting the present.

The element of *time* is of tremendous importance. Not all of the land surface now occupied by deciduous forest is of the same age. The land surrounding the Great Lakes has been available for plant occupancy for only about 20,000 years. Some of the seaward margin of the Atlantic Coastal Plain is no older. Areas affected only by earlier glacial advances have been available much longer, the Illinoian drift area for approximately 200,000 years, the Kansan for 600,000 years or more. The Atlantic and Gulf Coastal Plain (except the marginal strip where Pleistocene marine deposits are exposed) is older than any of the glaciated territory, but even it is youthful (in soils, in topography, in time available for vegetational development) as compared with the Appalachian and Ozarkian Uplands and the intervening low plateau on Paleozoic rock. These areas have been continuously available for plant occupation since before the origin of the angiospermous flora which dominates modern vegetation (Chapters 16, 17).

The key to an understanding of the Deciduous Forest Formation as a whole, and the spatial relations of its integral parts and the nature of these parts, is obviously to be sought in the old land areas, where ancient forest types have been able to continue existence in spite of vicissitudes in surrounding areas. The ancient Appalachian Upland, long recognized by taxonomists and students of species distribution as of paramount importance, is no less significant in ecologic thought.

A perusal of ecologic literature on the deciduous forest finds such literature largely confined to the peripheral areas. Except for a number of papers dealing with vegetational features of the Great Smoky Mountains and other adjacent mountain areas, there have been almost no studies other than those of the writer in the central, most critical area, which embraces much of Tennessee, Kentucky, and West Virginia. The ecologic features of the western margin of deciduous forest, of the generally secondary vegetation of the Piedmont, and especially of the great glaciated area north of the Ohio and Missouri rivers, and of New England have been worked with varying intensity. The glaciated area is young; its vegetation has been derived from that of the unglaciated land to the south. What lies to the south is *not* modified beech–maple; rather, the beech–maple to the north is derived from the older vegetation to the south. The intimate knowledge of the Beech–Maple or Maple–Beech association and the earlier lack of any information on the more complex central part of the deciduous forest is doubtless the reason for considering it "the typical association" of the deciduous forest (Weaver and Clements, 1938) and for the faulty interpretations of deciduous forest as a whole.

During the long period of existence of the deciduous forest, extending back certainly for some millions of years, there have been repeated changes in climate, either general in their scope or restricted to certain parts of the area. Each change has exerted a profound influence on the later extent of the forest. Each in turn has tended to stop migrations of the forest in certain directions, or to cause migrations in new directions, or by a selective action on species to lead to changes in numerical proportions of species and, therefore, to the development of new vegetation types. During this time, also, there have been repeated changes in the configuration of the land surface, some widespread, some local. Each of these changes has tended to affect forest migrations, to bring about regroupings of species, and, therefore, the development of new vegetation types. The modern regional segregation of the deciduous forest is, therefore, due not only to modern conditions of environment in the broadest sense, but also to the former environment.

THREE CENTURIES OF CHANGE

When the first settlers colonized the Atlantic seaboard, and later when hardy pioneers pushed westward over the mountains, they found an endless expanse of forest broken only by swamps and bogs, by cliffs or river bluffs too steep for forest, by windfalls and burns, by small grassy openings such as the serpentine barrens of the northern Piedmont, and by the prairie patches and "barrens" of the interior. The forest was a hindrance to travel, a hazard because of lurking Indians unfriendly to the invading white man, and an obstacle to homesteading, as land must be cleared for home and produce.

Such records as there are of the nature and appearance of the American forests of pre-colonization days are meager and fragmentary. They may serve to substantiate the evidences afforded by the remnants of today, but seldom do they enable one to visualize the original conditions. The accounts of travels by the early botanists were little, if any, better, for they were more concerned with the finding of new plants than with the broad vegetational features. Some of the early state geologic surveys were devoted to forest conditions or attempted to correlate forests and geologic features. These are valuable sources of information. The literature of forestry and, finally, ecologic papers (the latter usually dealing with quite circumscribed areas) yield information on vegetational conditions; however, by the time such contributions began to appear, but a small percentage of the Deciduous Forest Formation remained in anything like its original condition.

Very early in the twentieth century—but nearly 300 years after the white man started on his era of forest destruction on the Atlantic seaboard and 150 years after his penetration into the upper Ohio Valley region—John W. Harshberger (1911), recognizing the importance of leaving "a record of the original appearance of the country before the march of civilization has destroyed primeval conditions," published his work on the phytogeography of North America. This brought together for the first time information on the whole continent and attempted to characterize plant communities of the different phytogeographic regions recognized. Throughout, his treatment is illustrated by examples based chiefly upon his own observations during extensive travels. To this work the reader is referred for a history of botanical explorations.

Recent considerations of the entire Deciduous Forest Formation, such as are found in textbooks, give only a most general picture. The present volume, therefore, is an attempt to present information on the *original* forest pattern of eastern North America and on the composition of virgin forests, to analyze and compare climax communities, to trace the expansions and contractions of the Deciduous Forest Formation and its segregation into types, and to demonstrate the genetic relations of its several parts. The consideration of climax communities will of necessity begin with the forest of the central part—the ancient Appalachian Upland.

In the description of the forest regions, in Part II, the emphasis is placed upon original conditions. Statistical data are based upon as nearly natural stands as possible. Consideration of developmental stages is included, particularly where they constitute any considerable fraction of the original cover, or where they are significant in interpreting recent trends. After a picture of deciduous forest has thus been set forth, it will be possible to make some correlations with causal factors. The more important of these have already been suggested.

The material of this book is in part static or descriptive ecology, the description of vegetation and the presentation of statistical data; it is in part dynamic or interpretive ecology, seeking the causes for the observable vegetational features. In the search for causes, it is necessary to go back in time to the remote past, to draw from the geologic record. The past and the present are combined, in Part III, to build a story of the evolution of the present pattern of forest distribution and the relationships of deciduous forest climaxes.

The traveler today may see little to suggest the features which are described as characterizing a region, for untouched forests are seldom easily accessible. The complex pattern of secondary forest vegetation is but suggested, in the attempt to reconcile the apparent discrepancies between regional characteristics as stated, and what one sees while traveling—what remains today. The history of land utilization, as well as local habitat factors, affects the existing pattern. On land once covered by one climax type of forest, there may now be a score of unlike plant communities, at least some of which in no way suggest the original vegetation.

CHAPTER 2

Forest Ecology and Terminology

In the treatment of forest regions (Part II), the descriptive material is presented in such a manner as to emphasize its value in the interpretation of vegetation. The mere determination of what is present in the forest does not result in a knowledge of the forest. To know the forest we must recognize and understand significant vegetational features, the evidence of stability or of change, of forest development, the reactions of vegetation and their effects on forest environment and community continuance, the forest environment itself, and, finally, the factors determining and limiting forest regions. In order that such features (when presented in the descriptive text) may assume their proper significance in the reader's mind, a few fundamental ideas in the field of forest ecology will be briefly reviewed. This involves a limited amount of exact terminology, the use of terms which in general are ordinary English words but which have assumed more or less specific meanings in the field of ecology. Unfortunately, a decided lack of uniformity in usage is apparent in ecologic literature. Hence, it becomes necessary to define certain terms so that the sense in which they are used in this volume is clear. These are kept to a minimum and preference is given to more or less self-explanatory terms, when such meet the needs of good usage.[1]

FOREST VEGETATION

Vegetation may be thought of as made up of units or communities, some large and inclusive, some small and taking a place in the mosaic of vegetation everywhere apparent.

UNITS OF VEGETATION: COMMUNITY TERMS

Community is a general term used for any unit of vegetation regardless of rank or development.

The **formation** is the major vegetational unit; thus, deciduous forest is a

[1] For further definitions and discussion of terms applicable to the community, see Nichols, 1923; Clements, 1916, 1928, 1934, 1936; Clements and Shelford, 1939; Weaver and Clements, 1938; Toumey and Korstian, 1937, 1947; Cain, 1944; Society of American Foresters, 1944.

formation (see Chapter 3). Within it may be seen inclusions of other formations, for example, of the grassland formation and of the northern coniferous or spruce–fir formation.

Association is used here to denote a major climax unit of the formation, as Mixed Mesophytic association, Oak–Hickory association, Oak–Chestnut association, Beech–Maple association, and Maple–Basswood association. Each climax association possesses a certain unity throughout its extent which is due to (1) some uniformity in floristic composition, expressed primarily in the dominant species and secondarily in the accessory and subdominant species; (2) essential uniformity in physiognomy, a visible expression of the control of its characteristic dominant and subdominant species; and (3) historical or genetic origin. This does not mean that composition or even appearance are absolutely constant throughout the whole extent of the association. Differences are evident, due to variation in numerical representation of species, which in turn results in variation in physiognomy. Hence, no association can be defined in terms of proportions or of percentages of its constituents. Because there are these different expressions of the association, a number of terms have come into use.

Because no two dominants are exactly alike in behavior, in competition, in reaction, the components of an association segregate in all manner of ways, producing apparently more or less distinct communities, all of which are a part of the association. Such variants, due to shifting dominance or changing numerical importance of species of a complex association, as the Mixed Mesophytic, are **association-segregates.** In simpler associations, characterized by two or a few dominants, variants usually have but one dominant; these are **consociations,** climax units characterized by a single dominant. Consociations may appear among the association-segregates of complex associations. Geographical variations are evident in certain of the associations of the deciduous forest, correlated with range limits of species, with pairs of species (one dominant toward one geographic extreme, the other at the other extreme), or with pronounced differences in relative abundance of dominants. These are **faciations.** (Altitudinal faciations have also been recognized.) Local variations, generally due to edaphic causes, have been termed **lociations.**

It is necessary to distinguish between vegetational units in the *abstract* and in the *concrete* sense. Each forest stand which meets the specifications of a given association is a concrete piece of the association. The **concept** of the association, the **association-abstract,** is built up from the many concrete pieces of vegetation, just as is the concept of a taxonomic unit, the species, for example. All the individuals of a species are sufficiently alike to be recognized as belonging to that species; yet no two individuals are exactly alike in all respects. The individual stands of an association, although differing from one another, are sufficiently alike to be classed together. The stands which are described or whose composition is given (in Part II), represent the **association-concrete.** From these a general picture or concept emerges, which is the **association-abstract.** It is the association-abstract which is foremost in the chronological account of forest development (Chapter 17) and in the consideration of the relationships of climaxes (Chapter 18).

Succession and Climax

Succession is vegetational change, the replacement of one community by another as a result of natural and progressive (rarely, regressive) development. The series of stages together make up a **plant succession** or **sere.** Each stage or developmental unit may be assigned a specific term (**associes** or **consocies**), but here it is usually referred to as a **developmental** or **seral stage.** In much of the ecologic literature, these are referred to as "associations"—communities of more or less definite floristic composition and physiognomy, and in equilibrium with the habitat. Because of this, the association is here sometimes referred to as a "major association" or "climax association" to assure its status being understood. **Hydrarch succession,** or **hydrosere,** and **xerarch succession,** or **xerosere,** are terms used to designate the trend of succession, the former from hydric to mesic, the latter from xeric to mesic stages.

The replacement of one community by another depends upon, in fact is enforced by, changing environment. Environmental changes may be brought about by the reactions of the occupying vegetation alone, or by these along with topographic changes (the processes of erosion and deposition), or changes in climate. These are essentially the biotic, topographic, and regional successions long since defined by Cowles (1911).

Succession finally results in a **climax,** which is reached when, under existing climatic conditions, no further change is possible, because the reactions of the occupying dominants prevent the entrance of new forms. It is not necessary here to go into the problem of unity of the climax, of the so-called monoclimax and polyclimax concepts. That some degree of variation is admitted is indicated by the terms association-segregate, consociation, and faciation, which are different expressions of the climatic climax, or *climax.* The climax (climatic climax) occupies the average sites and is the result of long continuing climatic control. It is able to reproduce itself. However, under the impulse of gradually changing climates, and because of migrations induced by, but not necessarily keeping pace with, changing climate, climaxes may change. Any regional climax may be but a stage, a long pause, in a **climatic succession** or **clisere.**

In a forest, perpetuation of the existing type is indicated by the **essential accordance** of **canopy** and **understory** layers. For this reason, statistical data on forest composition should separate the canopy layer from the second layer, whose individuals will gradually replace the trees of the canopy, as these die off as a result of old age or accident. The proportions in lower layers are less significant, as competition will eliminate many individuals, and as different species respond differently to suppression. **Lack** of **accordance** of canopy and understory is always indicative of change, of development or succession. Some rhythmic variation in proportions of dominants is permissible within the concept of essential accordance.

Essential accordance of canopy and understory is a characteristic, not only of the climatic climax, but also of certain other self-perpetuating communities which occupy extreme sites, either topographic or edaphic. Such communities are sometimes referred to as **subclimax** and their sites as **subclimax** sites. The subclimax in this sense is not "the stage preceding the climax," but a community which, only theoretically, could be replaced by the climax. **Physiographic climax** is used for those self-perpetuating communities which can endure as long as the topographic form on which they occur (for example, the oak–hickory and oak–chestnut communities of ridge crests in the Mixed Mesophytic Forest region). **Edaphic climax** is occasionally used to refer to a self-perpetuating type dependent on an edaphic environment which persists with relatively little change through climatic and physiographic cycles; this term should be used with caution. **Postclimax** has been used to denote a community more mesophytic than the climax. Thus, local mixed mesophytic communities in the area where oak–hickory is climax may be referred to as postclimax communities. **Preclimax** bears the opposite relation to the climax, being more xeric than the climax. Both postclimax and preclimax communities are usually physiographic climaxes.

Recognition of the Climax

The recognition of a climax is not always possible if observations are confined to more or less local areas. While, in the forest, essential accordance of canopy and understory is indicative of a climax, this alone will not suffice for its recognition, as certain other communities (e.g., physiographic climax) simulate the climax in this respect. Convergence of seres, the tendency for the same community to develop as the result of unlike successions, as the end product in hydrosere and xerosere, in successions progressing under the impetus of biotic change (reactions) or of topographic change, is an excellent test of a climax which may be used in a more or less circumscribed area. The occupation of topographically mature sites is an attribute of a climax. The state of the soil, its maturity, its equilibrium with the occupying vegetation, and its general accordance with the regional type, ay aid in the recognition of the climax. A further test necessary for the l recognition of the climax is essential similarity over a large area.

Virgin and Second-Growth Forests

A **virgin forest** is one which has reached maturity through natural processes of development and has not been influenced by human activity. It is an old-growth stand but not necessarily a climax stand. The forest which has come in after disturbance is usually called a **second-growth** forest or **secondary** forest. In sections which have long been settled, the secondary forests are often third- or even fourth-growth stands. The second-growth forest is a stage in secondary succession.

A **secondary succession** (or subsere) is one beginning after disturbance on land which retains some influence of previous occupancy by vegetation. Some part of the previous vegetation may remain, as is always true after cutting. Cutting followed by fire is more destructive than cutting alone, not only because of the larger number of plants and propagules killed, but also because of the destruction of leaf litter and, in some instances, of humus. Complete denudation, plowing, or other agricultural utilization, although destroying all of the original vegetation, nevertheless does not remove all trace of former occupancy by vegetation, because the soil developed under the influence of the previous vegetation.

A **primary succession** (or prisere) is initiated when growth starts upon a previously unoccupied piece of land, a **primary bare area.** The most extensive primary bare areas which have been populated in the course of deciduous forest development are the great drift sheets of the North, land laid bare after the melting of the continental glaciers, and the newly emergent areas of the Coastal Plain, some of which appeared at various times in the Tertiary, some in the Pleistocene (Chapter 16).

Primary forest vegetation may be young, if the time since the initiation of succession is short, or it may be old. Some primary growth may remain in forests which have been influenced by human activity. A forest selectively logged is no longer virgin, although it may be largely primary. A forest heavily grazed is not a virgin forest, although its arboreal layers may be intact. Recovery after minor influences may be sufficient to obscure or almost obscure their former presence. No sharp line of demarcation can be made between the absolutely virgin forest and the old primary stand which, although lightly disturbed in the past, has lost all traces of disturbance. Windfalls and fire have occurred in forests from time immemorial; they have initiated secondary successions, small or large, depending upon the extent of the catastrophe. The mature forest which finally develops after the visit of a catastrophic agent has usually been considered a virgin forest (cf. the old white pine groves of the Unglaciated Allegheny Plateau in Pennsylvania, p. 402). Occasionally, in the older sections of the country (older from the viewpoint of time of settlement), second-growth forests are seen which resemble primary stands. Probably nowhere except on the Atlantic slope has there been time enough for a second-growth forest (started after the white man's entrance) to even approach an original forest in appearance.

The greater the disturbance, the more different is the second-growth stand from the original forest. Instances have been observed where a fence line separates a young hickory woodland from a climax beech–maple forest; a young white oak stand from a forest in which 50 per cent of the larger trees are beech; where, except for a few old beech trees (remnants of a former beech–white oak forest), scrub pines dominate the slope. The secondary forest not only may tell nothing about the nature of the original cover, but also it may even be very misleading in its implications, because of some resemblance to a well known or widespread primary forest type. The secondary slope forest is more xeric than the primary. *Disturbance sets back succession.*

An **all-age stand** is generally a primary stand, or in large part primary. The individuals have entered at various times, when and where space permitted. An **even-age stand** is always secondary; the individuals entered at approximately the same time, because suddenly much space was available. Any considerable gap in ages is indicative of a period of disturbance (as grazing) during which no trees could start. An even-age undergrowth in an old-growth forest may indicate a time at which disturbance ceased and abundant reproduction began, or a time when the young trees were killed (as by fire) and, hence, the lower layers temporarily opened up.

Dominant, Subdominant, and Accessory Species

The **dominants** in a forest are those trees of the canopy, or superior arboreal layer, which numerically predominate in the stand. Where several species are involved, these may be spoken of as **codominants.** The dominant species of lower layers of a forest are referred to as **subdominants,** i.e., dominants of a subordinate layer. Subdominants are species which belong to a different life-form than the dominants. For example, dogwood may be a subdominant in a forest where white oak is a dominant; dogwood and white oak, the former a small tree and the latter a large one, differ in life-form. In addition to the dominant and codominant species, there are usually other species which may be called **accessory** species. Some investigators call these species "secondary" because of their secondary importance, while the predominant species are referred to as the "primary" species. This usage is confusing, as it might seem to imply that certain of the species were remnants of a primary stand while others had entered, in secondary succession, after disturbance.

Competition

Competition is "a struggle between individuals for growing space, both above and below ground, and for light, water, and nutrients" (Toumey and

Korstian, p. 156, 1937). In a forest, trees are always close enough to one another for competition to take place. Its effects are evidenced in the form of trees, in the suppression of young individuals, and in the crowding out of weaker individuals. The effects of competition below ground may at first seem less apparent. Nevertheless, root competition is often severe and may prevent the entrance of seedling trees and undergrowth plants.[2] That is, invasion may fail to take place, not because the environment is fundamentally unfit, but because the competition of occupying vegetation is too severe. Competition is almost lacking in pioneer communities; as long as there is unoccupied space above and below ground, new individuals can enter; however, competition soon follows.

Suppression is one of the evidences of competition. A suppressed tree is one whose growth is greatly retarded, whose crown is overtopped by other individuals crowded about it. Annual increases in height are very slight and annual increment of wood is correspondingly small. In cross section, the period of suppression is readily discernible because of the slow growth. **Release** results when competition is reduced due to death (or cutting) of nearby large trees. The time of release is apparent in cross section because of the sudden increase in growth increment.

Tolerance

Observation will reveal that young individuals of certain species of trees are commonly seen only in the open while others are seen within the shade of the forests, that some are **intolerant,** others **tolerant.** "**Tolerance** may properly be defined as the ability of a tree to develop and grow in the shade of, and in competition with, other trees. In short, it is the ability of a tree to withstand competition and still maintain its growth" (Toumey and Korstian, p. 332, 1937). Ecologically, tolerance is of great importance, for differences in relative tolerance of species *in part* determine the progress of succession. Pioneers are generally intolerant species, and often are incapable of growing in their own shade, even though their crowns are open, as is usually the case with intolerant trees. Some less intolerant species will grow beneath them, and in time will replace the earlier growth. More and more tolerant trees appear with the progress of succession until finally, in climax communities, the canopy species are ones which can successfully reproduce and maintain the constancy of the community on a given site. Species of more or less intermediate tolerance may persist as suppressed individuals for many years.

The tolerance of a given species varies under different environmental

---

[2] Experimental investigations involving the use of trenched plots have shown that where root competition is removed by severing roots, abundant growth may enter where formerly the forest floor was almost devoid of plants (Toumey and Kienholz, 1931; Korstian and Coile, 1938).

conditions. Under more favorable conditions, tolerance increases. With higher temperatures, less light is required. Hence, it may be expected that a given species will not everywhere constitute a part of the same, or even comparable forest communities. The mere presence of a species in two geographically distant communities is not always indicative of similar communities. Chestnut, for example, has a wide range of tolerance in the Mixed Mesophytic Forest region, where it is a frequent constituent of mixed mesophytic climax communities; northward, it is less tolerant. Hophornbeam (*Ostrya virginiana*) is listed in published scales of tolerance as "very tolerant," and in the North it is an understory species in beech–maple communities. Yet, farther south it is more often a tree of open xeric slopes. Either its optimum environment is in northern latitudes, or, as is indicated by certain morphological characters, there are two distinct races which behave differently.

Reactions

Forest vegetation reacts upon its environment, gradually but constantly changing it. Progressive development of vegetation, from the time the first pioneer enters a bare area on through all seral stages, is in large part a result of reactions of occupying vegetation. The isolated pioneer casts some shade and by its decay adds some organic material to the soil. The cumulative effects, as vegetation increases in amount, bring about changes in light, temperature, relative humidity, rainfall interception, soil nutrients, soil texture, soil moisture—in the physical environment—which induce changes in vegetation. Only in climax and subclimax communities are the reactions favorable to the continuance of the occupying vegetation. A stage of equilibrium has been reached. The occupying type of vegetation may then continue on indefinitely, or until such time as the equilibrium is destroyed by topographic or climatic change.

Ecotones

Transitional areas, where characters of adjacent communities or regions are intermingled, may be referred to as **ecotones.** Broad ecotones result where controlling environmental conditions change gradually, narrow ecotones, where conditions change abruptly. Broad ecotones always present problems in the interpretation of boundaries because of the overlapping of vegetational features.

FOREST ENVIRONMENT

The forest environment is a complex of atmospheric, topographic, and soil influences, complicated by the reactions of the forest and its inhabitants, plant and animal, on the environment, and by the competition between the individuals of the forest. It is in part inherent in the physical organization of the site, in part created by the forest. Not only is environment far from uniform throughout a forest region, but also it varies from place to place within a given stand. In fact, adjacent individuals of unlike size or kind may occupy quite unlike local environments, due to their different heights—unlike exposure to atmospheric factors—and different root systems, which may inhabit soil layers differing in air, moisture, humus, and inorganic constituents.

No environmental factor operates independently of others. Each affects, and is affected by, other factors. Their action is concerted and simultaneous. "Controls" do not exist in nature. Raup (1942) and later Cain (1944) have emphasized the complexity of environment, the mutual or concerted action of environmental factors. Coincidences of agreement between vegetational boundaries and temperature or precipitation intensities (amounts, duration, extremes, etc.), or between vegetational boundaries and soil type boundaries cannot be interpreted as meaning that the factors separately are controlling factors. The correlations (real or apparent) which can be discovered between the boundaries of forest regions and climatic and edaphic factors must be accepted with reservations.

Environmental Factors

Environmental factors may be classified as climatic, physiographic, edaphic (soil), and biotic. Anthropeic and pyric might be included (Nichols, 1923). Some factors directly affect the physiologic activities of plants; others (particularly the physiographic factors) only indirectly affect them through their influence upon the direct factors.

Climatic or **atmospheric factors**—temperature, moisture, light—although fairly similar over considerable areas as far as measurements of their intensity indicate, may actually vary, as far as effectiveness is concerned, in circumscribed areas. The amount of rainfall, *per se,* is of less importance than the amount absorbed by and retained in the soil where it is available to plants. Differences in run-off due to differences in vegetation cover, soil porosity, topography, and rate of rainfall in adjacent areas may result in great differences in soil water. Even the "average" or "mesic" situations differ in ways incapable of measurement. The temperature of the air—at a weather station or at a particular spot in the forest—is not the temperature in all forest strata nor in all parts of the forest; it varies with the vegetative cover and with topography. Light, although varying primarily with latitude and altitude (angle of incidence of sun's rays and thickness and quality of the atmosphere penetrated), nevertheless varies locally because of cloudiness, topography (angle and direction of slope), and shade (vegetation). Thus each climatic factor is affected by the other groups of factors.

However, it is possible to utilize climatic maps—seasonal temperature, length of frostless season, temperature extremes, seasonal precipitation, precipitation effectiveness, duration of droughts, etc.—in the search for factors of importance in forest distribution, because the long-time control of climate over the larger vegetation units is a result of the usual rather than the extreme or exceptional conditions. The extremes are most effective near the margin of range of a community or a species, for the usual tolerance range of community or species is greater than the range of climatic conditions well within a vegetation region.

Extreme low temperatures in winter may result in the killing of young individuals of certain species, or of the young growth on older individuals, or in the killing of flower buds.[3] Such phenomena may limit the extra-regional occurrence of forest types and may in part determine the limits of forest regions.

The length of the frostless season—or the growing season—is a temperature factor of considerable importance, one which appears to be correlated with certain forest boundaries. A late spring freeze may affect individuals of one species while others are unharmed. The susceptible species is limited to regions where late frosts are infrequent. Its presence in one region and absence in another accentuates the differences between regions, even though it be only an accessory species of the climax, or a species of a seral stage.

The amount and seasonal distribution of precipitation, the precipitation-evaporation ratio, the duration and season of droughts, all are moisture factors affecting species distribution and, hence, forest distribution.[4]

Light (as a regional factor) may be significant in limiting ranges because of the effects of length of day on flowering and fruiting. Most studies of such effects have been directed at economic herbaceous species; only a few wild plants have been considered (Allard, 1932). It has been shown that a single species (e.g., *Bouteloua curtipendula,* a widespread species of the Grassland) may be made up of distinct races differing so greatly in their length of day requirements that flowering may be inhibited when plants are (experimentally) grown in regions of different day-length from whence they came (Olmsted, 1945). However, the wide latitudinal range of most of our tree species indicates that this factor is not critical.

Physiographic factors have to do with the nature of the topography. Physiographic development is progressive or cyclic under the influence of erosion and deposition. A physiographic cycle or **erosion cycle,** initiated by

[3] Nash (1943) lists beech as the most severely injured species, followed by white oak, in the extreme cold of the winter of 1942–43 in Maine. A number of observers noted injury to beech in the Adirondacks during the severe winter of 1935–36. The flower buds of redbud (*Cercis canadensis*) were killed in southern Ohio during the winter of 1935–36.

[4] Hursh and Haasis, 1931, show that different species in the Southern Appalachians react differently to protracted summer droughts.

uplift of the land, progresses through youth, maturity, and old age, to or toward the formation of a peneplain. Changes in vegetation, in the nature of communities and their extent, are related to progress in the erosion cycle (see Chapter 17). The local influence of physiographic factors is due to the local features of topography, most important of which is slope, including angle of inclination and direction of slope. **Slope exposure,** sometimes referred to as "aspect" by foresters, influences other factors more directly operative on the plant. A slope is designated according to the direction it faces. Slope exposure influences temperature, light, and wind exposure, and hence evaporating power of the air, the amount of moisture in the soil, the rate of decomposition of leaf litter, and very often the kind of humus formed. Angle of slope affects run-off, and hence soil moisture. Flatness of the landscape and a lack of stream dissection cause poor drainage and, hence, poor aeration, which in turn affect humus formation.

Edaphic factors are soil factors, including such features as soil moisture, soil texture, soil composition, soil temperature, soil gases, etc. Soil factors are complex and interrelated. Soil texture affects porosity, which in turn affects the amount of moisture and soil air, and hence the nature and rate of humus decomposition and incorporation. Humus content in turn affects soil porosity, water absorption and retention.[5] The ramifications and interactions are legion. Soils may differ physically and chemically in relation to underlying rock. Differences in composition are often evidenced by differences in alkalinity or acidity (pH). Soils differ with position on a slope.[6] They differ in profile and in humus layers, and also undergo development much as do plant communities. The regional type is climatically controlled, but the local variations are controlled by parent material, topography, and occupying vegetation. It is readily seen that edaphic factors are not independent of the other groups of factors.

Local differences in edaphic factors are responsible for many of the differences so evident in the vegetation of a single region, section, or smaller area. Many species are more or less limited in their distribution, or have a patchy or disjunct distribution, because of some soil factor.[7] Outstanding examples are seen in the serpentine barrens (p. 248), in the Cedar Glades (p. 131), in the remarkably sharp contrasts between the vegetation on contiguous calcareous and noncalcareous soils (p. 119), and in the influence or control of sandy soils (pp. 257, 343). That some forest boundaries

[5] Forest soils are more porous than field soils and absorb water more rapidly. Porosity, and hence water absorption are decreased by grazing and fire (Auten, 1933).

[6] Foresters and soil scientists sometimes refer to the **catena,** the series of soils from dry to wet or from ridge to valley, which is more or less comparable to the series of communities occupying the area.

[7] Mason (1946) emphasizes the influence of edaphic factors in narrow endemism.

approximate soil boundaries becomes evident in comparing the extent of forest regions and their subdivisions with soil maps.

Biotic factors include those influences associated with the activities or reactions of plants and animals. Most of these have already been considered under the topics competition, tolerance, and reactions. Biotic factors can not be separated from edaphic or regional factors in the development of soils.

SOILS[8]

Soils undergo development. The nature of this development is determined by climate, topography (which affects ground water), parent material, and vegetation (which is affected by climate and topography). Young soils partake largely of the character of the substratum or parent material. Mature soils reflect the influence of environment, they have acquired a **profile.** Horizons of the **soil profile,** including the humus layers, are shown in tabular form. Not all of the horizons are present in all soils. A water-logged or **glei** (or gley) horizon is present in some soils. It is near the surface in permanently wet soils (grading into the A_1 or A_2 horizon), somewhat deeper in periodically wet soils (grading into the B horizon, and distinguished by mottling), and deep, located in the parent material (C horizon), in certain moist but drained soils.

HORIZONS OF THE SOIL PROFILE

L Layer—The litter of freshly fallen leaves and forest debris.

A Horizon—A zone of extraction or eluviation, of removal of material by ground water, either in solution or in suspension.

A_0—Undecomposed and partly decomposed organic debris.

- **F Layer**—Undecomposed and partially decomposed forest litter, still recognizable as to origin.
- **H Layer**—The humified remains; principally organic material so decomposed that it is no longer recognizable as to origin. (H layer may be absent.)

A_1—Dark in color; a mixture of mineral material and humus (thick in some regions, thin or absent in others).

A_2—Light in color; leached and deficient in organic material and soluble salts (absent in some soils).

B Horizon—A zone of concentration or illuviation containing precipitated soluble salts and coagulated colloidal humus carried down from A horizon. Usually deeper in color than leached A_2 horizon. (Sometimes subdivided into B_1, B_2, B_3.)

C Horizon—Parent material, either weathered or unweathered.

The processes of humus decomposition and soil formation differ in different climatic regions and under different forest canopies. They differ

[8] For a more complete treatment of the topics included here, see Kellogg, 1936; Wilde, 1946; Lutz and Chandler, 1946.

within a single region because of the influence of substratum, slope exposure, and forest cover. Hence soil profiles vary in large part due to the nature of the humus layer.

The nature of the **humus layer** is largely determined by the forest cover, although topography, substratum, and ground water are modifying influences. Two main kinds of humus were recognized by Romell and Heiberg (1931), Heiberg (1937), and Heiberg and Chandler (1941). These are **mull** and **mor** (in the first paper called "duff"). **Mull** is a "mixture of organic matter and mineral soil, of crumbly or compact structure, with the transition to lower layers not sharp" (Heiberg, 1937). **Mor** is "a humus layer of unincorporated organic material, usually matted or compacted or both, distinctly delimited from the mineral soil unless the latter has been blackened by the washing in of organic matter" (Heiberg and Chandler, 1941). Several types of mull and of mor may be distinguished. One of these, "twin mull," combines certain characteristics of mull and mor.

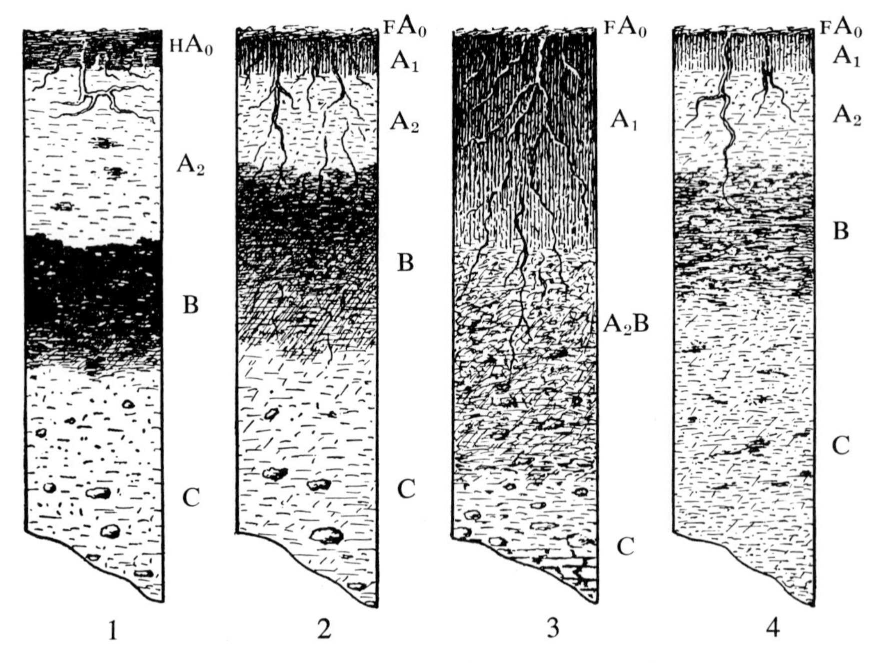

Representative profiles of mature soils from four zonal soil groups: (1) Podzol, showing thick accumulation of mor humus (H layer), strongly leached A_2 horizon, and hardpan development in the B horizon. (2) Gray-brown podzolic soil, showing an F layer and a thin A_1 horizon (where mineral material and humus are mixed) overlying the leached A_2 horizon. (3) Melanized soil, with a thin F layer (often absent), the deep and dark A_1 horizon which is mull humus, and the absence of clear differentiation of the A_2 and B horizons. (4) Red podzolic soil, lateritic soil, or red-erth of the South, with profile similar to that of the gray-brown podzolic soils, but differing greatly in color because of laterization. Depth of profiles about 5 feet.

Wilde (1946) considers that there are three basic types of forest humus: **earth mull, duff mull,** and **mor** or **raw humus.** Duff mull is the equivalent of twin mull; otherwise, the two classifications agree.

Mull humus commonly develops under mixed hardwood stands. It is seen at its best (as crumb mull or coarse mull) in rich mesophytic forests, particularly those of the Mixed Mesophytic Forest region. It occurs locally in mixed hardwood communities of other regions; in the North generally where maple, basswood, and ash (rather than beech) are abundant. Mull humus supports a rich herbaceous vegetation in which the spring flora is outstanding (p. 45). Characteristic mull humus species are numerous, including *Dryopteris Goldiana, Athyrium pycnocarpon, A. thelypteroides, Disporum lanuginosum, D. maculatum, Actaea pachypoda, Caulophyllum thalictroides, Dicentra* spp., *Dentaria diphylla, Viola palmata, V. canadensis, Hydrophyllum* spp., and *Phacelia bipinnatifida.* Large earthworms (*Lumbricus terrestris* L.) are common in coarse mull and have much to do with bringing about the mixing of organic and mineral material. Mull tends to retard podzolization. It is an essential constituent in the profile of melanized soils.

Duff mull, as defined by Wilde (1946), consists of an A_0 horizon of friable organic remains grading into a more or less developed A_1 horizon with incorporated humus, and seldom over six inches in depth. The podzol-forming tendency of duff mull is much less than that of mor, but there is some tendency toward podzolization. This type of humus layer is transitional in character. It develops under mixed hardwood–coniferous forests in the podzolic soil region; it is a common type under oak–hickory forests in the "prairie forest" or "grood soils" region, which is essentially the Northern Division of the Oak–Hickory Forest region. Both mull and mor plants occur in the herbaceous layer.

Mor humus commonly develops under conifers, and is particularly characteristic of the Northern Coniferous or Spruce–Fir forest, and of coniferous communities in the Hemlock–White Pine–Northern Hardwoods region. A hardwood type of mor (laminated mor or leaf duff) is seen frequently in oak and oak–chestnut forests on a noncalcareous substratum; it commonly supports an ericaceous shrub layer. The herbaceous flora is sparse and almost entirely different from that of mull; the rich vernal flora is lacking. Among the characteristic species are *Lycopodium* spp., *Maianthemum canadense, Trillium undulatum, Isotria* (*Pogonia*) *verticillata,* and *Cornus canadensis*. Mor is almost always strongly acid in reaction. A mor humus layer favors leaching of the soil, hence favors the formation of podzols.

Because of the differences between the prevailing soils of different geographic areas, **soil provinces** have been recognized in which the mature soils of average situations are essentially similar. The soil provinces reflect the combined influence of climate and vegetation. Because of their relation to climate, they have a somewhat zonal arrangement, hence may be referred to as "zonal soil groups." The several maps showing the soil provinces of eastern United States are not in complete agreement (Marbut, 1935; Kellogg, 1936; Wilde, 1946). Marbut and Kellogg recognize four zonal forest soils; **podzols, gray-brown podzolic soils, brown podzolic soils,** and **red** and **yellow podzolic soils.** The **prairie soils** province is to the west, with extensions penetrating into the forest region. Wilde recognizes five "soil-forest-provinces" in the East, and his classification is the only one

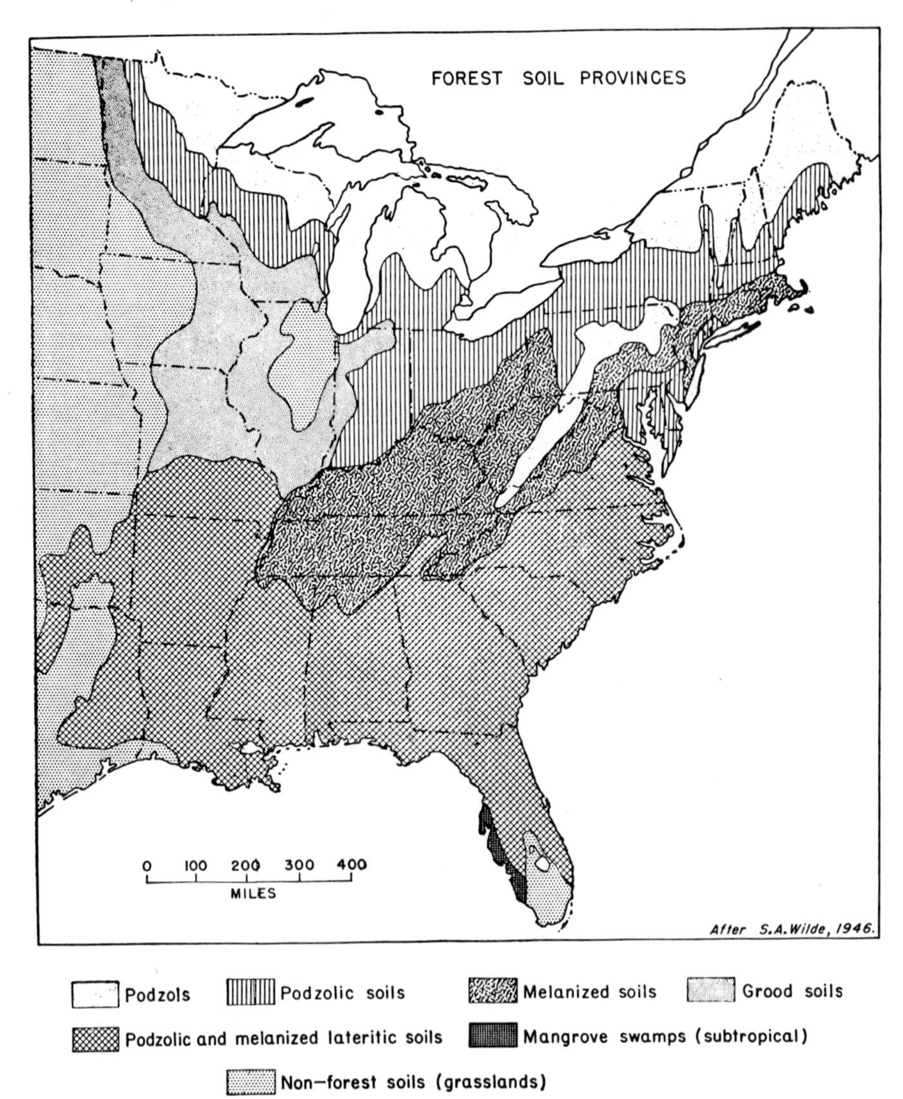

which takes cognizance of the typical dark soils of the mixed deciduous forests (see map, p. 24). The provinces and vegetational correlations which Wilde recognizes are:

1. Podzols—the area of northern coniferous forest.
2. Podzolic soils—the area of mixed coniferous and hardwood forest.
3. Grood soils—prairie-forest soils on which oaks and hickories predominate.
4. Melanized soils—the dark soils of the central deciduous forest.
5. Podzolized and Melanized Lateritic soils—the soils of the southern pine and hardwood forests.

To the west of these are the grassland soils. Kellogg maps a number of intrazonal soils, "great groups of soils with more or less well-developed soil characteristics reflecting the dominating influence of some local factor of relief, parent material, or age over the normal effect of climate and vegetation," and azonal soils, "soils without well-developed soil characteristics." Of these, planosols occupy a large part of the area of Illinoian glaciation, and lithosols occupy parts of plateau and mountainous sections (see map, p. 26).

The podzols, the dominant soils of the North, occupy about the same geographic area in the two classifications. The gray-brown podzolic soil region of the earlier classification is much more extensive than the podzolic soils region as mapped by Wilde, including almost all of the area of melanized soils and part of the grood soils. Marbut (1935) maps as "without normal profiles" extensive areas in the Appalachian Highlands where some of the best mulls are developed; these are melanized soils. The area of grood soils is partly within the prairie soils province, partly within the gray-brown podzolic province. The red and yellow soils (Marbut) and the podzolized and melanized lateritic soils (Wilde) are the soils of the South. The former name emphasizes color (which may be matched in other zonal soil areas), the latter emphasizes origin.

Podzols are produced by leaching or podzolization. In the typical podzol there is a thick mat of raw humus (mor) or undecomposed organic matter sharply delimited from the mineral soil beneath. The upper part of the mineral soil (A_2 horizon) is strongly leached and pale or ashy-gray in color (occasionally there is a thin slightly darkened A_1 layer just beneath the humus mat); the B horizon is reddish-brown and compacted or cemented. Podzols are the prevalent upland soils of the Northern Coniferous Forest and of dominantly coniferous communities throughout the North. They occur under conifers in other regions and sometimes develop in dry oak woods on a noncalcareous substratum, where their presence is usually indicated by a well-developed heath layer. **Podzolic soils** are moderately leached, and are transitional between the true podzols and the weakly podzolized soils where the A_0 horizon is thin, and the A_1 horizon, with

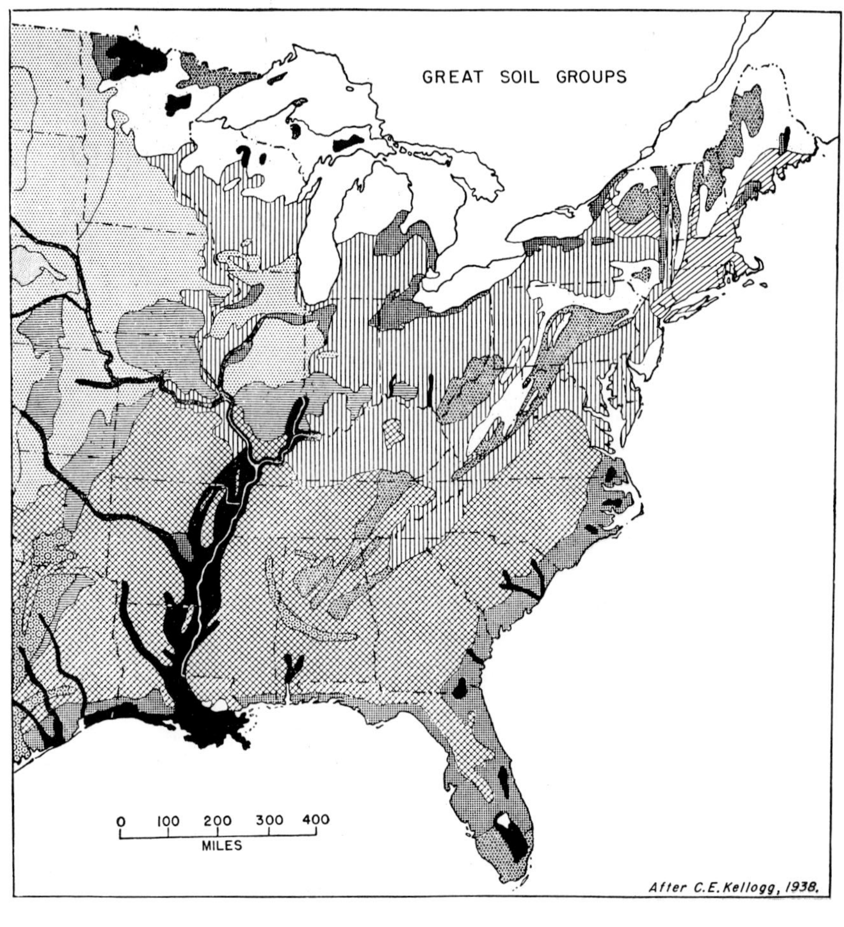

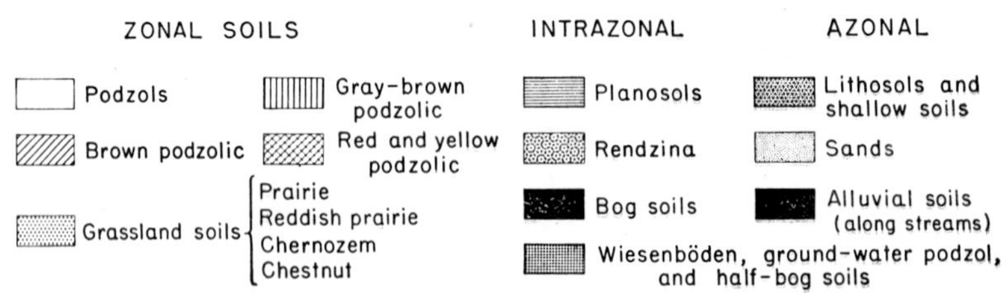

incorporated humus, is well developed. Included here are the **brown podzolic** soils, or brown forest soils, and the **gray-brown podzolic** soils, or gray-brown forest soils. Geographically, the podzolic soils border the podzols on the south. The podzolic soils belt as mapped by Wilde includes principally the Beech–Maple Forest region and part of the Hemlock–White Pine–Northern Hardwoods region. Forests of hardwoods, especially maple, beech, and yellow birch, with red spruce or hemlock, occupy the podzolic soils. As mapped by Kellogg, the podzolic soil province includes much of the area distinguished by Wilde as characterized by melanized soils.

Grood soils are transitional between prairie soils and forest soils. They have been described by Marbut (1935) as the gray-brown podzolic soils of the Prairie Region. Grood soils are more or less leached soils supporting oak–hickory forests in which a number of other species occur.

Melanized soils are humus-darkened soils, the dark or almost black mull humus soils of rich mesophytic forests. Geographically, they occupy a position between the podzols and podzolized soils of the North and the lateritic soils of the South. Leaching is limited to the most readily soluble compounds. The profile is made up of A_0, A_1, and C horizons (sometimes with a faintly differentiated A_2 or B horizon). The A_0 horizon may be a discontinuous F layer, with the H layer absent; the A_1 horizon is characterized by pronounced humus incorporation, and crumbly structure, growing gradually lighter in color downward, and up to 16 inches in depth. The area of melanized soils, as mapped by Wilde, corresponds well with the area of the Mixed Mesophytic Forest region and the Western Mesophytic Forest region, and includes much of the Oak–Chestnut Forest region. In the first of these regions, melanized soils predominate; in the second, they predominate in certain areas but are interrupted by areas of podzolic and lateritic soils; in the last, they occur only locally in areas of mixed forest and are lacking in extensive areas of oak forest. The melanized soils are among the most productive forest soils of the continent.

Lateritic soils are characteristic of moist warm climates. It is to these soils in our southern states that the terms "red- and yellow-erths" and "red and yellow soils" have been applied. In the warm climate of the South, chemical decomposition is rapid; winters are open, hence, leaching proceeds throughout the year, washing out the bases and removing part of the silica in drainage waters. Iron and aluminum oxides accumulate, imparting pronounced red and yellow colors to the soil. The lateritic soils display a range of podzolization comparable to that of the zonal groups to the north. Moderate leaching of lateritic soils results in **podzolized lateritic soils,** in which the horizon of incorporated humus is absent or poorly developed. These soils are strongly acid and are infertile. They support the pine and pine–oak communities of the Oak–Pine and Southeastern Evergreen Forest regions. Humus-enriched or **melanized lateritic soils** often develop over a basic substratum, producing fertile soils. The Orangeburg soil, derived from calcareous material, is representative (p. 283). Such soils support a variety of broad-leaved tree species; their forest communities contrast strongly with those of surrounding areas, where podzolized lateritic soils predominate.

Part II

THE DECIDUOUS FOREST FORMATION

CHAPTER 3

The Deciduous Forest as a Whole

The Deciduous Forest Formation of eastern North America is a complex vegetational unit most conspicuously characterized by the prevalence of the deciduous habit of most of its woody constituents. This gives to it a certain uniformity of physiognomy, with alternating summer green and winter leafless aspects. Evergreen species, both broad-leaved and needle-leaved, occur in the arboreal and shrub layers, particularly in seral stages and in marginal and transition areas. They are not, however, entirely lacking even in some centrally located climax communities.

Since vegetation develops in large measure in response to climate, the great climax unit, the formation, is the long-time expression of climatic control. Throughout the vast area occupied by this formation, the seasonal distribution of rainfall and the length of the growing season are sufficiently alike to favor the dominance of deciduous forest. Westward, with decreasing precipitation and longer and more frequent droughts, the character of the forest changes; it becomes more and more confined to ravine slopes and valleys, with the intervening upland occupied by prairie. Northward, the shorter growing season and the very low winter temperature extremes eliminate one after another of the deciduous forest species. Deciduous communities are there confined to more favorable sites.[1] The admixture of conifers in the forest and an alternation or mosaic of coniferous and deciduous communities characterize this northernmost section. Although this is often considered as a distinct formation—the Hemlock–White Pine–Northern Hardwoods—it must be considered with the deciduous forest because of its intimate relation to adjacent forests to the south and west. Eastward, in middle latitudes, deciduous forest extends to the Atlantic coast, except as interrupted by edaphic communities. Southeastward and southward, a more or less gradual increase in pines is apparent. Although deciduous forest communities extend to the south Atlantic and Gulf coasts and well into peninsular Florida, they are limited to better soil sites; much of this area is, or was, occupied by southern pines. Finally, the lack of a

[1] In some instances, range limits in the North do not appear to be related to climate; certain of the deciduous species cannot advance until soil development results in suitable edaphic situations (Hutchinson, 1918).

winter dormant season eliminates deciduous species and warm-temperate and sub-tropical broad-leaved evergreen species become dominant.

Topographic features are exceedingly varied. Although only indirectly affecting the nature of the climaxes, topography (particularly maturity in the erosion cycle) is important in determining the proportions of the surface occupied by climax and by developmental stages. The northern part of the deciduous forest area is glaciated territory; its topography, except in the few mountainous areas, is for the most part flat to rolling; its drainage systems are poorly developed, and wet and undrained areas are frequent and often extensive. Much of the unglaciated area is dissected upland; some of it is mountainous. Undrained areas are few, except where they have been formed on broad developing flood plains, or remain on upraised but undissected peneplains. Diversity in topography results in a wide variety of forest communities, most of which (within a given climax area) are developmentally related.

Unity of the Deciduous Forest Formation is emphasized not alone by the deciduous habit, but also by the presence of a number of wide-ranging genera and species, particularly *Quercus, Acer, Fagus,* and *Tilia.* The first two of these genera, however, have representatives in other formations as well. Additional characteristic genera abundantly represented in climax communities over a considerable portion of the deciduous forest include *Carya, Fraxinus, Ulmus, Betula, Liriodendron,* and *Castanea.* No tree species range throughout the deciduous forest except those regularly associated with early seral stages, as *Ulmus americana* and *Salix nigra.* The Deciduous Forest Formation has been called the *Quercus–Fagus* formation (Clements, 1916, 1928; Weaver and Clements, 1938), as *Quercus* is represented in most parts of the deciduous forest and *Fagus* in several of the climaxes.

The forest is a layered community. The canopy, the superior or arboreal layer, is made up of dominant and accessory species. Those trees which are used to designate the climaxes are, of course, dominants of the superior or arboreal layer. Yet subdominant trees and herbaceous vegetation are as much a part of the community as are the larger trees, and in some instances, because of more limited distribution, even more characteristic of the climaxes of which they are a part. Frequently, in disturbed areas, they are better indicators of forest type than are the trees of the secondary forest.

The "theory of forest types," which uses herbaceous plants as indicators of forest type, was elaborated by Cajander (1926), and applied by Heimburger (1934) in the Adirondacks. The idea has been used independently by a number of ecologists (Korstian, 1917; Gordon, 1937). The herbaceous layer is in part a result of reactions of arboreal layers; its features may be significant in exact local delimitations between communities because of the differences in composition (Potzger and Friesner, 1940; Braun, 1940).

Herbaceous vegetation also may be indicative of advance or state of development. The suggestion that lower layers (synusiae) are independent of the canopy was emphasized by Lippmaa in 1933 and 1935, and later discussed by Cain (1936) and by Gleason (1936). In the present consideration of deciduous forest communities, inferior or lower layers will be considered as a significant part of the forest.

The Deciduous Forest Formation is made up of a number of climax associations differing from one another in floristic composition, in physiognomy, and in genesis or historical origin. While the delimitation of associations may be made on a basis of dominant species, and it is from these that the climax is named, dominants alone do not suffice for the recognition of these units. To one familiar with all of them in the field, there is an intangible something in aspect, in physiognomy, which is always supported by concrete differences in composition of one or more layers of the forest; there are almost always soil profile differences. It will be shown (in Part III) that the climaxes differ in origin.

Since the climax association is an abstract concept based upon familiarity with numerous concrete examples, differing in minor details from one another, its delimitation in space is difficult, if not impossible. The respective positions of the several associations can be shown diagrammatically, but they cannot be mapped unless one wishes to map the location of each forest stand which meets the specifications of the concept. In order to clarify the concept of each association, and at the same time to demonstrate the variable character of the association, concrete examples of communities will be used, selected from among the many upon which the characteristics of the association in the abstract have been built.

Although the delimitation in space of an association is difficult, if not impossible, it is entirely possible to recognize and to map forest regions which are characterized by the prevalence of specific climax types, or by mosaics of types. These regions are natural entities, generally with readily observable natural boundaries based on vegetational features. Boundaries may in part be determined by the limits of the more or less continuous ranges[2] of characteristic species (there may be relic or advance disjuncts beyond the general range limits). Where boundaries are indistinct, either because of transitions or human disturbance, or the lack of sufficient detailed information on vegetation, it is necessary to use more or less arbitrary boundaries, sometimes approximating natural physiographic boundaries. Even where most distinct, boundaries can be correct only within the limits

[2] The taxonomist's maps of distribution of species, showing occurrence by counties, frequently do not show the real ecologic situation. The outlying county, where the species may be rare, shows up as prominently on such a map as does the county where the species is abundant. Neither are the number of herbarium records in any way indicative of abundance, as there is usually an over-emphasis, in collecting, of rare or disjunct species.

imposed by gradations between adjacent areas and by dovetailing of communities from adjacent regions.[3]

Forest regions must not be confused with climax associations. Even though a region is named for the climax association normally developing within it, it should not be assumed that the region is coextensive with the area where that climax can develop. Each of the several climaxes, although characterizing a specific region, nevertheless occurs in other regions. Although a specific climax association characterizes a region, many other communities, as seral stages, physiographic climaxes, and secondary communities, occur in the region, and in fact often far exceed the climax in areal extent. Such communities, whether primary or secondary, are as much a part of the regional vegetation as are the climax communities.

The determination of extent of any forest region is based upon physiognomy and similarity of composition of forest communities, which in turn result from environmental influences, present and past. Climatic control, although determining the relative positions of the several major climaxes, does not in general determine regional boundaries. Instead, most of these appear to be determined largely by historical factors—changing climates and physiography of past ages (see Part III). However, the influence of climate must not be ignored. Both to the north and to the south, temperature and length of the growing season are very important. Thus, the Hemlock–White Pine–Northern Hardwoods region (often considered as a formation of coördinate rank with the Deciduous Forest) occupies an area of short frostless season—generally less than 150 days. The limits of the Southeastern Evergreen Forest region, which is transitional to the Subtropical Evergreen Forest Formation, are in large part determined by the control of temperature over species range. That is, the several forest formations of eastern North America—Boreal Forest, Deciduous Forest, Subtropical Evergreen Forest, and their intermediates, the Hemlock–White Pine–Northern Hardwoods Forest, and the Southeastern Evergreen Forest—all of which develop in a humid climate, are determined by temperature factors. The subdivisions of these formations are in part due to differences in moisture factors. For example, the area of spruce–hardwoods in the Northeast (in the Northern Appalachian Highland division of the Hemlock–White Pine–Northern Hardwoods Forest) is at least in part controlled by the frequent development of "superhumid" climate.[4] Areas where an oak–hickory climax is general are areas where moist subhumid or even dry subhumid conditions occur more or less frequently. Areas where a mixed mesophytic climax

[3] The boundaries of the several regions have been determined chiefly by field observations; however, in drawing these boundaries, all available information has been used.

[4] Superhumid, humid, moist subhumid, and dry subhumid are terms used for the principal climatic types which occur in the Deciduous Forest area. The limits of these are determined by precipitation-effectiveness (Thornthwaite, 1941).

prevails are areas where a humid climate is normal and where anything drier is of very rare occurrence. The existence of this climax in drier regions is only possible where the compensating effects of the local environment offset the regional dryness.

The present pattern of forest distribution in eastern North America in terms of forest regions, as shown on the map of forest regions, is somewhat as follows:

1. The **Mixed Mesophytic Forest region,** centrally located, and occupying much of the unglaciated Appalachian Plateaus. This region is characterized by the prevalence of mixed mesophytic climax communities. It is the stronghold of the Mixed Mesophytic association—the climax association in which dominance is shared by a number of species, particularly beech, tuliptree, several species of basswood (but not *T. americana*), sugar maple, sweet buckeye, chestnut, red oak, white oak, and hemlock. This association develops only on moist but *well-drained* sites. Deeply melanized soil and a mull humus layer are characteristic features. This climax reaches its best development in the Cumberland Mountains where many variants occur. Although characteristic, the climax is not universally present. Certain physiographic climaxes (preclimaxes) resembling the climaxes of other regions and the many edaphic and secondary communities add to the diversity of the region.

2. The **Western Mesophytic Forest region,** between the Mixed Mesophytic Forest region and the bluffs of the Mississippi River. This includes the Interior Low Plateau[5] and some contiguous areas. Its vegetation is a mosaic of unlike climaxes and subclimaxes, and thus may be thought of as an ecotone. Representative examples of the Mixed Mesophytic association occur frequently in its eastern part, and more locally westward. Oak–hickory and prairie communities resembling the climaxes to the west and several intermediate types, as oak–tuliptree and beech–chestnut, take part in the mosaic. Edaphic and secondary communities further complicate the picture. Gray-brown podzolic, lateritic, and melanized soils, together with certain intra-zonal types (rendzinas, planosols), present a mosaic pattern comparable to that of the vegetation.

3. The **Oak–Hickory Forest region,** centering in the Interior Highlands (Ozark and Ouachita Mountains) and extending northward and northeastward onto glaciated territory. This region is characterized by the development of climax communities in which oaks and hickories are dominant. It is the center of development of the Oak–Hickory association, the climax of the drier western part of the deciduous forest.

4. The **Oak–Chestnut Forest region,** including most of the Ridge and Valley and Blue Ridge Provinces and extending across the Piedmont onto the Coastal Plain from northern Virginia northward, and onto glaciated

[5] For map of physiographic provinces, see p. 492.

territory in southern New England. This region is characterized by the former dominance of oak–chestnut forest[6] on most of its slopes, and by the dominance of white oak forest, particularly on the broad expanses of the Great Valley. This region is the center of development of the Oak–Chestnut association, a climax in which chestnut, red oak, chestnut oak, and tuliptree are the most frequent dominants, and of the white oak physiographic climax. Like other regions, it is diversified by inclusions of other climax associations, of which the mixed mesophytic cove forests are particularly well developed, and by altitudinal variations, because of its generally mountainous character. In much of the region, little primary forest remains and regional features are obscured by the prevalent secondary communities.

5. The **Oak–Pine Forest region,** occupying most of the Piedmont Plateau from Virginia southward and stretching across the Gulf States as a transition belt between more northern and more southern forest. This region might be thought of as an eastern oak–hickory region, for it is characterized by the development of an oak–hickory climax, in which white oak is the most abundant species, and of other drier types of oak and oak–hickory forest. However, the almost universal dominance of pines (particularly loblolly and yellow) in secondary forests and their widespread occurrence throughout the region has resulted in general usage of the term "oak–pine." This is a region of podzolized and melanized lateritic soils, the red and yellow soils of the South.

6. The **Southeastern Evergreen Forest region,** including some deciduous forest communities and confined to the Coastal Plain, extending from southern Virginia to Texas. Its most prominent vegetational features are the preponderance of evergreen trees and the great expanses of subclimax longleaf pine forest. However, successional trends point to the development of broad-leaved climax communities which are dominantly deciduous toward the north; while southward, because of the codominance of the evergreen magnolia, the communities are transitional to the Subtropical Evergreen Forest.

7. The **Beech–Maple Forest region,** lying to the north of the Mixed Mesophytic and Western Mesophytic Forest regions. Its southern boundary is the southern boundary of Wisconsin glaciation. This region is characterized by the development of the Beech–Maple association, in which beech and sugar maple are the dominant species. Because of the youth of the land surface, much of the area is occupied by developmental communities, the climax being limited to the better drained but mesic sites. The regional soils are gray-brown podzolic forest soils.

8. The **Maple–Basswood Forest region,** the smallest and most northwesterly region of the deciduous forest, occupying the Driftless area and extending a short distance northwestward onto young drift in southcentral

[6] For discussion of present status of oak–chestnut forest, see p. 192.

Minnesota. It is characterized by the development of a climax in which sugar maple and basswood (*Tilia americana*) are the dominant species and red oak a usual associate.

9. The **Hemlock–White Pine–Northern Hardwoods region,** often considered as a distinct formation, stretching across the north from Minnesota to the Atlantic Coast and occupying a position between the Deciduous Forest and the Boreal Forest. The region is characterized by the pronounced alternation of deciduous, coniferous, and mixed forest communities. Sugar maple, beech, basswood, yellow birch, hemlock, and white pine are climax dominants of the region as a whole, but not all occur in all parts of this vast region. Red pine and, in the east, red spruce are characteristic species of the region, the former in dominantly coniferous communities, the latter in dominantly deciduous communities.

The forest regions have, with two exceptions, been subdivided into sections. These sections, although characterized by the same climaxes, differ from one another in some of their vegetational features because of differences in topography, soil, climate, or history.

Of these forest regions, the Mixed Mesophytic, Western Mesophytic, Oak–Hickory, Oak–Chestnut, and Beech–Maple together comprise what is known to foresters as the "Central Hardwood" region; the Oak–Pine is transitional between this and the "Southern" region; and the Hemlock–White Pine–Northern Hardwoods approximates the "Northern" region as mapped by Mattoon, 1936.

The description of environmental features within any one of these regions is necessarily somewhat comparative. Prevalent features tend to set a standard. Thus, in a region of strong relief, small flattish areas along creeks are called "bottoms" or flats; hills are really hills. In a region of low relief where bottomlands are extensive, small flats on creeks are relatively insignificant; the term hill is applied to any recognizable elevation, an elevation so slight that it would be called flat in a rugged area. What may be called "swamp forest" in one region might be thought of as a comparatively dry type in another where deep swamps are extensive. The "rich woods" of an area where oak and pine woods prevail is not rich when compared with regions where mixed mesophytic forest prevails; it is generally comparable to the dryer phases of the mesophytic forests of such regions. Local usage must be followed in some instances; hence, it is not always possible to standardize the use of terms when they are nontechnical.

Species occur in different combinations in different regions; hence, the ecologic significance of a species in one region is often quite different from its significance in another. Also, similar combinations (of dominants) may be different in their ecologic status. A few examples will clarify these points. Chestnut, in the Mixed Mesophytic Forest region, displays (or displayed) great ecologic amplitude, occurring in typical mixed mesophytic communities, in the driest ridge-top communities as a dominant, and in many intermediate types. In the Oak–Chestnut region, it is normally codominant with oaks in mesic slope forests. In the Western Mesophytic Forest region, it may be codominant with beech. Tuliptree, also, normally occurs in a wide range of habitats and, hence, in quite unlike combinations in different regions. It is commonly a dominant in oak–chestnut–tuliptree communities of the Oak–Chestnut region; in mixed mesophytic communities of the Mixed Mesophytic region, the Western Mesophytic region, and the Oak–Chestnut region; in oak–tuliptree types of the Western Mesophytic Forest region; and is frequently mentioned as a constituent of bottomland forests in the Oak–Pine region. White oak has an even wider ecologic range. In the Mixed Mesophytic region, it is codominant with beech or with beech and hemlock in certain drier site segregates and occurs in the typical mixed mesophytic communities. It is generally the dominant of mesic climax and physiographic climax oak and oak–hickory forests in large parts of the Oak–Hickory, Oak–Chestnut, and Oak–Pine regions. It is a dominant in late developmental stages in hydrarch successions on alluvial lands and poorly drained till plains. Hophornbeam (*Ostrya virginiana*), a frequent subdominant in beech–maple climax communities of the Beech–Maple and Hemlock–White Pine–Northern Hardwoods regions, is southward often a tree of open xeric slopes.

Contrasts in the behavior of species in different regions may be noted. While chestnut sprouts freely after cutting in all parts of its range (most prolifically toward the northeast, however) and oaks, if not too large, in a considerable part of their area (again, especially northward), such mesophytes as beech and sugar maple, which give rise to sprout stands in the north, do not regenerate southward. Mesic forests, except in the north, if cut, will reproduce only through long secondary succession involving growth from seed. Hophornbeam, also, sprouts freely in the north, sometimes so prolifically as to suppress reproduction of other species (Diller and Marshall, 1937), while southward, sprouts are rare.

Secondary successions differ greatly in complexity and in number in different parts of the deciduous forest. Only in peripheral areas, in the regions of few climax dominants, is it possible to speak of *the* old-field succession. In areas where the Mixed Mesophytic association is climax, the early stages of subseres are legion.

The observable facts of present forest composition and distribution make up the bulk of the material of Part II. The evolution of the present pattern of forest distribution as determined from these facts, from past influences, and past distribution is considered in Part III.

CHAPTER 4

The Mixed Mesophytic Forest Region

In general, the Mixed Mesophytic Forest region is coextensive with the unglaciated Appalachian Plateaus except in the north, where a lobe of the Hemlock–White Pine–Northern Hardwoods region extends southward in Pennsylvania, and in the south, where the transition to the southeastern Oak–Pine region becomes evident before the physiographic limits of the plateau are reached. It includes, then, all of the Cumberland Mountains, the southern part of the Allegheny Mountains, all but the northeast arm of the Unglaciated Allegheny Plateau, and all but the southernmost end of the Cumberland Plateau.

The eastern boundary of this region lies approximately along the Cumberland and Allegheny fronts. Only along the middle part, near the Virginia-West Virginia line, is this boundary indistinct. Here, the features of the Mixed Mesophytic region lap over onto the adjacent ridges of the Ridge and Valley Province, which here are crowded close to the plateau escarpment. The vegetational boundary here, between the Mixed Mesophytic and Oak–Chestnut regions, is, like the physiographic boundary, more or less arbitrary, because of a transition area. The western boundary of the Mixed Mesophytic region is drawn along the western escarpment of the Cumberland and Allegheny Plateaus.[1] The northern boundary is irregular and indefinite, as tongues of mixed mesophytic forest extend northward across the plateau along major valleys and a lobe of northern forest extends southward on the high plateau of northwestern Pennsylvania and in the Allegheny Mountains. A broad transition belt intervenes between typical Mixed Mesophytic forest to the south and Hemlock–White Pine–Northern Hardwoods to the north (p. 86). The southern boundary, also, is indefinite and is drawn within the transition to the Oak–Pine region, but near the physiographic boundary.

The Mixed Mesophytic association, which characterizes this region, is the most complex and the oldest association of the Deciduous Forest Formation. It occupies a central position in the deciduous forest as a whole, and from it or its ancestral progenitor, the mixed Tertiary forest, all other climaxes of the deciduous forest have arisen (Chapters 17, 18).

The recognition of mixed mesophytic forest (although not under that

[1] For discussion of this boundary, see p. 123.

name) and its relation to the Tertiary forest should be credited to Harshberger who, in 1904, wrote of the "mixed deciduous forest formation" in southeastern Pennsylvania. He stated that this forest

may be looked upon as the northeastern extension of the forest found developed in its highest character in the region drained by the Tennessee river and its tributaries and by streams arising in the Allegheny mountains and flowing eastward into the Atlantic. Arbitrarily, a line drawn from a point where the Ohio joins the Mississippi river, east to the Cumberland mountains and thence along the Allegheny mountains to the west branch of the Susquehanna river in Pennsylvania, then to the Blue Ridge and along it to the Skuylkill river, following the hills on the south side of the Great Valley to the Delaware river, represents the northern limit during glacial times of the forest which during the Miocene period extended north into the Arctic regions.

His arbitrary northern limit approximates the northern boundary of the Western Mesophytic Forest region (in which mixed mesophytic communities are of frequent occurrence), of the Mixed Mesophytic Forest region, and of that part of the Oak–Chestnut region in which mixed mesophytic communities occur.

The term "mixed mesophytic" has been in use since 1916 (Braun, 1916; Sampson, 1930). It is the name applied to a climax association in which dominance is shared by a number of species; hence, generic designation is not satisfactory. It is not synonymous with mixed forest, although many mixed forests are, technically, mixed mesophytic forests. Not all mesophytic forests of mixed composition can be correctly termed "mixed mesophytic." For instance, Allard and Leonard (1943), in describing the original forest of Bull Run Mountain, Virginia, state that it was "largely mixed mesophytic forest with the white oak, chestnut oak, black oak, black gum, hickory, tulip poplar, white ash, and some beech." As the authors capitalize Oak–Chestnut and Oak–Hickory, and not "mixed mesophytic," they probably were not using mixed mesophytic in its restricted sense. Certainly such a community is not a part of the Mixed Mesophytic association.

The Mixed Mesophytic forest climax is a community in which the dominant trees of the arboreal layer are beech (*Fagus grandifolia*), tuliptree (*Liriodendron tulipifera*), basswood (*Tilia heterophylla, T. heterophylla* var. *Michauxii, T. floridana, T. neglecta*), sugar maple (*Acer saccharum, A. saccharum* var. *nigrum, A. saccharum* var. *Rugellii*), chestnut (*Castanea dentata*), sweet buckeye (*Aesculus octandra*), red oak (*Quercus borealis* var. *maxima*), white oak (*Q. alba*), and hemlock (*Tsuga canadensis*). In the Southern Appalachian outliers, silverbell (*Halesia monticola*) is also one of the dominants. Additional more or less abundant or local species are birch (*Betula lutea* var. *allegheniensis, B. lenta*), black cherry (*Prunus serotina*), cucumber tree (*Magnolia acuminata*), white ash

View of mountain slope clothed with mixed mesophytic forest. Note variety of crown forms, and preponderance of broad crowns, features of the Mixed Mesophytic association. White basswood (*Tilia heterophylla*), a characteristic species, is conspicuous from a distance because of the whitish color of the crowns of the trees. It is abundant on north and east slopes, and usually absent on south slopes. (Courtesy, *Am. Midland Naturalist,* **16:**549.) Photo by the author.

(*Fraxinus americana,* including var. *biltmoreana*), and red maple (*Acer rubrum*). Sour gum (*Nyssa sylvatica*), black walnut (*Juglans nigra*), and species of hickory (especially *Carya ovata* and *C. cordiformis*) occur in a large proportion of the stands but are never abundant. To the 25 species and varieties already mentioned, about a dozen others which sometimes appear in climax stands could be added. Data on forest composition (to be given later) will list many of the species. Not all of these species occur in any one area; some are not found everywhere that the Mixed Mesophytic climax association occurs.

Because of the large number of dominants of this climax, the composition and relative abundance of the dominants vary greatly from place to place. This has resulted in the development of apparently distinct climax communities, association-segregates, all of which are actually a part of one great climax unit (Braun, 1935). Some will prefer to call certain of these faciations or lociations, but this would necessitate distinguishing, for example, a sugar maple–basswood–buckeye faciation in one place and a sugar maple–basswood–buckeye lociation in another. The use of two vegetational unit terms for one community is not only objectionable, but also obscures the common origin and genetic relation of the two geographically separated pieces of the one community, sugar maple–basswood–buckeye. Segregation in the mixed mesophytic forest locally results in communities closely simulating other climaxes. For instance, oak–chestnut communities develop on some of the drier sites and oak–hickory communities on others. These are subclimaxes or physiographic climaxes in the area of the Mixed Mesophytic climax. They are not developmental steps in local seres, but their segregation from the typical mixed mesophytic forest, by gradual dropping out of the more mesophytic species and species of narrower community range, is readily demonstrable in extensive primeval areas of diverse topography (Braun, 1940). The recognition of this genetic relation within the Mixed Mesophytic Forest is of great importance, for it tends to clarify the relations existing between adjacent climaxes, and to explain the distribution of unlike climaxes in transitional areas. These relations will be considered after all climax communities and regions have been discussed (see Chapters 17, 18).

The general recognition of a climax with a large number of dominants, such as the Mixed Mesophytic, has been slow, although a number of writers have long since pointed out the mixed character of certain climax forest communities (Harshberger, 1904; Griggs, 1914). The idea that a climax is dominated by one or a very few species is widespread. It is suggested by Clements' statement that the climax "is reached when the occupation and reaction of a dominant are such as to exclude the invasion of another dominant" (1928, p. 105). This was later modified to read "occupation and reaction of the dominants or dominant are such as to exclude the invasion of other dominants" (Weaver and Clements, 1938, p. 81). It is used by Hough (1936) in comparing the East Tionesta Creek and Heart's Content forests, where he states that the former is higher in the scale of ecologic development "as indicated by the paucity of species reaching 5 per cent in abundance and 25 per cent in frequency." In this case, the regional climax has few dominants; hence, the conclusion reached is valid. But as a principle, this idea is untenable. The most luxuriant tropical selva is a climax in which a tree species may not be represented more than once in an acre (Richards, 1945). It is a recognized fact that the tendency toward dominance by fewer species increases with distance from equatorial regions. More northern forests, such as spruce–fir or beech–maple, have few dominants. The association of the Deciduous Forest which occupies the area of optimum moisture and temperature conditions of North America, is characterized by a larger number of dominants. In this it resembles lower latitude forests and displays its origin from the widespread mixed Tertiary forest of more equable climate.

Early ecologic familiarity with the Beech–Maple climax resulted in its being projected (mentally) southward, apparently as far as beech or sugar maple are abundantly represented. No one familiar with the true Mixed Mesophytic can fail to separate these two climaxes. Care must be taken, also, to distinguish mixed forest communities that are developmental stages from mixed communities that are climax (pp. 309, 318). Sampson (1930) clearly distinguished such communities in northeastern Ohio.

Although it is the canopy layer which primarily characterizes the Mixed Mesophytic climax, lower layers also have distinctive features. To the large number of canopy species must be added the lower trees, which seldom or never attain canopy position, as dogwood (*Cornus florida*), the magnolias (*Magnolia tripetala, M. macrophylla, M. Fraseri*), sourwood (*Oxydendrum arboreum*), striped maple (*Acer pensylvanicum*), redbud (*Cercis canadensis*), ironwood or blue beech (*Carpinus caroliniana*), hop-hornbeam (*Ostrya virginiana*), holly (*Ilex opaca*), and service-berry (*Amelanchier arborea*). Among the shrubs, *Lindera Benzoin, Hamamelis virginiana, Asimina triloba, Hydrangea arborescens,* and *Cornus alternifolia* are most generally present and abundant. *Rhododendron maximum,* a dominant shrub in many situations, is very local in, or entirely absent from, most of the more typical mixed mesophytic communities. Other species, more or less widespread, are *Viburnum acerifolium, Ribes Cynosbati, Pyrularia pubera, Stewartia ovata* (*S. pentagyna*), *Sambucus canadensis, Evonymus americanus, E. atropurpureus, Clethra acuminata,* and *Aralia spinosa.* Many more species, especially ericads, occur in subclimax situations. Woody lianes, neither numerous as to species nor abundant as individuals in the climax, include: *Parthenocissus quinquefolia, Vitis* spp., *Celastrus scandens, Bignonia capreolata* (more often not in climax communities), *Aristolochia durior,* and *Smilax hispida.*

It will be noted that among the woody plants many are southern in distribution, a few are endemic in the southern Appalachian Highlands, a very few are northern, and others are widespread species. Plants of the first two categories best serve to distinguish this association from the more northern Beech–Maple association. Perhaps the most characteristic canopy trees of the Mixed Mesophytic association are white basswood, *Tilia heterophylla,* and sweet buckeye, *Aesculus octandra,* both of which are southern in distribution. It should be noted that the basswoods of the Mixed Mesophytic Forest region are not the basswood of the northern forest. All of the more abundant species (*T. heterophylla,* including var. *Michauxii, T. floridana, T. neglecta*) are more southern in distribution than *Tilia americana* (*T. glabra*) of the Beech–Maple, Maple–Basswood, and Hemlock–White Pine–Northern Hardwoods regions. The more northern *T. americana* appears to occur only in the northern part of the Mixed Mesophytic region and in its western ecotone, the Western Mesophytic Forest

Sweet buckeye, *Aesculus octandra,* is one of the most characteristic species of the Mixed Mesophytic association, and a tree of large size (here 42 inches d.b.h.). In this forest, on the lower northwest slope of Pine Mountain, Kentucky, it is associated with sugar maple, basswood, and beech. (Courtesy, *Am. Midland Naturalist,* **16:**551.) Photo by the author.

region. The magnolias of the lower tree layer and a number of the shrubs and lianes are southern. Only hemlock among the canopy dominants and *Acer pensylvanicum* among the small trees are usually classed as northern. The latter is never abundant in the Mixed Mesophytic region and is local in distribution.

The category to which hemlock belongs is open to question. Because it is a dominant in the Hemlock–White Pine–Northern Hardwoods region, it is usually thought of as northern, although it does not extend far into the adjacent boreal forest. It is just as conspicuous and abundant in much of the lower elevation Applachian forest as it is in the North. To explain its occurrence in the mixed mesophytic forest by Pleistocene migration *from* the north is to overlook the general mixed character of the Tertiary forest, which by zonal segregation gave rise to the late Pliocene and Pleistocene banding of formations we recognize today. If hemlock was not a part of the undifferentiated forest of the Tertiary, of which the mixed mesophytic is a persisting remnant, how is the distribution of its root parasite, *Buckleya distichophylla,* to be explained? This rare endemic of the Southern Appalachians (p. 481) has, as have many other Tertiary genera, two close relations in Asia. Boreal species do not display this eastern America-eastern Asia relationship. This feature and the contemporaneous arrival of hemlock and mesophytic hardwood species in postglacial northward migration, as shown by pollen records (p. 465), point so strongly to the long indigenous nature of hemlock in the area of the Mixed Mesophytic climax that it seems best to omit mention of it as a "northern" species. If the entire flora of the Mixed Mesophytic Forest region be considered, developmental and edaphic communities as well as climax communities, it is apparent that southern plants far outnumber northern and that the endemic element is prominent. However, the floristics of the several climaxes is a separate study.

The emphasis thus far in this general consideration of the mixed mesophytic forest has been upon its woody constituents. The herbaceous vegetation is exceedingly rich and varied. The vernal flora is unexcelled in the deciduous forest and is accented by such showy flowers as *Trillium grandiflorum, T. erectum, Erythronium americanum, Cypripedium Calceolus* var. *pubescens* (*C. parviflorum* var. *pubescens*), numerous species of *Viola, Sanguinaria canadensis, Stylophorum diphyllum, Delphinium tricorne, Hydrophyllum* (four species), *Phacelia bipinnatifida, Phlox divaricata,* and *Synandra hispidula.* Among the numerous though smaller white, whitish, or greenish flowers are *Anemone lancifolia, A. quinquefolia, Anemonella thalictroides, Actaea pachypoda, Caulophyllum thalictroides, Claytonia virginica, C. caroliniana, Dicentra canadensis, D. Cucullaria, Cardamine Douglassii, Tiarella cordifolia,* etc. In summer, under the deep shade of the forest canopy, there are few flowering plants with showy flowers; ferns,

The vernal flora of mixed mesophytic forests is exceedingly rich and varied. Here, *Trillium grandiflorum* and *T. erectum* (the latter represented by red, white, and yellowish forms) are very abundant, while among the other species represented are *Delphinium tricorne, Stellaria pubera* var. *sylvatica, Dicentra canadensis, D. Cucullaria, Viola eriocarpa, V. canadensis, Disporum lanuginosum,* and *D. maculatum.* Ferns, which will later dominate the aspect, are just beginning to unroll. (Courtesy, *Ecol. Monographs,* **10:**213.) Photo by the author.

especially *Dryopteris Goldiana, Phegopteris hexagonoptera, Adiantum pedatum, Athyrium pycnocarpon, A. thelypteroides,* and *Osmunda Claytoniana* are very abundant in undisturbed areas. The late summer and fall flora contain a variety of mesophytic asters, goldenrods, and other composites (especially *Aster cordifolius, A. divaricatus, Solidago caesia, S. latifolia,* and *Eupatorium rugosum*), as well as a scattering of species from other families.

The interdependence of canopy and herbaceous layer is strongly marked. The leaf litter of typical mixed mesophytic forest is conducive to the formation of a mull humus layer. By far the larger number of the herbaceous plants are typical mull plants rooting in the porous and friable melanized soil. Few of them can grow in the ground layer conditions of oak–chestnut or oak–hickory forests. This contrast between the herbaceous plants of different forest types has been studied by Potzger and Friesner (1940), working in Indiana in adjacent oak–hickory and impoverished mixed mesophytic (called beech–maple by the authors). It is apparent in the under-

growth lists of mixed mesophytic and oak–chestnut communities in the Cumberland Mountains (Braun, 1935, 1940, 1942). The contrast with the herbaceous vegetation of the beech–maple forest is less marked, as some parts of that forest have a mull humus or humus of intermediate type, where the vernal flora is prolific. The variety and luxuriance is, however, greater in the mixed mesophytic forest. Southern species and endemics, represented by *Disporum maculatum, Astilbe biternatum, Phacelia bipinnatifida, Arisaema triphyllum* (as contrasted with *A. atrorubens*), etc., further differentiate the herbaceous layer of the mixed mesophytic forest from that of the beech–maple forest.

In the geographic area included in the Mixed Mesophytic Forest region, climax mixed mesophytic communities originally covered most of the land surface except the drier ridge-tops and upper southerly slopes, the flood plains, and a few peculiar edaphic situations. Mixed mesophytic forest communities are not, however, confined to this region, although beyond its limits they are less continuous and usually occupy only the better sites (Chapter 13, p. 444). Toward the limits of distribution of the Mixed Mesophytic association, a smaller proportion of the land was originally occupied by mixed mesophytic forest and a larger proportion by phases of other climaxes. However, excellent mixed mesophytic communities do occur in isolated areas. These outlying areas of mixed mesophytic forest will be considered in connection with the regions in which they occur.

From the central part of the region, in the Cumberland Mountains (the area of best development of mixed mesophytic forest), there is a gradual reduction in complexity outward toward the spatial limits of the Mixed Mesophytic climax. This change may be accompanied by dovetailing of mixed mesophytic communities and communities from adjacent climaxes, producing regions of the sort seen to the west of the Mixed Mesophytic region. It is generally apparent northward and northwestward in gradual impoverishment—decrease in the number of species represented—resulting in the development of communities suggesting beech–maple. This ecotone is not marked by pronounced dovetailing of communities. Eastward the limits of the true mixed mesophytic are more abrupt, because of the physiographic features of the Ridge and Valley Province. In the Southern Appalachians, true mixed mesophytic communities are well developed, but are separated from the Mixed Mesophytic Forest region by the Great Valley of the Ridge and Valley Province. Northeastward, impoverishment of the complex mixed mesophytic has there resulted in a somewhat comparable mixed climax in which oaks and (formerly) chestnut are abundant, together with some of the more mesophytic constituents of the Mixed Mesophytic association, especially beech. That forest is here included in the Oak–Chestnut region, although it has in part been referred to as Mixed Mesophytic (Gordon, 1937; Bromley, 1935).

Much of the area of the Mixed Mesophytic Forest region is now occupied by secondary forests. The best mixed mesophytic forest areas, if only once cut-over and not otherwise utilized, come back in mixed stands similar to the original mixed mesophytic, because of the rapid growth of trees of the lower layers. *Liriodendron* may occupy coves where once splendid mixed mesophytic forest grew. Many second-growth stands bear little or no resemblance to the original forest cover. This is particularly true of those which follow the segregates with lower water requirements. Site changes, due to increased insolation, more rapid run-off, and erosion, may result in development of communities suggesting oak–hickory rather than mixed mesophytic. Scrub pine (*Pinus virginiana*) may clothe slopes once occupied by the beech–white oak segregate of the mixed mesophytic forest. Only from primary remnants can the original type be ascertained.

The time which has been available for vegetational development in the Mixed Mesophytic region is so great (uninterrupted since Tertiary) that few seral stages of primary successions remain. Primary forest types on slopes in the area where the mixed mesophytic is climax may range from the most mesophytic of mixed forests to the driest pine or oak forests. These latter

River birch, *Betula nigra,* is abundant along streams in the southern half or more of the Mixed Mesophytic Forest region. Here, on Red Bird River, in the Rugged Eastern Area of the Cumberland Plateau, it is seen in the foreground on the right, and, with sycamore, lines the left bank of the stream.

forests are not mixed mesophytic communities, but are subclimaxes or seral stages related to extreme local conditions. In considering the Mixed Mesophytic Forest region, all these types must be included, as well as the climax types. This will be done by descriptive and statistical treatment of representative areas selected from the several sections of the Mixed Mesophytic Forest region.

Flood-plain successions, which are more or less uniform over greater areas of the deciduous forest, need only be mentioned briefly. In the Mixed Mesophytic Forest region, flood plains do not occupy any appreciable part of the whole area, nor are they, with the exception of that of the Ohio River, subject to protracted periods of overflow. The characteristic flood-plain vegetation is in many places little more than a band of streamside trees. Larger areas of cottonwood and silver maple, trees which will endure long-time submergence, are seldom seen. Willows, sycamore, sweet gum, and river birch line most of the streams, the last named being the most abundant species in the southern half or more of the region. Willow oak (*Quercus Phellos*) occurs on broad but local stream flats, and *Ulmus serotina* is occasional on abrupt stream banks.

Three sections of the Mixed Mesophytic Forest region are recognized: the **Cumberland Mountains,** the **Allegheny Mountains,** and the **Cumberland** and **Allegheny Plateaus.** The last of these is further divided into distinct subsections. The Cumberland Mountains are outstanding in the splendid development of mixed mesophytic forest. In the Low Hills Belt of the Cumberland and Allegheny Plateaus section, mixed mesophytic communities are much circumscribed.

CUMBERLAND MOUNTAINS

The Cumberland Mountains section of the Mixed Mesophytic Forest region is essentially coextensive with the physiographic section by the same name. This is a mountainous section, whose highest and most characteristic part is carved from a large fault block with more or less upturned edges forming the monoclinal bordering mountains—Pine Mountain along the northwest and Cumberland Mountain and Stone Mountain on the southeast. The more prominent mountains of the interior strongly dissected part are Black Mountain, Little Black Mountain, and Log Mountain with elevations of from 3000 to over 4000 feet. A mountainous area adjacent to the fault block section on the west is included in the Cumberlands, and is here referred to as "the outlying area" of the Cumberlands.

Deep melanized soils characterize most of the area, even the steep mountain slopes unless too dry for mixed mesophytic forest. Gray-brown podzolic soils or, in some instances, shallow podzols with a laminated or fibrous mor humus layer occur on dry slopes and ridges occupied by oak–

chestnut or oak–pine communities. Locally, where very abundant, hemlock has caused the formation of a deep mor layer which almost excludes herbaceous plants.

Mixed mesophytic forest of superlative quality covers, or originally covered, most of the slopes of this section. In number of tree species, in size of individuals, in variety of forest types, it ranks as one of the finest deciduous forest areas of North America.

The primary forest is a mosaic of climax and subclimax communities intimately interrelated and complexly variable. Developmental communities, except those of secondary successions, are few. Some of the climax communities accord well with the abstract concept of the Mixed Mesophytic association; others, characterized by fewer dominants, are association-segregates, communities resulting from shifts in dominance and local segregation of some of the dominants of the mixed forest. Some of these segregates are related to variations in altitude, some to differences in slope exposure and, hence, to the moisture factor, while some appear to be fortuitous variations. Diversity in topography and microclimates is largely responsible

Mixed mesophytic forest covers most of the slopes of the Cumberland Mountains. It is interrupted along ridge crests and on upper southerly slopes by less mesic forest types. (Courtesy, *Ecol. Monographs,* **12**:429.) Photo by the author.

for the complexity of the mixed mesophytic forest of the Cumberland Mountains.

If, from a distance, one views any mountain slope in the Cumberland Mountains (except the southeast slope of Pine Mountain and the northwest slope of Cumberland Mountain), the impression is gained that the whole slope of a thousand feet or more of vertical range is covered by a mixed forest almost homogeneous from bottom to top. Here and there, giant tuliptrees tower above the other trees; whitish blotches which are *Tilia heterophylla* dot the slope. In autumn, when fall coloration aids in distinguishing species from a distance, we may note that sugar maple is scattered throughout but tends to be more abundant on higher slopes; while beech, readily distinguished by its coppery color, does not extend to the higher elevations. In the lower outlying parts of the Cumberlands, where the altitudinal limits of this species are not reached, beech is seen to extend to the tops of slopes, to the limits of mixed mesophytic forest. Red oak (*Quercus borealis* var. *maxima*), although nowhere abundant, is seen to be well distributed on the slope and is easily identified as the bright green crowns among the many richly colored species. Dead snags here and there are chestnut. Occasional hickories may be recognized, and we know that in the mass are many other species lacking distinctive features visible from a distance. If we are viewing a south slope, we may miss the whitish *Tilia heterophylla* or see but a few of these trees; oaks will be somewhat more abundant, especially *Quercus montana,* always readily distinguished by color from a distance; white oak will be noted, more or less concentrated with beech on the lower slopes, below about 2000 feet in elevation. The highest southerly slopes and narrow ridge-tops will stand out in strong contrast to the general slope appearance, for there chestnut oak and chestnut prevail.

The dominant vegetational feature of both Black and Log Mountains is the splendid deciduous forest of the mountain slopes. Some altitudinal sequence is discernible, as beech and white oak drop out at about 2000 feet, and a sugar maple–basswood–buckeye, or sugar maple–basswood–buckeye–tuliptree segregate is prominent at middle elevations.

The most striking features of the inner (dip) slopes of Pine Mountain and Cumberland Mountain are the large amount of pine in parts of the forest where shallow sandy soils overlie the steeply dipping sandstone strata, and the high proportion of hemlock in some of the gorges. The outer (scarp) slopes are (or were until recently) clothed with deciduous forest, more mesic on Pine, less mesic on Cumberland Mountain, because of differences in slope exposure.

Mixed mesophytic forest prevails throughout the outlying area of the Cumberlands, except where occasional outcrops of massive sandstone favor the growth of pine and on ridge crests where subclimax oak–chestnut stands occur. Because this outlying area does not attain as high elevations as the central part of the Cumberlands, the forest on the slopes is more homogeneous (if a mixture may be called homogeneous); beech goes almost to the tops of the ridges and is a constituent of most of the mixed mesophytic communities; the sugar maple–basswood–buckeye segregate, so conspicuous above 2000 feet on Black and Log Mountains, is poorly represented or absent.

The most conspicuous subdivisions of the forests of this section are those related to composition: (1) the all-deciduous mixed mesophytic; (2) the hemlock–mixed mesophytic; and (3) subclimax communities. During the period of fall coloration, when the many species of the mixed mesophytic forest have assumed their varied hues, the areas of typical mixed mesophytic, of mixed mesophytic with hemlock, of oak–chestnut, and of pine communities stand out clearly. Then the small area occupied by oak–chestnut forest is most apparent and the areal dominance of the Mixed Mesophytic association is emphasized.

Representative examples of the three principal types of communities are selected from the three topographic divisions of the section, to illustrate forest composition, and as a basis for later comparisons with other parts of the Mixed Mesophytic Forest region, and with other climaxes.[2]

The All-deciduous Mixed Mesophytic Communities

The all-deciduous mixed mesophytic forest, considered in its entirety, is most closely representative of the abstract concept of the Mixed Mesophytic association. As exemplified in the Cumberland Mountains, this is made up of some 30 or more canopy species, of which about a dozen are more or less constantly present. Composition (Table 1) is demonstrated by percentages obtained from transects ascending or partly encircling the slopes. Seven samples are selected from areas of unlike altitudinal range and unlike exposure, to illustrate the wide range of variability in percentages of constituent species.

These seven sample areas, taken together, show the slope forest to be composed of about 35 species, some of which are generally present while others appear in only one, two, or three samples. Of the 33 species[3] listed in the seven sample areas of mixed forest (Table 1), nine are present in all samples, regardless of altitude or slope exposure. These are *Acer saccharum, Tilia heterophylla, Castanea dentata, Quercus borealis* var. *maxima, Liriodendron tulipifera, Fraxinus americana, Magnolia acuminata, Nyssa*

---

[2] For further information on these communities and their relations to, or gradation into, adjacent communities, the reader is referred to the papers here cited, and especially to the composition charts in those papers (Braun, 1935a, 1940, 1942).

[3] *Carya* sp. not counted; *Fraxinus americana* and var. *biltmoreana* counted as one; and *Tilia* counted as one.

Table 1. Composition of all-deciduous mixed mesophytic forest of the Cumberland Mountains in Kentucky.

A. Mesic slopes: *1,* Joe Day Branch, Black Mountain, Letcher County; *2,* Upper slopes of Log Mountain, Bell County; *3,* Northwest slope of Pine Mountain above Limestone Creek, Whitley County; *4,* Big Tree Branch, Lynn Fork of Leatherwood Creek, Perry County; *5,* Buck Branch of Jellico Creek, Whitley County.

B. South slopes: *6,* River Ridge, Black Mountain, Letcher County; *7,* Limestone Creek, Whitley County.

	A					*B*	
	1	*2*	*3*	*4*	*5*	*6*	*7*
Number of canopy trees	299	479	259	296	516	278	243
Acer saccharum	31.1	29.8	13.5	14.5	18.0	18.0	.4
Tilia heterophylla	20.1	14.6	10.8	5.7	6.2	2.2	.8
Liriodendron tulipifera	4.0	5.6	18.9	33.4	7.8	14.4	18.9
Fagus grandifolia	5.4	. .	10.0	14.9	29.1	7.9	15.6
Aesculus octandra	16.7	13.4	8.1	. .	5.4	.7	. .
Castanea dentata	9.4	13.6	8.9	8.8	2.9	19.1	11.9
Quercus borealis maxima	6.3	5.0	3.1	2.4	3.3	5.4	9.1
Quercus alba	. .	3.3	. .	6.4	7.0	.3	19.8
Fraxinus americana	1.3	3.3	7.0	.3	2.0	1.1	1.2
Magnolia acuminata	1.0	1.5	5.8	2.4	1.2	2.9	4.1
Acer rubrum	.7	. .	. .	2.4	.5	2.9	2.1
Betula allegheniensis	.7	. .	. .	1.0	. .	. .	. .
Nyssa sylvatica	.3	1.9	3.7	2.0	1.6	3.2	6.6
Magnolia Fraseri	.3	. .	. .	.3	. .	. .	. .
Juglans cinerea	.3	.2	1.1	.3	.4	. .	. .
Juglans nigra	.3	.8	.4	1.7	1.2	.3	.8
Quercus montana	.3	.8	1.9	1.0	1.3	11.9	. .
Prunus serotina	.3	.2	. .	. .	. .	.3	. .
Carya ovata	1.0	.4	1.1	3.3	. .	. .	. .
Carya tomentosa	. .	.2	.8	. .	. .	. .	1.6
Carya cordiformis	. .	.2	. .	. .	. .	.7	. .
Carya glabra	. .	. .	.4	. .	. .	4.0	. .
Carya spp.	.3	4.0	1.5	2.4	4.8	2.9	5.7
Sassafras albidum	. .	. .	1.1	. .	. .	. .	. .
Morus rubra	. .	. .	1.1	. .	. .	. .	. .
Cladrastis lutea	. .	. .	.8	. .	1.3	. .	. .
Ulmus americana	. .	. .	.4	. .	.5	. .	. .
Ulmus alata	. .	. .	. .	. .	.4	. .	. .
Liquidambar Styraciflua	. .	. .	. .	. .	1.0	. .	. .
Quercus velutina	. .	. .	. .	. .	.8	. .	.4
Robinia Pseudoacacia	. .	1.0	.4	. .	. .	1.4	. .
*Cornus florida**	. .	. .	. .	. .	. .	. .	.4
*Magnolia tripetala**	. .	. .	. .	. .	. .	. .	.4
Tsuga canadensis	. .	. .	. .	. .	. .	.3	. .

* Unusually large and approaching canopy position.

sylvatica, and *Juglans nigra.* Two, *Fagus grandifolia* and *Quercus montana,* are present in six of the seven samples. *Aesculus octandra, Acer rubrum, Quercus alba,* and *Juglans cinerea* are present in five samples; the last is absent from both south slope samples. Seventeen species are present in only one to three of the samples. Some of these, *Robinia Pseudoacacia, Quercus velutina, Ulmus americana, Ulmus alata,* and *Liquidambar Styraciflua,* more properly belong to other communities and are relicts from preceding developmental stages or accidentals from nearby unlike communities. Some, though infrequent, belong in the mixed mesophytic forest. *Cladrastis lutea,* of low frequency and present in only two of the samples, is confined to mixed mesophytic forest in the Cumberland Mountains, although in central Kentucky and Tennessee it is more abundant on river bluffs. It is represented by individuals in all layers; the largest do not attain the size reached by other species of the community. Some species, *Cornus florida, Magnolia Fraseri,* and *M. tripetala,* are regularly understory trees rarely attaining a semi-canopy position.

Degree of presence, however, is not an indication of dominance, for of those species present in all samples, *Juglans nigra* is always a minor constituent of the forest, and *Fraxinus, Nyssa,* and *Magnolia acuminata* never assume a dominant role. *Fagus,* present in six of the seven samples, is one of the dominants of the mixed mesophytic forest; its absence from one of the samples is due to altitude, for *Fagus* is not found in the Cumberland Mountains between about 2000 and 4000 feet elevation. *Quercus montana,* with the same degree of presence as *Fagus,* assumes a dominant role only in the oak–chestnut communities of this area; in the mixed mesophytic communities, it is always a minor constituent. *Aesculus octandra* and *Quercus alba,* present in five of the seven samples, are dominant species in the sugar maple–basswood–buckeye and white oak–beech association-segregates respectively (see Tables 2, 4).

In general, the more abundant species, those which together may be called dominants of the typical mixed mesophytic forest of the Cumberland Mountains, are *Acer saccharum, Tilia heterophylla, Liriodendron tulipifera, Castanea dentata, Fagus grandifolia, Aesculus octandra, Quercus borealis* var. *maxima,* and *Q. alba.* These eight species comprise from 68 to 76 per cent of the canopy of the southerly slope mixed mesophytic forest (Table 1, areas 6, 7), and from 73 to 93 per cent of the mixed mesophytic forest of the other more mesic situations (Table 1, areas 1–5). Variations in relative abundance of the dominants and accessory species may be noted by comparing the seven areas (Table 1).

In connection with the above seven samples of the entirely deciduous mixed mesophytic forest, no mention has been made of lower layers. Evergreen species are infrequent; *Rhododendron maximum,* if present, is generally confined to the immediate vicinity of a stream or localized near

Tuliptree, beech, and sugar maple are the most abundant species in this stand of mixed mesophytic forest. The luxuriant undergrowth and the many ferns are characteristic features of the all-deciduous mixed mesophytic forests. Big Tree Branch of Lynn Fork of Leatherwood Creek, Perry County, Kentucky. (Within Area 4 of Table 1.) (Courtesy, *Nature Magazine,* **27**:238.) Photo by the author.

rock outcrops. The trees of the canopy are represented in lower layers and with them may be *Carpinus caroliniana, Cornus florida* (abundant), *Cercis canadensis, Amelanchier arborea, Ostrya virginiana* (infrequent), *Oxydendrum arboreum,* and one or more species of *Magnolia.* The herbaceous layer is distinctive, more luxuriant, and more continuous than in any other type of forest, and includes a great number of species.

The picture of composition of the mixed mesophytic forest just given is one obtained by observation of extensive areas and from traversing large forest tracts. However, careful study reveals the tendency toward segregation of species in some situations. Many variations are found, among which are a few well-marked association-segregates which must be treated separately in any consideration of this part of the Mixed Mesophytic Forest region.

THE SUGAR MAPLE–BASSWOOD–BUCKEYE FOREST

In the higher Cumberland Mountains, this segregate forms an altitudinal band extending from about 1500 or 2000 feet in elevation upward for about 1000 feet or more. In its lowest part, beech may be abundant, but in most of the community, it is entirely absent. The tuliptree is almost always present and is in places as abundant or more abundant than any one of the three characteristic species. Toward the upper altitudinal limits, the proportion of sugar maple increases, and on the highest slopes of Black Mountain, above 3500 feet, birch (*Betula lutea* var. *allegheniensis*) increases in amount and becomes one of the dominant species; basswood and buckeye are proportionately reduced in number.

In the outlying section of the Cumberland Mountains and in other sections of the Mixed Mesophytic Forest region, this association-segregate develops only locally, usually on north slopes, and with no relation to altitude. Closely related forest communities occur in the cove and ravine forests of the Great Smoky Mountains, there complicated by the addition of silverbell tree (*Halesia monticola*), black cherry, and, sometimes, hemlock.

The sugar maple–basswood–buckeye or sugar maple–basswood–buckeye–tuliptree forest is one of the most magnificent communities of the whole mixed mesophytic forest. Its trees are large (mostly three to four feet, d.b.h.[4]) with towering columnar trunks, often rather widely spaced. Its herbaceous layer is exceedingly luxuriant, with large ferns and a wealth of spring flowers, among which *Trillium grandiflorum, T. erectum* (red, white and yellow forms), *Delphinium tricorne, Phacelia bipinnatifida, Uvularia grandiflora* (orange-yellow), and *Disporum maculatum* are perhaps most showy. The last named is here limited to this community. The soil is deep and dark, with good crumb mull humus. Again the interrelations of canopy species, soil development, and herbaceous layer are emphasized. The features displayed by this community—wide spacing of trees and continuous

[4] The abbreviation d.b.h. is commonly used for diameter breast high.

In the sugar maple–basswood–buckeye forest these three species usually comprise 80 per cent or more of the canopy. Black Mountain, Letcher County, Kentucky. (Within Area 1 of Table 2.) (Courtesy, *Ecol. Monographs,* **12**:423.) Photo by the author.

luxuriant herbaceous layer—are in keeping with the findings of Lutz (1932), who states that

at any given age, after competition begins, good forest sites usually support a smaller number of tree individuals per unit area than poor sites. Larger size, early expression of dominance, and eliminative competition largely account for the relatively small number of individuals on good sites . . . On good sites there appears to be a larger number of shrubby and herbaceous individuals per unit area than on poor sites.

Typical examples and variations of the sugar maple–basswood–buckeye forest of Black Mountain, Log Mountain, and Pine Mountain are selected to demonstrate canopy composition (Table 2). The three dominants of this

Table 2. Percentage composition of the sugar maple–basswood–buckeye and sugar maple–basswood–buckeye–tuliptree forest types in the Cumberland Mountains, Kentucky: *1,* Slopes of Black Mountain, Letcher County; *2,* Slopes of Log Mountain, Bell County; *3,* Northwest slope of Pine Mountain above Limestone Creek, Whitley County; *4,* Northwest slope of Pine Mountain near Scuttlehole Gap, Letcher County.

	1	*2*	*3*	*4*
Number of trees	162	216	147	129
Acer saccharum	35.8	39.3	20.4	18.6
Tilia heterophylla	23.4	21.7	10.9	20.9
Aesculus octandra	24.7	29.7	12.9	16.3
Liriodendron tulipifera	4.3	1.4	21.8	18.6
Fraxinus americana	2.5	3.7	6.7	2.3
Quercus borealis maxima	1.8	.5	3.4	7.8
Magnolia acuminata	.6	2.0	4.1	1.5
Castanea dentata	1.2	7.0	4.7	7.0
Fagus grandifolia	. .	. .	3.4	. .
Carya ovata	. .	. .	2.0	. .
Carya tomentosa	. .	. .	1.4	.8
Carya glabra	. .	. .	. .	3.9
Carya spp.	3.0	1.0	2.0	. .
Juglans nigra	.6	.5	.7	2.3
Juglans cinerea	.6	.5	. .	. .
Betula allegheniensis	.6	. .	. .	. .
Nyssa sylvatica	. .	.5	1.4	. .
Magnolia Fraseri	.6	. .	. .	. .
Quercus montana	. .	. .	1.4	. .
Sassafras albidum	. .	. .	1.4	. .
Cladrastis lutea	. .	. .	1.4	. .

association-segregate (areas 1 and 2) comprise over 80 per cent of the canopy (84 per cent in an area on Black Mountain and 83 per cent in an area on Log Mountain). Not only is this an extensive community, but one remarkably uniform in composition.

On Pine Mountain, tuliptree is regularly associated with sugar maple, basswood, and buckeye in a codominant role (Table 2, areas 3, 4). The

Tuliptree is usually a constituent of the sugar maple–basswood–buckeye forest, and sometimes one of the dominants of this forest. Pine Mountain, Whitley County, Kentucky. (Within Area 3 of Table 2.) (Courtesy, *Ecol. Monographs,* **12**:430.) Photo by the author.

four dominants make up 66 per cent of the stand in one area and 75 per cent of the stand in another area 60 miles away. It will be noted that, in both of these examples of the sugar maple–basswood–buckeye–tuliptree forest, the four dominants compose a smaller per cent of the stand than do the three dominants of the typical sugar maple–basswood–buckeye association-segregate. This is a nearer approach to typical mixed mesophytic forest in which dominance is spread over many species.

COMMUNITIES IN WHICH BEECH IS DOMINANT

Beech may be dominant in a mixed forest which develops near the foot of mountain slopes and in ravines. In the Cumberland Mountains, this segregate, approaching the status of a beech consociation, is never extensive. In other parts of the area where mixed mesophytic forest is climax, it may cover valley flats. Three examples are selected (Table 3, areas 1–3) to illustrate the type; of these, beech comprises, respectively, approximately 70,

Table 3. Percentage composition of ravine communities of the mixed mesophytic forest of the Cumberland Mountains of Kentucky in which beech is the dominant species: *1,* Limestone Creek, Whitley County (lower on slopes than area *3* of Table 2); *2,* Laurel Fork, Bell County, valley floor; *3,* Lower slopes, Oldhouse Branch, Perry County; *4,* Buck Branch, Whitley County; *5,* Cave Branch, Harlan County.

	1	*2*	*3*	*4*	*5*
Number of canopy trees	210	127	94	71	76
Fagus grandifolia	70.5	58.3	72.3	56.3	48.7
Acer saccharum	6.2	.8	7.4	19.7	15.8
Liriodendron tulipifera	1.4	7.9	3.2	5.6	10.5
Tilia heterophylla	5.2	1.6	7.4	5.6	7.9
Aesculus octandra	1.4	..	3.4	1.4	5.2
Quercus borealis maxima	1.4	.8	..	1.4	..
Quercus alba	.5	9.4	1.1	..	..
Magnolia acuminata	1.4	.8	1.1	..	1.3
Castanea dentata	1.9	6.3	..	1.4	5.2
Acer rubrum	1.0	7.1	..	..	..
Carya ovata	3.8	.8	1.1	..	..
Carya sp.	1.0	..	..	..	1.3
Nyssa sylvatica	.5	4.8	..	..	1.3
Fraxinus americana	..	..	1.1	1.4	..
Juglans cinerea	..	..	1.1	..	..
Ulmus americana	1.0	..	..	1.4	..
Quercus montana	.5	..	..	..	..
Magnolia tripetala	.5	..	1.1	..	..
Tsuga canadensis	.5	1.6	..	..	..
Cladrastis lutea	..	..	..	4.2	..
Ulmus alata	..	..	..	1.4	..
Betula allegheniensis	..	..	..	..	1.3
Betula lutea	..	..	..	..	1.3
Carpinus caroliniana	1.4	..	..	..	..

58, and 72 per cent of the canopy. It may rarely form as much as 80 per cent of the forest, but only over small areas.

This beech type of forest has a less continuous herbaceous layer than the typical mixed mesophytic or the sugar maple–basswood–buckeye forest. This is correlated with the greater amount of slowly decomposing beech leaf litter.

In a few places in the Cumberland Mountains, and only over small areas, sugar maple may be second in abundance to beech. This grouping is so unusual and of such small extent that it would not be recognized as a segregate if it were not for the fact that this combination of dominant species characterizes the great expanse of the beech–maple forest to the north of the mixed mesophytic forest. It, like the beech type, is a community of ravines and lower north slopes. Only two examples were found among the many areas of forest studied (Table 3, areas 4, 5). Neither area was large, as indicated by the small number of trees (71 and 76) included in the samples. It will be noted that both contain six of the usual dominants of the mixed mesophytic forest and that both contain the two most characteristic species of the mixed mesophytic forest.

Beech is dominant in the mixed mesophytic forest along valleys and on the lower slopes of ravines. Tree to right, four feet in diameter. Lower slopes of Pine Mountain, Whitley County, Kentucky. (Within Area 3 of Table 3.)

Not all forests in which beech prevails are lower north slope and ravine areas such as these just considered. Beech may be abundant in forests in which the principal species are chestnut, oaks, and hickories, together with some of the usual dominants of the mixed mesophytic forest.

BEECH–WHITE OAK OR WHITE OAK–BEECH FOREST

A forest of beech and white oak, in which either species may predominate, is the usual southerly slope forest type at low elevations in the Cumberland Mountains (Table 4). *Liriodendron* is usually present and may

Table 4. Percentage composition of the white oak–beech forest type and its variations in the Cumberland Mountains of Kentucky: *1,* Jakes Creek, Letcher County; *2,* Buck Branch of Jellico Creek, Whitley County; *3,* Oldhouse Branch, Perry County; *4,* Lynn Fork of Leatherwood Creek, Perry County; *5,* Line Fork, Letcher County.

	1	*2*	*3*	*4*	*5*
Number of canopy trees	153	135	55	86	242
Quercus alba	41.8	25.2	23.6	11.6	28.1
Fagus grandifolia	30.1	21.5	27.0	39.5	23.2
Liriodendron tulipifera	5.2	19.2	21.8	4.6	3.7
Acer saccharum	. .	.7	5.4	22.1	7.9
Castanea dentata	6.5	.7	7.3	4.6	.8
Acer rubrum	6.5	2.2	5.4	. .	6.6
Nyssa sylvatica	2.6	5.2	1.8	2.3	2.5
Quercus borealis maxima	1.3	3.7	1.8	1.2	3.3
Tsuga canadensis	. .	. .	. .	. .	16.1
Carya ovata	. .	2.2	. .	. .	. .
Carya cordiformis	. .	. .	. .	. .	1.7
Carya spp.	2.6	9.6	. .	1.2	. .
Betula allegheniensis	.6	. .	. .	1.2	2.1
Fraxinus americana	. .	1.5	. .	. .	. .
Tilia heterophylla	. .	.7	3.6	7.0	. .
Aesculus octandra	. .	. .	. .	1.2	.4
Juglans nigra	. .	. .	. .	2.3	. .
Juglans cinerea	. .	. .	1.8	. .	1.2
Magnolia acuminata	. .	. .	. .	1.2	.8
Magnolia Fraseri	. .	. .	. .	. .	.4
Oxydendrum arboreum	. .	. .	. .	. .	1.2
Ulmus alata	. .	.7	. .	. .	. .
Quercus velutina	2.6	3.0	. .	. .	. .
Quercus montana	. .	3.7	. .	. .	. .

share dominance with white oak and beech. Hemlock may or may not be present, but when present sometimes assumes a dominant position, then suggesting the hemlock–white oak type (see Table 6, D). The white oak–beech association-segregate is a well-marked forest type, extensively developed throughout the Mixed Mesophytic Forest region. The white oak type of the Appalachian Valley (p. 238) appears to be genetically related. White oak–beech communities which are developmental occur in certain areas, for example, on the till plains of southwestern Ohio (Braun, 1936). These must not be confused with the climax community under consideration.

Because of its location on lower slopes, often adjacent to cultivated valleys, and because of the high commercial value of white oak and its ready accessibility in such locations, good examples of the white oak–beech community are not often seen. The rather small number of species (10) in one of the samples (area 1) and the appearance of black oak with chestnut and red maple suggest a possible transition to oak–chestnut forest. This, however, is not supported by the general composition of white oak–beech communities, which have little in common with the oak–chestnut. In another sample (area 2), white oak, beech, and tuliptree are codominant in a community of 15 species, most of which are common species of the mixed mesophytic forest. The higher proportion of beech and fair representation of sugar maple in area 3 suggest the possibility of transitions to the beech–sugar maple ravine type. Such transitional communities are not uncommon, but seem best considered as variations of the white oak–beech forest type, even where sugar maple is more abundant than the white oak (area 4). As hemlock may be an abundant species in the white oak–beech forest, one example (area 5) of this variation is included, although this is not an all-deciduous community.

In the five examples of white oak–beech forest and its variations presented, these two species comprise from 47 to 72 per cent of the forest canopy. Twenty-three species are included in the five samples, as compared with 33 species in the samples of typical mixed mesophytic forest (Table 1). Six species, *Quercus alba, Fagus grandifolia, Castanea dentata, Liriodendron tulipifera, Quercus borealis* var. *maxima,* and *Nyssa sylvatica,* are present in all samples of the white oak–beech community. All of these are species of high presence in typical mixed mesophytic communities. The most mesophytic species are generally absent or infrequent in the white oak–beech forest, although sugar maple is present in four of the five samples, *Tilia heterophylla* in three, and *Aesculus octandra* in two. That the white oak–beech is an integral part of the Mixed Mesophytic climax is evident; its designation as an association-segregate emphasizes its relation to the mixed forest.

CHESTNUT–SUGAR MAPLE–TULIPTREE FOREST AND VARIATIONS

A series of communities may be noted in which two or more of the following are most abundant: beech, chestnut, sugar maple, tuliptree, and sometimes white oak and hickory (Table 5). In the Cumberland Mountains, the chestnut–sugar maple–tuliptree association-segregate is most distinct. It is well expressed on many of the higher slopes of westerly and southwesterly direction if they are not too steep or are not convex slopes (area 1). This community (here composed of 16 species) is a variation of the mixed

Table 5. Percentage composition of the chestnut–sugar maple–tuliptree forest type and its variations in the Cumberland Mountains of Kentucky: *1,* Upper westerly slopes, Left Fork of Colliers Creek, Letcher County; *2,* Upper slopes of Log Mountain, Bell County; *3,* Head of Limestone Creek, Whitley County; *4,* Nolan Branch, Clay County.

	1	*2*	*3*	*4*
Number of canopy trees	158	109	62	87
Castanea dentata	26.0	23.8	19.3	13.8
Acer saccharum	26.6	20.2	4.8	5.7
Liriodendron tulipifera	13.9	8.3	16.1	12.6
Quercus alba	. .	5.5	17.7	16.0
Quercus borealis maxima	6.3	11.0	1.6	2.3
Quercus montana	6.3	. .	6.4	2.3
Fagus grandifolia	1.3	. .	. .	28.7
Tilia heterophylla	6.3	5.5	4.8	2.3
Aesculus octandra	3.2	5.5	. .	. .
Nyssa sylvatica	.6	5.5	. .	4.6
Magnolia acuminata	2.5	1.8	3.2	1.2
Carya ovata	.6	.9	. .	. .
Carya glabra	. .	. .	1.6	. .
Carya spp.	1.3	9.2	16.1	5.7
Acer rubrum	3.2	. .	. .	1.2
Fraxinus americana	. .	1.8	3.2	. .
Betula sp.	. .	. .	. .	1.2
Juglans nigra	.6	. .	3.2	. .
Juglans cinerea	. .	. .	1.6	. .
Prunus serotina	.6	. .	. .	. .
Quercus velutina	. .	. .	. .	1.2
Robinia Pseudoacacia	.6	.9	. .	1.2

mesophytic forest characterized by the local dominance of the three species named. Seven of the eight most abundant species of more typical mixed mesophytic forest are present, including the two most characteristic species; white oak is absent as the elevation is too high. The prominence of chestnut further distinguishes this from more representative mixed mesophytic forest. The high percentage of chestnut in any composite picture of mixed mesophytic forest in the Cumberland Mountains (*cf.* Braun, 1940, p. 237) is due largely to its dominance in this segregate. Variations in the chestnut–sugar maple–tuliptree type occur; in some, white oak, red oak, or hickory may be relatively more abundant. The grouping (in some of the examples) of chestnut, oaks, and hickories emphasizes the relation of the oak–chestnut and oak–hickory forests and helps to explain why, with the dying of chestnut, many former oak–chestnut communities now appear to be oak–hickory communities. It helps to explain the alternation of subclimax oak–chestnut and oak–hickory communities on ridges in the outer part of the Mixed Mesophytic Forest region and in ecotone regions, where the character of underlying rock (soil pH) may determine which of the two communities develops. At lower elevations beech is generally present, often as one of the dominants, together with chestnut, tuliptree, and white oak (area 4). This variation of the chestnut–sugar maple–tuliptree segregate is significant because, with some modifications, it is widely represented outside of the area where the Mixed Mesophytic climax prevails. It is suggestive of the mixed mesophytic communities of ravines in the Oak–Chestnut region to the east and of the chestnut–beech–tuliptree and oak–chestnut–tuliptree communities extensively developed in the Western Mesophytic Forest region.

In all of these variations of the chestnut–sugar maple–tuliptree forest, the undergrowth is similar to, but less continuous and less luxuriant than, that of the more typical mixed mesophytic and sugar maple–basswood–buckeye communities. This forest type is not a transition community but an integral part of the Mixed Mesophytic climax. There are, of course, transitions to the oak–chestnut of the driest sites of the Cumberland Mountains; these generally lack the most characteristic species of the Mixed Mesophytic association and contain some of the distinctive oak–chestnut features, especially in the undergrowth where ericaceous plants are more or less well represented.

A beech–chestnut or beech–heath type is somewhat distinct. In this, beech and chestnut are two of the most abundant species, usually with more or less oak. Ericaceous shrubs—*Rhododendron, Kalmia, Vaccinium,* and sometimes *Azalea*—are abundant; shrubs and herbaceous plants of the more typical mixed mesophytic forest are also present. No extensive undisturbed stands have been seen, but numerous fragments indicate the frequent occurrence of this type. It is found locally in many places in the Mixed Mesophytic Forest region.

The composition of all communities containing chestnut has been modified by the dying of chestnut due to blight. Where chestnut was a dominant, as in the chestnut–sugar maple–tuliptree forest, radical changes have taken place in light and in root competition, and hence in growth and composition of lower layers. In many places, it was still too early at the time the studies were made to determine what species would replace chestnut in the future forest; generally, the release results in more rapid growth of oak and sugar maple saplings. In some areas, sugar maple will increase greatly in abundance.

The Hemlock–Mixed Mesophytic Communities

Forests containing hemlock are confined to the lower elevations in these mountains. In different facies, they occupy mesic or dry slopes. The abundance of hemlock in the deciduous forest ranges from a mere scattering of individuals to over 50 per cent of the stand. As long as the abundance does not exceed about 30 per cent, little modification in the composition of lower layers of the forest is noted, although the herbaceous layer is less continuous and a few species more or less dependent on the hemlock leaf-mold appear, particularly *Mitchella repens* and *Viola rotundifolia,* species not common in purely deciduous mixed mesophytic communities, although often in communities transitional to oak–chestnut.[5] When a greater proportion of hemlock is present, the lack of sunshine in the spring and the development of a distinctive hemlock mor humus layer exclude many of the characteristic shrub and herb species of the more typical all-deciduous mixed mesophytic forest and encourage the growth of ericaceous plants. *Rhododendron maximum* and, on drier sites, *Kalmia latifolia* are frequent constituents of the shrub layer. Where abundant, they exclude almost all herbaceous plants. The forests with a high percentage of hemlock and a Rhododendron layer are generally confined to deep sandstone gorges where they might be interpreted as physiographic climaxes. Elsewhere, they are very limited in extent, being nothing more than local patches in an otherwise mixed deciduous forest.

To show the composition of mixed mesophytic forests containing hemlock, representative examples (Table 6) are selected from the three principal topographic subdivisions of the Cumberland Mountains—from the lower slopes of Black Mountain, from the outlying area, and from the inner (dip) slope of Pine Mountain.

The hemlock–mixed mesophytic forest of mesic ravine-slopes (Table 6, A, B) is comparable to the typical mixed mesophytic forest, as it contains in its varied canopy representatives of most of the dominant species and species of high presence in the typical mixed mesophytic. However, sugar maple, basswood, and buckeye are usually poorly represented. Hemlock, beech, and tuliptree are usually most abundant, comprising between 45 and 70 per cent of the canopy. Red maple, and sometimes white oak, may be codominants. The forest of sandstone gorges of Pine Mountain (Table 6, C) is similar in specific content but contains a much higher percentage of hemlock, whose dominance is reflected in the composition of shrub and ground layers.

In some drier slope forests with hemlock (Table 6, D, area 8), hemlock, white oak, and red maple are most abundant, together comprising, in the sample given, 81 per cent of the canopy. This type always occurs below an elevation of about 2000 feet, the upper altitudinal limit of white oak. Hemlock may be codominant with chestnut in a forest suggesting oak–chestnut rather than mixed mesophytic (Table 6, D, area 9). In the sample given in the table, these two species comprise 60 per cent of the canopy; scarlet oak and white oak, comprise 26 per cent. The dominance of evergreen heath

[5] For a comparison of the herbaceous layers of the hemlock–mixed mesophytic and all-deciduous mixed mesophytic forests, see "Ecological Transect of Black Mountain," Table 3, areas I and II, p. 210 (Braun, 1940); "Forests of the Cumberland Mountains," Fig. 14 and Table 1 (Braun, 1942); and "Vegetation of Pine Mountain, Kentucky," R, 3, pp. 542–544 (Braun, 1935).

Table 6. Percentage composition of canopy of hemlock–mixed mesophytic forest of the Cumberland Mountains, Kentucky.

A. Mesic slopes of Black Mountain in Letcher County: *1,* Lower part of Joe Day Branch; *2,* Lower part of Staggerweed Hollow; *3,* Slopes of wide valley of Colliers Creek.

B. Outlying area of the Cumberlands: *4,* Lynn Fork of Leatherwood Creek; *5,* Clover Fork of Leatherwood Creek, Perry County.

C. Gorges on the southeast (dip) slope of Pine Mountain: *6,* Bad Branch, Letcher County; *7,* Kentucky Ridge State Forest near Pineville, Bell County.

D. Hemlock–white oak and hemlock–oak–chestnut segregates of drier slopes. Black Mountain, Letcher County: *8,* South-facing cove slopes, Colliers Creek; *9,* Westerly slopes, Meeting House Branch.

	A			*B*		*C*		*D*	
	1	*2*	*3*	*4*	*5*	*6*	*7*	*8*	*9*
Number of canopy trees	198	90	115	283	226	146	293	64	103[1]
Tsuga canadensis	30.3	14.4	13.4	29.7	17.7	56.2	45.7	42.2	31.0
Fagus grandifolia	14.6	41.0	16.5	26.5	37.6	2.7	12.3	4.7	1.0
Liriodendron tulipifera	11.1	10.0	17.4	3.5	12.8	11.6	9.0	..	..
Acer rubrum	11.6	4.4	10.4	4.6	.4	5.5	4.0	10.9	3.9
Quercus alba	7.5	..	6.9	10.9	2.2	..	10.6	28.1	11.6
Tilia heterophylla[2]	..	10.0	..	1.4	9.3	3.4	.2	..	..
Castanea dentata	5.0	3.3	2.6	5.7	.4	2.7	3.7[6]	1.6	29.0
Acer saccharum[3]	4.5	3.3	..	3.5	8.0	..	..	..	..
Quercus borealis maxima	2.5	2.2	5.2	1.8	2.2	1.4	.6	6.2	..
Betula alleghenienesis[4]	2.0	1.0	1.7	7.1	4.4	1.4	5.8	1.6	1.0
Betula lutea	..	..	..	..	..	9.6	..	..	..
Aesculus octandra	1.0	2.2	.9	..	.9	..	..	1.6	..
Nyssa sylvatica	3.5	3.3	2.6	..	..	1.4	.4	..	1.0
Magnolia acuminata	1.0	1.0	..	..	.9	.7	.4	..	..
Magnolia Fraseri	2.5	..	2.6	.7	..	1.4	..	3.1	1.0
Fraxinus americana[5]	..	1.0	..	..	.4	..	.2	..	..
Juglans nigra	.5	1.0	2.6	..	..	..	..	..	..
Juglans cinerea	1.0	..	3.5	..	..	..	..	..	..
Carya spp.	.5	1.0	.9	1.4	.4	.7	1.4	..	..
Quercus montana	.5	..	..	..	..	1.4	2.7	..	..
Quercus coccinea	..	..	..	..	..	..	..	..	14.5
Quercus velutina	..	..	..	..	..	..	..	..	3.9
Magnolia macrophylla	..	..	..	..	..	..	.2	..	..
Sassafras albidum	..	..	..	..	..	..	.2	..	..
Pinus rigida	..	..	..	..	..	..	..	..	1.9

[1] Composition based on three samples, together composed of 103 trees; for composition of the three samples separately and contrast with mixed mesophytic forest of opposing ravine slopes, see Braun, 1940, Fig. 16 and Table 5, pp. 218, 219.

[2] Some *T. floridana* and *T. heterophylla* var. *Michauxii* may be included; *T. heterophylla* is most abundant.

[3] Varieties of *Acer saccharum* are not distinguished; both var. *nigrum* and var. *Rugellii* are present.

[4] Some *Betula lenta* probably included.

[5] *Fraxinus biltmoreana* not distinguished from *F. americana*.

[6] As dead chestnut had been cut close to the ground, there may have been more trees than indicated here.

The hemlock–mixed mesophytic forest of Lynn Fork, Perry County, Kentucky. (Within Area 4 of Table 6.) (Courtesy, *Ecol. Monographs,* **12**:438.) Photo by the author.

shrubs, *Rhododendron* and *Kalmia,* is further evidence of the oak–chestnut relationship of the community.

The status of hemlock in deciduous forest communities has often been discussed. Its probable antiquity in the southern Appalachian Highland has already been mentioned (p. 45). By definition (because of its different life-form), it has been excluded from the deciduous forest (Clements, 1934, 1936; Weaver and Clements, 1938). However, its frequent occurrence in a variety of deciduous forest communities—in communities made up largely of the usual dominants of the mixed mesophytic and in white oak and oak–chestnut segregates—is apparent in the field and is shown statistically in Table 6. Its interpretation as "postclimax," because it "requires the protection afforded by both mesocline and canyon-like valley" (Clements, 1934) is refuted by its frequent occurrence on west-facing bluffs, and by its association with scarlet oak, black oak, and pitch pine, three of the most xeric species of the Mixed Mesophytic Forest region. Hemlock should be considered as one of the dominants of the Mixed Mesophytic association, and the hemlock, hemlock–mixed mesophytic, hemlock–white oak, and hemlock–chestnut communities as association-segregates. The southern aspect of some of the hemlock forests is emphasized by the large number of, and abundance of, southern species. For example, a forest on Pine Mountain (area 7 of Table 6, C), in addition to having 13 or 14 of the common tree species of the mixed mesophytic forest (at least some of which range little if at all northward of this region), contains a large representation of distinctly southern small trees and shrubs in its lower layers, as *Magnolia macrophylla, M. tripetala, Aralia spinosa, Oxydendrum arboreum, Cornus florida, Ilex opaca, Rhododendron maximum, Clethra acuminata,* and *Stewartia ovata* (*S. pentagyna*). The broad-leaved evergreens and the large proportion of southern species in the low-tree and shrub layers, as well as the southern species in the canopy, emphasize the distinctness of the hemlock–mixed mesophytic forest from the hemlock–northern hardwoods forest. The two have little in common except the abundance of hemlock.

Subclimax Communities

Although mixed mesophytic communities cover (or originally covered) the greater part of the Cumberland Mountains, the drier ridge-tops, and upper southwesterly slopes, certain rocky and shallow soil areas are covered with subclimax or physiographic climax communities. The nature of these subclimaxes and their relations to the Mixed Mesophytic Forest climax are significant in the interpretation of relationships of the major associations of the deciduous forest.

In the highly variable and complex mixed mesophytic forest of the Cumberlands, with its tendency toward segregation into more or less distinct forest types, it is possible to find gradations from typical mixed mesophytic forest to a forest of very different aspect and composition, the oak–chestnut forest.

OAK–CHESTNUT FOREST

Both of the dominant species of this forest, *Castanea dentata* and *Quercus montana,* are present in the typical mixed mesophytic forest (Table 1). The former is one of the eight most abundant species of the mixed mesophytic forest and a dominant in certain of its segregates (Table 5). Its dominance is not indicative of oak–chestnut forest. *Quercus montana,* codominant with *Castanea* in many stands, and sometimes entirely replacing it, is never abundant in the mixed mesophytic forest. This combination of dominant species is characteristic of many oak–chestnut stands. The most characteristic and most mesophytic species of the mixed mesophytic forest—basswood, buckeye, sugar maple, ash, walnut, and butternut—are absent or represented only by an occasional tree in marginal areas. Most of the species of the oak–chestnut forest are species which occur in the mixed mesophytic forest, thus emphasizing the relation of the two forest types. In fact, if only composition of the forest canopy be considered, it is impossible to make more than an arbitrary boundary between oak–chestnut and mixed mesophytic forest. The lower layers are, however,

The chestnut oak woods of narrow ridge crests is similar to chestnut oak forests of the Oak–Chestnut region. View on a narrow but flat-topped stretch of the summit of Pine Mountain in Bell County, Kentucky. (Courtesy, *Ecol. Monographs,* **12:**427.) Photo by the author.

distinctive. The reaction of forest canopy on germination and growth of accessory and subordinate species is so pronounced that concrete differences between the lower layers of oak–chestnut and of mixed mesophytic forest are evident. Deep oak leaf litter, slow humus decomposition, and intense light in spring (due to late leafing of dominants), coupled with consequent more rapid water loss, are features of the oak–chestnut forest. The delicate herbaceous plants of the mixed mesophytic forest, which thrive in deep mull humus, are absent. The open herbaceous layer includes a much smaller number of species, among which *Cypripedium acaule, Heuchera longiflora, Lysimachia quadrifolia, Aureolaria* (*Gerardia*) *laevigata, Pedicularis canadensis, Coreopsis major,* and *Hieracium venosum* are frequent. Ericaceous plants may be conspicuous and usually are, but not always, a part of the oak–chestnut forest.

Seven samples of the oak–chestnut forest of the Cumberland Mountains are selected to demonstrate the composition of this type (Table 7). In five of these (areas 1–5), chestnut oak and chestnut are by far the most abundant species, together comprising from 58 to 85 per cent of the stand.

Table 7. Percentage composition of oak–chestnut subclimax communities of the Cumberland Mountains of Kentucky: *1,* Joe Day Ridge, Black Mountain, Letcher County; *2,* Lynn Fork of Leatherwood Creek, Perry County; *3,* Pine Mountain, Letcher County; *4,* Buck Branch of Jellico Creek, Whitley County; *5,* Nolan Branch, Clay County; *6,* Pine Mountain, Letcher County; *7,* Pine Mountain, Bell County.

	1	*2*	*3*	*4*	*5*	*6*	*7*
Number of canopy trees	244	74	60	87	116	69	56
Quercus montana	46.7	59.4	56.6	66.6	31.0	13.0	44.6
Castanea dentata	32.0	25.7	18.3	10.3	26.7	36.2	7.1*
Quercus alba	1.2	1.4	1.6	5.7	. .	17.4	10.7
Quercus coccinea	4.9	. .	3.3	. .	. .	13.0	1.8
Quercus velutina	. .	. .	3.3	2.3	. .	3.0	. .
Quercus borealis maxima	.4	. .	. .	. .	.9	. .	1.8
Nyssa sylvatica	. .	. .	. .	4.6	5.2	1.5	. .
Liriodendron tulipifera	.4	1.4	1.6	3.4	8.6	6.0	3.5
Acer rubrum	1.6	8.0	1.6	1.1	3.4	6.0	. .
Carya spp.	2.5	1.4	. .	3.4	. .	. .	18.0
Magnolia acuminata	. .	. .	. .	1.1	6.0	. .	. .
Betula sp.	. .	. .	. .	. .	7.7	. .	. .
Fagus grandifolia	7.0	. .	. .	. .	5.2	. .	. .
Acer saccharum	. .	. .	. .	. .	2.6	. .	. .
Tilia heterophylla	. .	. .	. .	. .	.9	. .	. .
Oxydendrum arboreum	.4	. .	. .	. .	.9	1.5	. .
Robinia Pseudoacacia	. .	. .	. .	1.1	.9	. .	. .
Tsuga canadensis	1.2	. .	. .	. .	. .	. .	1.8
Pinus echinata	. .	. .	5.0	. .	. .	1.5	10.7
Pinus rigida	1.6	2.7	8.3	. .	. .	1.5	. .

* *Castanea* may have been more abundant as dead trees had been cut close to the ground and hence may have been overlooked.

In three of these five, pines are present, while the other two contain no conifers. The number of deciduous species ranges from 10 to 13, which is less than in most mixed mesophytic stands. In the sixth example (area 6), chestnut is the most abundant species, and chestnut oak less abundant than white oak; the chestnut and oaks together here make up 83 per cent of the stand. A few areas of oak–chestnut somewhat suggest oak–hickory; such a community on Pine Mountain is represented by a sample of 56 trees (area 7). The small number of trees in the sample areas of oak–chestnut forest (except area 1), as compared with mixed mesophytic forest samples, is a result of smaller size of oak–chestnut communities. They never occupy large continuous areas but may be narrow strips following dry ridges (area 1). However, because of the smaller number of species in the community as a whole and its greater uniformity, smaller samples are adequate.

In the seven sample areas, 20 species are represented, of which three—*Quercus montana, Castanea dentata,* and *Liriodendron*—are present in all. Two species—*Quercus alba* and *Acer rubrum*—are present in six of the seven samples. Red maple is a usual constituent of the oak–chestnut forest, while sugar maple is generally absent or represented by few individuals. Of the remaining 15 species, *Quercus coccinea* and *Q. velutina* are frequent in drier oak–chestnut areas and are never a part of the mixed mesophytic forest (either may appear rarely as an accidental, probably entering in a wind-fall opening). The pines (*Pinus echinata, P. rigida*), more properly a part of the pine and pine–oak physiographic climaxes, are frequent constituents of the oak–chestnut communities on Pine Mountain, where oak–pine and oak–chestnut communities may be adjacent to one another and, hence, intergrade to some extent.

In the oak–chestnut communities of the Cumberlands, the dying off of approximately a quarter of the canopy trees (because of blight) has resulted in a very open forest. Abundant light reaching the lower layers has encouraged rapid growth of many of the understory trees and saplings and, in some places, the development of dense tangles of blackberries and Smilax. In all of the drier sites where *Quercus montana* is abundant in the canopy, this species seems to be the one favored in the undergrowth by the changed conditions. Often, there is a continuous layer of young individuals. The hickories also, especially *Carya glabra* and *C. ovata,* are increasing and, in some places, there are indications that an oak–hickory forest will replace the original oak–chestnut stand. In the driest areas, and especially if fires have occurred, *Quercus coccinea* is increasing. In the more mesic situations, sugar maple or white oak are growing rapidly.

PINE AND PINE–OAK COMMUNITIES

On shallow sandy soils over sandstone, pines (*P. rigida, P. echinata,* and, in places, *P. virginiana*) prevail (Table 8). This physiographic climax type is extensively developed on Pine Mountain and on Cumberland Mountain, where dipping sandstone strata are exposed on ridge crests and but shallowly covered with sandy soil for some distance down the dip slope. Elsewhere in the Cumberland Mountains, it is limited to narrow, dry, rocky summits. Secondary pine communities (usually *P. virginiana* is the dominant species) are frequent on eroded slopes, particularly if the under-

Open pine woods with a shrubby layer made up principally of ericaceous plants on a ridge where the underlying sandstone is near the surface. The contrast between such stands and the prevailing mixed mesophytic forest is striking. (Courtesy, *Am. Midland Naturalist,* **16:**535.) Photo by the author.

lying rock is noncalcareous shale. The primary forests are open stands, more or less park-like in appearance, with a continuous grassy herbaceous layer, through which many forbs and low blueberry bushes are scattered, or with a more or less continuous shrubby layer. Beds of Cladonia occupy a position at the edge of outcropping sandstones. Such pine forests may grow where there are only a few inches of soil over the bed rock. The

Table 8. Percentage composition of pine–oak subclimax forest type; summit of Pine Mountain near Scuttlehole Gap, Letcher County, Kentucky (114 trees).

Pinus rigida	43.0	
Pinus virginiana	16.6	63.6
Pinus echinata	4.0	
Quercus montana	17.5	23.5
Quercus coccinea	6.0	
Castanea dentata	12.0	
Nyssa sylvatica	.9	

deciduous species which are present—usually *Quercus montana, Q. velutina, Q. coccinea, Nyssa sylvatica, Castanea dentata,* and *Oxydendrum arboreum*—are small in stature. They increase in proportion and in size where the soil is deeper, resulting in transitions to chestnut oak or oak and chestnut communities.[6] In some situations, fire is a contributory factor in maintaining the pine communities. It is not, however, the primary cause.

Contacts and Transitions

The Cumberland Mountains, a distinct physiographic section, have been treated as a vegetational unit distinct from surrounding areas. We must recognize, however, that vegetational boundaries are seldom as sharp as physiographic boundaries. The southeastern boundary of the Cumberland Mountains, formed by the escarpment which here is Cumberland and Stone Mountains, is the approximate limit of the area where mixed mesophytic forest prevails. In the adjacent Ridge and Valley Province to the southeast, mixed mesophytic forest is confined to certain favorable sites, while phases of oak and oak–chestnut forests are the regional types. The most characteristic features of the mixed mesophytic forest, its decidedly mixed aspect, its small proportion of oaks (except on southerly slopes), and, especially, its buckeye and white basswood, do not extend southeastward except in isolated areas. These isolated areas will be considered in connection with the discussion of the regional vegetation.

Surrounding the Cumberland Mountains on all except the southeast are the Cumberland and Allegheny Plateaus. Mixed mesophytic forest extends

[6] For further information on the pine and pine–oak communities, see Braun, 1935.

into and even beyond the plateau; the vegetational boundary between the Cumberland Mountains and the plateau is one of degree rather than kind. The pronounced segregation of dominants which has been strongly emphasized in the discussion of the Cumberland Mountains is limited to the Cumberland Mountains section; hence, beyond its limits the mixed mesophytic forest is more uniform in character. Through the high maturely dissected portion of the plateau to the northeast in Virginia and southern West Virginia, the forest retains, in a large degree, the aspect of the forest of the Cumberland Mountains. About 175 miles to the northeast, the southern end of the Allegheny Mountains forms a wedge between the plateau (here called Allegheny Plateau) and the Ridge and Valley Province.

ALLEGHENY MOUNTAINS

This section includes the southern part of the Allegheny Mountain physiographic section of the Appalachian Plateaus Province and some contiguous mountainous area to the south. Since the vegetational boundary is eastward of the physiographic boundary in West Virginia, a strip of the Ridge and Valley Province where ridges are closely crowded (the most prominent of these is Allegheny Mountain) belongs in the Mixed Mesophytic Forest region. This strip is here included in the Allegheny Mountains.

The Allegheny Mountains section is an area in which underlying rocks are mildly folded (in contrast to the more or less horizontal strata of the Allegheny Plateau on the west and to the intensely folded and sometimes metamorphosed rocks of the Ridge and Valley Province to the east. This structural feature is responsible for certain of the topographic features, which in turn affect the vegetation. Mountain ridges tend to be longitudinal and separated from one another by more or less longitudinal valleys, in places connected by narrow water-gaps crossing the ridges. The ridges have developed on the more resistant rock, the valleys, sometimes broad and plateau-like, on the weaker strata. The Greenbrier River Valley above Marlinton, West Virginia, one of these longitudinal valleys, skirts the eastern base of Back Allegheny Mountain, which is technically the eastern boundary of the Allegheny Mountains. However, the eastern boundary of this forest section follows the east slopes of Allegheny Mountain from its southern terminus near White Sulphur Springs northeastward (essentially along the West Virginia-Virginia border), continuing, along the Allegheny Front, across Maryland (between Cumberland and Frostburg) into Pennsylvania.[7] The western boundary extends northward from the wedge-like southern tip of

[7] The regional boundary in southern Pennsylvania is very prominent when approached from the east over the Pennsylvania Turnpike. The whitish blotches formed by *Tilia heterophylla* are particularly conspicuous in the mixed forest, which contrasts strongly with the oak forests of slopes to the east.

the section, joining the western boundary of the physiographic section northwest of Bald Knob.[8]

Northward, the Allegheny Mountain section broadens. In western Maryland and southern Pennsylvania, its broad plateau-like valleys are high and extensive and its ridges rather low ranges (relative to these valleys) extending in a southwest-northeast direction. Valleys formed in the present erosion cycle contrast with broader and older valleys on weaker rocks, sometimes flat and sometimes gently rolling, which represent the Harrisburg peneplain. There are vegetational correlations with these surfaces of different ages and with soils.

This area is included in the Mixed Mesophytic Forest region because of the character of its lower elevation slope forests. Altitudinal and latitudinal transitions from mixed mesophytic to more northern forests of beech, birch, and maple, and of spruce occur in the Allegheny Mountains and afford opportunity of demonstrating the relations of these forest climaxes. Northward in the section, a larger proportion of the land is at a high elevation. Northern vegetation, therefore, becomes more general and mixed mesophytic communities are confined to lower elevations, especially to slopes of valleys cut into the more or less plateau-like valleys and to slopes of the Allegheny Front. Along the latter, mixed mesophytic forest extends far northeastward as a narrow band between the Ridge and Valley section of the Oak–Chestnut region and the southern lobe of Hemlock–White Pine–Northern Hardwoods forest on the Appalachian Plateaus.

Mixed Mesophytic Communities

No good areas of mixed mesophytic forest at lower elevations in the Allegheny Mountains were found. All have been more or less modified by cutting. It is, perhaps, the absence of good primary areas of mixed mesophytic forest along the more traveled routes that in part accounts for its lack of general recognition. However, existing remnants give unmistakable evidence of former occupancy of most of the valley slopes and lower mountain slopes (surfaces of the present erosion cycle) by mixed mesophytic forest. Its exact composition cannot be determined from such remnants and more or less secondary stands.

The mixed mesophytic forest of the valley slopes of the southern Allegheny Mountains apparently contained a large proportion of *Acer saccharum* (including var. *nigrum*), *Fagus grandifolia*, and *Quercus borealis* var. *maxima*. With these were *Liriodendron tulipifera*, *Magnolia acuminata*,

[8] On the map showing province boundaries (Fenneman, 1938), the southern terminus of the Allegheny Mountain section is shown just north of Bald Knob. Actually, Back Allegheny Mountain (the eastern boundary of the section) continues some 6 or 8 miles south of this point. The writer includes in the Allegheny Mountain forest section this part of Back Allegheny Mountain, the southern end of Cheat Mountain, and the southern end of Allegheny Mountain.

Fraxinus americana, and, in some areas, *Tilia heterophylla, Prunus serotina, Ulmus fulva, Quercus alba, Q. montana, Nyssa sylvatica, Betula lenta, Juglans nigra, J. cinerea,* and, locally, *Aesculus octandra. Cornus florida* is usually present in the southern part of the area. Remnants of this mixed mesophytic forest occur along the valleys of the Tygart, Cheat, and Youghiogheny Rivers, and also on the eastern slope, in valleys cutting or indenting the Allegheny Front, and sometimes on talus slopes.

This forest varies in composition from place to place but does not display the marked segregation of its dominants that is a feature of the Cumberland Mountains. In flatter areas or in the rolling belt which sometimes borders the lower mountain slopes, sugar maple may be dominant, often with white oak, and also tuliptree, chestnut, cucumber tree, and hickories.

On southerly slopes, white oak and red oak are more abundant than the other species, but the forest retains its mixed character. In an area of southeast slope in Tygart Valley south of Huttonsville, where white oak is abundant, the canopy includes: white oak, red oak, chestnut oak, tuliptree, beech, sugar maple, chestnut, basswood, buckeye, walnut, hickory, black cherry, and sour gum. In general, in the mixed woods in which oaks prevail, white oak, red oak, sugar maple, tuliptree (if elevation not too high), and others of the usual constituents of the mixed mesophytic forest are present. This type of mixed woods is (with some variations) one of the prevalent communities of slopes at lower elevations in the Allegheny Mountains.

Shreve (1910), in discussing the slope forests of the "Mountain Zone" of Maryland (a discussion based largely upon observations of second-growth stands), states that white oak is the most abundant species. With it, the six commonest subordinate species, about equally abundant, are linden, cucumber-tree, sugar maple, sweet birch, red oak, and shagbark hickory. He states that "the abundance of these six species is one of the principal characteristics of the forests of the Slopes." The slope forests of this part of the Allegheny Mountains, although not called mixed mesophytic, have always been known as mixed forests. Certainly they are not beech–maple, oak–chestnut, or oak–hickory. Although they may contain the dominants of any one of these, they usually contain the dominants of all three. This mixed forest of slopes in which oaks prevail is distinct from the white oak community of the Harrisburg peneplain areas and straths.

Chestnut is frequently an abundant constituent of these mixed oak woods. A representative example with fairly old trees, but obviously not entirely primary, was seen on the Virginia slope of Allegheny Mountain; in this sample, oaks (*Q. borealis* var. *maxima, Q. montana,* and *Q. alba*) comprise almost half, and chestnut nearly 20 per cent of the stand (Table 9, A). The admixture of basswood and sugar maple relates this to the Mixed Mesophytic association.

Table 9. Composition of variants of mixed mesophytic forest in which oaks and chestnut or chestnut and maple comprise 60 per cent or more of the canopy.

A. Old-growth (but not primary) forest on Virginia slope of Allegheny Mountain.

B. Primary forest on slopes of Back Allegheny Mountain above Linwood, West Virginia.

C. Primary forest on Point Mountain between Valley Head and Webster Springs; high dissected Allegheny Plateau adjacent to Allegheny Mountains; elevation about 3000 ft.

	A	*B*	*C*
Number of canopy trees	61	128	77
Castanea dentata	18.0	46.1	37.6
Quercus borealis maxima	23.0	11.7	3.9
Quercus montana	13.1	..	..
Quercus alba	11.5	2.3	..
Acer rubrum	5.0	14.0	..
Acer saccharum	5.0	7.8	22.1
Prunus serotina	..	.8	16.9
Fagus grandifolia	..	.8	13.0
Tilia spp.	10.0	1.6	..
Magnolia acuminata	6.5	6.2	2.6
Fraxinus americana	..	3.9	1.3
Betula lenta	1.5	3.1	..
Betula lutea	..	..	1.3
Carya spp.	6.5	.8	..
Robinia Pseudoacacia	..	.8	..
Magnolia Fraseri	..	..	1.3

In another variation of the slope woods, chestnut and sugar maple (or sugar maple and red maple) are the most abundant species. A primary area in which chestnut comprises nearly 46 per cent and sugar maple and red maple 22 per cent of the canopy (Table 9, B) is seen in the forest above Linwood, West Virginia, on the slopes of Back Allegheny Mountain. Curran (1902) called this the chestnut subtype of the slope forest, stating that it occupied the steep slopes above the streams. The composition, taken from his data, is given in Table 10, A; here chestnut formed 36 per cent and sugar maple 20 per cent of the stand (trees over 12 inches d.b.h.). This chestnut–sugar maple variation of the mixed mesophytic forest of slopes occurs also in the adjacent higher parts of the Allegheny Plateau (Table 9, C). Release of understory individuals of sugar maple (due to dying out of chestnut) has resulted in their rapid growth, indicating an increase in the proportion of sugar maple in the future forest.

The abundance of chestnut and the presence in some examples of tuliptree and cucumber magnolia emphasize the southern relationship of these slope forests, which are similar to the chestnut–sugar maple–tuliptree association-segregate of the Cumberland Mountains (p. 63). The yellow birch and *Tilia americana* suggest the transition to more northern forest types. The northern aspect becomes more noticeable in Maryland, where the

Table 10. Composition of four forest types of slopes and ridges of western Maryland. (Data as given by Curran, 1902.)

A. "Chestnut subtype" of "Slope Timber," average of 15 acres, trees 12 inches and over d.b.h.

B. "White oak subtype" of Garrett County.

C. "Ridge Timber," average of 33 acres, trees 12 inches and over d.b.h.

D. "Hemlock subtype," Youghiogheny River, Garrett County.

	A	*B*	*C*	*D*
Chestnut	36.06	10.75	47.34	.12
Sugar maple	20.51	. .	. .	2.45
Red oak	9.16	4.03	20.04	.06
Basswood	8.75	. .	.24	1.18
Yellow birch	3.99	. .	. .	6.60
Sweet birch	3.34	.14	3.88	.19
Beech	2.83	. .	.09	1.58
White oak	2.51	81.12	10.73	.33
Chestnut oak	.23	1.11	6.43	. .
Red maple	. .	2.29	5.34	.64
White pine	. .	.21	. .	.25
Hemlock	. .	. .	.29	86.04
Other species	12.62	.35	5.62	.56

general absence of tuliptree and dogwood may be noted, and where a few northern species appear, as *Diervilla Lonicera, Maianthemum canadense,* and *Aralia nudicaulis.*

Secondary groves of tuliptree dot many of the more mesic disturbed slopes where mixed mesophytic forest formerly prevailed. On all southerly slopes there is little doubt that the proportion of oaks and hickories is now greater than in the original stand.

Because of the high altitude of many of the valleys (mostly above 2000 feet except near major streams) and the more northerly latitude of this section, the area of mixed mesophytic forest is altitudinally limited. Upward, the change to a more northern forest type in which beech, maple, and birch prevail is apparent. This altitudinal sequence will be considered later (p. 82).

White Oak Communities

White oak forest occupies some of the flat areas (straths) in valleys of a former erosion cycle. Such white oak communities are more prevalent in the valleys of the Ridge and Valley Province (Chapter 7), but occur locally in the Allegheny Mountains, as they do also in the Allegheny Plateau to the west. All such areas have been greatly disturbed by human occupancy, so that the original composition cannot now be determined. There are places on the plateau to the west where it appears that beech originally accompanied the white oak, suggesting the derivation of this type from the white oak–beech segregate of the Mixed Mesophytic association. Small areas

Virgin white oak near Anthony Creek in the Monongahela National Forest, West Virginia. (Courtesy, U. S. Forest Service.)

in valleys of the Alleghenies of West Virginia are referable to the white oak or white oak–hemlock type. In the upper Greenbrier Valley (elevation about 2700 feet), there are small white oak areas,[9] and secondary areas, both in the valley and on adjacent slopes, where white pine is mingled with the white oak.[10] These secondary white oak stands are a conspicuous feature of western Maryland, there also sometimes with white pine. A white oak forest type (about 80 per cent white oak) was recognized as one of the virgin forest types of western Maryland (Curran, 1902). The composition of this forest type as determined by Curran is given in Table 10, B.

On the Virginia slope of Allegheny Mountain, white oak as a dominant is usually associated with red oak, chestnut oak, and some tuliptree and birch. This forest type is extensively developed in the Warm Springs district

[9] White oak grows at higher elevations in the Alleghenies of West Virginia and Maryland than in the Cumberland Mountains farther south.

[10] This is what has been called the white pine area of West Virginia (map, West Virginia State Planning Board, 1943).

of the George Washington National Forest, where it will be considered later (in the Ridge and Valley section of the Oak–Chestnut Forest region, p. 229).

Oak–Chestnut Communities

On convexities of valley slopes and on higher rocky slopes and ridges, chestnut was usually abundant. As in the Cumberland Mountains, chestnut oak may be the dominant species. *Betula lenta* is usually a constituent of this community. Ericaceous shrubs, as *Kalmia, Azalea,* and species of *Vaccinium* are abundant in rocky woods. The chestnut and oak–chestnut areas are in most places much disturbed by cutting and fire, and the chestnut is killed by blight. Primary forests of this type no longer exist; they were probably similar to those of the Cumberland Mountains. This is the sort of forest referred to by Curran (1902) as "ridge timber," and by Shreve (1910) as "the ridge type of forest." Table 10, C gives the composition of original stands of this type in Maryland where chestnut predominated. Shreve states that in second-growth stands of this forest, chestnut, chestnut oak, and red oak together form 75 to 95 per cent of the stand.

Chestnut Ridge, which forms the western boundary of the Allegheny Mountains from near Morgantown, West Virginia, northeastward nearly half-way across Pennsylvania, is one of the ridges where the oak–chestnut forest type prevails. Jennings (1927) calls this the "chestnut–chestnut oak–black oak association" and uses as an example the forest on "the uppermost slopes and undisturbed top of Chestnut Ridge, above Hillside, Westmoreland Co., Pa." The dominant trees listed are chestnut, chestnut oak, black oak, and pitch pine; with these occur red maple, black birch, sour gum, tuliptree, sassafras, white oak, and basswood. Ericaceous shrubs and ground plants (*Kalmia, Azalea, Vaccinium, Epigaea,* and *Gaultheria*) are abundant and characteristic of this community. The vegetation of Laurel Hill to the east of Chestnut Ridge is similar (Jennings, 1927).

Forests of the Higher Elevations

In many places, the Allegheny Mountains of West Virginia reach elevations of more than 4500 feet. Spruce Mountain is over 4500 feet for five miles and 4860 feet at its highest part, Spruce Knob. The valley of North Fork at its east base is about 1800 feet, giving a local relief of 3000 feet. Shavers Mountain is approximately 4000 feet in elevation for 10 miles and has many higher knobs. At its southern end (there called Back Allegheny Mountain), the highest knob is over 4800 feet in elevation. This mountain, then, rises over 2000 feet above Tygart Valley toward the west and about 1500 feet above the upper Greenbrier Valley to the east.

In Maryland and southern Pennsylvania, the old valley floors are higher and more extensive, but the ridges do not attain as high an elevation. The average elevation of much of the surface is between 2700 and 2800 feet. Narrow younger valleys are cut deeply into the old valleys.

In order to show the relation of the mixed mesophytic forest of the lower slopes to the middle elevation forests, a few examples of altitudinal sequences will be given.

On lower slopes (chiefly valley slopes of the present erosion cycle), there is mixed forest in which sugar maple, beech, red oak, and tuliptree are most abundant, together with cucumber magnolia, birch, ash, hickory, chestnut, red maple, basswood (*T. americana* and *T. heterophylla*), cherry, red elm, and, among the smaller species, dogwood and *Aralia spinosa*. This is mixed mesophytic forest of the type already described. The ground cover has the rich and varied appearance so universal in mixed mesophytic forest. Variations are typical of the mixed mesophytic forest—white oak locally more abundant on some of the lower slopes and chestnut dominant on convexities of south slopes. Higher, tuliptrees drop out and birch increases in abundance; soon beech, sugar maple, and birch become the dominant species. On the highest slopes and summits, red spruce (*Picea rubens*) appears.

On many of the high slopes, the original forest cover has been entirely removed. Young birch or birch and pin cherry (*Prunus pensylvanica*) may now cover the slopes. In places, sprout beech and red maple make very dense stands, sometimes with birch. Whether primary or secondary, the shrub and herbaceous growth will readily distinguish this altitudinal band from the mixed mesophytic forest below. *Diervilla Lonicera, Viburnum alnifolium, Oxalis montana, Maianthemum canadense, Circaea alpina, Hydrophyllum virginianum, Corydalis sempervirens,* and other more or less northern species occur in these higher elevation beech–maple–birch woods.

The vegetation of the slopes and summits of Spruce Mountain and of the successions in burned-over areas near Spruce Knob have been described by Core (1929). Here, the range of elevation is about 3000 feet. The lowest slopes were originally covered by mixed forest, remnants of which still exist. Although Core did not distinguish the low belt of mixed mesophytic forest, his list of species of the foothills clearly defines it. Higher, beech and maple are abundant. A higher band of oak–hickory distinguished by Core appears to be the equivalent of the more general oak–chestnut of steep rocky slopes. The chestnut is of course dead, which gives to the oaks and hickories higher proportional percentages. The summit shows remnants of the primary spruce–birch forest and a number of secondary communities.

A virgin forest area some three miles east of Linwood at the southern end of Back Allegheny Mountain (where it joins the southern end of Cheat Mountain) afforded opportunity of studying the variations of the middle elevation forest (3300 to 4000 feet) in relation to slope exposure and contour. In the forest as a whole, the most abundant trees are beech, sugar maple, red maple, sweet birch (*Betula lenta*), and basswood (*Tilia americana*), which together comprise 75 per cent of the canopy. Eight other species, most of them more or less generally distributed in the forest, make up the remaining 25 per cent of the canopy (Table 11, A). In the lowest

Table 11. Altitudinal sequence of forest types in Allegheny Mountains, West Virginia.

A. Forest of middle elevations (3300–4000 ft.) on southern end of Back Allegheny Mountain near Linwood; composition based on 597 trees distributed in five adjacent areas traversed in ascending mountain.

B. Forest of higher elevations (4200–4400 ft.) on Shavers Mountain (18 miles north of *A*) showing transition from beech–maple to spruce forest.

	A						*B*		
	Total	*1*	*2*	*3*	*4*	*5*	*1*	*2*	*3*
Number of canopy trees	597	154	209	99	91	44	173	52	136
Fagus grandifolia	23.8	31.2	24.9	7.1	27.5	22.7	33.5	28.8	8.8
Acer saccharum	16.7	24.0	20.1	5.0	13.2	9.1	31.2	5.8	..
Acer rubrum	11.9	10.4	12.9	16.2	6.6	13.6	1.1	..	13.2
Betula lenta	11.9	7.1	11.9	21.2	9.9	11.4	3.5	9.6	9.5
Tilia americana	10.9	9.7	10.0	20.2	9.9	..	3.5	..	..
Fraxinus americana	7.0	5.2	6.2	5.0	14.3	6.8	1.7	..	..
Magnolia acuminata	5.4	1.9	4.3	3.0	9.9	18.2	1.1	..	..
Quercus borealis maxima	4.4	2.6	1.0	13.1	3.3	9.1	..	..	..
Betula lutea	3.8	..	5.2	5.0	4.4	6.8	4.0	13.4	12.5
Castanea dentata	2.8	5.2	1.4	4.0	1.1	2.3	..	..	..
Prunus serotina	.7	2.6	..	..	..	..	11.6	3.8	..
Carya sp.	.5	..	1.4	..	..	..	..	..	..
Magnolia Fraseri	.2	..	.5	..	..	..	..	..	..
Picea rubens	..	..	..	..	..	..	8.7	38.5	55.9

A. *2, Betula lutea* enters; *3,* steeper, rocky north-northwest slope and hence more *Betula lenta; 4,* North slope (*Virburnum alnifolium* abundant); *5,* flat.

B. All slopes gentle and northerly.

part of this forest, beech and sugar maple comprise 55 per cent of the canopy; the five most abundant species of the forest as a whole here make up 82 per cent of the stand. In the undergrowth, all of the canopy species are represented, with beech predominating. In addition, *Acer pensylvanicum, Amelanchier* sp., *Crataegus* sp., and *Hamamelis virginiana* are present. The herbaceous layer contains a large number of species, for the most part typical mull plants. This layer is essentially that of the mixed mesophytic forest; typically northern species are absent.

Slightly higher, yellow birch enters and the proportion of beech and sugar maple decreases. Here, also, *Viburnum alnifolium* appears. As in the forest lower down, the canopy species are represented in the undergrowth, together with *Acer pensylvanicum, Hamamelis, Azalea* sp., *Viburnum aceri-*

Beech and sugar maple are the most abundant trees in mesic forests of middle elevations in the Allegheny Mountains of West Virginia. The undergrowth of these forests is essentially that of mixed mesophytic forests. In the late summer aspect, the narrow-leaved spleenwort (*Athyrium pycnocarpon*), and composites (*Aster divaricatus, Solidago caesia, Eupatorium rugosum*) are conspicuous. (Courtesy, U. S. Forest Service.)

folium, and *V. alnifolium.* Many of the herbaceous plants of the lower elevation are here, but the presence of *Maianthemum canadense, Oxalis montana, Trillium undulatum,* and *Hydrophyllum virginianum* in considerable quantity emphasizes the change to a more northern type of forest. The forest is essentially the beech–maple–birch forest of the North. Differences of slope exposure and steepness result in variations in composition but, except for steep convex slopes and a flat summit gap, the mesic slope forest prevails. *Betula lenta* is the most abundant species on very steep rocky slopes which resemble wooded rock slides. Chestnut, red oak, red maple, and cucumber magnolia are most abundant on convexities of the slope. The slopes to the higher knobs had been cleared for pasture; hence, the sequence to the spruce forest is interrupted.

On Shavers Mountain near the Gaurdineer Fire Tower (about 18 miles north of the Linwood forest), a virgin forest (at an elevation above 4000 feet) shows the transition from beech–maple to spruce forest. Cut-over

At higher elevations, yellow birch is a constituent of the forest, and occasional red spruce occur. Hobble-bush (*Viburnum alnifolium*), more common at higher elevations, or farther north, is present. Monongahela National Forest, West Virginia. (Courtesy, U. S. Forest Service.)

forests on the west slope of adjacent Cheat Mountain toward Tygart Valley at Huttonsville (2000 feet) indicate the entire forest sequence from the sugar maple consociation of the valley through the mixed mesophytic forest of the lowest slopes, the beech–maple–birch–basswood forest of middle elevations, to beech, birch, and spruce on the highest slopes—the forest type which on Shavers Mountain was still represented by virgin stands at the time of observation (September, 1938).

In the lower part of this virgin forest, beech and sugar maple are dominant, comprising 65 per cent of the canopy, with black cherry and spruce (*Picea rubens*) next in abundance. Slightly higher, the proportion of spruce increases. The highest part of the area is essentially spruce forest similar to that of the Adirondacks. This transition from the deciduous forest with a small amount of spruce to a forest in which spruce is dominant is shown in Table 11, B. In the dominantly deciduous part of this forest, beech predominates in the undergrowth. Northern and mountain species prevail in the shrub layer: *Acer spicatum, Acer pensylvanicum, Ilex montana, Sambucus pubens* (*S. racemosa*), and *Viburnum alnifolium.* The herbaceous layer contains many of the species common to lower elevations; it is very similar to that of the beech–birch–maple forest at Linwood. Where spruce forms a high proportion of the canopy, yellow birch and spruce are conspicuous in the reproduction. The herbaceous layer has a decidedly northern aspect, with an abundance of *Dryopteris spinulosa, D. noveboracensis, Lycopodium* spp., *Oxalis montana,* and other northern and mountain species, such as *Trillium undulatum, Clintonia borealis, Maianthemum canadense,* and *Aster acuminatus,* together with a smaller representation from the forest of lower elevations.

The forests of the higher parts of the Allegheny Mountains of West Virginia, although within the area of the Mixed Mesophytic Forest region, are outliers of more northern vegetation types. The higher elevations of these mountains, as compared with those of the Cumberland Mountains, together with their more northern position and the continuity of ridges northward, readily explain the northern character of the higher ridges.

Northern Transition

Farther north in the Allegheny Mountains, the proportion of area originally occupied by more northern vegetation increases. Much of the high tableland of western Maryland and adjacent Pennsylvania can best be thought of as a southern arm of the Hemlock–White Pine–Northern Hardwood Forest. The vegetation of the glades and swamps of this area is decidedly northern, with *Picea mariana, Larix laricina, Pinus Strobus,* and *Tsuga canadensis* as the principal trees (see Shreve, 1910). The hemlock forests of rocky slopes are southern extensions of the northern type of hemlock forest, in which yellow birch is abundant, and northern shrubs—

Taxus canadensis, Diervilla Lonicera, Sambucus pubens—are frequent. The composition of a representative example of hemlock forest is given in Table 10, D, based upon Curran (1902). That author also distinguishes a white pine subtype, early destroyed by cutting. To a large extent, these northern communities are edaphic and owe their persistence to the submature topography of the area.

The northern part of the Allegheny Mountains is in the Hemlock–White Pine–Northern Hardwoods region. Even there, the forest of valley slopes of the present erosion cycle is a mixed forest (for example, along the Susquehanna River) made up of sugar maple, beech, tuliptree, red oak, basswood, red maple, red elm, ash, cherry, shellbark hickory, sweet birch, chestnut, chestnut oak, walnut, and an occasional white pine and hemlock. Such dovetailing of northern and southern forest types is to be expected in transition areas.

CUMBERLAND AND ALLEGHENY PLATEAUS

Much of the area of the Cumberland Plateau and the Unglaciated Allegheny Plateau is within the Mixed Mesophytic Forest region. There is, of course, pronounced variation from north to south, for this plateau area extends through nine degrees of latitude—from 33° to 42°. In its middle latitudes, the forests are prevailingly mixed mesophytic, except as modified by edaphic conditions. The prevailingly mixed mesophytic forest extends to the western boundary of the Allegheny and Cumberland Plateaus in Ohio and Kentucky. Northeastward, there is a gradual transition to northern forest types with the predominance of beech, sugar maple, hemlock, and white pine; the high plateau of northern Pennsylvania is beyond the limits of the Mixed Mesophytic Forest region. Southward, oaks increase; southern pines dominate certain edaphic communities; and true mixed mesophytic communities are more or less confined to valley slopes.

The nature of prevailing forest is also seen to vary in relation to major topographic features of the plateau. Four subdivisions of this section are recognized. The **Rugged Eastern area,** lying immediately to the west of the Cumberland Mountains and between the Cumberland Mountains and the Allegheny Mountains, is a maturely dissected area of strong relief. Most of the slopes are covered with mixed mesophytic forest. Adjacent to this is the **Low Hills Belt,** an area of low relief and relatively gentle slopes. It is an irregular band extending from southern Kentucky (where it is very narrow) northeastward to about Pittsburgh, and is widest in southeastern Ohio and adjacent Kentucky and West Virginia. Here, there was a larger proportion of oak in the original mixed forest, and much of the second growth strongly suggests oak–hickory forest. A third subdivision is the **"Cliff Section,"** so named for the bold cliffs which characterize it. From about London, Kentucky, southwestward across Tennessee and into Alabama, the plateau surface is submaturely dissected; the broad, flat uplands are but shallowly indented by valleys of a former erosion cycle or here and there are cut by deep, narrow, young gorges. Forests of beech and sweet gum and open swampy glades occupy the poorly drained areas, while mesa-like remnants of the plateau between deep gorges are clothed with oak forest. The rugged and ragged western edge of the southern half of the plateau (the west-facing escarpment of the basal Pottsville sandstone) presents conditions unlike all the rest of the plateau. Ecologically, this is an area of great interest because of the occurrence here of endemic species and of disjuncts, an area where rich mesophytic forests in the gorges alternate with pine-clad promontories and dry oak uplands. In northern Kentucky and locally in Ohio, the sandstone outcrop lies well within the margin of the plateau. This is a narrow band similar to the western edge farther south. These together will be referred to as the "Cliff Section." The **Knobs Border area** forms the western portion of the plateau in northern Kentucky and southern Ohio. Here, Devonian shales form much of the slopes and Mississippian rocks cap the hills.

These distinct topographic and vegetational subdivisions of the plateau will be more easily understood if the geologic features which are responsible for them are explained. The general plateau level is an uplifted peneplain, for the most part much dissected.[11] The plateau is underlain chiefly by Carboniferous strata dipping gently eastward. Of these, the basal sandstone or conglomerate of the Pottsville series (called Rockcastle in Kentucky and Walden and Lookout in Tennessee) is most significant. This is the most resistant rock in the plateau and reaches a maximum thickness of about 700 feet in Tennessee. It forms the rugged western margin; there the escarpment stands as cliffs above the weaker rocks to the west and is indented by many narrow gorges. The basal Pottsville is overlain by easily eroded beds, mostly shales. In Tennessee these have been largely removed, leaving the sandstone as the surface rock (see A of accompanying Fig.). This is the area shown as "lithosols" on the soil map (p. 26). A band some 20 or 25 miles wide along the western margin of the plateau in southern Kentucky (northward to about London) is the stripped surface of this sandstone (B of Fig., p. 89). Farther east, where the resistant sandstone dips beneath the surface, the plateau is finely dissected hill country. To the west these hills are low and the slopes gentle; eastward, as summit levels gradually rise and the resistant basal Pottsville becomes deeply buried, the country is more rugged. Farther north, the resistant basal Pottsville sandstone is thinner; the western escarpment of the plateau is along the Mississippian outcrop, which (with the underlying Devonian shale) forms a hilly marginal

[11] For a discussion of peneplains represented, see Chapter 16 and Fenneman, 1938, pp. 293–301.

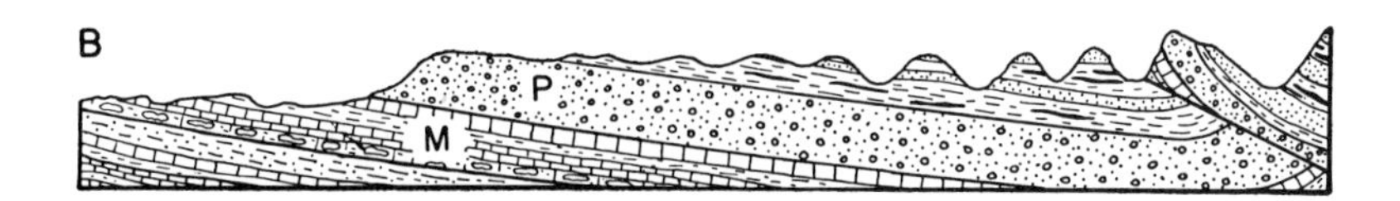

Generalized diagrammatic sections across the Cumberland and Allegheny plateaus: *A,* in Tennessee (Southern district); *B,* in southern Kentucky (Middle district); *C,* in northern Kentucky; *D,* in southern Ohio and West Virginia. Letters within the diagrams designate age of strata: *P,* basal Pottsville (Pennsylvanian); *M,* Mississippian; *D,* Devonian.

belt, the "Knobs Border area." The band in which the Pottsville sandstone outcrops is back from the plateau margin (C of Fig., p. 89). Here it caps some of the hills as cliffs or forms the upper walls of deep and narrow valleys and gorges. Just to the east, again, is the Low Hills Belt, and then the more rugged and deeply dissected portion. Still farther north, across eastern Ohio and West Virginia, or eastern Ohio and Pennsylvania, the shallow synclinal structure of the plateau and the prevalence of easily eroded strata result in a more uniformly dissected surface varying somewhat in relation to the underlying structure and material (D of Fig., p. 89). In general, there is an area of low relief in the vicinity of the principal preglacial rivers and a fairly rugged western border, the Knobs.

Near the western border of the northernmost part of the Unglaciated Allegheny Plateau, the topographic and vegetational subdivisions distinguished farther south are absent. The dissected plateau of moderate relief contrasts strongly both topographically and vegetationally with the adjacent till plains to the west. Parts of the area resemble the Low Hills Belt; in fact, it may be convenient to consider the Low Hills Belt as extending westward in lobes to the western border of the plateau. In the latitude of Zanesville, Ohio, and extending from the western border of the plateau eastward almost to the Ohio River is an area of more mesophytic vegetation. This was prevailingly mixed mesophytic, although in the culled forest, beech is now most prominent. Many small areas preserve remnants of the mixed forest. The secondary slope woods are largely oaks, except on northerly and easterly slopes, where secondary groves of tuliptrees are further indication that the original vegetation was a mesophytic type. This area more or less coincides with the headwaters area between the preglacial Newark River and opposing tributaries of the Teays drainage.[12] Just as the oak woods of the northern division of the Low Hills Belt apparently is correlated with the reduced topography best developed nearer the major drainage lines of late Tertiary times, so also are the intervening areas of prevailingly mixed mesophytic forest correlated with the less reduced interstream sections which must always have been more or less hilly. Such correlations, which exist in many places, will be considered more fully later (Chapter 17).

Rugged Eastern Area

No definite limits can be given to this maturely dissected area of high relief, which forms more or less of a band from the Allegheny Mountains at about the Pennsylvania border southwestward almost to Corbin, Kentucky, and is widest in West Virginia, where it extends from the southeastern boundary of the plateau to the vicinity of Charleston. Throughout this area, mixed mesophytic forest prevails, except on dry slopes and ridge-tops.[13] In composition, the forests are essentially similar to those of the outlying section of the Cumberland Mountains already considered (p. 51 and Table 1, areas 4, 5), or, in their northern part, to the adjacent lower and middle elevation forests of the Allegheny Mountains (p. 76). Farther from the eastern boundary, summit elevations decrease and relief becomes less, about 500 to 700 feet. In areas where the underlying rock is shale and, hence, soils infertile and subsoils impervious, there is an increase in the proportion of beech and white oak and a decrease in the other dominants of the mixed mesophytic forest. The forest is, however, mixed forest, made up of the same species as the more typical mixed mesophytic forest.[14]

In general, most of the rocks of the plateau are shales and sandstones; limestone is a minor element. Where these different kinds of rock outcrop on slopes in the typical maturely dissected topography of this section, lithological character of the strata has little effect on the vegetation except

[12] For a map of preglacial hydrographic basins of the upper Ohio Valley, see Tight, 1903, Plate I.

[13] Circumscribed areas, stratigraphically different (as the shale flats near Princeton, W. Va.) depart from the regional vegetation type. Where limited in extent, they must be omitted in a general treatment. So also must such relic communities as Cranberry Glades in Pocahontas County, W. Va., be omitted.

[14] For an example of this type of mixed mesophytic forest, see Braun, 1942, p. 437 and Fig. 34.

as it may determine the presence or absence of a few species. For example, heath shrubs are more abundant along a sandstone outcrop; red cedars (*Juniperus virginiana*) may grow in a band along a limestone outcrop on a cleared slope. A number of herbaceous species might be considered as indicators of calcareous or of noncalcareous rock.

Where limestone is the surface rock over any considerable area, its influence is reflected in the aspect of vegetation. The region around Lewisburg, West Virginia, may serve as an illustration. Here a rolling plateau with sinks and underground drainage has developed. White oak is the dominant tree; with it are chinquapin oak (*Q. Muhlenbergii*) and black oak, hickories, walnut, cherry, ash, tuliptree, sugar maple, cucumber magnolia, and basswood. The composition suggests oak–hickory forest; yet many of the trees of the Mixed Mesophytic association are present. Ravine slopes in this area support remnants of typical mixed mesophytic forest. In general aspect, this is no longer a forested region; due to the fertile limestone soil and rolling topography, most of it is utilized. It is sometimes referred to as the Bluegrass of West Virginia, for today the dominant ground cover is *Poa pratensis.*

In certain areas where noncalcareous shales are the surface rock, broad, poorly drained flats have developed. In such areas, pin oak and red maple are now the dominant trees; sedge meadows, shrub communities of *Spiraea tomentosa* and *S. alba,* or alder thickets are developmental stages of the hydrosere.

Very characteristic areas of the Rugged Eastern area may be seen in the vicinity of Hazard and Pikeville, Kentucky, where the relief is about 900 feet, in the northwestern angle of Virginia (Dickenson and Buchanan counties), in the vicinity of Williamson and Charleston, West Virginia, and along the New River and the Gauley River in West Virginia. Although a large part of the area is well wooded and all of the characteristic trees of the Mixed Mesophytic association are present, few entirely primary areas remain. In general, all slopes are covered by mixed mesophytic forest, with *Tilia heterophylla* most conspicuous on east and north slopes and beech proportionately more abundant on south and west slopes where oaks (particularly white oak) are conspicuous. Hemlock is seen in some ravines, and when present is usually accompanied by Rhododendron. *Magnolia tripetala* is a conspicuous ravine tree, while *M. macrophylla* sometimes forms the dominant growth in young secondary stands. Where massive strata outcrop on the mountain sides, occasional pitch pines are seen. The uppermost slopes and ridge-tops are occupied by oak–chestnut or oak–hickory communities, or, more often, by a combination in which oaks are dominant and chestnut and hickory equally abundant. Chestnut oak (*Q. montana*) is the most abundant oak in the oak–chestnut communities, white oak and black oak in the oak–hickory communities. Post oak (*Q. stellata*) is sometimes a constituent of the driest summit forests. Pines

Strongly dissected country along the Gauley River, West Virginia, where mixed mesophytic forest prevails. (Courtesy, U. S. Forest Service.)

(*P. rigida* and *P. echinata*) may be scattered on the ridge-tops or occasionally localized in small groups. In the saddles of ridges, the mixed mesophytic forest comes to the summit, the beech, basswood, buckeye, sugar maple, and cucumber magnolia of the most mesophytic slopes here mingling with white oak and hickory. Sweet gum (*Liquidambar Styraciflua*) is occasionally scattered in the mesophytic forest of the lower slopes. If present in a locality, it is apt to be abundant in the young second growth of dry, shaly slopes. Secondary forests in this section vary greatly, the degree of disturbance largely determining the nature of the stand. The more mesic slopes, if they have not eroded badly, are growing up in mixed forest; while south and west slopes usually have a high proportion of white oak. Eroded slopes—and there are many in this area of mountainside corn "patches"—may grow up in scrub pine, sassafras, persimmon, and other xeric pioneer species.

Low Hills Belt

The Low Hills Belt, which belies the name "mountains" so generally applied to all the plateau in Kentucky and West Virginia, begins in southern

Kentucky as a narrow wedge between the higher and more rugged land to the east and the plateau margin to the west. Here its topography is determined by the resistant Pottsville sandstone, which is overlain by weak shales. These have been partly eroded away, forming low hills. The valleys between them are narrow or broad, their depth determined by the resistant underlying stratum which establishes a local base level for the smaller streams. The bottoms of these valleys are moist or even swampy. Bands of red maple and sweet gum, with patches of alder and swamp herbs, grow in the moist soil. On the adjacent low hills, beech, white oak, black oak, and hickory prevail in secondary stands. Some of the wider valleys are swampy and the adjacent hills reduced to low swells. This is the topography of a former erosion cycle. Nearer the margin of the plateau these streams are deeply entrenched as a result of rejuvenation after uplift of the peneplain (Schooley peneplain) on which the broad valleys and low hills were formed. The forest of the wettest parts of such old valley flats is mostly pin oak, sweet gum, and red maple, with considerable beech if not too swampy, and some white oak, tuliptree, and occasional chestnut[15] in the drier parts. Where a sequence can be traced in remnants of primary communities, the developmental series appears to be pin oak–sweet gum–red maple → beech–red maple–sweet gum → beech. A large number of shrub and herb species occupy small swampy openings. Many of these are coastal plain species, relics of the old-age topography of the Schooley peneplain (see Chapter 15, p. 481). This topography merges with the flat borders of the "Cliff Section." Some of the larger streams have trenched the rolling plateau, extending features of the Cliff Section far back into the plateau.

Farther to the northeast, where the outcrop of Pottsville sandstone lies back of the margin of the plateau, the topography of the Low Hills Belt changes. There are no broad flats, although narrow, swampy ravine bottoms do occur. All of the evidence indicates that the prevailing original forest type of slopes of this part of the Low Hills Belt was white oak–beech, giving way on the upper slopes to oak or oak–hickory. Mixed mesophytic forest occupied the coves and lower ravine slopes. Cutting has resulted in the expansion of the upper slope type, so that today oak or oak–hickory forest prevails. The existing forest may, therefore, be misleading in the impression given as to regional vegetation, especially if the trees of the secondary forest have reached considerable size. A small area seen in Carter County, Kentucky, demonstrates the difference between primary and secondary forest. The small patch of primary forest (35 trees) high on a slope is 50 per cent beech and 20 per cent white oak, with the remaining 30 per cent made up of six other species. Immediately adjacent, and

---

[15] Chestnut was killed by blight in these swamp areas long before it was affected in the more mountainous land to the east. One standing dead chestnut measured a little over 7 feet in diameter.

separated only by a fence, is a second growth stand which is dominantly white oak—a woods of the sort generally seen in this area.

A broad area of low relief and lower summit elevations entirely independent of the influence of the Pottsville sandstone lies adjacent to the Ohio, Kanawha, lower Big Sandy, and preglacial Teays river valleys (D of Fig. on p. 89). The narrow belt farther to the southwest merges with this broad area without pronounced topographic or vegetational break, although for a short distance between the two the features are ill-defined. This broader and lower area represents the Harrisburg peneplain or, according to some, the Allegheny peneplain, which was developed over much of the Allegheny Plateau (see Chapter 16). Throughout, the low rounded hills are typical; in many places, local relief is not more than 200 or 250 feet. Northward in Ohio and Pennsylvania, the Low Hills Belt expands to include a large part of the Unglaciated Allegheny Plateau. No section of the plateau has suffered more from cutting, burning, and attempted agricultural utilization. Ease of access resulted in early exploitation. The old Midland Trail and old National Highway traverse it; the Chesapeake and Ohio, Baltimore and Ohio, and Pennsylvania railways go through it for many miles. In a large part of the area, timber was cut for charcoal burning, the charcoal to be used in early local iron furnaces which were numerous here. Much forest was cut and burned in the clearing of the land for farms. As most of this area has a heavy subsoil (derived from shales), surface wash and gullying is often extreme. This, together with the type of land utilization which the section has suffered, has resulted in pronounced change in aspect from the original condition.

An old "report on the timber growth of Greenup, Carter, Boyd, and Lawrence counties in eastern Kentucky" (Crandall, 1876) is valuable in reconstructing a picture of the original vegetation. Tables accompanying that report "show approximately the relative abundance of the more common species of trees." Observations were made

> first, in the bed of the valley, including also, in most cases, about an equal area of slope; second, the side hill at that part of the slope which appeared on all accounts to be most nearly a medium between hill-top and valley; and third, the top of the hill or ridge, including more or less slope . . . the counts chosen being regarded as representative of this part of Eastern Kentucky.

Percentage composition of the forests of eight localities is given in Crandall's table of "Old Forest Growth." Composition is given for valley, hillside, top of hill, and for the three combined. Also, the percentages of all eight localities together are given (reproduced as Table 12). It is obvious from these figures that the oaks were more numerous than other trees on the hillsides and hilltops (46 and 66 per cent respectively), and hickory rather abundant, especially in the Lawrence County area. The most meso-

Table 12. Forest composition determined from "counts of old forest trees" in Greenup, Carter, Boyd, and Lawrence counties in eastern Kentucky; data based upon eight localities and divided into forest of valley (*V*), hillside (*H*), and top of hill (*T*). (Selected from Table I of Crandall's report, 1876.)

	V	*H*	*T*	*All*
Number of trees	704	646	665	2015
White oak	14.9	23.7	12.9	17.1
Black oak	6.1	12.9	16.4	11.7
Chestnut oak	. .	4.4	29.7	11.2
Post oak	. .	. .	1.8	.7
Other oaks[1]	1.7	4.6	5.5	3.8
Beech	27.8	6.9	. .	11.9
Maple[2]	11.6	4.6	.5	5.7
Chestnut	2.0	3.2	5.3	3.5
Hickory	3.0	12.0	10.0	8.2
Yellow poplar	6.4	5.8	3.5	5.2
Gum	3.1	4.6	2.6	3.4
Ash[3]	.7	1.8	.6	1.0
Linden	2.6	1.5	. .	1.4
Sycamore	6.2	. .	. .	2.2
Buckeye	2.1	. .	. .	.7
Elm[4]	3.7	2.1	. .	2.0
Black walnut	2.6	2.3	.3	1.7
White walnut	.7	. .	.2	.3
Hemlock	4.7	.9	. .	1.9
Pine[5]	. .	6.7	10.6	5.6

[1] "Other oaks"—red oak, Spanish oak, laurel oak, and blackjack oak.

[2] "The sugar tree or rock maple . . . makes up a large proportion of the maples. Along the banks of streams the white maple . . . is common, while an occasional red maple . . . is found, as also the ash-leaved maple . . ."

[3] Mostly white ash.

[4] Represented by several species—*U. americana, U. fulva, U. alata,* but mostly the first.

[5] Several species of pine; yellow pine the common species; white pine in Buffalo Creek, Carter Co.; scrub pine more common in second growth.

phytic species prevailed in the valleys; the forests were mixed mesophytic, with beech, white oak, sugar maple, and tuliptree (yellow poplar in Crandall's table) most abundant. A few of Crandall's remarks about the trees, especially as to size, are of interest, for one rarely sees a *big* tree in this section today.

White oak has a wider range and a greater development in numbers than any other species. In size, it ranks with the largest of the hard wood trees, often reaching a diameter of three and a half feet. . . . The chestnut oak often predominates on the ridges, extending its range downward in rapidly decreasing proportion, rarely being found in the valleys.

A number of species of oaks are grouped under "other oaks," among which is red oak, which he says "reaches dimensions scarcely less imposing than those of the white and black oak." Beech is "found mostly along the foot of the hills. It sometimes becomes prominent well up on the slope, and not unfrequently occurs in scattered growth along the highest ridges. It often shows a diameter of three feet." Chestnut "is found in all localities, and in such size as to give it a prominence much greater than is shown by its percentage in the tabular view." . . . "The yellow poplar, the tulip tree or whitewood . . . ranks in size above all the other trees in Eastern Kentucky . . . ranges in size from two to five feet in diameter, having a cylindrical trunk of great length." While linden (*Tilia*) comprises only 1.4 per cent of the total number of trees of all localities, Crandall states that it "is abundant in some shaded valleys and on some moist slopes. In the tables it falls below its proportional numbers . . . from the difficulty of selecting average localities for all the species." Additional figures were given by Crandall for second-growth stands, some of them quite young (only 22 years). Because bottomlands and lower slopes are generally under cultivation after once cleared, the data refer only to sides and tops of hills. The proportion of oaks on the hillsides is greater, on the tops of the hills about the same, in second-growth as in old-growth stands. Even in these sites, most of the more mesic species were represented. The second-growth of the middle nineteenth century started immediately after clearing and before erosion had removed the humus. Subsequent changes have still further reduced the later second- and third-growth stands to a more xeric type—the prevailing oak and oak–hickory communities now seen.

There is evidence that mixed mesophytic forest was poorly represented, even in the original forest cover, in some parts of the Low Hills Belt. A primary forest area[16] near Gallipolis, Ohio, in a section of very low relief and low summit elevations may be considered oak–hickory forest (Table 13). In the forest area as a whole, including ridges, southerly and northerly slopes, oaks and hickories made up about 62 per cent of the canopy; on the ridges and upper slopes, 98 per cent; on the northerly slopes, only 29 per cent. Sugar maple was present in the understory, even on the ridges, and dominant in the understory of northerly slopes, especially in a tuliptree cove. Beech, although not a constituent of the canopy of the ridges or southerly slopes and poorly represented in the forest as a whole (5.5 per cent of the total, 19 per cent in ravine), was present in the understory of south slopes. There were indications, then, of tendencies toward increasing mesophytism, with perhaps the ultimate replacement of oak–hickory except on the ridges.

Comparable features can be traced as far as the Pittsburgh area. Jennings (1927) states that the forest of the rolling uplands or rounded hills is dominantly white oak, with *Carya ovata, Acer rubrum, Quercus imbricaria,*

---

[16] Morton's Woods; when visited in 1934 lumbering operations had been started. Trees there attained the sizes mentioned by Crandall (1876). White oak on south slope, 39 inches d.b.h.; tuliptree, 42.6 inches d.b.h.

Table 13. Percentage composition of forest of ridge, southerly slope, northerly slope, ravine, and area as a whole, near Gallipolis, Ohio.

	Ridge	*South*	*North*	*Ravine*	*Area as a Whole*	
Number of canopy trees	107	67	143	67	384	
Quercus alba	40.2	47.8	12.6	44.8	32.0	50.2
Quercus montana	24.3	5.9	4.2	7.5	10.7	
Quercus velutina	13.1	1.5	. .	7.5	5.2	
Quercus borealis maxima	.9	1.5	4.9	. .	2.3	
Liriodendron tulipifera	. .	4.5	34.9	7.5	15.1	15.1
Carya ovata	4.7	7.5	4.2	3.0	4.7	12.2
Carya cordiformis	2.8	5.9	2.8	6.0	3.9	
Carya tomentosa	7.5	1.5	. .	. .	2.3	
Carya glabra	4.7	. .	. .	. .	1.3	
Fraxinus americana	.9	20.9	11.9	4.5	9.1	9.1
Fagus grandifolia	. .	. .	5.6	19.4	5.5	5.5
Juglans nigra	.9	. .	4.9	. .	2.1	7.8
Acer saccharum	. .	1.5	4.2	. .	1.8	
Ulmus fulva	. .	. .	4.2	. .	1.5	
Aesculus octandra	. .	. .	2.8	. .	1.0	
Nyssa sylvatica	. .	. .	2.1	. .	.8	
Celtis occidentalis	. .	1.5	.7	. .	.5	

Q. coccinea, Q. prinus (*montana*), *Q. velutina, Q. rubra* (*borealis maxima*), *Castanea,* and *Prunus serotina.* He correlates this forest type with remnants of the Harrisburg peneplain.

The occurrence of this rather extensive area of oak and oak–hickory in the Low Hills Belt, immediately adjacent to and surrounded by more mesophytic communities, raises some questions as to causal factors. The oak–hickory is not here a climatic climax community. Soil, underlying rock, topography (erosion cycles), and Pleistocene drainage changes, all may be contributory. There is evidence that all of the area from which forest composition examples are taken (Tables 12, 13) was within a great pro-glacial lake of early Pleistocene time, formed when the then northward-flowing major rivers were ponded (Wolfe, 1942). The discussion of the problems involved is reserved for later treatment (Chapters 15, 17).

Cliff Section

The name "Cliff Section" is used because of the bold cliffs of Pottsville sandstone or conglomerate which are a prominent feature throughout most of its extent. In much of the area, the Pottsville sandstone is the surface rock of the plateau, in which case the soil is sandy. Elsewhere, it is overlain by shales, the nature of whose weathering in part determines the topography and soils; even here the resistant Pottsville establishes the temporary local base-level and is hence responsible for the numerous swampy places near streams which have not yet felt the effects of rejuvenation.

From near Berea, Kentucky, southward, this section lies along the western margin of the Cumberland Plateau, until in Tennessee and Alabama it expands to the entire width of the plateau; there the Pottsville (Walden and Lookout) sandstone is the surface rock over all the plateau. The west-facing escarpment, together with any small adjacent benches on Mississippian limestone, the plateau, and its gorges are all included in the Cliff Section. Northward from about Berea, the Cliff Section lies back of the western margin of the plateau. It necessarily lacks the prominent westward-facing escarpment seen to the south, for the plateau escarpment now follows the Mississippian outcrop farther to the west. Caps of Pottsville sandstone appear on the narrow interstream areas, and cliffs of this sandstone form the walls of gorges and box canyons of streams which cut across the Pottsville outcrop. Overhanging cliffs, with great semicircular recesses sometimes extending far back under the cliffs above, are known as "rock-houses." These features are particularly well displayed along the Red River (tributary to the Kentucky). From south of Portsmouth, Ohio, northward, the continuity of the Cliff Section is broken and it is represented only by small isolated areas. In Hocking County, Ohio, the typical features of the Cliff Section, here due to the Black Hand sandstone of Mississippian age, occur

A typical view in the Cliff Section where Pottsville sandstone cliffs rim the deep valleys. Mixed mesophytic forest occupies the valley slopes; pine, oak, or oak–pine communities occupy the plateau. Red River, Cumberland National Forest.

near the western escarpment of the plateau. From Berea, Kentucky, northward to Hocking County, Ohio, the Knobs Border area lies to the west of the Cliff Section.

Extreme diversity of topography and, hence, of habitats is responsible for the great variety of unlike communities of this section. Its slope forests are among the finest of mixed mesophytic forests, resembling in many respects those of the Cumberland Mountains. However, not all of the Cliff Section is (or was) occupied by mixed mesophytic forest; some parts are too dry and some too wet. Southward, there is a transition to the southern oak and oak–pine types. The middle part of the section is most representative and will be used as the basis for the present discussion, with which the northern and southern ends may be compared.

MIDDLE OR MOST REPRESENTATIVE DISTRICT

In southern Kentucky, the plateau is flat to rolling, its surface but shallowly indented by small streams whose valleys retain the old-age features reached on an earlier peneplain (the Schooley peneplain) now uplifted. A stream may head near the margin of the plateau in a swamp or boggy depression remarkable in its assemblage of disjunct species—mostly from the Coastal Plain (see p. 481, Chapter 15). Or, coming from somewhat farther east, it flows sluggishly in an old-age valley whose broad flat is occupied by swamp forest, in which pin oak and sweet gum are most abundant. Red maple, sour gum, swamp white oak (*Q. bicolor*), shingle oak (*Q. imbricaria*), and cow oak (*Q. Prinus*) may be scattered in the swamp forest. In less wet areas which are not subject to occasional periods of standing water, beech is always abundant, and sometimes white oak. River birch (*Betula nigra*) follows along the watercourse or, with alder and other shrubs, is scattered over the whole swamp in secondary communities. The undergrowth contains a large number of swamp shrubs—*Cephalanthus occidentalis, Spiraea tomentosa, Aronia* (*Pyrus*) *melanocarpa, Rhododendron* (*Azalea*) *nudiflorum, Itea virginica, Lyonia ligustrina, Ilex verticillata, I. opaca,* and *Viburnum cassinoides*—together with more mesic species. The herbaceous growth is a mixture of hydro-mesophytes and mesophytes, the latter occupying small mounds and hummocks. *Lygodium palmatum* is abundant. Coastal plain species are present, as in the swampy stream heads, and help to explain the vegetational history of the area (see Chapters 15, 17). The soil is pale with iron mottling (a gley soil) as a result of the wet, poorly aerated conditions. On the low, almost imperceptible slopes which border these valleys, beech prevails, with white oak, sweet gum, and tuliptree. In most instances, the primary forest of the swamp gives way on either side to secondary oak or oak–pine communities. The original forest of the interstream areas apparently contained a large proportion of white oak and considerable tuliptree; some areas were largely beech.

The stream finally becomes somewhat entrenched, although the topography as a whole may not be greatly changed. The swamp forest is, of course, restricted and secondary oak, oak–hickory, oak–tuliptree, oak–pine, or pine forests cover most of the area. The usual constituents are *Quercus alba, Q. velutina, Q. stellata, Carya ovata, C. tomentosa, Liriodendron tulipifera, Cornus florida, Pinus echinata,* and *P. rigida.* Other oaks and hickories, *Nyssa, Acer rubrum,* and *Pinus virginiana* may be present. Various expressions of the older secondary forests on the rolling interstream uplands are seen. White oak is dominant in some areas, but more often white oak, black oak, and tuliptree together predominate. Yellow pine may be associated with the oaks or alone may dominate. Dogwood is almost always the dominant tree in the understory, sometimes forming a continuous layer beneath the pines. In spring, this dogwood layer is very conspicuous. These several types of woods have a dry aspect. Shrubs are few, except for scattered patches of *Vaccinium vacillans, V. stamineum, Rhododendron* (*Azalea*) *nudiflorum, Smilax glauca, Vitis* spp., and, locally, of *Gaylussacia brachycera,* and occasional *Viburnum acerifolium.* Among the widely spaced herbaceous plants are *Pteridium aquilinum, Polystichum acrostichoides, Uvularia perfoliata, Tipularia discolor, Cimicifuga racemosa, Podophyllum peltatum, Oxalis violacea, Geranium maculatum, Viola hirsutula* and, sometimes, *V. pedata, Chimaphila maculata, Obolaria virginica, Houstonia caerulea, Antennaria solitaria,* and *Solidago* spp. Leaf litter tends to accumulate on the surface and humus incorporation is slow. The podzolic soils of these communities contrast with the melanized soils of the mixed mesophytic communities and are unsuitable for most of the species of mull humus. These oak, oak–tulip, or oak–pine woods have a xeric aspect, very different from that of a mixed mesophytic forest.

Toward the margin of the plateau (either along the western escarpment or at a margin above one of the deeply entrenched rivers), the stream, which thus far has been not far below the upland level, begins to cut much more deeply and then suddenly drops as a waterfall over the sandstone cliff into a gorge below. These marginal interstream areas are the driest parts of the plateau. Small isolated areas stand as flat-topped pinnacles, the tops always pine-clad. Narrow promontories extend out irregularly from the general margin and narrow isthmuses connect almost separated pieces of the plateau. Some of these are 500 or 600 feet vertically above the valley below and perhaps not over 10 feet wide. Within a few feet of the margin of the sandstone cliff the ground is covered by moss and *Cladonia* cushions. A little farther back, heath shrubs prevail, usually *Kalmia latifolia, Gaylussacia baccata,* and *Vaccinium vacillans* (with *Epigaea* and *Gaultheria* underneath), but in some areas the box huckleberry (*Gaylussacia brachycera*) is the dominant shrub. Pines (*P. rigida, P. virginiana,* and, sometimes, *P. echinata*) grow nearly at the cliff margin; scarlet oak and chestnut oak (*Q.*

Streams drop suddenly from the plateau into deeply entrenched valleys. Yahoo Falls (75 feet in height), near South Fork Cumberland River, Cumberland National Forest.

coccinea and *Q. montana*) often mingle with the pines. Instead of this pine–heath or pine–oak–heath community, some of the promontories are occupied by open pine woods (the three species of pine) with a grassy layer of *Andropogon scoparius, A. glomeratus,* and *Sorghastrum nutans,* in which are a few scattered forbs. Fires have modified most (perhaps all) of these pine summits, although the abundance of large *Cladonia* mats is an indication that there has been no fire for many years. The pine and pine–oak of the plateau margin are stable communities which can be considered as physiographic climaxes. Reduction of the sandstone margin of the cliff is infinitely slow.

The contrast between the dry pine or oak forest of the plateau and the forest of the slopes below the cliffs (over which one can look from the edge of the plateau above) is extreme. Every such slope, unless at some time completely cleared, is occupied by mesophytic forest. Mixed mesophytic forest occupies all except extreme situations, its composition varying with slope exposure. Hemlock is the dominant tree in the narrowest gorges and reentrant angles where streams drop from the plateau above and on north-facing talus slopes, if these are composed mostly of large sandstone fragments. In such places, Rhododendron often forms impenetrable thickets. Hemlock also occurs as a dominant with white oak and beech in ravines opening widely to the south, thus suggesting the white oak–beech–hemlock and hemlock–white oak segregates of the Cumberland Mountains. Beech is the dominant tree of the gradually sloping valley floor, which develops as soon as the gorge widens or the stream emerges from between the bordering cliffs. As in the Cumberland Mountains, white oak and beech are most abundant on many southerly slopes; while the more representative mixed mesophytic forest communities occupy the more mesic slopes. Few virgin areas remain, although much of the forest of slopes in deep valleys and gorges is in part primary. Inaccessibility delayed, but did not prevent, the logging of these rich forests.

A number of specific examples of forests, their composition and environment, will illustrate the general features which have been mentioned. These are grouped, in Tables 14 to 17, as representative of all-deciduous mixed mesophytic, hemlock–mixed mesophytic, and beech–mixed mesophytic.

Extensive and very fine forest areas were located in Pickett County, Tennessee, near the Kentucky line, on the property of the Stearns Coal and Lumber Company.[17] In the forest of the plateau, white oak, chestnut, and chestnut oak were most abundant, with red oak, black oak, tuliptree, cucumber tree, sour gum, and occasional yellow pine (*Pinus echinata*). This plateau type, a physiographic climax, was reproducing itself, as shown by essential agreement of canopy and understory. Dogwood was abundant in this primary forest, as it is also in many secondary forests. The Heaven-

[17] The last virgin tract in their forest land was being logged in 1947.

Basswood (right), tuliptree, and hemlock in Heavenridge tract, Pickett County, Tennessee.

ridge tract of 6500 acres was a magnificent mixed forest. The canopy species of this slope forest were tuliptree, hemlock, basswood, walnut, cucumber tree, sugar maple, red maple, chestnut, chestnut oak, sour gum, shellbark hickory, bitternut hickory, buckeye, and white oak, of which the first two were most numerous. Tuliptrees were about 3½ feet d.b.h. and 60 to 70 feet to the first branch. The undergrowth, in addition to the species of the canopy, contained American holly, umbrella magnolia, great-leaf magnolia, sweet birch, dogwood, and papaw. Hemlock was generally local, occurring in rocky parts of the valleys, high on the slopes in crevices of the cliffs, and in the rocky talus at the foot of the cliffs. Thus, as in the Cumberland Mountains, the mixed mesophytic forest displays two unlike phases, the all-deciduous mixed mesophytic and the hemlock–mixed mesophytic. An example of the former, in which beech, tuliptree, and basswood make up about 50 per cent of the canopy, is illustrated by Table 14, A; an example of the latter, in which beech and hemlock make up over 75 per cent of the canopy, by Table 15, A.

Pronounced variations in composition occur, although association-segregates, except those of south slopes, are usually not well defined. Beech–white oak, hemlock–white oak, and beech–hemlock–white oak–

Beech–sugar maple–white oak forest of lower slopes, Heavenridge tract, Pickett County, Tennessee.

Table 14. Composition of canopy of representative mixed mesophytic forests of the "Cliff Section."

A. "Ring Pine" Valley, Rock Creek, Pickett County, Tenn.

B. Steep east slopes of Cumberland River, Pulaski County, Ky.

C. Carter Caves, Carter County, Ky.

D. Cove on Tygarts Creek near Cascade Caves, Carter County, Ky.

	A	*B*	*C*	*D*
Number of canopy trees	74	97	131	69
Fagus grandifolia	14.9	34.1	30.5	23.2
Liriodendron tulipifera	23.0	18.6	17.5	15.9
Tilia neglecta+heterophylla	10.8	3.1	11.4	18.8
Acer saccharum	2.7	3.1	13.0	13.0
Quercus borealis maxima	4.0	5.1	2.3	5.8
Fraxinus americana	1.4	2.1	4.6	4.3
Quercus alba	6.8	3.1	1.5	2.9
Castanea dentata	5.4	1.0	6.1	1.4
Magnolia acuminata	4.0	8.2	. .	. .
Carya ovata	5.4	8.2	2.3	1.4
Carya tomentosa	4.0	4.1	. .	. .
Carya sp.	4.0	. .	.8	. .
Nyssa sylvatica	4.0	6.2	2.3	1.4
Aesculus octandra	4.0	. .	.8	1.4
Betula lenta	4.0	1.0	3.0	. .
Acer rubrum	1.4	. .	.8	. .
Juglans nigra	. .	2.1	2.3	. .
Ulmus fulva	. .	. .	. .	1.4
Oxydendrum arboreum	. .	. .	.8	. .
Tsuga canadensis	. .	. .	. .	8.7

tuliptree communities, with a number of the usual constituents of richer mixed mesophytic forests, occupy southerly slopes. The second and third of these are well represented on valley slopes of southward-flowing tributaries of No-Business Creek, Scott County, Tennessee (Table 15, C, D). In such valleys, the more mesic hemlock–mixed mesophytic communities, with occasional basswood and buckeye, are confined to the lowest ravine slopes.

The tremendous difference between primary and secondary slope forest was demonstrated by a small secondary area adjacent to the mixed mesophytic forest illustrated by Table 14, A. Erosion had removed all humus, exposing clay and shale; the trees were scrub pine and oaks, with a band of sweet gum along a seepage zone. Certainly such a community gives no indication of the luxuriant forest growth once occupying these situations.

Forest communities in the vicinity of Cumberland Falls, Kentucky, illustrate the variety of types possible in a limited area: pine and oak–pine uplands, mixed mesophytic slope forests, hemlock communities, and other variations. Although many large trees remain and some parts are essentially virgin, large untouched areas are lacking. Here holly (*Ilex opaca*) reaches canopy size in the mixed mesophytic forest with hemlock. The Cumberland

Table 15. Composition of hemlock–mixed mesophytic forests of gorges, southerly slopes, and northerly slopes of the "Cliff Section."

A. Yahoo Creek, McCreary County, Ky.

B. Rock Creek Natural Area, Cumberland National Forest, Laurel County, Ky.

C. and *D.* Slopes of No-Business Creek, Scott County, Tenn.*

E. Glen Eden, Lee County, Ky.

F. Cove forest near Carter Caves, Carter County, Ky.

	Gorges		*Southerly Slopes*		*Northerly Slopes*	
	A	*B*	*C*	*D*	*E*	*F*
Number of canopy trees	93	117	349	115	90	88
Tsuga canadensis	25.8	38.5	51.9	25.2	23.3	35.2
Fagus grandifolia	51.6	6.0	3.1	30.4	15.5	31.8
Quercus alba	. .	. .	25.2	12.2	. .	. .
Liriodendron tulipifera	6.4	19.2	4.0	12.2	21.1	6.8
Tilia neglecta+heterophylla	4.3	. .	. .	. .	13.3	14.8
Acer rubrum	1.1	17.2	2.9	.9	4.4	. .
Acer saccharum	6.4	. .	2.3	8.7	4.4	3.4
Quercus borealis maxima	1.1	.9	.9	2.6	. .	1.1
Betula lenta	1.1	6.8	1.7	. .	. .	. .
Aesculus octandra	2.1	. .	. .	. .	1.1	4.5
Ilex opaca	. .	4.4	. .	. .	. .	. .
Castanea dentata	. .	3.4	. .	. .	4.4	. .
Nyssa sylvatica	. .	2.5	6.0	2.6	. .	. .
Magnolia acuminata	. .	. .	.9	2.6	3.3	. .
Carya ovata and *tomentosa*	. .	. .	.6	. .	. .	1.1
Fraxinus americana	. .	. .	. .	. .	. .	1.1
Quercus montana	. .	.9	.3	. .	. .	. .
Oxydendrum arboreum	. .	.9	. .	. .	. .	. .
Magnolia macrophylla	. .	.9	. .	. .	6.7	. .
Ulmus americana	. .	. .	. .	. .	1.1	. .
Pinus Strobus	. .	. .	.3	.9	. .	. .
Juglans nigra	. .	. .	. .	1.7	. .	. .

* The writer is indebted to Dr. Royal E. Shanks for these data.

River flows in a gorge, and some distance below the Falls becomes entrenched in the Mississippian limestone below the Pottsville sandstone. Where the forest is continuous from the plateau to the river, as in a few places farther down stream in the Cumberland National Forest, mixed mesophytic forest covers the slopes on sandstone and limestone alike (Table 14, B). Rock outcrops of course have rock plants, many of which are generally limited to one or the other substratum. In the forest canopy of the slopes, certain minor differences in composition are discernible. Chestnut, a usual constituent of the mixed mesophytic forest, is generally absent on limestone soils; hence, the upper forest (here three-fourths of the slope) contains chestnut, while the lower does not. Sugar maple is often more abundant on limestone soils and here is prominent in the canopy of the lower forest and not in the upper (although it is in the undergrowth in both

Beech–white oak forest of lower slopes, Ring Pine Valley, Pickett County, Tennessee.

parts). The differences in aspect of the forest and its undergrowth on the two substrata are so slight that a separation into two communities is not desirable. Climax Mixed Mesophytic forest develops irrespective of the nature of underlying rock.

Some of the small tributaries of the Cumberland and Rockcastle Rivers, which join these streams in their gorges, have cut deep, narrow gorges a mile or two back into the plateau. The Rock Creek Natural Area in the Cumberland National Forest (19 miles southwest of London) occupies such a gorge with precipitous walls, which trends in a direction slightly north of west. In it, hemlock, tuliptree, and red maple are the dominant species, comprising 75 per cent of the canopy, in which nine additional species make up the remaining 25 per cent (Table 15, B). The forest is an all-aged stand, in which all of the canopy species are represented in the undergrowth. Hemlock is proportionately less abundant than in the canopy; holly and bigleaf magnolia are abundant. The larger holly trees, which attain canopy status, are 16 inches d.b.h. *Rhododendron maximum* forms an almost continuous layer from cliff to cliff; *Clethra acuminata* is scattered. Herbaceous species are almost excluded; scattered individuals only were seen of *Dryopteris spinulosa, Dennstaedtia* (*Dicksonia*) *punctilobula, Lycopodium lucidulum, Medeola virginiana, Cypripedium acaule, Goodyera* (*Epipactis*)

Holly is present in the undergrowth in this mixed mesophytic forest, and occasionally reaches canopy size. Here, it is associated with beech, tuliptree, hemlock, ash, red maple, and hickory. Photo in early spring, before deciduous trees were in leaf. Cumberland National Forest, McCreary County, Kentucky.

pubescens, and *Tipularia discolor,* with a few additional species confined to the cliffs or mossy rocks. Narrowness of the gorge, its precipitous walls, and its direction combine to produce this extreme type of hemlock–mixed mesophytic forest.

Not all parts of the valleys of rivers cutting across the Cliff Section are gorges. In the occasional wider parts of valleys there are alluvial terraces and graded talus accumulations presenting a habitat topographically very unlike the slopes, but supporting a similar forest community (Table 17, A). That is, mixed mesophytic forest has developed on slopes where they have become sufficiently reduced and on depositional areas when they have been sufficiently built up. Forests in these situations always have a high percentage of beech, but the typical species of the Mixed Mesophytic association are present. They may also contain an occasional white elm, relict of an earlier developmental stage, and bamboo (*Arundinaria gigantea*), generally most abundant near streams. The herbaceous growth is exceedingly luxuriant and like that of rich slope woods.

An excellent sample of slope forest (on soil derived from limestone) is

In the Red River Valley. A beech consociation occupies the valley floor on the left; the slopes on the right (which face nearly south) are occupied by rich mixed mesophytic forest. Above this are the cliffs which rim the valley, with pines and oaks growing to the very edge. Cumberland National Forest.

preserved in the Natural Bridge area of Powell County, Kentucky, in the Red River drainage basin (Table 16). Again the fact that the Mixed Mesophytic association is independent of substratum is demonstrated. The absence of chestnut and the presence of *Quercus Muhlenbergii* are related to the calcareous soil. Chestnut (now dead) is present higher on the slope. A variant of the slope forest is illustrated by an area in Lee County, Kentucky (Table 15, E). The essential similarity of these forests and the mixed mesophytic forests containing hemlock in the Cumberland Mountains is evident (cf. Table 6, areas 2, 7). In this part of the Cliff Section, white pine (*Pinus Strobus*) is a constituent of some mixed mesophytic communities of gorge heads. Such a forest may contain hemlock, white pine, beech, tuliptree, sugar and red maple, sweet birch, white basswood, chestnut, hickory (shellbark and pignut), and white oak. Although this may suggest the "hemlock–white pine–northern hardwoods" forest, it is quite different in aspect. The magnolias in the understory (*M. macrophylla, M. tripetala*), the holly, and the wealth of Rhododendron give a distinctly southern aspect to the community.

In the Carter Caves region of Carter County, Kentucky, which lies close

Table 16. Composition of representative area of mixed mesophytic forest of slopes to demonstrate essential accordance of canopy and second layer (trees 8–12 inches d.b.h.); Natural Bridge, Powell County, Ky. (sample of 107 canopy trees; 57 second layer trees).

	Canopy	*Second layer*
Tsuga canadensis	21.5	21.0
Fagus grandifolia	15.9	19.3
Tilia heterophylla	15.9	5.3
Liriodendron tulipifera	10.3	10.5
Acer saccharum	9.3	8.8
Betula lenta	3.7	7.0
Juglans nigra	2.8	. .
Acer rubrum	2.8	5.3
Carya cordiformis	2.8	3.5
Quercus Muhlenbergii	2.8	5.3
Quercus borealis maxima	2.8	3.5
Quercus alba	2.8	. .
Magnolia acuminata	1.9	3.5
Nyssa sylvatica	1.9	1.7
Fraxinus americana	.9	3.5
Juglans cinerea	.9	. .
Quercus montana	.9	. .
Ulmus fulva	. .	1.7

to the northern end of the more or less continuous band of the Cliff Section and immediately adjacent to the Low Hills Belt, a wide range of forest communities may be seen within a few miles radius. The Pottsville sandstone is not as thick as farther south (about 90 feet) and the plateau area is more continuous. Cavernous Mississippian limestone forms the valleys, and all the larger streams cut below the sandstone into the limestone. The upland removed from the vicinity of gorges is in that part of the Low Hills Belt which is developed above the Pottsville sandstone horizon. Its vegetation, almost entirely secondary and dominantly oak, gives no suggestion of the variety or luxuriance of vegetation nearby. On gradual slopes of the plateau above the cliffs, beech is dominant, usually with red maple, oaks, and, in places, chestnut (Table 17, C). Such a forest has a poor undergrowth of a type indicative of leached and acid soils: *Mitchella repens, Gaultheria procumbens, Epigaea repens, Tipularia discolor,* etc. It is very different from the beech consociation of the valleys. Along the driest cliff margins, oaks, hickories, and pines prevail, with occasional hemlocks arising from clefts. Far below, on the flood plain of Tygarts Creek, the flood-plain hydrosere presents a strong vegetational contrast. Succession progresses from the white elm, sycamore, silver maple, river birch, and box elder community of the muddy river flats, through gradual changes with loss of river margin species and addition of mesic species, to a beech consociation with associated characteristic species of the Mixed Mesophytic climax (Table 17, B), and finally to typical mixed mesophytic forest on the ag-

Table 17. Composition of mixed mesophytic forest in which beech forms half or more of the canopy.*

A and *B*. Forests of stream flats: *A*, On Rockcastle River near Livingston, Ky.; *B*, On Tygarts Creek and tributaries near Carter Caves, Carter County, Ky.

C. Drier upland slope forest, slopes above cliffs; composition based on four areas in Carter and Elliott counties, Ky.

	Stream Flats		*Uplands*
	A	*B*	*C*
Number of canopy trees	72	78	137
Fagus grandifolia	59.7	78.2	49.6
Acer saccharum	1.4	7.7	.7
Liriodendron tulipifera	5.5	1.3	1.5
Aesculus octandra	2.8	1.3	1.5
Quercus alba	1.4	2.6	8.0
Quercus borealis maxima	2.8	. .	3.6
Tilia spp.	2.8	1.3	. .
Acer rubrum	2.8	. .	9.5
Tsuga canadensis	. .	3.8	5.8
Nyssa sylvatica	. .	1.3	1.5
Ulmus americana	7.0	1.3	. .
Juglans nigra	5.5	. .	. .
Juglans cinerea	1.4	. .	. .
Fraxinus americana	4.2	. .	. .
Carya ovata	1.4	. .	. .
Morus rubra	1.4	. .	. .
Platanus occidentalis	. .	1.3	. .
Castanea dentata	. .	. .	8.8
Quercus velutina	. .	. .	5.8
Carya glabra	. .	. .	2.2
Oxydendrum arboreum	. .	. .	1.5

* Both types contain representatives of most of the usual dominants of the Mixed Mesophytic association; the forest of stream flats contains a few representatives of the flood-plain hydrosere.

graded inner margin of the flood-plain area which is continuous with the reduced adjacent slopes. Mixed mesophytic forest (Table 14, C, D) occupies the intermediate slopes between the plateau communities and the valley communities. It was in such a situation that the "Carter Caves virgin forest" was located.[18] In this area, beech, tuliptree, sugar maple, and basswood comprise 73 per cent of the canopy, with the remaining 27 per cent made up of 12 species. Lower layers of the forest contain all of the species of the canopy and, in addition, red elm, mulberry, redbud, umbrella magnolia, dogwood, ironwood (*Carpinus*), and papaw. This forest well illustrates the typical aspect of the best mixed mesophytic forests, punctuated by the large

[18] This forest was logged in 1944. The area is now within the newly established Carter Caves State Park.

size of trees,[19] the all-age character of the stand, and the exceedingly luxuriant herbaceous undergrowth with quantities of ferns.[20] Hemlock is abundant in restricted areas (coves and deep hollows) in this forest, locally comprising 30 to 40 per cent of the canopy (Table 15, F). In these hemlock groves, Rhododendron is dominant in the undergrowth; the number of herbaceous species is less and *Dryopteris spinulosa* is more abundant.

NORTHERN DISTRICT

Northward, where the Cliff Section begins to be discontinuous, remnants of the Pottsville sandstone cap parts of the ridge-tops; between these cliffy caps are low saddles where the Mississippian limestone is the underlying rock. Along such ridge-tops, communities characterized by chestnut oak (with heath shrubs and, usually, chestnut) alternate with open oak–hickory communities (with no heaths, but often with prairie plants). These communities are sharply delimited and are correlated with sandstone and limestone substrata, respectively. This correlation between the composition of ridge-top communities and underlying rock is indicative of their subclimax status.

In this discontinuous band of Cliff Section, there are a number of more or less isolated areas where the characteristic features of the Cliff Section recur. The more southern species drop out one by one, as range limits are reached, and a few northern species appear (*Sambucus pubens* and *Maianthemum canadense* in the Hocking Valley area). The mixed character of the slope forests is evident. Among the northern outliers of the Cliff Section are High Rock, near Haverhill, on the Ohio River, gorges near Jackson, Ohio, and the Hocking Valley areas (on Black Hand sandstone) now included in Ohio's Forest Parks system. The communities occurring in the last mentioned areas have been described by Griggs (1914).

SOUTHERN DISTRICT

In the southern part of the Cumberland Plateau, from northern Tennessee to the Warrior Tableland of Alabama, the Pottsville sandstone is the surface rock over the entire plateau; the term "Cliff Section" might then be applied to the whole plateau, for wherever streams have cut gorges, cliffs are present. As a whole, the plateau is little dissected; the topography is submature and level or rolling upland prevails, representing the Schooley peneplain.

[19] Beech, 33 inches; buckeye, 37 inches; red oak, 44 inches; chestnut, 65 inches, d.b.h.

[20] In a sample strip through the forest, 14 species of shrubs and woody climbers, 19 species of ferns (including one *Lycopodium* and four rock ferns), and 81 species of herbaceous flowering plants were listed.

The vegetation of this old and little-modified surface is very different from that of the slopes. Today, little of the original upland vegetation remains. The poor secondary growth gives little indication of the former forest. The poorly drained spots are swampy and red maple now prevails, although such areas were probably similar to those farther north already considered (p. 99). Communities of post oak and blackjack oak (*Q. stellata* and *Q. marilandica*) occupy very shallow dry soil areas of the plateau where the sandstone is close to the surface. Over the greater part of the area, one sees mixed oak, oak–hickory, and oak–pine communities made up of *Quercus alba, Q. montana, Q. coccinea, Q. falcata, Q. stellata, Q. borealis* var. *maxima, Carya glabra, C. tomentosa, Nyssa sylvatica, Oxydendrum arboreum, Diospyros virginiana, Cornus florida, Pinus echinata, P. virginiana,* and other xeric species. Not all of these are associated in any one place; the general impression is of oak woods. The large number of showy summer-blooming herbaceous plants suggests that the forest of the upland was not everywhere exceedingly dense, that there always were sunny spots in which this now prevalent herbaceous growth found suitable habitats. Plateau margins above gorges support pine–heath communities as in the more typical Cliff Section farther north.

A semi-virgin plateau forest may be seen at Falls Creek Falls, Van Buren County, Tennessee (elevation 2000 feet). Here white oak comprises 83 per cent of the canopy; associated with it are black oak, hickory, tuliptree, chestnut, sour gum, and occasional basswood. The older inhabitants state that "the forest on top was mostly oak, with some poplar [tuliptree], gum, and hickory." The tuliptrees were long since removed for timber from this area of primary white oak.

Some 75 miles farther south in Alabama, where the summit of Lookout Mountain (a part of the Cumberland Plateau) spreads out to a wide tableland at an elevation of about 1150 feet, the original forest was evidently similar to that at Falls Creek. Mohr (1901) says that "the mountain was recently covered with a fine hard-wood forest, chiefly of oaks, and was noted for the abundance of white oak timber (*Quercus alba*) and tan-bark oak [*Q. montana*]." Farther west, on the Warrior Tableland above 900 or 1000 feet, "the tan-bark or mountain oak largely prevails, associated with black oak, occasionally with scarlet oak. . . ; also with mockernut, pignut hickory, and fine chestnut trees . . . , and with white oak (*Quercus alba*) and highland gum (*Nyssa sylvatica*)" (Mohr, 1901). At lower elevations, yellow pine (*P. echinata*) formed 20 to 30 per cent of the timber growth, with loblolly pine (*P. Taeda*) in areas of deficient drainage. Nearer the Tennessee border and north of the Tennessee River, where the plateau is much dissected—the area referred to by Mohr as detached spurs of the Cumberland Mountains—"the table-lands of comparatively limited extent, support a more varied and heavier tree growth than the table-lands of

Oak forest, a physiographic climax of the plateau at Falls Creek Falls, Tennessee. Dogwood and pink Azalea are common in the undergrowth.

the Warrior basin, differing chiefly by the total absence of pines and the appearance of species common also to the forests of the Ohio Valley . . . Oaks form the predominating forest growth of these highlands—white oak, mountain oak, and fine black oak." Other trees mentioned for this area as growing in richer spots include sugar maple, cucumber tree, "silver-leaf linden" (*Tilia heterophylla*), and sweet buckeye (Mohr, 1901).

There is no doubt that oak or oak–hickory forest prevails over the large area of submaturely dissected plateau (old peneplain surface) in Tennessee and Alabama. But a submature topography does not support the regional climax. This forest type, although areally dominant, no more represents the climatic climax than do the small patches of this same community on the ridge-tops of the plateau farther north. It is a physiographic climax maintained by topography (old) and soil (sandy and leached or lithosol) and frequently affected by fire. Where the plateau is dissected, on the slopes of valleys cut into the plateau and of ravines indenting it, even along the bold westward-facing escarpment of the Cumberland Plateau, mixed mesophytic forests are seen. If the whole plateau were dissected, as it is farther north, mixed mesophytic forest would be the prevailing type. Although areally limited, it is the climax of this area; hence, almost all of the Cumberland Plateau is included in the Mixed Mesophytic Forest region.

A few examples of composition of slope forests will emphasize the contrast between these and the pre-erosion forests. As compared with the plateau soils, the soils of the slopes are deeper and richer, darker in color (melanized), and, on forested slopes, have a mull humus layer.

Table 18. Composition of mixed mesophytic forest in southern part of the Cliff Section.

A. Lower northwest slopes of Cane Creek, Van Buren County, Tenn.

B. Middle east, southeast, and south slopes of Cane Creek.

C. West-facing escarpment of Cumberland Plateau east of Cowan, Tenn.

	A	*B*	*C*
Number of canopy trees	76	206	74
Liriodendron tulipifera	17.1	17.5	21.6
Fagus grandifolia	28.9	.5	. .
Tilia heterophylla +	11.8	.5	10.8
Acer saccharum	2.6	1.9	8.1
Aesculus octandra	2.6	2.4	8.1
Magnolia acuminata	. .	7.3	8.1
Nyssa sylvatica	6.6	4.4	2.7
Quercus alba	6.6	16.0	5.4
Quercus borealis maxima	3.9	8.7	12.2
Quercus montana	1.3	8.2	. .
Quercus Muhlenbergii	. .	. .	8.1
Quercus velutina	. .	1.0	. .
Carya ovata	6.6	16.5	6.8
Carya tomentosa	. .	8.2	. .
Carya spp.	2.6	.5	6.8
Acer rubrum	3.9	.5	. .
Fraxinus americana	2.6	1.9	1.3
Juglans nigra	1.3	.5	. .
Ulmus alata	. .	3.4	. .
Tsuga canadensis	1.3	. .	. .

East of Spencer, Tennessee, on the lower northwest-facing slopes of Cane Creek (a wide, not gorge-like valley), the forest is mixed mesophytic, made up of 15 species of canopy trees of which beech, tuliptree, and basswood comprise nearly 60 per cent of the stand (Table 18, A). Higher, on east, southeast, and south slopes, oaks and hickories are more abundant; tuliptree, shellbark hickory, and white oak comprise 50 per cent of the canopy of a mixed forest in which 18 species are represented in the sample (Table 18, B). Buckeye, basswood, sugar maple, and beech are present, but in small numbers. In deeper valleys or gorges, as in the Falls Creek Falls area in Tennessee, hemlock is abundant in a mixed forest in which tuliptree, basswood, buckeye, red oak, sugar maple, red maple, yellow-

The valley of Falls Creek. The contrast between the hemlock–mixed mesophytic forest of the valley slopes below, the open pine woods of cliff margins, and the oak woods of the plateau (as shown on p. 114) is pronounced.

wood, and birch (*B. lenta, B. lutea*) are prominent; Rhododendron and magnolia are abundant in the understory.

The same conditions prevail in the gorges in the plateau in Alabama. Mohr (1901) describes secluded valleys in the tableland where the forest growth is of great luxuriance and variety. In addition to many of the trees of the upland, a number of mesophytic species occur: beech, elm (*U. americana*), butternut, cow oak (*Q. Prinus*), basswood (*T. heterophylla, T. americana*),[21] cucumber-tree, umbrella magnolia, and large-leaf magnolia, the last here reaching a trunk diameter of 20 inches. In a few of these gorges, hemlock is found, even as far to the southwest as the border of the Coastal Plain, where the larger trees associated with it are, in order of abundance, *Ilex opaca, Liriodendron tulipifera, Liquidambar Styraciflua, Fagus grandifolia, Castanea dentata, Nyssa sylvatica,* and *Quercus borealis* var. *maxima* (Harper, 1937). With a similar group of species, hemlock extends into the adjacent Tennessee Valley area of the Highland Rim, where, on Little Mountain, Alabama, the Hartselle sandstone (Mississippian) affords habitats similar to the Pottsville sandstone (Harper, 1943a).

[21] Although referred to by Mohr as *T. americana,* it is almost certain that this should be referred to one of the other almost glabrous species.

The west-facing escarpment of the Cumberland Plateau in southern Tennessee rises abruptly above the mostly level to rolling lower plateau which is the Highland Rim. Its slopes are covered by mixed forests in which *Tilia heterophylla* is prominent. The composition of a forest area on the escarpment east of Cowan, Tennessee, is given in Table 18, C. Tuliptree, red oak, and basswood are the three most abundant species, and buckeye and sugar maple are well represented. Beech is abundant on the lower slopes, and basswood and sugar maple more abundant on less exposed slopes than shown in Table 18, C.

SECONDARY FORESTS

As much of the Cliff Section is submarginal land and includes most of the Cumberland National Forest, problems of reforestation and improvement of existing secondary stands are important. With a knowledge of primary forest types of the several distinct habitats and of developmental trends in each, the forest possibilities of the area may be envisioned. White oak and tuliptree were among the most abundant species of the plateau forest; both are valuable timber trees. White oak is a constituent of many second-growth stands; tuliptree, however, is not reproducing abundantly on the plateau. Soil deterioration and fire have set back development so that more xeric (and often less valuable) species prevail. Tuliptree is reproducing abundantly on some of the mesic slopes, particularly in east and northeast coves. The secondary forests of slopes vary more in relation to slope exposure than do the primary forests. Southerly and westerly exposures, particularly where they have been visited by fire, are often reduced to scrubby oak and pine (*P. virginiana, P. rigida*) communities. With repeated disturbance, scarlet oak assumes dominance on upland and xeric slopes. On more mesic slopes, the variety of secondary communities in the Cliff Section seems infinite. Where disturbance has been slight, mixed forest prevails, often a result of growth of young trees already in the area. Slopes which have been completely denuded (and usually also subjected to fire) have fewer species; pioneer secondary communities may be dominated by one or a few trees—sometimes tuliptree, but more often by other species; *Magnolia macrophylla* is in places the dominant growth. One of the most striking features of secondary vegetation of the plateau is the white pine, which is more or less abundant from the vicinity of Natural Bridge, Kentucky, northeastward across the Red River Valley and into the Licking River drainage, and locally almost to Portsmouth. It is also abundant farther south, in the northern Tennessee area of the Cliff Section. White pine is not limited to any particular site. It may be associated with oaks (usually white oak) on the low hills of the plateau, or on dry west and south slopes below the plateau level. It may be a constituent of the marginal pine–heath or even dominate in this community. It is frequently associated with tuliptree on the mesic slopes of gorges, or sometimes with a variety of mesophytic species. The study of natural tendencies in the many secondary communities and their correlation with stable primary communities should yield information of economic value.

Knobs Border Area

In northern Kentucky and southern Ohio, the westernmost part of the Allegheny Plateau lies to the west of the Pottsville sandstone outcrop. Mississippian and Devonian shales, limestones, and sandstones are the underlying rocks. This rather narrow band is included in the plateau. In its southern part, from Berea to Frenchburg, Kentucky, this area is lower than the Pottsville escarpment, strongly dissected, and includes a considerable proportion of limestone soils (C of Fig. on p. 89). Farther north in Kentucky and in southern Ohio, the western escarpment of the Mississippian rocks is as high as that of the Pennsylvanian (D of Fig., p. 89). This northern part of the rim, which is higher than the plateau farther east, represents the Schooley peneplain, as does the Pottsville margin farther south. Many of the hills of this border area are more or less conical, their slopes cut on soft shales (on the Devonian black shale in much of the area), their summits sometimes capped by more resistant strata (limestone or sandstone of Mississippian age). It is the shape of many of these hills which gives the name "Knobs" to the area. Between the hills, in parts of this Border area, broad valleys extend in from the adjacent Bluegrass section of the Western Mesophytic Forest region. Many of these valleys have two levels, a lower swampy flat and a higher flat to rolling terrace which is a tongue of the Lexington plain (a correlative of the Harrisburg peneplain) extending in between the higher hills and knobs. Thus, a variety of topographic and soil conditions are offered in the Knobs Border area.

More than any other part of the Mixed Mesophytic Forest region, except Pine Mountain in the Cumberlands, this area demonstrates the interrelations between vegetation and underlying rock and between vegetation and physiographic history. In its edaphic communities—on the black shale slopes, on the high and low valley flats, on certain of the limestone outcrops—it is distinct from other parts of the plateau. In its mixed mesophytic forests, it is similar to the plateau to the east. The contrast between this area and the adjacent lower land of the Bluegrass to the west is pronounced.

A large proportion of the slopes of the Knobs Border area was originally occupied by mixed mesophytic forest. Mixed mesophytic forest with beech, tuliptree, basswood, red oak, buckeye, sugar maple, walnut, shellbark hickory, white ash, cucumber tree, red elm, bitternut hickory, white oak, redbud, dogwood, and other accessory species is seen in many places. Sugar maple may be absent, and its place taken by red maple; chestnut is then a constituent of the canopy. This variation is related to underlying rock and occurs on mesic slopes on the noncalcareous Ohio black shale. Sugar maple is abundant in some of the less rugged areas where limestone is the underlying rock. The dominant species of the secondary forest communities of these limestone areas are red cedar and redbud. Locally, the sugar maple–basswood–buckeye–tuliptree segregate of the Mixed Mesophytic association develops. In such a community on Easter Run, Adams County, Ohio, these four species comprise 88 per cent of the canopy in a forest of only nine species. Not only is the canopy of this stand similar to that of this segregate in the Cumberland Mountains, but the herbaceous layer also is comparable in luxuriance. There is no altitudinal correlation here as there is in the Cumberland Mountains. Beech is absent from this community (again as in the Cumberland Mountains), although beech is a dominant in mixed forests in comparable nearby situations.

The proportion of oaks and/or of chestnut increases on the upper slopes and, finally, near the summit of the ridge or knob, mixed mesophytic forest gives way to chestnut oak (*Q. montana*), to chestnut, to oak–chestnut, to oak–hickory, to oak–chestnut–hickory, or to pine. Any of the subclimax communities of ridge-tops of the Cumberland Mountains may be seen here. Pines frequently cap the knobs and extend some distance down steep southwest slopes. These summit communities, when composed of pine, oak, and chestnut, usually have a heath layer of *Kalmia, Gaylussacia,* and species of *Vaccinium,* and if the soil is sandy, such herbs as *Viola pedata, Iris verna, Chrysopsis nervosa, Aster linariifolius,* etc. Such summit communities are correlated with noncalcareous underlying rock. The oak–hickory communities (without heaths) more often occur on limestone ridge-tops. The differences between communities at different levels on the slopes are correlated with soil moisture, humidity, and (on upper slopes) character of the underlying rock, not with elevation, since the total relief is generally not over 500 or 600 feet, often less.

Secondary vegetation on the shale slopes is very different from the primary. Few areas of secondary mixed forest occur, except where all cutting has been selective, and where fires have been absent or few and light. Usually, erosion has removed all humus, in fact all of the A horizon of the soil, leaving either an impervious subsoil—slightly weathered shale—or a slope strewn with rock fragments and supporting little herbaceous vegetation beneath a scrubby forest growth. The more badly eroded slopes have been occupied largely by pines; other slopes are mostly oak. In areas where there has been little disturbance for many years (perhaps a century), oak woods are well established; often they have a shrub layer of *Vaccinium* and *Gaylussacia.* Development is exceedingly slow and there is seldom indication of progress beyond this oak or oak–blueberry stage. In fact, it is only by comparison with unlike forests of apparently similar sites that the probable secondary character of much of the oak woods is indicated.

High on the slopes of many of the knobs, there is a chestnut oak community unlike the oak woods lower down. There is little except chestnut oak—in every layer. The ground is either covered by a deep accumulation of these decay-resisting leaves, beneath which is a layer of root-mor overlying the almost unchanged mineral soil; or is windswept and bare; or is partially covered by moss and lichen cushions, in which are scattered small blueberry and huckleberry bushes. When these become established, leaf accumulation begins, and many bushes are almost buried. The bare areas are completely devoid of humus or soil; the term "oak barrens" seems particularly fitting. One herbaceous plant, *Silene caroliniana* Walt. var. *Wherryi* (Small) Fern., is characteristic of these areas. Although areally unimportant, these oak barrens are such a conspicuous feature of the Devonian black shale slopes as to be visible for long distances. There is no indication as to whether this community is primary or secondary; certainly, there is no indication of what may have preceded it; and there is no indication of change.

On many of the lowest slopes of wider valleys, a white oak–beech community once prevailed which appears not to be related to slope exposure, as it occurs alike on northerly and southerly slopes, differing only in the somewhat higher proportion of beech and admixture of other species on northerly slopes and in the specific content of its herbaceous layer. Few

In the watershed of Kinniconnick Creek, three distinct levels are visible: the wide valley floor, seen at the right, which is trenched by the tree-bordered stream; the high hills bordering the valley, which here form the sky-line; and an inner dissected plateau represented by the low ridge in the foreground. This intermediate level (of Harrisburg peneplain age) is clad in white oak forest, here with dogwood forming a lower layer. Lewis County, Kentucky.

of these areas remain even in relatively undisturbed condition; all occupy soils which erode badly and dry out rapidly after logging. Scrubby oak thickets and groves of pine (*P. virginiana*), which give no suggestion of the original forest, now occupy many of these slopes. Young beech trees occur in the undergrowth of some of the better secondary oak stands, suggesting the developmental trend toward the return of a white oak–beech forest.

The high-level valley floors which extend between the hills merge with the lower white oak–beech slopes. Wherever forest remains on this level or rolling surface, it is dominantly white oak, with scattered black oak and post oak, and occasional other species. Secondary white oak forest occupies most of the area not in pasture or in cultivation. A correlation, noted in the Allegheny Mountains (p. 79) and in the Low Hills Belt of the plateau (p. 97), between this white oak community and the Harrisburg peneplain is again suggested.

Wide swampy valley flats are either alluvial (in the glacially ponded streams) or are underlain by the impervious Ohio black shale. The swamp forest of either type of valley is made up principally of pin oak, sweet gum, and red maple, with swamp white oak, shingle oak, and occasional beech and southern red oak (*Q. falcata*). Alder and Spiraea (*S. tomentosa*) are usually abundant. Drier parts of the shale flats are occupied by a type of oak woods (*Q. alba, Q. falcata, Q. stellata, Q. marilandica, Q. velutina*) indicative of poor soil and impervious subsoil.

Relic prairie communities and scattered prairie species occur along the belts of limestone in the Knobs Border area. These are most extensive and best developed along the Silurian dolomite outcrops at the west base of the hills in Adams County, Ohio. Although a significant feature in connection with the interpretation of vegetational migrations, they need not be considered here (Braun, 1928a).

Only the most prominent of the many edaphic communities of the Knobs Border area have been mentioned. A detailed study of any part of this area will disclose many features which do not accord with the broad picture here given.[22] However, this area is a part of the Mixed Mesophytic Forest region because of the prevalence of and variety of mixed mesophytic forest communities. That it is distinct from the adjacent area to the west, physiographically, vegetationally, and floristically, is apparent from even casual observation.

[22] One such treatment, the "Vegetation of the Mineral Springs Region of Adams County, Ohio" (Braun, 1928) may be referred to.

CHAPTER 5

The Western Mesophytic Forest Region—A Transition Region

The Western Mesophytic Forest region has as its eastern boundary the western escarpment of the Cumberland and Allegheny Plateaus. Its western boundary lies along the loess bluffs of the Mississippi River at the eastern limit of the Mississippi alluvial plain. The region extends from northern Alabama and Mississippi northward to the southern boundary of Wisconsin glaciation in Ohio and eastern Indiana, and to the southern boundary of Illinoian glaciation farther west. In Illinois, it includes the Ozark Hills.

This region is essentially the area known to physiographers as the Interior Low Plateau (see map, p. 492), and includes in addition, the area of Illinoian glaciation in southwestern Ohio and southeastern Indiana, and the area of coastal plain sediments in western Kentucky, western Tennessee, and northern Mississippi.

This broad and irregular band 100 to over 200 miles in width lying west of the Cumberland and Allegheny Plateaus is here designated as a transition region. Unlike the preceding region, this region is not characterized by a single climax type. The major vegetation types occurring in this region form a complex mosaic which is the resultant of present influences and of past influences operative within recent enough time that their effects on vegetation have not yet been eliminated. The pronounced influence of underlying rock, early recognized by geologists,[1] is evidence of the lack of complete climatic control. Substratum, by its influence on ground water, soil fertility, and topography, may augment or in part counteract climatic influences, may favor persistence of vegetation types no longer in accord with climate, may modify, or may retard vegetational development. The mosaic pattern of climax vegetation types[2] in the Western Mesophytic Forest region is in strong contrast to the situation in the Mixed Mesophytic Forest region to the east, where a single climax type (with its segregates) dominates.

[1] In the old Kentucky Geological Surveys, 1856–61, reference is made frequently to the apparent correlation between forest composition and geologic horizon.

[2] Climax is here used in a broad sense, to include the diverse physiographic climaxes and relic climaxes occurring in the area.

The forests of this transition region are in general less luxuriant in aspect than are those of the Mixed Mesophytic Forest region. In its eastern part, mixed mesophytic forests are of frequent occurrence; westward they become more and more limited in extent and more closely dependent on very favorable habitat conditions. The forest communities of some parts of this transition region, although mixed forest, can hardly be called mixed mesophytic forest. There is a greater tendency toward dominance of a few species. Some of these mixed forests suggest certain of the segregates of the Mixed Mesophytic association which develop in the Cumberland Mountains. This is indicative of the genetic relations involved. Because of the many mixed forest communities, and of the gradual change from east to west in extent of mixed mesophytic communities and in composition of forest, and of the increasing frequency of communities in which oaks are dominant, this is considered a transition region.

In addition to the wide variety of forest types of the uplands in the Western Mesophytic Forest region, there are extensive alluvial swamps, especially in the lower Green River and lower Wabash River basins. Vegetationally, these are extensions from the alluvial lands of the Mississippi embayment. Phases of swamp forest occur locally and sometimes over extended areas on the undissected Illinoian Till Plains of southwestern Ohio and southeastern Indiana, and on flatter parts of the Mississippian Plateau section where they are often called flatwoods. The Cedar Glades of Tennessee, and the cedar barrens of limestone slopes in the "Barrens" of Kentucky are interesting edaphic communities. The prairies or barrens of the western part of the region together with the small local patches of prairie which extend northward through the Knobs into Indiana and around the southern side of the Bluegrass into the Knobs Border area of the Mixed Mesophytic Forest region are of value in reconstructing vegetational history and migrations in the region.

The transition from extensive mixed mesophytic communities in the east to extensive oak and oak–hickory communities in the west is well-marked, in spite of the mosaic pattern of communities, and the many and large prairie areas. This change is further emphasized by the dropping out westward of the most characteristic trees of the Mixed Mesophytic Forest—*Tilia heterophylla* and *Aesculus octandra*.

The regional boundary on the east is emphasized by the differences in specific content of adjacent communities on either side of the boundary. Many Appalachian species extend beyond the western escarpment of the plateau only locally, if at all.[3] Heaths are almost absent from the extended limestone areas of the Western Mesophytic Forest region, thus their ranges are abruptly interrupted at the regional boundary. (Some of them recur in the Knobs and Western Coal Fields.) *Quercus montana,* a dominant in

[3] For examples of such species, and known ranges in Kentucky, see Braun, 1943.

many subclimax communities of the Mixed Mesophytic Forest region, is absent from large areas of the transition region. Here, on the limestone soils, *Q. Muhlenbergii* is its ecologic equivalent. *Cladrastis lutea,* which occurs only sparingly in mixed mesophytic communities in the Cumberland Mountains and Cliff Section of the Cumberland Plateau (and farther south in the Great Smokies), is an abundant tree on river bluffs in the eastern part of the transition region, where it extends from southern Indiana (very rare) south through Kentucky and Tennessee. The reputedly rare *Pachysandra procumbens* is an abundant plant of mesophytic woods of this region, rarely extending slightly east of the regional boundary, in valleys which are little more than tongues of the Low Plateau. *Trillium recurvatum,* unknown in the Mixed Mesophytic Forest region, is abundant in many of the mesophytic woods of the Western Mesophytic Forest region. *Trillium Gleasoni* replaces *T. grandiflorum* in mesophytic woods in many parts of the region. *Hypericum dolabriforme* is an endemic of this transition region. *Helenium tenuifolium,* a very common weed in grazed and much disturbed open places in Missouri, Arkansas, eastern Oklahoma, western Tennessee, and western Kentucky, rapidly decreases in abundance eastwardly as that part of the transition region is reached where mixed mesophytic communities are more or less frequent. It perhaps might be called an indicator of oak–hickory forest. *Coreopsis major,* although not a plant of mixed mesophytic communities, is common in the Mixed Mesophytic Forest region and extends west in decreasing amounts, dropping out completely in the Oak–Hickory region. Examples could be multiplied indefinitely which illustrate the distinctness of the boundary between the Mixed Mesophytic Forest region and the Western Mesophytic Forest region, or which emphasize the transitional nature of the region under consideration.

The greater degree of agricultural utilization (except in the Knobs) as compared with that in the Mixed Mesophytic Forest region has resulted in a much more fragmentary forest vegetation. The "Bluegrass Region" of Kentucky has suffered most, the Pennyroyal and Nashville Basin very greatly. The picture of original vegetation must be reconstructed from mere fragments and with the aid of very inadequate old accounts.

Six sections are distinguished in the Western Mesophytic Forest region. To the northeast is the **Bluegrass section** (known also as the Lexington plain, and in popular usage as the Bluegrass Region), the most anomalous of all vegetation areas of eastern United States. Similar in some respects is the **Nashville Basin** of Tennessee. The **area** of **Illinoian glaciation** in southwestern Ohio and southeastern Indiana is vegetationally distinct from the Bluegrass to the south and the younger glaciated land to the north. The **Hill section** is distinguished vegetationally by the large amount of mixed mesophytic forest on its slopes, and floristically by its strong Appalachian element. This section includes: (1) the Knobs of Kentucky which surround the Bluegrass on the south and west and extend northward into Indiana as the Norman Upland; (2) the Shawnee physiographic section, which is bordered on the south and east by the Dripping Springs escarpment and extends northward to the Illinoian glacial boundary in Illinois and Indiana (thus including the Ozark Hills of Illinois and the unglaciated parts of the Wabash Lowland and Crawford Upland in Indiana); and (3) the Mitchell Plain in Indiana which lies between the Norman and Crawford uplands (thus including all of what is sometimes referred to as "the knobs" or "chestnut oak upland" of Indiana (Deam, 1940). Under the name **Mississippian Plateau,** is included a large area underlain by rocks of Mississippian age, and extending from the Cumberland Plateau escarpment westward to the Tennessee River, and southward into northern Alabama and northeastern Mississippi. It surrounds the Nashville Basin and divides the Hill section into a narrow eastern lobe (the Knobs) and a large western lobe (the Shawnee physiographic section). It includes the Eastern Highland Rim, the Pennyroyal, and the area south and west of the Nashville Basin, and is essentially the Highland Rim section of physiographers. The area west of the Tennessee River, the **Mississippi Embayment section,** is physiographically a part of the Coastal Plain. As here limited, it does not include the alluvial plain of the Mississippi River. It is an area of Cretaceous and Tertiary sediments along whose western border lies the belt of Loess Hills.

BLUEGRASS SECTION

The Bluegrass section, popularly known as the "Bluegrass region," is a roughly circular area nearly 100 miles in diameter, underlain by Ordovician limestones, and situated in north-central Kentucky,[4] west of the Appalachian Plateau. This area is surrounded except on the north (where the glacial margin forms its boundary) by the belt of "Knobs," the eastern arm of which has already been considered as a part of the Mixed Mesophytic Forest region (p. 118). From any part of the Knobs, the Bluegrass appears as an extensive basin with rolling terrain. The plain, which is the floor of this apparent basin, and known to physiographers as the Lexington Plain, is not uniform throughout. The central part, known as the Inner Bluegrass, is surrounded by two concentric bands, the Eden Shale belt and the Outer Bluegrass.[5]

The Inner Bluegrass is a fertile and prosperous section of rolling land with a mild karst topography, interrupted near its southern boundary by

[4] A small angle extends into Ohio at the west base of the Appalachian Plateau escarpment.

[5] The recognition of such subdivisions in a broad vegetational treatment may appear unnecessary. The general concept of the "Bluegrass Region" is about equally applicable to the Inner and Outer Bluegrass; the Eden Shale belt is in vegetation, in soil, and even in culture, unlike the other parts.

the gorges of the Kentucky River and its tributaries. Except in these gorges —which are in no sense representative of the Bluegrass—no areas of natural vegetation remain. Old forest trees, widely spaced in a carpet of Kentucky bluegrass, dot the extensive estates. From these it is not possible to construct a picture of the original condition. No comparable combination of species occurs elsewhere in the deciduous forest (except small areas, equally affected by man's occupation, in the Nashville Basin). Some of the species occur together in forest communities bordering the Prairie Peninsula in west-central Ohio. Blue ash and bur oak may, perhaps, be thought of as most characteristic. Many species are represented among the old trees still remaining, including: *Quercus macrocarpa, Q. Muhlenbergii, Q. alba, Fraxinus quadrangulata, F. americana, Celtis occidentalis, Acer saccharum, Juglans nigra, Prunus serotina, Gleditsia triacanthos, Gymnocladus dioicus, Carya ovata, Ulmus americana, Aesculus glabra,* and *Morus rubra.* Beech trees are not present. The absence of undergrowth—of shrubs and native herbs—is striking; hence no clue to original conditions may be obtained from this source. The bluegrass (*Poa pratensis*) from which the section gets its name, is generally conceded as not indigenous here; it was not a part of the original vegetation.

The Outer Bluegrass appears to have been very similar to the Inner. Underdrainage is less pronounced, and sinks few. Perhaps related to this ground water condition, beech trees are present, along with the species of the Inner Bluegrass. Semi-natural areas are almost lacking, except on valley slopes, and at the margins of the area.[6]

The Eden Shale belt is a hilly belt, underlain, as suggested by the name, by shales, mudstones, and arenaceous layers. It is agriculturally inferior to the land on either side, and because of its hilliness has deteriorated badly. Casual observation suggests that this is an oak or oak–hickory belt, for in the woods of today these trees generally prevail. About 1850, some of the land was considered very good tobacco land. Utilization of formerly wooded slopes for tobacco, extensive sheep-grazing (to the point of bad overgrazing), and subsequent erosion and removal of top soil have greatly modified this belt. Fragments of mixed forest remain in some of the deeper ravines, but the dominance of oak is indicated by the scattering of old trees near houses and public buildings, and by the occasional grazed oak woods.

A few vegetational features, differing from those outlined above, may be observed. In the Eden Shale belt, a broken ring of red cedar is locally prominent; this may be correlated with the effects of bison, as red cedar is very abundant in places once frequented by these animals. In the gorges

[6] The Lloyd Preserve near Crittenden in Grant County occupies a topographic situation intermediate between conditions of the Outer Bluegrass and Eden Shale belt. It will be used in the reconstruction of vegetation cover.

of the Kentucky River and its tributaries, communities range from the most mesophytic on deep talus of sheltered slopes, to dry and open red cedar communities on exposed cliffs.

Original Condition

What was the original condition of the Bluegrass section? What did the early colonists from Virginia find in the last quarter of the eighteenth century? Perhaps the earliest general description is that of Daniel Boone, published in 1784,[7] where we read:

> We . . . from the top of an eminence, saw with pleasure the beautiful level of Kentucke.[8] . . . We found everywhere abundance of wild beasts of all sorts, through this vast forest. The buffaloes were more frequent than I have seen cattle in the settlements, browsing on the leaves of the cane, or cropping the herbage on those extensive plains, fearless, because ignorant, of the violence of man.

Filson states:

> The country in general may be considered as well timbered, producing large trees of many kinds, and to be exceeded by no country in variety. Those which are peculiar to Kentucke are the sugar-tree, which grows in all parts in great plenty, . . . the honey locust . . . , the coffee-tree . . . , the pappa-tree does not grow to a great size . . . , the cucumber-tree is small and soft, with remarkable leaves . . . Black mulberry-trees are in abundance. The wild cherry is here frequent, of a large size, . . . Here also is the buck-eye . . . and some other kinds of trees not common elsewhere. Here is plenty of fine cane, on which the cattle feed, and grow fat. . . . There are many cane brakes so thick and tall that it is difficult to pass through them. Where no cane grows there is abundance of wild rye, clover, and buffalo-grass, covering vast tracts of country, and affording excellent food for cattle. The fields are covered with abundance of wild herbage not common to other countries. The Shawanese salad,[9] wild lettuce, and pepper-grass, and many more, as yet unknown to the inhabitants. . . . Here are seen the finest crown-imperial in the world, the cardinal flower, so much extolled for its scarlet colour; and all the year, excepting the three winter months, the plains and valleys are adorned with variety

[7] For this and other descriptive material of this early date, see Filson's "Kentucke," by Willard Rouse Jillson, 1930, which contains a facsimile reproduction of the original Wilmington edition of 1784.

[8] The term "Kentucke" seems sometimes to have been applied only to the land lying to the west and north of the Appalachian Plateau escarpment, while all the rugged land from Virginia and Carolina to what is now the Bluegrass was known as the "Wilderness."

[9] Shawanese salad, is, in Van Wyck, "A dictionary of plant names," said to be *Hydrophyllum virginianum*. As there is no definite record of this species in the Bluegrass, it is more likely one of the other species of *Hydrophyllum,* all three of which do occur. *H. canadense* is still used by some of the hill people, who have probably continued the custom established by early settlers.

of flowers of the most admirable beauty. Here is also found the tulip-bearing laurel-tree, or magnolia, . . .

That the Bluegrass was originally a forest region appears evident. That it was unlike any existing forest is also evident. The great quantity of cane, mentioned repeatedly in early publications, must have presented an aspect not now found in any deciduous forest. The cane was originally so abundant that Filson's map (1784) shows on it areas marked by such terms as "Fine Cane Land," "Abundance of Cane," and "Fine Cane." These occurred in both the Outer and Inner Bluegrass; on his map they are shown between the North Fork and main Licking rivers, in the upper South Fork of the Licking, in the region about Elkhorn Creek, and between "Bards Town" and "Harrods Town." It is difficult to picture a forest land, "covered with cane, wild rye, and clover."[10] These cane lands and wild rye lands were not treeless and must have contrasted strongly with true grasslands, which are distinguished on this same early map. "Natural meadow" is shown where now are the prairie areas of Adams County, Ohio; between the Green and Salt Rivers, where today in Hart, Hardin, and Meade counties, there are remnants of prairie. Filson's map states: "Here is an extensive tract call'd Green River Plains, which produces no Timber, and but little Water; mostly fertile, and covered with excellent Grass and Herbage."[11] The forest of the Bluegrass could not have been dense, if it was "adorned with variety of flowers of most admirable beauty" in all but the three winter months. The lack of showy flowers during all the summer months is characteristic of our denser deciduous forests.

No mention is made by Filson of the hilly belt now known as the Eden Shale belt, except that he says part of the land is hilly.

The accounts of travels of the early botanists give almost no data on appearance of the country. Individual species are mentioned, but there is little material of ecologic significance. André Michaux, in 1793, mentions "portions of forest lands" between Lexington and Danville (Thwaites, vol. I, 1904). In Bradbury's Travels (Thwaites, vol. V, 1904), the country near Limestone (now Maysville) was described as "almost wholly covered with trees, many of which are of great magnitude." This is a typical part of the Outer Bluegrass.

The early geologic surveys of Kentucky (David Dale Owen, 1857–1861) more than three-quarters of a century after settlement began, give data on specific locations in a country by that time largely utilized. In all, about

[10] Cane is *Arundinaria gigantea,* which now occurs in limited quantity in many parts of Kentucky; wild rye is doubtless a species of *Elymus;* clover is of course some legume, and as early geologic surveys speak of cane and pea vine, it is probable that the "clover" of Filson is the "pea vine" of early surveys. "Pea vine" is *Amphicarpa* (see Wood's Botany, 1869).

[11] Much of the Green River Barrens are south of the Green River, but "barrens" did exist in the area shown on the map.

25 species of trees are mentioned in accounts of the Inner Bluegrass. Sugar maple, walnut, bur oak, white oak, and several species of bristle-tip oaks, hickories, ash (blue, white and "black"), wild cherry, black locust, honey locust, and mulberry are frequently mentioned; hackberry and coffee nut, relatively prominent today, are less often mentioned. In a few locations, poplar (*Liriodendron*) is mentioned, generally with white oak. It seems evident that there were several more or less distinct forest communities.

One of these communities, composed of bur oak, blue ash, sugar maple, honey locust, walnut, and some additional species, appears to have occupied the "cane land." Remnants of this community, without any cane, can still be seen. A few quotations from early surveys will illustrate this community. "The growth in Fayette was originally pignut hickory, large overcup oak, white-oak, ash, black[12] and honey locust, buckeye,[13] and mulberry, much the same, indeed, as in Woodford, and in both counties, in the early settlement of the country, there was a luxuriant growth of cane and pea vines,[14] which appears also to have been the case over large tracts of the richer portions of the blue limestone formation of Kentucky, even on high ridge land" (Ky. Geol. Surv., vol. III, pp. 66–67, 1857). In Franklin County, a situation is described which "is emphatically ash, locust, walnut, and burr-oak land. The original forests of these trees had a dense undergrowth of spice-wood and cane. The black locust attains here a great size—three feet through—and the honey locusts are very numerous in the primitive woods. . . . Both blue and black ash[15] flourish here; also sugar tree, spanish,[16] and chincapin oak, buck-eye, coffee-nut, and hackberry" (Ky. Geol. Surv., vol. II, p. 59, 1857). In Bourbon County, there is mentioned "the huge burr-oak timber which forms a marked feature in the growth of timber of this part of the country, associated with sugar-tree, honey-locust, buck-eye, and box-elder." The soil was deep, with (in 1856 or 1857) a "luxuriant growth of blue grass," where "in the early settlement of the country an abundant, general, large undergrowth of cane gave name to the Cane Ridge." (Ky. Geol. Surv., vol. II, pp. 63–64, 1857). It will be noted that beech is not mentioned. It is known to have occurred only locally on what were called "sobby beech flats," small poorly drained spots on the headwaters of Elkhorn Creek in Fayette County, and on the divide between the Licking and Kentucky rivers in Bourbon County.

[12] Black locust is frequently mentioned, and its large size commented on. Although not generally considered to be indigenous here, it should be noted that trees 3 feet in diameter were growing here along with native forest trees, only 75 years after the first settlers penetrated the wilderness.

[13] The buckeye of the Inner Bluegrass is *Aesculus glabra.*

[14] Pea vine is *Amphicarpa bracteata* (*A. monoica*).

[15] Black ash does not occur in Kentucky.

[16] Spanish oak, *Quercus falcata,* does not occur in the Inner Bluegrass; some other bristle-tip oak, perhaps *Q. Shumardii,* should be substituted.

The Outer Bluegrass is, and was, very similar to the Inner, as is evidenced by Geological Survey statements: "locust, black walnut, black and blue ash, wild cherry, and some white oak; undergrowth of cane" (Bath County, east of Sharpsburg; Ky. Geol. Surv., vol. III, p. 133, 1857). "The growth is sugar-tree, hickory, ash, and some walnut" (Boyle County). Beech is present in some communities of the Outer Bluegrass, and is, locally, a dominant tree, especially in the northern tip of the Outer Bluegrass, and in parts of the outer periphery. Its former greater prominence is indicated by many statements in the early surveys. Tuliptree, also, was more frequent than it is today. A forest in which beech, tuliptree, sugar maple, white oak, and red oak were most abundant, originally occupied much of the more rolling topography near the periphery.

The contrast between the Eden Shale belt and the Inner Bluegrass on the one side and the Outer Bluegrass on the other, when a forest cover was more prevalent, was apparently as noticeable as it is today. For example, the west half of Mercer County is in the Eden Shale belt, the east half in the Inner Bluegrass. "To the west of this line the growth is white and black oak, hickory, sugar-tree, dogwood, and some poplar. The soil is, for the most part, shallow near the rock. . . . To the east the prevalent growth is sugar-tree, ash, black, and white walnut; this is emphatically, the blue-grass region of Mercer County" (Ky. Geol. Surv., vol. III, p. 81, 1857). The contrast between the Bluegrass (either Inner or Outer) and the "oak land" is frequently mentioned. The oak land was good tobacco land, which probably accounts for the excessive denudation it has suffered. On the lower slopes, beech and sugar maple mingled with the oaks.

The Lloyd Preserve near Crittenden, Kentucky, illustrates the type of mixed forest on the less steep slopes of the area along the boundary of Outer Bluegrass and Eden Shale belt. Here oaks (mostly white and black) form about 30 per cent of the canopy, beech 22 per cent, and sugar maple 15 per cent, in a mixed forest of 16 canopy species (Table 19). That this area is representative of the more mesophytic situations of the Eden Shale belt, can be judged by comparison of its composition with the several lists of trees

Table 19. Mixed forest of slopes at the boundary between the Outer Bluegrass and Eden Shale belt; Lloyd Preserve, near Crittenden, Grant County, Ky. (sample of 167 trees).

Fagus grandifolia	22.2	*Ulmus fulva*	3.0
Quercus alba	16.2	*Quercus Muhlenbergii*	2.4
Acer saccharum	15.0	*Quercus Shumardii Schneckii*	1.8
Juglans nigra	9.6	*Nyssa sylvatica*	1.8
Quercus velutina	9.0	*Carya glabra*	1.2
Fraxinus americana	6.6	*Carya* sp.	.6
Tilia neglecta	5.4	*Celtis occidentalis*	.6
Carya ovata	4.2	*Fraxinus quadrangulata*	.6

given in old surveys. The oak forest of ridges, tablelands, and middle slopes varies somewhat in composition, although white oak is always abundant, ranging from 25 to 80 per cent of the canopy. A representative area of oak forest, in which oaks form 70 per cent of the canopy, is illustrated by Table 20. In all of the oak forests, dogwood is a conspicuous species of the understory.

Table 20. Representative area of oak forest of the Eden Shale belt; south slope, Twelve-mile Creek, Campbell County, Ky. (sample of 285 trees).

Quercus alba	37.9	*Ulmus fulva*	1.0
Quercus Shumardii Schneckii	17.5	*Ulmus americana*	1.0
Quercus Muhlenbergii	13.7	*Gymnocladus dioicus*	.7
Carya ovata	8.1	*Quercus imbricaria*	.4
Fraxinus americana	8.1	*Quercus macrocarpa*	.4
Juglans nigra	3.8	*Fagus grandifolia*	.4
Acer saccharum	3.8	*Morus rubra*	.3
Carya glabra	2.4	*Nyssa sylvatica*	.4

NASHVILLE BASIN

The Nashville Basin of middle Tennessee somewhat resembles the Bluegrass section of Kentucky, in form, in topography, and in some of its vegetational features. It is a more or less oval area approximately 60 miles wide and 120 miles long in a north-south direction, enclosed by the Highland Rim (Mississippian Plateau), dissected and reduced outliers of which form hills within the Basin. Its topography is rolling to hilly, with numerous flat areas on the floor of the basin where the underlying Ordovician (Lebanon and Ridley) limestone is exposed or but shallowly covered with soil.

The "Cedar Glades" of the Nashville Basin are its most distinctive feature. These are open to dense stands of red cedar (*Juniperus virginiana*), which occupy the flats on the Lebanon limestone. Scattered individuals of deciduous trees (*Quercus stellata, Q. Muhlenbergii, Carya ovata, Cercis canadensis, Bumelia lycioides,* and *Ulmus alata*) among the cedars, and shrubs (*Forestiera ligustrina, Rhus aromatica,* and *Rhamnus caroliniana*) do not greatly modify the aspect given by the dominant cedars. In spring, the open cedar glades are carpeted with flowers, many of them short-lived annuals. *Arenaria patula* is most abundant, often forming continuous white expanses; *Sedum pulchellum* covers many of the rockiest parts. Among other abundant and showy species are *Phlox bifida, Verbena canadensis, Psoralea subacaulis* (which very much resembles a lupine), *Oenothera triloba,* and several species of *Leavenworthia* (*L. uniflora, L. torulosa, L. aurea,* and *L. stylosa*).[17] In summer these cedar glades are very dry, and most of the

[17] The species of Leavenworthia were the subject of a cytogeographic study by J. T. Baldwin (1945).

herbaceous plants disappear. Patches of prickly pear (*Opuntia*) and other pronounced xerophytes (as *Cheilanthes lanosa, Agave virginica, Croton capitatus,* and *C. monanthogynus*) are then much more conspicuous.[18] Of particular interest from the floristic viewpoint is the presence of a number of endemic species (*Lesquerella Lescurii, Leavenworthia stylosa, L. aurea, L. torulosa, Psoralea subacaulis*) and species of but slightly less restricted range (*Petalostemon foliosum, P. Gattingeri,* and *Lobelia Gattingeri*). Western species, also, are well represented. The unique character of the Cedar Glades lends interest to this small area within the deciduous forest. Comparison with other rock-outcrop vegetation was made by Harper (1926).

Indication of successional development is seen in some of the denser red cedar stands where there is deeper soil. Hickories and oaks are conspicuous among the hardwood invaders and sugar maple is sometimes present. More heavily wooded areas in which cedars have been shaded out have a sparse herbaceous layer of typical woods flowers. Other "glades," known as "hardwood glades" occur on outcrops of more massive limestones (Galloway, 1919). On lower rocky hills and in the hardwood glades, oaks (white and bristle-tip species), hickories, winged elm, hackberry and blue ash form open stands. Secondary cedar communities, lacking the Cedar Glades flora, follow cutting of the hardwood species.

Rolling parts of the Nashville Basin, where the soil is deep, originally supported a forest of large trees. Here white oak and tuliptree are the principal species on the knolls, with sugar maple most abundant on the slopes. Accessory species include *Celtis laevigata, Ulmus americana, U. alata, Juglans nigra, Quercus borealis* var. *maxima, Q. Muhlenbergii, Q. macrocarpa, Carya ovata, Nyssa sylvatica, Prunus serotina,* and *Liquidambar Styraciflua.* Lower spots between the knolls lack the white oak, tuliptree, and sugar maple. This rolling part of the Nashville Basin resembles the Bluegrass section of Kentucky. Higher hills, which are the reduced remnants of the surrounding Highland Rim, support mixed mesophytic forest on their sheltered slopes. In these communities, beech is an abundant tree; thus, the vegetation of the Highland Rim remnants within the Nashville Basin resembles the Rim vegetation rather than that of the Basin.

AREA OF ILLINOIAN GLACIATION

Adjacent to the Bluegrass section on the north, in southwestern Ohio and southeastern Indiana, is an area covered by Illinoian drift, but not reached by later glaciers. This area is included in the Western Mesophytic Forest region, primarily, because of the mixed mesophytic forest communities oc-

[18] The seasonal aspect societies of the cedar glades have been described by Freeman (1933).

cupying dissected portions of the area, and because of the vegetational contrasts between this area of old drift and the younger land to the north. Furthermore, the occurrence of the southern sweet gum (*Liquidambar Styraciflua*) as a dominant in developmental forest stages of undissected portions of the Illinoian drift area, and its almost complete absence north of this line, suggests relationship with the more southern forest region.

Flats

Much of the area of Illinoian drift is flat; hence, the names "flats" and "Illinoian till plain" so frequently applied. Regional soil maps (p. 26) designate this as an area of planosols. The impervious soils of these plains (Avonburg, Blanchester, and Clermont silt loam) accentuate the poor drainage conditions of the flat and undissected surface. These conditions afford favorable environment for hydro-mesophytic forest communities. Pin oak, sweet gum, red maple and white elm separately, and in various combinations, together with some accessory species as swamp white oak, sour gum, white oak, shellbark hickory, and beech compose the developmental forest stages. In secondary forests, pin oak and/or sweet gum frequently dominate. The hydrarch succession of the undissected flats terminates in a beech forest, which here is a physiographic climax. Sugar maple is not a part of this community. Late developmental stages may be dominated by white oak and beech, by sweet gum and beech, or by the three species together. Various combinations of the dominants of early and late developmental stages have resulted in a variety of related forest types (Braun, 1936; Chapman, 1942). Soil water, determined by minor variations in topography, and light are important factors controlling forest composition. Sweet gum is more prevalent in the Indiana part of the area than in Ohio; this has led to the recognition there of the beech–sweet gum forest type on these flats (Gordon, 1936).

In transitional bands between the flats and the slopes of the area, where drainage and aeration are better and yet dissection is not apparent, sugar maple appears with the beech. This beech–maple forest is a seral community which is replaced by mixed mesophytic forest wherever dissection becomes more evident.

Dissected Areas

Dissected parts of the area of Illinoian glaciation are to be seen near any of the larger streams which traverse the area. The drier slopes and exposed river bluffs display remnants of an oak–ash–maple forest, which was the prevailing primary forest type of such xero-mesophytic situations. On less dry slopes and in maturely dissected places, forest development has progressed to a Mixed Mesophytic Forest climax (Braun, 1916, 1936). Such mixed mesophytic communities are essentially like those of the Mixed

Mesophytic Forest region, displaying a comparable luxuriance of the herbaceous layer. The most characteristic trees of the Mixed Mesophytic Forest region—*Aesculus octandra* and *Tilia heterophylla*—are among the dominants of these communities. In some areas, the larger size of the beech trees and their decadent character suggest that the mixed mesophytic forest has succeeded a community in which beech alone was dominant (Table 21). The continuance of the mixed mesophytic community is in-

Table 21. Compostion of forest in Ault Park, Cincinnati, Ohio: Percentages of canopy trees (all over 18 inches d.b.h.) and, separately, trees 30 inches and over and trees 18–30 inches d.b.h.; and percentages of middle size or second-layer trees (10–18 inches d.b.h.). Demonstrates trend from beech forest to climax mixed mesophytic forest.

	Canopy Trees over 18 in. d.b.h.			*Second-layer*
	Total	*30" +*	*18"–30"*	*10"–18"*
Number of trees	138	86	52	152
Fagus grandifolia	54.3	83.7	5.8	7.9
Tilia heterophylla	16.6	4.6	36.5	23.7
Aesculus octandra	8.0	4.6	13.5	11.8
Fraxinus americana	5.8	1.2	13.5	13.1
Acer saccharum and *nigrum*	3.6	3.5	3.8	11.2
Prunus serotina	2.9	. .	7.7	4.6
Celtis occidentalis	1.5	. .	3.8	6.6
Ulmus americana	1.5	. .	3.8	7.8
Quercus Muhlenbergii	1.5	1.2	1.9	.6
Quercus borealis maxima	1.5	. .	3.8	2.0
Juglans nigra	.7	1.2	. .	4.6
Juglans cinerea	.7	. .	1.9	. .
Ulmus fulva	.7	. .	1.9	2.6
Quercus Shumardii Schneckii	.7	. .	1.9	.6
Aesculus glabra	. .	. .	. .	.6
Sassafras albidum	. .	. .	. .	.6
Fraxinus quadrangulata	. .	. .	. .	.6
Gleditsia triacanthos	. .	. .	. .	.6

dicated by the composition of the second layer of this forest. Time available for forest development, even in this area of old glaciation, has been much less than in the unglaciated Mixed Mesophytic Forest region, from which the vegetation must have migrated. Locally, development and segregation of communities have reached conditions comparable to those attained in the Mixed Mesophytic Forest region (Cobbe, 1943).

In most of the mixed mesophytic forest communities of this section of the Western Mesophytic Forest region, beech forms approximately 50 per cent of the canopy. That this forest is, however, distinct from the beech physiographic climax of undissected areas of the Illinoian drift, may be seen readily by a comparison of accessory species of the two communities. In the mixed mesophytic communities, tuliptree, sugar maple, basswood, walnut and white ash are almost always present together with some six or eight species of lesser frequency. In the beech physiographic climax, the species just mentioned are generally absent or infrequent, and white elm, red maple, and shellbark hickory—all of which are more abundant in earlier seral stages—are the usual accessory species.[19] While the mesophytic forests of the southern part of this section (Table 21) are not distinguishable from those of the Mixed Mesophytic Forest region, usually containing *Tilia heterophylla, Aesculus octandra, Liriodendron tulipifera,* and other species, those of the northern part suggest the transition character of the section, for neither *Aesculus octandra* nor *T. heterophylla* extend far within the Illinoian glacial boundary. Compare the composition of forests illustrated by Table 22, in which A is an area a few miles within the margin of glaciation, and B, an area in the northern part of the Illinoian drift area.

The dissected zone bordering the Ohio River and its major tributaries

Table 22. Canopy composition of forests in southern and northern parts of area of Illinoian glaciation in Indiana.

A. Mixed mesophytic community in Clifty Falls State Park.

B. Beech–mixed mesophytic community near Batesville.

Note that *Tilia heterophylla* and *Aesculus octandra* are absent in the more northern area.

	A	*B*
Number of canopy trees	80	251
Fagus grandifolia	27.5	62.5
Acer saccharum	20.0	3.2
Tilia heterophylla	15.0	. .
Liriodendron tulipifera	8.8	12.7
Quercus borealis maxima	10.0	1.2
Quercus alba	. .	8.8
Fraxinus americana	8.8	2.0
Aesculus octandra	6.3	. .
Ulmus fulva	. .	3.6
Juglans nigra	1.2	2.4
Celtis occidentalis	1.2	.4
Quercus Muhlenbergii	1.2	.4
Quercus velutina	. .	.8
Nyssa sylvatica	. .	.8
Carya ovata	. .	.4
Juglans cinerea	. .	.4
Fraxinus quadrangulata	. .	.4

[19] Data on forest composition of the till plains, and a comparison with composition of mixed mesophytic communities of slopes, are given by Braun, 1936. Such data as given by Keller (1946) for Klein's Woods in the "flats" of southeastern Indiana entirely obscure the mosaic of developmental communities which every stand on the flats displays and erroneously suggest a forest of mixed composition made up of mesophytes and hydro-mesophytes, of shade-tolerant and intolerant species intermingled.

in the area of Illinoian glaciation and adjacent northern Bluegrass is contiguous at the east with the typical mixed mesophytic forests of the Knobs Border area of the Mixed Mesophytic Forest region. This dissected zone affords favorable sites for mixed mesophytic forest which here extends westward to the Knobs as a narrow band.

HILL SECTION

The Knobs form the eastern lobe of the Hill section and extend northward into Indiana as the Norman Upland; the Shawnee physiographic area forms the western lobe and extends northward to the Illinoian glacial boundary in Indiana and Illinois. These two areas are included in one section because of the similarity of their forests. At the south, the eastern and western lobes are widely separated by an arm of the Mississippian Plateau. They converge northward, and here are separated by the narrow Mitchell Plain which is included in this vegetation section. The Hill section, then, includes the Knobs of Kentucky, all of unglaciated southern Indiana (parts of the Norman Upland, Mitchell Plain, Crawford Upland, and Wabash Lowland), the Ozark Hills of Illinois, the Western Coal Fields area of Kentucky together with its rugged border, the Dripping Springs escarpment.

Knobs of Kentucky

The Knobs extend around the southern and western borders of the Bluegrass. For part of its length, this hilly belt is bordered on the outside by Muldraugh's Hill, whose bold escarpment looks inward toward the Bluegrass. The Knobs area is in part underlain by calcareous strata, in part by noncalcareous shales and sandstones. The rugged topography, marked by numerous conical hills, is in strong contrast with the rolling topography of the Bluegrass on the one side, and the flat to strongly karst topography of the Mississippian Plateau on the other. On the south and southwest the Knobs merge with a hilly and dissected part of the Mississippian Plateau which extends more or less east and west a little to the south of the Knobs and toward the Dripping Springs escarpment.

A wide variety of topographic situations and unlike soils in the Knobs afford habitats ranging from the most mesophytic of deep ravines and lower sheltered slopes to the most xerophytic of shaly knobs, sandy ridge crests, and limestone ledges. Extensive mixed mesophytic forest communities occupy (or originally occupied) the more favorable sites. Oak, oak–hickory, and oak–chestnut communities occupy many of the drier slopes and uplands; while mesophytic uplands sometimes support a chestnut–beech–tuliptree type. An oak–tuliptree type is represented on areas of low relief. Pine woods (*P. virginiana*), more extensive in secondary than in primary growth, cover

A white oak–tuliptree or white oak–black oak–tuliptree community is a common forest type of the Hill section, represented in all its subdivisions. This white oak–tuliptree stand is in the Shawnee National Forest in the Ozark Hills of southern Illinois. (Courtesy, U. S. Forest Service.)

many of the dry and barren Devonian shale slopes. In the driest situations where limestone outcrops, xerophytic red cedar communities and occasional typical prairie patches interrupt the deciduous forest cover.

The vegetation of the Knobs, in contrast with that of the Bluegrass, suggests that of the Appalachian Plateau. Many of its plants are Appalachian species, and essentially absent from the Bluegrass, as for example, *Pinus virginiana, Quercus montana, Castanea dentata, Oxydendrum arboreum, Vaccinium vacillans, V. stamineum, V. arboreum,* and *Gillenia stipulata.* Here is (geologically) an interrupted extension of the Appalachian Plateau, its highest summits capped with residual Pottsville material. It is, physiographically, a much reduced and dissected remnant of the Schooley peneplain (the reduced summit level of the Appalachian Plateau). This physiographic correlation with the Schooley peneplain may be significant in explaining the mixed mesophytic outliers, which then are relics of an older and once more extensive forest type (see Chapter 17).

Some of the outlying relics of mixed mesophytic forest contain its characteristic trees, *Tilia heterophylla* and *Aesculus octandra,* and its wealth of herbaceous plants, including the larger species of *Athyrium* (*A. pycnocarpon,* and *A. thelypteroides*). Some, situated higher on ravine slopes and on the rolling upland levels, recall certain of the segregates of the mixed mesophytic forest which occur in the Cumberland Mountains. In these, chestnut was an abundant constituent of the original forest, often comprising over 30 per cent of the canopy. Its death (years before it disappeared from the forests of the Cumberland Mountains) has affected forest composition, resulting in an increase in tuliptree and sugar maple. Oak–hickory communities occupy some of the higher slopes, especially on the more separated hills of the Knobs.[20]

In a forest area including flat to rolling upland, ravines, upper easterly and northerly slopes, the influence of site factors on relative abundance of the constituent species is well illustrated (Table 23). On the flat to rolling upland, chestnut, beech, and tuliptree together comprise about 78 per cent of the canopy. The chestnut trees were by far the largest trees, about 5 ft. d.b.h. Wherever there were two or more dead chestnut trees close together, several moderately large tuliptrees are now growing, thus suggesting replacement of chestnut by tulip. Beech, although appearing in the table as codominant with chestnut and tuliptree, actually is smaller than the standing dead chestnut and may have been a second layer in a chestnut–tuliptree forest, a type comparable to one of the segregates of the mixed mesophytic forest of the Cumberland Mountains. The smaller size of beech and the somewhat layered character of the forest suggest a successional trend toward dominance of beech with elimination of upland and development of

[20] "Primitive forest growth black and white oak, hickory, ash, dogwood, &c" (near top of the Salt River hills, Hardin Co., Ky.). Ky. Geol. Surv., vol. III, p. 285, 1857.

Table 23. Composition of canopy of forest eight miles north of Campbellsville, Taylor County, Ky., in strongly dissected area at southern border of the Knobs; influence of site factors on relative abundance of constituent species demonstrated by dividing area into (1) flat to rolling upland, (2) upper west slope, (3) northerly slopes, (4) ravines.

	Area as a Whole	*Flat to Rolling Upland*	*Upper West Slope*	*Northerly Slopes*	*Ravines*
Number of trees	428	200	51	121	56
Fagus grandifolia	33.2	26.5	33.3	38.0	46.4
*Castanea dentata**	22.4	34.0	11.8	17.4	1.8
Liriodendron tulipifera	15.2	18.0	5.9	19.8	3.6
Acer saccharum	4.9	2.5	5.9	5.0	12.5
Nyssa sylvatica	3.5	.5	7.8	7.4	1.8
Carya ovata	3.3	1.5	9.8	2.5	5.3
Quercus alba	2.8	1.0	11.8	3.3	..
Carya spp.	2.3	1.5	5.9	..	7.1
Quercus borealis maxima	2.1	2.5	..	1.7	3.6
Aesculus octandra	2.1	..	..	..	16.1
Quercus coccinea	2.1	4.5	..	..	..
Juglans nigra	1.4	3.0	..	..	..
Fraxinus americana	1.4	..	1.9	3.3	1.8
Quercus velutina	.9	1.5	1.9	..	..
Sassafras albidum	.7	1.5	..	..	..
Acer rubrum	.7	.5	3.9	..	..
Ulmus fulva	.2	.5	..	..	..
Carya tomentosa	.2	..	..	.8	..
Prunus serotina	.2	..	..	.8	..
Morus rubra	.2	.5	..	..	..

* Chestnut all dead.

gentle slopes. This is substantiated by the greater size and greater abundance of beech on the slopes. In the prominence of white oak with beech on upper westerly slopes, the beech–white oak type of the Mixed Mesophytic Forest region is suggested (compare with Table 4). Upper northerly slopes are more mesophytic, but with the same dominants as the uplands; beech, tuliptree, and chestnut make up 75 per cent of the canopy. Sugar maple, although forming only 5 per cent of the canopy, is abundant in the understory, and is apparently thriving because of release by dying of chestnut. *Pachysandra procumbens* is conspicuous in the herbaceous layer, which is richer than that of the upland community. The forest of the ravines resembles the typical mixed mesophytic communities of the Mixed Mesophytic Forest region, but has fewer species and a higher percentage of beech. The greater abundance of beech is a feature of this Transition region; beech generally comprises about 50 per cent of the canopy of its mixed mesophytic communities. This in part may account for the less rich herbaceous growth, as compared with more typical mixed mesophytic communities. Where the forest of deeper ravines has a lower percentage of beech and more of the other mesophytic species, whose leaf-litter decomposes more rapidly, the undergrowth is richer.

Along the border of the Knobs facing the Bluegrass, areas underlain by Silurian and lower Devonian limestones are vegetationally distinct. Red cedar is dominant on the driest rocky slopes. Gently rolling areas are generally utilized, but forest remnants and statements in early surveys[21] indicate the former prominence of tuliptree, usually with beech, sugar maple, oaks, and hickories. Sometimes (if underdrained) the more mesophytic species are absent and the oaks are codominant with tuliptree. The oak–tuliptree type is illustrated by a forest area southeast of Louisville, Ky., where black oak, white oak and tuliptree compose about 60 per cent of the canopy of a forest of thirteen species (Table 24, A). The possible

Table 24. Oak–tuliptree forest type.
A. Cedar Creek woods, Jefferson County, Ky.
B. Donaldson's Woods, Spring Mill State Park, Indiana.

	A	*B*
Number of canopy trees	97	308
Liriodendron tulipifera	28.9	17.9
Quercus alba	13.4	43.5
Quercus velutina	17.5	10.1[1]
Quercus borealis maxima	5.1	
Carya ovata	9.3	5.8[2]
Carya glabra and *cordiformis*	8.2	3.9
Fagus grandifolia	4.1	7.1
Acer saccharum	5.1	3.2
Juglans nigra	4.1	. .
Fraxinus americana	1.0	3.2[3]
Nyssa sylvatica	1.0	2.9
Prunus serotina	1.0	1.3
Quercus Muhlenbergii	1.0	. .
Acer rubrum	. .	.6
Ulmus americana	. .	.3

[1] *Q. borealis maxima* and *Q. velutina* are present in proportion about 2:3.
[2] Some *C. laciniosa* may be included, as Cain (1932) lists this species.
[3] All small trees in this area are the pubescent form referable to "*F. biltmoreana.*"

early cutting of white oak from this forest, thus allowing the tuliptree to assume greater prominence, is suggested by a comparison with a forest (on the Mitchell Plain) in Spring Mill State Park, Indiana, where white oak is most abundant (Table 24, B). A similar type occupies rolling upland in the Shawnee section (Table 27, area 3).

[21] "Associated with the poplar timber . . . are beech, white oak, red hickory, and sugar tree." Also, "the growth is beech, large poplar, large white and black oak, hickory, and large black walnut." (Nelson Co. Ky.; Ky. Geol. Surv., vol. III, p. 94.)

Flats on the Devonian black shale are usually small in extent. The vegetation is similar to that of comparable situations in the Knobs Border area of the Mixed Mesophytic Forest region (p. 121).

THE HILL SECTION IN INDIANA

In Indiana, the term "Knobs" is frequently applied to the entire unglaciated highland area (Parker, 1936), or area of "chestnut oak upland," as it is called by Deam (1940) in his map of the floral areas of Indiana. Here, the northward extensions of three physiographic sections, more distinct vegetationally and topographically farther south, merge in one hilly area whose three subdivisions—Norman Upland, Mitchell Plain, and Crawford Upland[22]—need not be distinguished in a broad treatment. Vegetationally, all of this hilly area may be considered as belonging to the Knobs.

In its extension into Indiana, the Hill section is striking because of the contrasts with the Illinoian till plain to the east. Its aspect at once suggests mixed mesophytic forest, because of the nature of the slope forests. Vegetationally, this is an area of mixed forests—usually some phase of mixed mesophytic forest on northerly slopes, and of oak or oak–hickory forest on drier slopes and ridges (for map, see Gordon, 1936). The influence of micro-climates is well illustrated in this area, which lies close to the regional boundary of the Oak–Hickory Forest region (Friesner and Potzger, 1937). The more mesophytic forest types of ravines and northerly slopes are best considered as attenuated northern extensions of the Mixed Mesophytic association, displaying some of the luxuriance of herbaceous layer (but with fewer species). In number of tree species, they resemble the typical mixed mesophytic forest; but in the greater abundance of beech and sugar maple, and in the absence of or low frequency of buckeye and basswood, they suggest a transition to the more northern Beech–Maple Forest region. As Potzger and Friesner (1940a) have pointed out, the Indiana Knobs (together with other topographically similar areas of south-central Indiana) is an area of "merging phenomena which are sensitively balanced." As is to be expected in a transition region, the forests display features of the three adjacent climaxes—Mixed Mesophytic, Beech–Maple, and Oak–Hickory. Floristically, the Indiana Knobs area is marked by an attenuated Appalachian flora (Parker, 1936).

Numerous vegetational studies in the Hill section of Indiana illustrate the variety of communities in this area of diversified topography (Cain, 1932; Friesner, 1937; Potzger, 1935; Potzger and Friesner, 1940, 1940a; Potzger, Friesner and Keller, 1942; Friesner and Potzger, 1937). Locally,

[22] For the extent and location of these three physiographic areas, and their relation to the Knobs as here defined, see maps in Malott, 1922; Fenneman, 1938; Parker, 1936.

hemlock is a constituent of the forest, again emphasizing the Appalachian relationship of this area. It is usually confined to steeper slopes and canyon rims (Friesner and Potzger, 1932, 1932a, 1936, 1944).

A white oak forest type (white oak with tuliptree, black and red oaks, and a variety of other species) is well represented by Donaldson's Woods, a virgin forest tract in Spring Mill State Park, Indiana (Cain, 1932). Here, 43 per cent of the large trees are white oak; this species and tuliptree, black and red oaks, and hickories, make up approximately 80 per cent of the canopy (Table 24, B). In appearance, this is a white oak forest. Most of the very large trees are white oak and tulip; a very few of the larger trees are beech. However, beech is abundant among the larger understory trees, and sugar maple in the smaller size classes. This type of forest does not appear to be climax for this region, as there are few white oaks in the undergrowth; the indications are toward an increase in beech and sugar maple, which,

A virgin white oak forest, Donaldson's Woods, Spring Mill State Park, Indiana. (Courtesy, U. S. Forest Service.)

together with the several species of oaks and hickories, may make up a mixed forest of a character more or less transitional between mixed mesophytic and oak–hickory. This forest occupies an under-drained area on the Mitchell Plain.

The mixed mesophytic forest type is represented by the Cox Woods, south of Paoli, in Orange County, Indiana. Potzger, Friesner and Keller (1942) consider this to be "a fine example of the typical climatic climax forest of Indiana," where "the ultimate under optimum mesophytic soil moisture conditions is the mixed mesophytic forest with great prominence of Fagus and Acer." This area lies along the boundary between the Mitchell Plain and Crawford Upland, along the northward extension of the Dripping Springs escarpment. It includes easterly slopes of a Chester sandstone ridge, and typical karst topography of the adjacent lower Mitchell Plain. In the forest as a whole, beech, tuliptree, sugar maple, and walnut compose about 80 per cent of the canopy in which 17 species are present (Table 25). The

Table 25. Cox Woods (Indiana Pioneer Mother's Memorial Forest), Orange County, Indiana.

A. Canopy composition based on sample strips traversing forest as a whole, *W;* ridges and southerly slopes, *1;* and mesophytic slopes, *2.*

B. Composition determined from quadrats of 100 square meters each: *W,* 50 quadrats, the forest as a whole; *1,* 20 quadrats, 10 on ridge and 10 on upper slopes; *2,* 20 quadrats, 10 on mid-slope and 10 at foot of slope. (Data for *B* from Potzger, Friesner, and Keller, 1942, Tables I and III.)

	A *Canopy Trees*			*B* *Trees 10" + d.b.h.*		
	W	*1*	2	*W*	*1*	2
Number of trees	313	83	230	83	30	33
Fagus grandifolia	40.6	34.9	42.6	33.7	16.6	42.4
Liriodendron tulipifera	15.3	4.8	19.1	18.1	13.3	18.2
Acer saccharum and var. *nigrum*	14.0	8.4	16.1	15.7	10.0	9.1
Juglans nigra	10.2	. .	13.9	13.2	10.0	15.2
Quercus alba	6.7	22.9	.9	4.8	20.3	. .
Quercus borealis maxima	3.8	7.2	2.6	1.2	. .	. .
Carya ovata	1.9	3.6	1.3	2.4	3.3	6.1
Fraxinus americana	1.6	2.4	1.3	3.6	3.3	9.1
Quercus velutina	2.2	8.4	. .	3.4	3.3	. .
Carya glabra	.9	3.6	. .	. .	10.0	. .
Sassafras albidum	.9	3.6	. .	2.4	. .	. .
Ulmus fulva	.6	. .	.9	. .	3.3	. .
Celtis occidentalis	.3	. .	.4	. .	. .	. .
Carya cordiformis	.3	. .	.4	. .	. .	. .
Platanus occidentalis	.3	. .	.4	. .	. .	. .
Gymnocladus dioicus	. .	. .	. .	1.2	. .	. .
Quercus Muhlenbergii	. .	. .	. .	1.2	3.3	. .
*Fraxinus biltmoreana**	. .	. .	. .	. .	3.3	. .

* Not distinguished from *F. americana* in *A*.

tulip and walnut, although represented in all of the mesophytic communities of Cox Woods, are, nevertheless, more abundant locally. The more general dominance of beech and sugar maple is in accord with the usual condition in the Western Mesophytic Forest region, and in contrast with that in the Mixed Mesophytic Forest region. Local environmental factors in the Cox Woods have resulted in the development of more or less distinct communities related to soil moisture as determined by slope exposure and

Large walnut in Cox Woods, south of Paoli, Orange County, Indiana. (Courtesy, U. S. Forest Service.)

position on slope.[23] On mesophytic ravine slopes, the four dominants of the forest as a whole comprise 92 per cent of the canopy; while on ridges and southerly slopes they make up less than 50 per cent of the canopy. Walnut is absent except locally. The beech–white oak forest type is represented on the latter sites, with these two species comprising nearly 60 per cent of the canopy; locally on the driest ridges, an oak–hickory community is present. The higher ridges of the Chester sandstone escarpment, where oaks probably dominated, have been denuded of forest. The mixed mesophytic character of Cox Woods is emphasized by the number of plants of the herbaceous layer, where some 40 species were listed in a single traverse in early spring. *Trillium recurvatum,* absent from the Mixed Mesophytic Forest region, is abundant here.

To the west of the Crawford Upland in Indiana, mixed mesophytic communities are of rare occurrence. Wetter parts of the Wabash Lowland support phases of southern swamp forest, drier parts are typically oak–hickory, while beech and some of its associates occupy the most favorable situations (Gordon, 1936). Studies by Potzger and Friesner (1934) in Mauntel's Woods in southwestern Dubois County illustrate a typical oak–hickory forest of that area in which there is no indication of successional replacement by beech and maple (Table 26). Here, about 75 per cent of the

Table 26. Composition of Mauntel's Woods, Dubois County, Indiana, an oak–hickory forest; percentages based on 70 trees over 10 inches d.b.h. (Data from Potzger and Friesner, 1934.)

Quercus alba	47.1		*Fraxinus americana*	2.9	
Quercus velutina	22.9	75.7	*Platanus occidentalis*	2.9	8.6
Quercus borealis maxima	5.7		*Nyssa sylvatica*	1.4	
Carya ovata	5.7		*Betula nigra*	1.4	
Carya glabra	5.7	15.7			
Carya tomentosa	2.9				
Carya laciniosa	1.4				

trees over 10 inches d.b.h. are oaks (47 per cent white oak) and 15 per cent hickories; *Cornus florida* is conspicuous in the understory.

Ozark Hills

Farther west, in the Ozark Hills of southern Illinois, the more hilly terrain affords favorable habitats in its ravines for more mesophytic forest. Here there is good evidence of the occurrence of mixed mesophytic communities (Chapman, 1937). The upland forests are largely composed of oaks

[23] Potzger, Friesner and Keller (1942) distinguish ridge, upper slope, mid-slope, and foot of slope (their Table III), but make no mention of slope exposure. If their data for ridge and upper slope be combined, and for mid-slope and foot of slope, a comparison with the writer's data may be made (see Table 25, A and B).

Upland and ravine forests of the Ozark Hills of Illinois. In the former, oaks and hickories prevail; in the latter, beech and other mesophytes. Shawnee National Forest. (Courtesy, U. S. Forest Service.)

(mostly black oak and white oak); while on sheltered slopes, there is a variety of species, as beech, sugar maple, tuliptree, cucumber tree, wild black cherry, ash, elm, mulberry, and species of hickory (Miller and Fuller, 1921; Kuenzel and Sutton, 1937). As in the Wabash bottoms of Indiana, flood plain and swamp forests (with *Taxodium distichum*) occupy alluvial bottoms; these are extensions from the Southeastern Evergreen Forest region (p. 150). The floristic affinities of the Illinois Ozarks with the south and the Appalachian Upland are indicated by such species as *Magnolia acuminata, Quercus montana, Pinus echinata, Vaccinium arboreum,* and *Rhododendron* (*Azalea*) *nudiflorum* (Gleason, 1923); *Aesculus octandra,* recorded in one county, occurs only as a low shrub.

Dripping Springs Escarpment and Western Coal Fields

The western lobe of the Hill section in Kentucky includes a more or less circular area of rolling plateau partly encircled by a rugged belt whose most prominent feature is the outward-facing Dripping Springs escarpment. Valleys in the central part are deeply alluviated and frequently swampy; while near the borders the plateau is locally cut by deep and sometimes gorge-like valleys.

Secondary oak or oak–hickory forest prevails on the rolling plateau, and there is evidence, from remnants of original vegetation, and from statements in early surveys,[24] of the dominance of oaks in much of the original forest. Locally, there were flat areas known as "post-oak glades" (Ky. Geol. Surv., vol. III, p. 402, 1857). However, beech occurs on most ravine slopes, accentuating the transitional status of the region. On some of the better soils, the original forest contained beech, tuliptree and sugar maple, as well as oaks, hickories and other trees.[25]

Mesophytic forests, at least some of which are best classified as mixed mesophytic, characterize mesic slopes of the rugged bordering belt. Where streams cut gorges in the Pottsville sandstone, hemlock is locally abundant, and with it may be associated a number of Appalachian and mixed mesophytic forest species, including white pine, yellow birch (*Betula lutea* var. *macrolepis*), beech, oaks (*Q. alba, Q. montana*), chestnut, shellbark hickory, butternut, mulberry, tuliptree, magnolia (*Magnolia macrophylla, M. tripetala*), red maple, sugar maple, holly (*Ilex opaca*), sour gum, sourwood (*Oxydendrum*), and white ash. *Kalmia* is conspicuous in the undergrowth. That this is a disjunct community is evident. In aspect and floristic composition it resembles communities of gorges of the Cliff Section at the western margin of the Mixed Mesophytic Forest region. *Silene rotundifolia, Heuchera parviflora* var. *Rugellii,* and *Thalictrum clavatum* on the sandstone cliffs accentuate the similarity. *Rhododendron maximum,* common in such situations at the edge of the Appalachian Plateau, is lacking here.

Ravine slopes and the deep soil of talus accumulations in both sandstone and limestone areas of this rugged belt are occupied by mixed mesophytic forest. As is general throughout this region, beech usually is the most abundant species, while sugar maple (and/or black maple), tuliptree, white ash, red oak and white oak are its principal associates. Where a valley floor has developed in ravines, sweet gum and sycamore may be present, and butternut more abundant than on the slopes. Such mesophytic forests contrast strongly with the upper slope and plateau forests where oaks and hickories, or oaks and tuliptree are dominant. Composition of a series of these forest communities is given in Tables 27 and 28.

In the Mammoth Cave area, where slope forests on sandstone and on limestone soils occur, well-preserved forest remnants illustrate a variety of forest types. The "Big Woods" of Hart County in Mammoth Cave National Park illustrates: (1) the beech consociation of the ravine flats and adjacent lowest slopes where beech forms about 60 per cent of the canopy;

[24] For example, "an average specimen of Coal Measures soil, supporting a growth of white oak, hickory, ash, and poplar." Ky. Geol. Surv., vol. II, p. 30, 1857.

[25] Statements from early surveys give indication of various sorts of mesophytic forest: "a fine body of arable land," where the soil is fine loam, "supporting a heavy growth of Beech, intermixed with Poplar and Hickory, with an undergrowth of Pawpaw" (Ky. Geol. Surv., vol. I, p. 148, 1876); "primitive forest growth, yellow poplar, much sugar-tree, black oak, hickory, sweet and black gum, elm, some beech, and black walnut" (Daviess Co.: Ky. Geol. Surv., vol. III, p. 241, 1857).

Table 27. Composition of canopy of forest communities in the "Big Woods," Mammoth Cave National Park, Kentucky: *1,* Beech forest of ravine flats and adjacent lowest slopes; *2,* Mixed mesophytic forest of ravine slopes; *3,* White oak–black oak–tuliptree type of slopes of upland.

	1	2	*3*
Number of canopy trees	39	95	210
Quercus alba	. .	2.1	31.4
Quercus velutina	. .	. .	17.1
Liriodendron tulipifera	. .	11.6	23.8
Castanea dentata	2.5	1.0	8.1
Quercus borealis maxima	. .	5.3	4.3
Fraxinus americana	. .	6.3	3.3
Nyssa sylvatica	. .	3.1	1.9
Carya glabra	. .	2.1	1.9
Juglans nigra	. .	1.0	1.4
Quercus Muhlenbergii	. .	4.2	. .
Gleditsia triacanthos	. .	1.0	. .
Carya tomentosa	. .	. .	.5
Morus rubra	. .	. .	.5
Ulmus fulva	2.5	. .	. .
Carya cordiformis	2.5	. .	. .
Carya ovata	2.5	2.1	1.9
Juglans cinerea	10.2	1.0	3.3
Acer saccharum and var. *nigrum*	20.5	14.7	. .
Fagus grandifolia	59.0	44.2	.5

(2) the mixed mesophytic forest of ravine slopes where beech forms 44 per cent, tuliptree and sugar maple 26 per cent of the canopy in a forest of fifteen canopy species; and (3) the prevailing forest of slopes of the upland, a white oak–black oak–tuliptree type (here with chestnut) in which the three dominants comprise 72 per cent of the canopy (Table 27). In this last area, three species of hickory taken together are only about half as abundant as was chestnut. However, with the elimination of all chestnut from the area, the hickories may increase proportionately. This forest is similar to the oak–tuliptree areas illustrated by Table 24, except for the chestnut of the canopy, sourwood in the understory, and *Vaccinium pallidum* in the shrub layer, all of which are correlated with the sandstone substratum. Chestnut also appears on leached and under-drained ridge crests in the region, even though the underlying rock is limestone, suggesting that it may have a place in the climax forest, which resembles some of the oak–tuliptree–chestnut communities of the Oak–Chestnut Forest region as much or more than it does the forests of the Oak–Hickory region.

Forests of more luxuriant aspect occupy the limestone soils of slopes in the vicinity of the Green River, where a deep mull humus favors a rich herbaceous layer (Table 28, areas 1 and 3). In some of these slope forests, beech and sugar maple are dominant, forming over 50 per cent of the canopy, while in others the mixed character of the forest is more evident.

Table 28. Forests of the Mammoth Cave area:

A. Limestone bluffs of Green River: *1,* Mixed mesophytic forest of lower slopes; *2,* Oak–hickory forest of upper slopes.

B. Ravine slope forests: *3,* Mixed mesophytic forest of lower slopes; *4,* Transitional middle slope forest; *5,* Ravine slopes of upland showing forest type similar to that in the "Big Woods," Table 27.

	A				*B*			
	1		*2*		*3*	*4*	*5*	
Number of canopy trees	102		109		90	70	121	
Fagus grandifolia	31.4		2.7		20.1	7.1	9.9	
Acer saccharum and var. *nigrum*	23.5	54.9	5.5	8.2	10.0	8.6	6.6	
Fraxinus americana	9.8		10.1		8.8	2.9	4.1	
Fraxinus quadrangulata	2.9		1.8		. .	. .	. .	
Liriodendron tulipifera	3.9		5.5		22.2	25.7	28.9	
Quercus alba	2.9		16.5		11.1	14.3	24.8	60.3
Quercus borealis maxima	1.0		9.2		12.2	17.1	5.8	
Quercus Muhlenbergii	3.9	7.8	6.4	33.9	5.6	. .	.8	
Quercus velutina	. .		.9		. .	. .	. .	
Quercus montana	. .		.9		. .	. .	. .	
Carya ovata and *laciniosa*	8.8		14.7		2.2	. .	4.1	
Carya glabra and *tomentosa*			18.3	33.0	2.2	12.8	7.4	
Ulmus americana	2.9		3.7		. .	. .	. .	
Ulmus fulva	3.9		. .		1.1	2.9	. .	
Celtis occidentalis	2.9		. .		. .	. .	. .	
Aesculus glabra	1.0		. .		1.1	. .	. .	
Platanus occidentalis	1.0		. .		. .	2.9	. .	
Juglans nigra	. .		.9		. .	1.4	.8	
Juglans cinerea	. .		.9		. .	. .	1.6	
Prunus serotina	. .		.9		. .	. .	. .	
Tilia floridana	. .		.9		. .	. .	. .	
*Castanea dentata**	. .		. .		2.2	1.4	.8	
Nyssa sylvatica	. .		. .		1.1	1.4	. .	
Gymnocladus dioicus	. .		. .		. .	1.4	. .	
Sassafras albidum	. .		. .		. .	. .	1.6	
Oxydendrum arboreum	. .		. .		. .	. .	.8	
Morus rubra	. .		. .		. .	. .	.8	
Ulmus alata	. .		. .		. .	. .	.8	

* Chestnut may have been more abundant, as unsightly dead trees have mostly been removed.

There is an increase in the percentage of oaks and hickories on higher slopes (Table 28, areas 2 and 4), and a decrease in the luxuriance and continuity of the herbaceous layer. The community illustrated by Table 28, area 2 is typical oak–hickory forest; oaks and hickories make up 67 per cent of the canopy and 35 per cent of the second layer (middle size trees). A great increase in the abundance of maple (*Acer saccharum* var. *nigrum*) in this layer and the presence of most of the more mesophytic species of lower slopes suggest a possible areal extension of the mixed mesophytic forest type at the expense of the oak–hickory type. This is in accord with the trend indicated by the invasion of nearby prairies by woodland (p. 155). Where valleys cut less deeply into the upland, mixed mesophytic forest does not develop. The oak–tuliptree type, containing representatives of most of the mesophytes of lower valley slopes, prevails (Table 28, area 5).

Not all upland communities of the Mammoth Cave region belong to the oak–tuliptree type, although this seems to have been the prevailing type on the better soil areas.[26] Some communities are more xeric. On flatter uplands with poor soil, post oak (*Q. stellata*), together with southern red oak (*Q. falcata*), black oak (*Q. velutina*), and blackjack oak (*Q. marilandica*), characterize a forest community whose relationships are with the Ozark upland. Scrub pine (*P. virginiana*) along some of the Pottsville sandstone cliff margins, and red cedar on limestone and Cypress sandstone (Mississippian) cliff margins form open stands, each with its characteristic herbaceous ground cover.

Upland vegetation today is, except for some of the cliff margin communities, for the most part, secondary. An old-field community, in which *Andropogon virginicus* (or *A. virginicus* and *A. Elliottii*) is dominant together with *Smilax glauca* and *Solidago* spp., is invaded by winged sumac, sassafras, persimmon, red cedar, dogwood, and oaks, especially white oak (not all of these in any one old-field; hence, early secondary woody communities differ greatly from one another). A red cedar–dogwood community may become established before the oaks gain control, or, these species may be absent, and oaks follow the early woody invaders, or even invade the herbaceous community.

Edaphic Communities

Certain edaphic communities of the Hill section which occupy considerable area or which contrast in appearance with the prevailing vegetation must be mentioned. Almost throughout the central part of the western lobe of the Hill section (the Shawnee physiographic section), valleys have wide flat silt-filled floors in which are numerous swamps. In the valleys of the Wabash and Green Rivers, this feature is particularly pronounced. Swamp forests, usually of broad-leaved species, but in places dominantly bald cypress (*Taxodium distichum*), contain a large proportion of southern species. One of the more northern of the cypress swamps of this section—Hovey Lake in the Wabash bottoms—was studied by Cain (1935). Conspicuous among the broad-leaved trees of the swamp forests are *Populus heterophylla, P. deltoides, Salix nigra, Carya Pecan, Quercus*

[26] In the Kentucky Geologic Survey of 1876 (Shaler Survey), the statement is made that "the best soil of the table land of Edmonson County supports a growth of white and black oak, hickory, poplar and dogwood."

palustris, Q. bicolor, Q. Phellos, Q. Prinus, Celtis laevigata, Betula nigra, Acer saccharinum, A. rubrum, and *Liquidambar Styraciflua.* These forests of the alluvial valleys are extensions from those of the Mississippi alluvial plain which reaches the lower Ohio Valley (pp. 281, 291). Upstream, the more southern species gradually drop out and the flood-plain vegetation is similar to flood-plain vegetation in much of the deciduous forest area.

On the rims or slopes of sinks, on the hills of the outfacing escarpment which bounds the western lobe of the Hill section on the south, where the Chester limestone (Mississippian) outcrops, red cedar is the dominant tree. The cedars are widely spaced, and accompanied by an herbaceous flora with a high proportion of prairie species. These "cedar barrens" bear some resemblance to the "Cedar Glades" of the Nashville Basin, to the "glades" of Missouri and Arkansas, and to the cedar groves on limestone outcrops in northern Alabama. They are found both within and without the boundary of this section, being particularly abundant and well-developed in the Mammoth Cave area. As they are closely related to the "barrens," a large and significant vegetational feature of the Mississippian Plateau, and as they also occur there, they will be considered in connection with that section.

MISSISSIPPIAN PLATEAU

The Mississippian Plateau does not everywhere have the appearance of a plateau or plain. The Eastern Highland Rim (adjacent to the Cumberland Plateau) is strongly dissected with deep valleys. Outlying spurs of the Cumberland Plateau rise above it. In the eastern half of the section, near larger streams (especially near the Cumberland River), stream dissection has cut the plateau into an area of irregular ridges and slopes. In the stream headwaters region, stretching from northern Tennessee (north of the Nashville Basin) northeastward toward Danville, Kentucky, there is much hilly country. In the western and southern parts of the plateau, near the Tennessee River, the land is generally hilly. Elsewhere, a rolling plateau surface prevails. In the north (especially in the Pennyroyal district) it is dotted with sinks, and surface streams are few.

The primary vegetation pattern is a mosaic of unlike communities. In the dissected and hilly areas of the eastern half of the Mississippian Plateau, mixed mesophytic forest in which beech is dominant occupies many of the slopes; oak, oak–hickory, oak–chestnut, and related forest types occupy the drier slopes and ridges. Westward, mixed mesophytic forest becomes more circumscribed. Phases of oak forest prevail over much of the rolling plateau surface, generally on red or red-brown soil areas, and oak–hickory forest occupies many isolated hills. Because of the dominance of oak forest over much of the Mississippian Plateau, and the increasing limitation from east to west of mixed mesophytic forest to the most favorable habitats, this section particularly demonstrates the transition character of the region. Furthermore, extensive areas of prairie in the original vegetation cover (called "barrens" by the early settlers) emphasize the western relationship. Cedar barrens on some of the drier slopes, and swamp forests with their included herbaceous and shrub communities in depressions and on wet flats, add to the number of distinct vegetation types occurring in this section.

Dissected Eastern Highland Rim

Adjacent to the Cumberland Plateau escarpment, the nature and extent of mixed mesophytic forest is such as to make this dissected part of the Eastern Highland Rim essentially a part of the Mixed Mesophytic Forest region. Only the abundance of *Pachysandra* in the herbaceous layer is distinctive. In a few places, hemlock is an abundant constituent of the forest (Table 29, A). More often, beech is the dominant tree (Table 29, B).[27] This dominance of beech in the ravine slope forests continues westward essentially to the borders of the "Barrens." Beech is more or less

Table 29. Mixed mesophytic forests of the strongly dissected eastern part of the Mississippian Plateau:

A. Easterly slopes of Beaver Creek, Wayne County, Ky.

B. Slopes of northwesterly flowing creek tributary to the Obey River, Pickett County, Tenn.

	A	*B*
Number of canopy trees	77	114
Fagus grandifolia	24.7	64.9
Tsuga canadensis	18.2	. .
Acer saccharum	7.8	12.3
Liriodendron tulipifera	11.7	. .
Tilia heterophylla	7.8	.9
Aesculus octandra	6.6	1.7
Fraxinus americana	6.6	1.7
Quercus borealis maxima	3.9	. .
Quercus alba	2.6	1.7
Quercus Muhlenbergii	. .	2.6
Ulmus americana	. .	3.5
Ulmus alata	. .	3.5
Ulmus fulva	1.3	. .
Magnolia acuminata	2.6	1.7
Carya ovata	2.6	1.7
Carya sp.	. .	.9
Juglans nigra	1.3	.9
Juglans cinerea	2.6	. .
Nyssa sylvatica	.	.9
Liquidambar Styraciflua	. .	.9

[27] It is probable that in some places, the dominance of beech has been accentuated by early logging of more desirable species of a mixed forest.

abundant in ravine forests to the western edge of the Mississippian Plateau at the Tennessee River. Southward, in northern Alabama, slope forests of mixed mesophytic character occur in deep limestone soil of dissected areas adjacent to the Tennessee Valley.

Limestone Hills

Where residual limestone hills (Chester limestone) rise above the plateau, slopes lack the shelter afforded in ravines. Drier types of forest, differing in relation to slope exposure and shelter, are particularly pronounced in the western part of the section. In the east, mixed mesophytic forest (with a large proportion of beech) occupies any sheltered slopes and ravines in these hills; forests in which beech and white oak are abundant occur on some of the more southerly slopes. The beech and white oak communities have a xeric aspect, accentuated by the local occurrence of limestone xerophytes where rock outcrops on the slope. Oaks and hickories increase in proportion on higher slopes. The abundance of oaks and of hickories in the second-growth stands gives an oak–hickory aspect to these slopes.

Limestone ledges in most of these hills afford suitable habitats for a variety of xerophytic herbs and shrubs, at least some of which are plants commonly occurring in prairie communities (*Andropogon scoparius, Agave virginica, Euphorbia corollata, Gaura filipes, Aster oblongifolius,* and *Silphium trifoliatum*). A red cedar–prairie community (the "cedar barrens") forms a band around many of these limestone hills. In the Pennyroyal district, and along the Dripping Springs escarpment, this feature is well marked; here prairie areas are extensive, and prairie species numerous. In the eastern part of the plateau, the cedar barrens community appears always to have been more local.

After lumbering, the slopes eroded badly; most of the A horizon of the soil washed away. Conditions became entirely unsuitable for the reproduction of shade-tolerant mesophytes. The xerophytes of the limestone ledges spread. Red cedar took over many of the driest slopes, often forming broken bands around the hills. The more xerophytic oaks and hickories, together with redbud, persimmon, dogwood, and a few shrubs (*Rhus aromatica, Celtis pumila, Rhamnus caroliniana,* and *Bumelia lycioides*) may mingle with the cedars. Older secondary communities are largely of broad-leaved trees, in which the oaks and hickories prevail; a scattering of more mesophytic species suggests a possible gradual increase in mesophytism.

Plateau

The general plateau level is far from uniform; parts of it are rolling and underdrained; parts are flat or marked by shallow depressions and very poorly drained. The trees of the plateau vary from place to place, apparently in relation to underdrainage. Oak forest was formerly widespread. A mixed oak–tulip–chestnut type, with beech, hickory, and sugar maple, occupies some of the better drained, but moist habitats. Swamp forests occur on poorly drained flats and in shallow depressions; oak swamp in the wettest spots (*Q. palustris, Q. bicolor, Q. Prinus,* and *Q. Phellos*); mixed swamp, with oaks, sweet gum, red maple, sour gum, and beech, merging into wet beech woods on the wet to moist areas. Openings in these forests have a variety of swamp shrub and herb species, a considerable number of which are southern or coastal plain species.

The oak forest of the rolling plateau varies in composition. In general, white oak, Spanish or southern red oak, and black oak are the most abundant species, the white oak occurring in the most mesophytic situations. Accessory species include sugar maple, beech, tuliptree, chestnut, hickories, white ash, and occasional other species. An example in which oaks form 50 per cent of the canopy is given in Table 30. Here, although

Table 30. Forest of rolling plateau surface in Barren County, Ky., just east of the "Barrens." Percentages based on sample of 54 canopy trees.

Quercus alba	27.8	50.0	*Fraxinus americana*	5.5
Quercus Muhlenbergii	9.3		*Nyssa sylvatica*	3.7
Quercus borealis maxima	7.4		*Carya ovata*	1.8
Quercus velutina	5.5		*Carya tomentosa*	1.8
Acer saccharum	13.0		*Carya cordiformis*	1.8
Fagus grandifolia	11.1		*Juglans cinerea*	1.8
Castanea dentata	7.4		*Ulmus fulva*	1.8

tuliptree is not a canopy species, it is present in the undergrowth, which contains all of the canopy species, as well as dogwood, redbud, mulberry, and wild cherry. Sugar maple is reproducing abundantly. Westward and southward, post oak, and sometimes blackjack oak may be present; usually in these areas beech and sugar maple are absent, and white oak less abundant. The oak woods of the plateau almost always has an understory of dogwood; the herbaceous layer is very open. As a large proportion of the plateau is utilized agriculturally, remaining forest tracts are too small in extent, too isolated from one another, and too disturbed either by cutting or grazing or both, to make it possible now to ascertain what may have been the variations in composition and just how these were related to variations in topography.

Much of the oak forest of the plateau is secondary. At least some of it has come into the barrens (prairie) since the advent of the white man. It is impossible now to determine the original limits of forest, so complete has been the utilization of the land. Old maps, statements in old surveys, scattered relic prairie species, and even more scattered relic prairie communities are all that is left from which to reconstruct the original pattern of vegetation—a pattern significant in the interpretation of vegetational migrations. Young oak forest outside of the area of karst topography, if on red soil, is apparently secondary and replacing previous oak forest; if on gray or gray-brown soil, it may be a stage in the secondary return of a more mesophytic forest type.

Barrens

Within the area of karst topography the situation is more complicated, for here were the extensive prairie communities. Although not a part of deciduous forest vegetation, the prairies of the Mississippian Plateau are within the deciduous forest region. They were referred to as "barrens" by early settlers, who considered their treelessness to be indicative of lack of fertility of the soil. The largest of these prairie areas lay south of the Green River, extending in a northeast-southwest band about to the Tennessee border. Another, almost as large, reached northward almost to the Ohio River.[28] Other smaller areas extending westward from the first tract completed a semi-circular arc around the western lobe of the Hill section. Fifty miles farther east, in Wayne County, Kentucky, is another area which once was "barrens."[29] The only remaining evidence of this is the abundance of prairie species on roadsides and dry slopes. The prairie areas or "barrens" are closely correlated with areas of karst topography underlain by cavernous Mississippian limestones.[30] All early accounts of the "barrens" emphasize the tall grasses in the originally open country.[31] Prairie communities, except in the cedar barrens of slopes, have been almost completely destroyed. Pastures and cornfields cover most of the area. The prairies occupied the fertile red calcareous soils so conspicuous in the plowed fields today. Young stands of oak on red soil may, then, have displaced prairie, in which case they are not secondary forests, but stages in a climatic succession leading to the establishment of deciduous forest after a period of grassland dominance. A considerable part of the former prairie is now occupied by scrubby oak thickets, in which post oak and blackjack oak are abundant, and by black oak woods, some with an understory of dogwood. The rapidity of vegetational change in historic times, and the absence of a distinct grassland soil in some areas suggest a relatively short period of grassland occupancy. The prairies of the Mississippian Plateau, the cedar barrens of hills on the plateau and bordering escarpment, and the compa-

[28] This, without boundary, was indicated on Filson's map of "Kentucke" (1784) by the statement: "Here is an extensive Tract call'd Green River Plains, which produces no Timber, and but little Water, mostly Fertile, and covered with excellent Grass and Herbage" (Jillson, 1930).

[29] Ky. Geol. Surv., vol. IV, p. 492, 1861.

[30] For map of karst and sink-hole regions see Jillson (1927) or Fenneman (1938).

[31] Ky. Geol. Surv., vol. II, p. 30, 1857; vol. III, pp. 162, 167, 316, 1857. McInteer (1942) has quoted from a number of early accounts in a paper on the Barrens, to which the reader is referred.

rable communities of the Knobs are the remaining evidence of a former very different pattern of vegetation.

Tennessee Valley

Along the western border of the Mississippian Plateau, in the hilly belt along the Tennessee River, forest communities are uninterrupted by grassland. White oak–hickory woods which prevails on most slopes, gives way upward to drier slope types of forest in which white oak is absent or uncommon. The ridge tops are similar in appearance to those at the dissected margin of the Allegheny Plateau; chestnut oak, black oak, hickories, post oak and blackjack oak make up an open forest in which the shrubs and herbs also recall the Allegheny Plateau margin. Post oak–blackjack oak ridges in this area have much the aspect of similar situations in the Ozarks. The ravine communities, where beech, tuliptree, and sugar maple mingle with the white oak have something of the mixed mesophytic aspect.

In the Tennessee Valley of northern Alabama and northeastern Mississippi, an area of broken topography presents a variety of habitats, which, coupled with the southern position, accounts for the variety of communities of the area. The abundance of pines (*Pinus echinata* and *P. Taeda*) in the drier noncalcareous habitats emphasizes the transition to the Oak–Pine region to the south. The dominance of oaks in many communities, and the presence of mixed mesophytic communities on some of the ravine slopes make this a part of the Western Mesophytic Forest region. Harper (1942, 1943) distinguishes three subdivisions of the Tennessee Valley region of Alabama: the chert belt, the valley proper, and Little Mountain (to the south of the Tennessee River).[32]

In the chert belt, oaks (*Quercus alba* and *Q. falcata*) and locally pines (*Pinus echinata* and *P. Taeda*) dominate in the upland forests. Chestnut was formerly a constituent of the forest. Mohr (1901) referred to this belt as "barrens," because of areas of stunted trees (mostly blackjack oak), and of shrubby willow (*Salix tristis*). On slopes and uplands where there are limestone flats, red cedar is the dominant tree. Mohr states that "the red cedar forms extensive woods, of pure growth, interrupted only by bare openings where the rocky ground scarcely affords a foot-hold to shrub or herb." These Alabama "cedar glades or cedar brakes" resemble the cedar glades of the Nashville Basin more than they do the cedar barrens of the Mississippian Plateau in Kentucky.

Little remains of the natural vegetation of the valley proper. A red clay soil prevailed over much of the typical valley country; on this the southern red oak or Spanish oak (*Q. falcata*) was probably the dominant species (Harper, 1943).

Little Mountain, to the south of the Tennessee River, is capped by the

[32] For extent and location of these areas, see map, p. 274.

Hartselle sandstone (of Mississippian age, and equivalent to the Cypress sandstone of the Dripping Springs escarpment at the south edge of the Shawnee section). The soil of its level summit and gentle southward slope is sandy; pines and oaks are the dominant trees. Its steep scarp slope, which faces north, is cut by deep gorges in which are rich mesophytic woods. Hemlock occurs in one of these gorges, where it is associated with beech, tuliptree, sugar maple, and holly (Harper, 1943a).

In all of these subdivisions of the Tennessee Valley section of northern Alabama, the forests of protected ravine slopes are made up of a variety of mesophytic species, as beech, tuliptree, sugar maple, red oak, shellbark hickory, black walnut, basswood, white ash, red elm, and, locally, sweet buckeye. The occurrence of the two most characteristic species of the Mixed Mesophytic association (*Tilia heterophylla* and *Aesculus octandra*) at the southern margin of the Transition region is of particular interest.

The Tennessee River hills of northeastern Mississippi comprise an area which, in topography and geology, is more or less a transition between plateau and coastal plain. Its forests, in which oaks and pines prevail, are essentially a part of the Oak–Pine region. However, the forests on the richer soil of limestone slopes in the ravines are more mesophytic in character; they contain cucumber tree, tuliptree, walnut, butternut, mulberry, and chinquapin oak (*Q. Muhlenbergii*). Two of the characteristic herbaceous plants of the Western Mesophytic Forest region, *Trillium recurvatum* and *Pachysandra procumbens,* are found in these rich woods (Lowe, 1921).

MISSISSIPPI EMBAYMENT SECTION

This section comprises an area of coastal plain sediments extending from the Tennessee River on the east to the alluvial plain of the Mississippi River on the west. It is physiographically a part of the Mississippi embayment, a northern extension of the Gulf Coastal Plain. Topographically, this is an area of low relief, generally rolling, but locally hilly. Except in northern Mississippi, there is little tendency toward geologic control of topography such as is seen farther south on the Coastal Plain, and which results in belts differing vegetationally. Most of the larger streams flow sluggishly in broad alluvial valleys which are arms of the Mississippi alluvial plain. Along its western border lies the belt of Loess Hills, which rise more or less abruptly about 125 to 250 feet above the flood plain of the Mississippi River. This is an area of slopes and ravines quite unlike the low rolling upland which extends eastward to the Tennessee River.

In common with the Mississippian Plateau section to the east, the Mississippi Embayment section displays a mosaic of unlike vegetation types, of prairie, oak–hickory forest, swamp forest, and mixed mesophytic communities.

Oak–hickory forest occupies a considerable proportion of the upland; southward, yellow pine mingles with the oaks, thus suggesting the transition to the Oak–Pine region. In the northern and in the southern parts of the section (Kentucky, northern Tennessee, and northern Mississippi) prairie areas or "barrens" are of frequent occurrence and sometimes extensive (now mostly in cultivation). The broad alluvial valleys are occupied by phases of swamp forest, extensions from those of the Mississippi alluvial lands. The vegetation of the loess hills contrasts strongly with that of the remainder of the section; its ravine slopes are occupied by mixed mesophytic forest communities. It is because of these mixed mesophytic communities of the loess hills that this section is included in the Western Mesophytic Forest region rather than in the Oak–Hickory region.

The oak or oak–hickory forest communities of the rolling and moderately dissected uplands vary in composition in relation to topography and soils. White oak is generally an abundant species, becoming dominant in ravines and between the knolls of rolling areas. Southern red oak (*Q. falcata*) is often the dominant species on the low hills. A number of other oaks—post oak, blackjack oak, black oak, and locally, chinquapin oak—are found in the oak woods; the first of these is sometimes a dominant species on flat areas. Southward, yellow pine mingles with the oaks on hills in strongly dissected and sandy areas. At the southern limits of the section, loblolly pine (*P. Taeda*) occurs with the yellow pine, thus emphasizing the transition to the Oak–Pine region. Hickories, in greater or less abundance, are almost always present. Tuliptree is a frequent constituent of the white oak communities. Only rarely do beech or sugar maple occur, and then only in the lowest part of the white oak woods, or on immediate stream slopes. Westward, in and adjacent to the loess hills of the northern half of the section, these species become common. The undergrowth of the oak woods is relatively poor in species; in addition to young individuals of the canopy species, dogwood, wild black cherry, winged elm, sour gum, persimmon, mulberry, white ash, sassafras, and sometimes holly, may be found. Shrubby species include *Aralia spinosa, Sambucus canadensis, Corylus americana, Forestiera ligustrina, Symphoricarpos orbiculatus, Rhus radicans,* and *Parthenocissus quinquefolia.* The poor herbaceous layer includes a few widespread mesophytes (especially in the white oak areas) and a few of the xerophytes of open oak woods of the plateau to the eastward. The soil of much of the oak woods is red or reddish, sometimes sandy or gravelly, and sometimes clayey.

The prairie communities of the Mississippi Embayment section are similar to those of the Mississippian Plateau (exclusive of those of the dry limestone slopes where cedar barrens occur). They are (or were) most extensive in the north—the "barrens" of the Purchase area of western Kentucky, with outliers to the south in northern Tennessee. In northern

Mississippi, prairies occupy the northern end of the Black Belt, an area of deep black residual limestone soil (Lowe, 1921; Fenneman, 1938).

The broad alluvial valleys, sometimes but little lower than the rolling oak uplands, are occupied by luxuriant and dense forests. The principal tree species are *Quercus Phellos, Q. Prinus, Q. palustris, Q. lyrata, Q. nigra, Populus heterophylla, P. deltoides, Ulmus americana, U. alata, Celtis laevigata, Betula nigra, Carya Pecan, Platanus occidentalis, Acer rubrum, Liquidambar Styraciflua, Salix nigra, Acer saccharinum, A. Negundo,* and *Taxodium distichum.* These may be mingled in swamp forests, or segregated into more or less distinct communities related to period and depth of submergence. The bald cypress may occur alone in deep swamps; the black willow, cottonwood, silver maple and river birch may border streams. These alluvial swamp forests are extensions of the forests of the alluvial plains of the Mississippi River, and merge with them along the western margin of the section (Hazard, 1933).

Along the indefinite southern margin of the section in northern Mississippi, the belts of the Gulf Coastal Plain, here extending in a north-south direction, are vegetationally, as they are geologically and topographically, distinct (Lowe, 1921). On the poorer and sandy soils of Pontotoc Ridge and the eastern part of the North Central Plateau, and on the heavy gray clay of the Flatwoods belt, pines (*P. echinata* and *P. Taeda*) are abundant; this part of the Mississippi Embayment section could well be included in the Oak–Pine region. The boundary as here drawn agrees with one shown by Shantz and Zon (1924).

Loess Hills

The most interesting and surprising part of the section is its western border, the belt of Loess Hills which extends from the Ohio River southward into Louisiana, hence at the south passing out of the region under consideration. The loess deposit is as much as 100 feet thick along its western margin, while eastward it becomes thinner. The loess bluffs which rise abruptly from the Mississippi alluvial plain are well dissected with numerous ravines. The soil is generally gray-brown in color (yellow below where the loess is unmodified), with a mull humus layer, and contrasts with the prevailingly red soils of the oak lands to the east. The most characteristic and striking features of the Loess Hills belt are to be found along its deeply dissected western margin.

Mixed mesophytic forest covers the ravine slopes, and the flat or rounded divides between ravines. Along narrow ridge tops, the proportion of oaks increases. In fact, the forest vegetation of the dissected loess bluffs is very similar to that of dissected glacial deposits in the Illinoian Drift Area of the Western Mesophytic Forest region. It is unlike that of other parts of the Mississippi Embayment section, or of the western half of the Mississippian

Plateau. Beech is the most abundant species of these mixed mesophytic forests, often comprising about 50 per cent of the stand; common accessory species are tuliptree, red oak (*Quercus borealis* var. *maxima*), hickories, cucumber tree, sugar maple, and basswood (*Tilia neglecta, T. floridana*). The composition of a primary forest on the bluffs near Reelfoot Lake, Tennessee, is given in Table 31. In this particular area, sugar maple is not abundant, and basswood is not present.

Table 31. Composition of canopy and second layer of mixed mesophytic forest on slopes of the dissected loess bluffs of the Mississippi River near Reelfoot Lake, Tennessee.

	Canopy	*Second layer*
Number of trees	104	55
Fagus grandifolia	46.2	43.6
Liriodendron tulipifera	9.6	18.2
Quercus borealis maxima	8.6	5.4
Carya spp.	7.7	3.6
Magnolia acuminata	6.7	3.6
Nyssa sylvatica	4.8	. .
Fraxinus americana	3.8	. .
Quercus alba	3.8	9.1
Quercus velutina	2.9	. .
Juglans nigra	2.9	1.8
Acer saccharum	1.0	1.8
Liquidambar Styraciflua	1.0	1.8
Sassafras albidum	1.0	7.3
Juglans cinerea	. .	3.6

The undergrowth is made up of species of the canopy and, in addition, dogwood, redbud, mulberry, devil's walking stick (*Aralia spinosa*), papaw, ironwood (*Carpinus caroliniana*), and hophornbeam (*Ostrya virginiana*). The shrubs (as *Hydrangea arborescens, Evonymus americanus, Staphylea trifolia, Sambucus canadensis, Lindera Benzoin*), woody climbers (*Parthenocissus quinquefolia, Rhus radicans, Celastrus scandens, Bignonia capreolata, Vitis* spp.), and bamboo (*Arundinaria gigantea*) add to the luxuriant aspect. The herbaceous layer is made up of widespread mesophytes, most of which are found in mesophytic forest communities throughout the Mixed Mesophytic Forest region and Western Mesophytic Forest region. Ferns (*Adiantum pedatum, Athyrium pycnocarpon, A. thelypteroides, A. angustum, Phegopteris hexagonoptera, Polystichum acrostichoides,* and *Botrychium virginianum*) are abundant and emphasize the luxuriance of this forest.

A few flood-plain species usually occur in the mesophytic forest of the lowest slopes of bluffs adjacent to the flood plain of the Mississippi River, and of ravine flats in the bluffs.

Toward the eastern margin of the loess hills, where the loess mantle becomes thin, the topography is rolling. Beech, sugar maple, and tuliptree prevail except in depressions; while white oak, white elm, winged elm, walnut, shellbark hickory, sour gum, and sweet gum are more or less haphazardly distributed over swells and depressions.

Rich hardwood forests are a feature of the loess hills to their southern limit in eastern Louisiana. The vegetational transition, unlike that seen along the rest of the southern margin of the region, is not marked by the entrance of southern pines. Broad-leaved evergreen southern species, especially *Magnolia grandiflora,* and the abundance of the Spanish moss (*Tillandsia usneoides*), mark the transition to the Southeastern Evergreen Forest region.

No consideration of the loess bluffs along the east side of the Mississippi River flood plain can be concluded without reference to their physiographic counterpart to the west of the river—Crowley's Ridge. This ridge extends 200 miles across Missouri and Arkansas, through the center of the alluvial plain of the Mississippi River. "The sedimentary rocks which compose the ridge are of the same age as those in the Bluffs on the east, even to the late Tertiary surficial gravel which covered the plain before entrenchment" (Fenneman, 1938). Although far removed from the loess hills on the east side of the Mississippi River, the forest vegetation of Crowley's Ridge is more like that to the east of the Mississippi than that in the Ozark upland to the west.

On the north slopes, and in deep gulleys, there occurs a [forest] type which is more closely related to the forests of the western Appalachian Mountain region than to any type found elsewhere in Arkansas . . . Species occurring in it are as follows: White, black, northern red, chestnut, and burr oaks; shagbark, white and bitternut hickories; white and green ashes; hard and red maples; wild cherry; walnut, butternut; elms; basswood; chinquapin; red and black gums; beech and yellow poplar (Turner, 1937).

Here is an outlier of mixed mesophytic forest, so situated that it cannot be included in the Western Mesophytic Forest region, although similar to the forest of the western border of that region. Neither is it a part of the Oak–Hickory region centering in the Ozark Upland to the west. Rather, evidence indicates that it is a relic of a former more widespread mixed mesophytic forest (see Chapter 17).

CHAPTER 6

The Oak–Hickory Forest Region

The most westerly forest region of the deciduous forest is one in which oaks are dominant and hickories usually more or less abundant. This region extends from Texas to Canada; it varies in width from place to place and is frequently interrupted by prairie.

Oak–hickory forest is best developed and most continuous in the Interior Highlands—the Ozark and Ouachita Mountain region. It is widely distributed north of that region, where it extends over the older drift of the upper Mississippi Valley and into the Driftless area, and, in Minnesota, over the younger drift as a narrow band at the western margin of forest. It also extends northeastwardly over the Wisconsin drift in the region of the Prairie Peninsula where, as far as southern Wisconsin, southern Michigan, and west-central Ohio, it is represented by large but isolated tracts. Along the western border of the Oak–Hickory Forest region, and southwestward in Texas, decreased precipitation results in a transition to prairie. Here are savannah forests, scrubby oak communities, and tongues of woodland along the streams in the Prairie region. In the unglaciated land east of the Mississippi River, oak–hickory communities are well represented in the Western Mesophytic Forest region and in the Oak–Pine region of the Inner Coastal Plain and Piedmont (Chapters 5 and 8).

The boundaries of the Oak–Hickory Forest region are very irregular and indefinite, except along its contact with the Mississippi alluvial plain in Arkansas and Missouri. Northeastward from the Mississippi River, the Oak–Hickory region is in contact with the Western Mesophytic Forest region until it meets the Beech–Maple Forest region in Indiana. Lobes of oak–hickory penetrating into the Beech–Maple region in Indiana and southern Michigan, and detached areas of oak–hickory within the Beech–Maple region, make a definite boundary impossible. All of Illinois,[1] except the Ozark Hills in the south (included in the Hill section of the Western Mesophytic Forest region), the hilly northwestern corner (included in the Driftless section of the Maple–Basswood region), and the immediate vicinity of

[1] See Vestal (1934) for a "Bibliography of the ecology of Illinois."

Lake Michigan at the northeast (included in the Beech–Maple region) is within the Oak–Hickory Forest region; much of it, however, is prairie. The Driftless area of southwestern Wisconsin and adjacent Illinois, Iowa, and Minnesota is included in the Maple–Basswood region; although here, and northward in Wisconsin to the boundary of the Hemlock–White Pine–Northern Hardwoods region, oak–hickory and maple–basswood communities are both represented in what could be designated as a transition region (p. 329). Northwestward in Minnesota, a narrow strip of the Oak–Hickory region skirts the western border of the Hemlock–White Pine–Northern Hardwoods region. The sinuous western boundary, from Canada to Texas, must be drawn arbitrarily for woodland penetrates far up the streams into the Prairie region. In Texas, bands of post oak–blackjack oak forest or savannah extend southward from the Red River Valley.

Unlike most of the other forest regions, the Oak–Hickory region extends over land areas of very unlike ages. Its center, and area of best development is on one of the oldest land areas in the United States. Its southwestward extensions in Oklahoma and Texas occupy younger land areas, but, nevertheless, ones which date back to middle Tertiary or earlier. Its northward extensions are, except for isolated patches in the Driftless area, on glaciated land, in part on early Pleistocene deposits, and in part on young drift of Wisconsin age. The vegetation of this part of the Oak–Hickory region has been derived from the older area by Pleistocene and post-Pleistocene migrations. The genetic relations of the several parts of the Oak–Hickory region, and of this region to other forest regions, will be discussed later (Chapters 17, 18).

The Oak–Hickory Forest region is made up of two principal divisions, a **Southern Division,** including the Interior Highlands—Ozark and Ouachita Mountains—together with their western and southern borders, and a **Northern Division,** lying north of the glacial boundary. Oak–hickory communities of the unglaciated land east of the Mississippi River—in the Western Mesophytic Forest region and in the Oak–Pine region—might be considered as constituting a third division. The Southern Division is characterized by the prominence of southern species of oaks and hickories, as *Quercus stellata, Q. marilandica, Q. Shumardii* (and var. *Schneckii*), and *Carya Buckleyi* (var. *arkansana* and var. *villosa*), and in the bottomlands, *Q. nigra, Q. lyrata, Carya Pecan, C. myristicaeformis,* and *C. aquatica.* Although some of these species extend northward into the glaciated area, they are almost confined to the older drift areas. In the Northern Division, *Quercus macrocarpa* assumes great prominence, and *Q. ellipsoidalis* is confined to that division. The more widespread species of oaks and hickories (among them *Quercus alba, Q. borealis maxima, Q. velutina, Carya cordiformis,* and *C. ovata*) are present in both divisions.

SOUTHERN DIVISION

The southern division of the Oak–Hickory Forest region occupies unglaciated land. Its approximate northern boundary is the Missouri River. Two sections are distinguished, the **Interior Highlands,** and the bordering woodlands or **Forest–Prairie Transition** area to the south and west in Texas, Oklahoma, Kansas, and western Missouri. In the Interior Highlands, oak–hickory forest reaches its best development and greatest diversity in composition. Here the largest number of species of oaks and hickories occur. From this area, the oak–hickory communities of the glaciated territory were largely derived.

INTERIOR HIGHLANDS: OZARK AND OUACHITA MOUNTAINS

The prevailing forests of the Ozark and Ouachita Mountains are composed principally of oaks of several species. These are found in various combinations on different sites, and may or may not be accompanied by hickory. In the Ouachita Mountains, more so than in the Ozarks, yellow or shortleaf pine (*Pinus echinata*) may be codominant with the oaks or locally form pure stands.[2] The forests of the most mesophytic slopes usually contain sugar maple and an admixture of other mesophytic species. In a few situations on slopes, these mesophytic species prevail. Prairie openings, limestone glades, and balds locally interrupt the forest cover. The aspect of the area as a whole is, however, that of oak–hickory forest.

This vegetation area corresponds more or less closely with the physiographic provinces by the same names. These, like the Appalachian Plateaus and folded Appalachians, are ancient land areas which have had geologic and physiographic histories similar to their eastern relatives. The Ozark Plateaus Province (popularly called Ozark Mountains) consists of two somewhat distinct parts, a northern dissected plateau area, commonly called the Ozark Plateau (including the Springfield and Salem Plateaus), and a southern slightly higher and more rugged area, the Boston Mountains. The former represents a peneplain "believed to have been completed in late Tertiary time;" while the higher levels of the Boston Mountains represent an older peneplain formed "at some time in the Tertiary, perhaps at a time concurrent with the Schooley cycle in the Appalachians" (Fenneman, 1938). It will be of interest to note that in the Boston Mountains there are relic mixed mesophytic communities closely related to those of the Appalachian Plateaus where the reduced Schooley peneplain surface is pre-

[2] Liming (1946) has pointed out that shortleaf pine has been eliminated from some areas where it formerly grew, and is invading other areas once without pine.

View from the overlook at the White Rock Recreation area, looking across the broad Arkansas Valley toward the higher land beyond. Ozark National Forest about 25 miles south-southeast of Fayetteville, Arkansas. (Courtesy, U. S. Forest Service.)

served. The Ouachita Province also comprises two more or less distinct subdivisions, an eastern one (to which the term "Ouachita Mountains" is sometimes limited) marked by "a series of low hills separated by narrow rocky valleys," and a western one, partly in Arkansas and partly in Oklahoma, where one finds

> broad open spaces and rivers flowing not through wide alluvial valleys, except in the case of the Arkansas, but through level uplands of moderate elevation, many miles in extent, sparsely wooded . . . while on all sides can be seen on the horizon the outlines of more or less distinct ranges or isolated peaks and domes of conical or long canoe-shaped mountains (Palmer, 1924).

Two peneplains are represented in the Ouachita Province also; the younger is the late Tertiary peneplain. In the western section of the Ouachita Province, as in the Boston Mountains, there are mesophytic forest communities in coves on north slopes, and isolated occurrences of eastern species.

The prevailing rocks of the Ozark Plateau are limestones, cherts, and dolomites of Cambro-Ordovician and Mississippian age; the prevailing rocks of the Boston Mountains are sandstones and shales of Pennsylvanian age. The prevailing rocks of the eastern division of the Ouachita Province are early Paleozoic shales, sandstones, quartzites and cherts; of the western division, thickbedded sandstones of Pennsylvanian age, alternating with thin layers of shale. These lithologic features influence topography and the character of the vegetation. Steyermark (1940) believes that "in the Ozarks, the basis for the appearance and difference in the oak–hickory or oak–pine association on the one hand and the sugar maple association on the other, lies in the acidity and chemical nature of the soil." Turner (1935) distinguishes seven forest communities in northwestern Arkansas, which differ in degree of mesophytism.

Ozark Plateau

Some phase of oak–hickory or oak–pine forest prevails over most of the Ozark Upland, but in such an area of diverse topography, more xeric and more mesic communities are not uncommon. Broad uplands or narrow upland ridges represent the general level. Rocky barrens and limestone glades are local features of the upland. The larger streams, as the White, Current, Osage, Meramec and Gasconade rivers, have cut deep valleys; cliffs and talus slopes are frequent. There are numerous springs, some among the largest in the United States; the upper courses of streams flowing from these may be deeply entrenched in narrow valleys whose slopes afford mesophytic habitats.

On drier ridges with sandy soil, and steep more or less southerly slopes, the post oak–blackjack oak community commonly develops. This is an open community composed largely of *Quercus stellata* and *Q. marilandica.*

The Current River in the Clark National Forest in Missouri. Willows and sycamores border the stream; mesic slope forests in which sugar maple is usually abundant occupy the talus slopes, while more xeric communities are seen higher on the slopes. (Courtesy, U. S. Forest Service.)

Locally, either one may dominate or, with yellow pine (*Pinus echinata*), form a pine–oak community. Additional species commonly present are hickory (*Carya Buckleyi* var. *arkansana,* and var. *villosa*), winged elm (*Ulmus alata*), and persimmon (*Diospyros virginiana*). Black oak (*Q. velutina*) frequently enters into the post oak–blackjack oak community, suggesting its replacement by the black oak–hickory forest type. On the flatter to rolling parts of the Ozark Plateau and toward its western margin, the post oak–blackjack oak community is more widespread than in the rugged parts.

Although white oak is generally more abundant on northerly slopes, it enters into some of the ridge and plateau forests, associated with black oak, post oak and yellow pine (Table 32, area 1). Dogwood is a common understory tree. In all of these upland communities, species of *Vaccinium* are abundant, *Ceanothus americanus* and *Rhus aromatica* common. The herbaceous layer contains a variety of species, many of which are common to similar situations on the Appalachian Plateau. Among these are *Danthonia spicata, Ascyrum hypericoides, Viola pedata, Pycnanthemum*

Table 32. Composition of ridge and slope forests of the Interior Highlands: *1,* Ridge and upper slopes; *2,* Adjacent to *1,* but lower on slope, Shannon County, Mo.; *3,* North and northeast slopes, Hedges, Stone County, Ark.; *4,* North slope on limestone, Pivot Rock, Carroll County, Ark.; *5,* Northeast slope in limestone ravine, Big Spring, Carter County, Mo.; *6,* Limestone bench on south-southeast slope on White River, Carroll County, Ark.

	1	*2*	*3*	*4*	*5*	*6*
Number of trees	74	33	93	83	52	20
Quercus stellata	20.3	..	..	..	..	..
Quercus velutina	25.7	33.3	18.3	..	..	..
Quercus alba	25.7	51.5	58.1	36.1	..	..
Quercus borealis maxima	..	..	1.1	26.5	38.5	15.0
Pinus echinata	20.3	6.1	2.1	3.6	..	..
Carya Buckleyi	5.4	6.1	6.4	..	..	..
Nyssa sylvatica	1.3	..	8.6	7.2	1.9	..
Quercus coccinea	1.3	3.0	..	..	..	..
*Castanea ozarkensis**	..	..	5.4	2.4	..	..
Quercus Muhlenbergii	..	..	..	7.2	15.4	10.0
Acer saccharum	..	..	..	4.8	3.8	55.0
Juglans nigra	..	..	..	8.4	7.7	..
Juglans cinerea	..	..	..	..	3.8	..
Ulmus alata	..	..	..	1.2	..	10.0
Tilia floridana	..	..	..	..	9.6	..
Fraxinus americana	..	..	..	1.2	5.8	5.0
Fraxinus quadrangulata	..	..	..	..	..	5.0
Sassafras albidum	..	..	..	..	1.9	..
Morus rubra	..	..	..	..	3.8	..
Ulmus fulva	..	..	..	..	1.9	..
Carya cordiformis	..	..	..	1.2	5.8	..

* A smaller tree than the oaks.

flexuosum, Cunila origanoides, Aster patens, Aster linariifolius, and a number of legumes—*Lespedeza procumbens, Desmodium rotundifolium, Cassia nictitans, Tephrosia virginiana, Schrankia uncinata, Stylosanthes biflora,* and *Psoralea psoralioides.* The occasional chestnut in the oak forests of the southern Ozarks, although of a different species (*Castanea ozarkensis*), suggests an ancient Appalachian relationship. The general aspect of these uplands is similar to the poorer soil uplands of the Appalachian Plateaus; *Coreopsis major,* so abundant there, is absent here. Its place is taken to some extent by *Rudbeckia hirta.* The prairie species (as *Lithospermum canescens, Asclepias verticillata, Euphorbia corollata, Phlox pilosa, Eryngium yuccaefolium, Liatris squarrosa* and *L. scariosa*) in the open and park-like forests of some of the uplands may be relics of prairie openings invaded by forest.

White oak is an important constituent of many slope forests. It may have black oak as its principal associate, or, in more mesic situations, red oak (Table 32, areas 2, 3, 4). As in the drier oak forests, dogwood is an abundant understory species. A number of small trees and tall shrubs occur in the understory. Among these, *Cercis canadensis* (on limestone soils chiefly), *Ostrya virginiana, Carpinus caroliniana, Morus rubra, Amelanchier arborea, Asimina triloba,* and *Bumelia lanuginosa* are conspicuous in some areas. These oak forests have a less varied summer flora than the more open communities of ridges and upper slopes. They contain some shrubs common to mesophytic forests over a wide area—*Hydrangea arborescens, Parthenocissus quinquefolia, Rubus occidentalis, Smilax hispida,* and *Lindera Benzoin*—as well as some of those of the summit forests. The open herbaceous layer contains some of the species of the open oak forests and a number of widespread mesophytes conspicuous among which are *Smilacina racemosa, Uvularia grandiflora, Geranium maculatum, Sedum ternatum, Passiflora lutea, Cynoglossum virginianum, Solidago caesia,* and *Eupatorium rugosum.*

The more mesophytic slope forests contain a considerable amount of red oak. Sugar maple is a constituent of many slope forests, particularly on limestone soils, where it may be associated with white oak, red oak, chinquapin oak, basswood, and bitternut hickory. It may even become a dominant tree (Table 32, areas 4, 5, 6). Steyermark (1940) recognizes an "*Acer saccharum–Quercus alba* associes" of limestone slopes. These more mesophytic slope forests have a rich and luxuriant undergrowth resembling that of mesophytic forests farther east; they differ greatly from the prevailing oak forests of the region. Most of the shrubs and herbaceous plants are widespread species.

Lowlands of the region, exclusive of the alluvial lands of larger rivers, up which tongues of the Mississippi embayment forests extend, support stages of hydrarch successions leading to the establishment of an "*Acer sac-*

charum–Carya cordiformis associes." Steyermark describes the developmental stages in different parts of the Missouri Ozarks. Silver maple, sycamore, black willow, cottonwood, white elm, river birch and green ash are the principal trees of earlier stages. The edaphic climax is similar to the "*Acer saccharum–Quercus alba* associes" of slopes. Farther south, addi-

A virgin stand of white oak, Clark National Forest, Missouri. Note the relatively small size of trees, low branching of crowns, and open stand admitting abundant light to the undergrowth, and compare with more eastern forests. (Courtesy, U. S. Forest Service.)

tional species enter into the composition of lowland forests. Sweet gum is abundant in many places, and often associated with white and winged elm, walnut, river birch, and sycamore. An even more complex forest type which occupies moist soils in smaller bottoms is essentially similar to that on the adjacent ravine slopes; red oak, sweet gum, and mockernut hickory are important species used in the forest type name (Society of American Foresters, 1940).

The extremely xerophytic communities of cliffs, rocky limestone slopes, and balds are distinctive features of the Ozarks, although relatively unimportant in the forest vegetation as a whole (Palmer, 1921; Erickson, Brenner, and Wraight, 1942; Palmer and Steyermark, 1935; Steyermark, 1940). Red cedar (*Juniperus virginiana*) is the principal tree invader of rocky limestone areas. The glades, which occur on south and southwest slopes where thin-bedded dolomites outcrop, are characterized by a xerophytic herbaceous flora. They may be treeless, although more often there are scattered individuals of red cedar, oaks, dogwood, redbud, *Bumelia,* etc. The glades are considered to be an edaphic climax. Pioneer forest communities are, however, encroaching on many of the originally open areas of the Ozarks, and there is evidence that at the time of settlement, the land was more open than it is now (Palmer, 1921).

Boston Mountains

As in the Ozark Plateau, oak–hickory forest is the prevailing forest type of the Boston Mountains. Pine and pine–oak communities often occur. In these, sassafras, persimmon, and dogwood are common understory species. The oak forests of ridges and upper slopes often contain Spanish oak (*Quercus falcata*) and chestnut (*Castanea ozarkensis*), in addition to the more abundant post oak, blackjack oak, black oak, and white oak. The white oak–black oak forest type, well represented by a fine tract of forest at Hedges (at the northern edge of the Boston Mountains) in the Ozark National Forest (Table 32, area 3), occupies many of the slopes, both in the Ozark Plateau and Boston Mountains.

The Boston Mountains are of particular interest vegetationally, both because of the considerable number of southern and mesophytic Appalachian species occurring there, and because of the relic mixed mesophytic forests of protected slopes of deep ravines. Here beech may be a dominant species (Table 33). In such an area near Cass, Franklin County, Arkansas, beech extends well up the slope onto the rock outcrop where it meets and mingles with pine. Sweet gum, dominant on the ravine flat below, is a constituent of the slope forest, as it is locally on the Appalachian Plateaus. The aspect is quite similar to the mixed mesophytic forests with high percentage of beech such as are seen in the Western Mesophytic Forest region. Other areas have a more typically mixed mesophytic aspect. On

Spring in the Ozark National Forest, Arkansas, with dogwood abundant in the oak–pine forest. Slopes and ridges of this aspect are encountered frequently in the more strongly dissected parts of the Interior Highlands, and locally in regions to the east. (Courtesy, U. S. Forest Service.)

the slopes of a very deep valley near Ponca, Newton County, Arkansas, a massive sandstone forms cliffs near the top of the slope. Above, is the black oak–white oak–hickory type of forest. Below the cliffs, on the talus slope, is a mesophytic forest of sugar maple, red oak, white oak, chinquapin oak (*Q. Muhlenbergii*), basswood, and hickory; dogwood, redbud, and umbrella

Table 33. Composition of mesophytic forest on ravine slopes near Cass, Franklin County, Arkansas (sample of 71 canopy trees).

Fagus grandifolia	63.4	*Nyssa sylvatica*	5.6
Quercus borealis maxima	9.9	*Ulmus alata*	2.8
Quercus alba	8.4	*Ulmus americana*	1.4
Liquidambar Styraciflua	7.0	*Tilia floridana*	1.4

magnolia (*M. tripetala*) are conspicuous in the understory. Lower on the slope, the forest is dominantly beech, with red oak, sugar maple, cucumber tree (*Magnolia acuminata*), basswood (*Tilia floridana*), sour gum, and hickories. A similar mixed forest, of beech, sugar maple, chinquapin oak, white oak, Kentucky coffee tree, walnut, sweet gum, white ash, elm, and hickories, occupies the low banks immediately adjacent to the stream. The aspect of these slope forests is very luxuriant and strongly suggestive of mixed mesophytic forests of the Cumberland and Allegheny Plateaus.

These isolated mixed mesophytic communities are related to past forest migrations. Their preservation here, in a region whose physiographic history is similar to that of the Cumberland Plateau, is significant (see Chapter 17).

Western Border of the Ozarks

Toward the western border of the Ozark Plateau, prairie areas are larger and more frequent than in the interior. Here prairie often occupies the flatter, more moist sites and oak–hickory forest the rolling sites. Soil color and texture differences between the two topographic and vegetational situations are striking; dark fine-textured prairie soil contrasts strongly with the yellow to reddish coarser soils of forest areas.

In the western border area, deep ravines and gorges afford favorable habitats for mesophytic communities; sugar maple, white oak and red oak are abundant. The contrast between the open xeric plateau forest of low-statured trees, beneath which is a sparse herbaceous layer containing dry prairie species, and the comparatively luxuriant gorge forest with a mesophytic undergrowth, is pronounced. In such a situation at Dripping Springs, Oklahoma, the forest of gentle slopes above the gorge walls is made up largely of blackjack oak, post oak, white oak, black oak, and hickory in open stand; while the forest of the gorge slopes is dominantly white and red oaks, with a number of accessory species including red maple, red elm, mulberry, sour gum, hickory and chinquapin oak, and a lower layer in which are sugar maple, dogwood, hophornbeam, ash, and wild cherry (Table 34). Such widespread mesophytes as *Hydrangea arborescens, Parthenocissus quinquefolia, Rubus occidentalis, Impatiens biflora, Desmodium acuminatum, Polymnia canadensis,* and *Eupatorium rugosum,* occur in the shrub and herb layers. Much farther west, and beyond the

Table 34. Composition of plateau and gorge forest at Dripping Springs, Oklahoma, in the western part of the Ozark Plateau.

	Plateau	*Gorge Slopes*	
		Canopy	*Understory*
Quercus marilandica	39.7	. .	. .
Quercus stellata	15.1	. .	. .
Quercus velutina	9.6	. .	. .
Quercus alba	17.8	40.6	x
Carya sp.	15.1	6.3	. .
Pinus echinata	2.7	. .	. .
Quercus borealis maxima	. .	28.1	x
Quercus Muhlenbergii	. .	6.3	. .
Nyssa sylvatica	. .	6.3	x
Morus rubra	. .	3.1	. .
Ulmus fulva	. .	3.1	. .
Ulmus alata	. .	. .	x
Acer rubrum	. .	6.3	. .
Acer saccharum	. .	. .	x
Ostrya virginiana	. .	. .	x
Amelanchier arborea	. .	. .	x
Prunus serotina	. .	. .	x
Cornus florida	. .	. .	x
Fraxinus americana	. .	. .	x
Castanea ozarkensis	. .	. .	x

x, species present in understory.

limits of the Oak–Hickory region, isolated forest communities in which sugar maple is dominant, occur in deep canyons. The vegetation of such canyons in Caddo County, Oklahoma, where the forest is characterized by tall straight sugar maples one to two feet d.b.h. (and as much as 27 inches) and 50 to 75 feet tall, was described by Little (1939). Such isolated communities are relics of a former more extensive forest.

Ouachita Mountains

The Ouachita Mountains, so far as forest flora is concerned, are essentially similar to the Ozarks. Yellow pine is more abundant, hence this section is often mapped as oak–pine (Shantz and Zon, 1924). Here numerous though not extensive pure stands of pine occupy the poorer, rocky or sandy soils. Bruner (1931) considers that these "represent the northwestern extension of the southern pine forest." On better soils, pine is mixed with oaks and hickories which may ultimately replace the pine.

The pine–oak forest type is the most widespread type of the Ouachita Mountains and extends over the adjacent coastal plain area on the south. It "is an open forest of *Pinus echinata* with several species of oaks and hickories, such as *Quercus stellata, Q. marilandica, Q. velutina, Q. alba, Carya Buckleyi* and *C. alba*" (Little and Olmsted, 1936). *Quercus Shumardii* largely replaces the northern red oak in parts of the Ouachita Mountains. In a few of the most favorable sites, beech, sugar maple, and basswood occur. In speaking of the "Southeastern Oklahoma Protective Unit of the Oklahoma Forest Service," Little and Olmsted state that

the climax vegetation of the Unit is the *Quercus alba* climax forest association, a luxuriant deciduous forest of limited extent on the north-facing slopes of the higher mountains and on lower upland slopes near streams. *Quercus alba* is the dominant species, and other characteristic trees are *Tilia floridana, Carya alba, Quercus Shumardii,* and *Acer saccharum.*

Among other species listed as occurring in the "*Quercus alba* climax forest" are *Carya cordiformis, C. ovata, Juglans nigra, Quercus borealis* var. *maxima, Quercus Muhlenbergii, Magnolia acuminata, Asimina triloba, Hamamelis macrophylla, Cercis canadensis,* and *Fraxinus americana.*

The lowland forests of southeastern Oklahoma contain a number of distinctly southern and coastal plain species. They are

mixed forests of *Quercus Schneckii, Q. nigra, Q. Phellos, Hicoria* [*Carya*] *ovata,* and *H. myristicaeformis.* The more moist areas are clothed with mixed stands of *H. myristicaeformis* and *Q. nigra* or *Q. Phellos,* while the somewhat drier soils are occupied by *Q. Schneckii* and *H. ovata* (Bruner, 1931).

These lowland forests are comparable to those of the Mississippi Embayment section of the Western Mesophytic Forest region.

Local areas within the Ouachita Province are of particular interest because of the great diversity of communities in a small area, because of the extremely mesophytic character of some of the communities, and the occurrence in them of disjunct eastern species. Palmer (1924) writing on the "ligneous flora of Rich Mountain," briefly describes the vegetation of the opposing south slope of Blackfork Mountain and the north slope of Rich Mountain, and of the valley of Big Creek between these two mountains, a valley not over a quarter of a mile in width. Four communities are distinguished in the valley. A community of shrubs occupies the rocky channel, where *Cornus obliqua* and *Salix longipes* var. *Wardii* dominate. A dense growth of shrubs and herbaceous plants covers the wide low margin of gravel or coarse rubble where sand and silt accumulate. Here are found *Rhus radicans, Vaccinium stamineum, Lyonia ligustrina, Rhododendron roseum, R. oblongifolium, Hypericum prolificum, Callicarpa americana,* and a few small trees—*Carpinus, Ostrya,* and occasional individuals of *Quercus Phellos* and *Ilex opaca.* More elevated rocky surfaces, subject, however, to frequent floods, support *Celtis laevigata, Platanus, Liquidambar, Nyssa, Cornus florida, Chionanthus, Ptelea trifoliata, Bumelia lanuginosa,* etc. In the fertile loam soil of the valley floor above the level of ordinary floods there is a fine forest growth in which the most important species are yellow pine, mockernut and Arkansas hickories, white, black, Spanish, and post oaks, sweet gum, black gum, southern linden (*Tilia floridana*), and sugar maple. A few beech trees occur along the margins of the creek.

The contrast between the forests of the south and north slopes is emphasized by Palmer. On the south slope, there is a thin forest of pines and deciduous species (mostly oaks and hickories) which extends to the top of the mountain, except for scattered open areas where rock is exposed; on the north slope, the forest is denser, and composed almost entirely of deciduous trees, many of which attain large size. In the open forest of the south slope, which covers

> the greater portion of the mountain side, where the slope is not too steep for the accumulation of some soil . . . pine (*Pinus echinata*) attains the largest size and generally far overtops the deciduous trees. Amongst the commonest species of the latter class here are *Carya alba, C. Buckleyi* var. *arkansana, Quercus alba, Q. stellata, Q. velutina, Q. marilandica, Q. borealis* var. *maxima,* and *Robinia Pseudoacacia.*

Steeper slopes are covered by thickets of shrubs and stunted trees—locust, post oak, blackjack oak—except in the open rocky glades where few woody species occur. The dense forest of the north slope varies with angle of slope and nature of substratum. Along the deep ravines which indent the lower and middle slopes are forests of red and sugar maple, linden, oaks, and hickories; cucumber tree is abundant higher on the slope, as are also white and black oak, linden, and black gum. Small coves on the mountain slope, where deeper soil has accumulated and where there is a constant supply of moisture, support a particularly luxuriant forest in which,

> in addition to many of the trees and shrubs previously mentioned, are found fine specimens of the Black Walnut, Sugar Maple, White Ash and rarely the Umbrella-tree (*Magnolia tripetala*). Growing as a second layer or in the more open spots are the Papaw, Spicebush, Bladdernut, Southern Witch-hazel (*Hamamelis macrophylla*), Tear-blanket (*Aralia spinosa*) and silverbell-tree (*Halesia monticola* var. *vestita*).

The rich herbaceous vegetation of these coves contains many plants which might be seen in comparable situations in the Mixed Mesophytic Forest region.

> In the richer spots there are great beds of the Maidenhair-fern, Beech-fern and Christmas-fern, besides more rarely the Fragile Fern, Rattlesnake-fern and others. Bloodroot, Dutchman's-breeches, Wild Ginger, Wake-robin, Bellwort,

March-lily (*Erythronium americanum*), Spider-wort and Violets, yellow and blue, are amongst the common spring flowers. Here also, as the season advances are found the blue Cohosh, Black Snakeroot and White-fruited Actaea, and scores of other plants of similar association (Palmer, 1924).

The comparatively level rocky summit is occupied by the post oak–blackjack oak type of forest which is general on the uplands of the Ouachita and Ozark region.

The large assemblage of eastern species almost at the western limit of deciduous forest is remarkable. Some are far removed from the general area of their ranges, i.e., are disjuncts; others are rather widespread species of more mesophytic forest regions. Beech, abundant in a few spots in the Boston Mountains, is found in only two localities in southeastern Oklahoma. Most of the disjunct species occur in both the Rich Mountain section of the Ouachitas and in the Boston Mountains; a few are also found in the White River district of the Ozark Plateau. The occurrence of such species as *Magnolia tripetala, M. acuminata, Robinia Pseudoacacia, Oxydendrum arboreum, Chionanthus virginicus,* and *Halesia monticola* var. *vestita* emphasizes the Appalachian relationship of the forests of the Interior Highlands. The presence of endemic species and varieties, as *Hamamelis vernalis, Carya Buckleyi, Magnolia acuminata* var. *ozarkensis, Aesculus glabra* var. *leucodermis,* and *Castanea ozarkensis,* related to eastern species, is evidence of the long continued separation of the forests of the Interior Highlands (Ozark Plateaus and Ouachita Mountains) from those of the Appalachian Highlands.

Secondary Vegetation

Disturbing influences—lumbering, fire, and grazing—have modified most of the Ozark and Ouachita Mountain region. The post oak–blackjack oak community has been least affected by the first of these influences, due to the low commercial value of these trees. Pine and pine–oak communities have been severely modified; all good pine stands and most of the better pines from mixed communities have been cut. Almost all of the forests in which the commercially valuable oaks are abundant have been logged. Secondary (and modified primary) communities, in a small but representative ravine area at Gray Summit near St. Louis, were studied by Brenner (1942). Oak–hickory, white oak–sugar maple, chinquapin oak–sugar maple, open glade, and closed glade communities were distinguished. Drew (1942), working at the northern margin of the Ozark region (in Boone and Calloway counties, Missouri), traces stages in revegetation of abandoned cropland to the invasion of forest species and concludes that, 30 to 35 years after abandonment,

typical forest species, though then dominant, are not at all comparable in frequency and density to those of second-growth woodlands in the vicinity, thus suggesting a long interval of time prerequisite for reestablishment of such forest conditions.

FOREST–PRAIRIE TRANSITION

To the south and west of the Interior Highlands is an area occupied in part by oak–hickory forest, in part by oak and oak–hickory savannah, and in part by prairie. Throughout this area, post oak and blackjack oak are the dominant tree species of upland woods. Woodlands are usually not continuous; intervening spaces are occupied by prairie vegetation. This section may be considered as a forest–prairie transition. It extends southwestward from the Interior Highlands over the Coastal Plain of Texas. To the west, it extends into the Osage physiographic section of the Central Lowland. In eastern Kansas and western Missouri, west of the Ozark Plateau, prairie vegetation dominates; woodlands are confined to valley and ravine slopes.

Oak–hickory forest merges with the oak–pine of the Coastal Plain in eastern Texas. Southward and westward in Texas, it changes to post oak–blackjack oak savannah, or alternates with tall-grass prairie. Oak–hickory woodland in Texas regularly occupies more or less rolling or hilly land where the soil is sandy and porous (Tharp, 1926). It therefore occurs in bands whose position is controlled by outcropping of sandy formations. The widest band of oak–hickory forest extends in a southerly direction from the Red River in the vicinity of the Oklahoma-Arkansas boundary. Farther west, two strips of woodland, the Eastern Cross Timbers and the Western Cross Timbers, extend south from the Red River along the outcrops of the Woodbine and Trinity sands. In all of the oak–hickory communities of Texas, *Quercus stellata* is the dominant species; *Q. marilandica* and *Carya Buckleyi* are usual associates. Conspicuous among the small trees or larger shrubs are *Vaccinium arboreum, Viburnum rufidulum, Rhus copallina, Crataegus apiifolia, C. spathulata, Ilex decidua,* and *I. vomitoria* (Tharp, 1926). In the eastern part of the Texas oak–hickory area, persimmon and sassafras are more or less prominent in old fields. The numerous prairie species which occur in the oak–hickory communities are indicative of the transitional position of these communities. The more western oak–hickory woodlands are broken and patchy, and better considered as savannah forests. In the extreme west (and mostly beyond our limits), southwestern species, particularly cacti and mesquite, characterize the fringe of the Western Cross Timbers (Dyksterhuis, 1948).

North of the Red River, the oak–hickory savannah occupies a broad band stretching northward across Oklahoma into southern Kansas. This coincides approximately with the areas known as the "Sandstone Hills" and "Chautauqua Hills," where the underlying rock is sandstone. As is true in the "Cross Timbers" of Texas, the hills are occupied in part by oak savannah.

In southern Oklahoma, oak–hickory savannah is continuous with the oak–hickory forest of the Ouachita section, and extends westward across the Arbuckle Mountains. Northward, it is separated from the oak–hickory of the Ozarks by a broad expanse of *Andropogon* prairie. The dominants of the oak–hickory savannah are *Quercus marilandica, Q. stellata,* and *Carya Buckleyi.* The oaks are usually small and scrubby, with broad low crowns. The blackjack oak indicates dry, sterile soils. Although a shrubby undergrowth is present in some places, in others the undergrowth is of grasses, *Bouteloua gracilis* in drier places, and taller grasses in moist soils (Bruner, 1931). Some of the shrubs and herbaceous plants of this section are Ozarkian species which reach the limits of their ranges in eastern Oklahoma and southeastern Kansas.

Along the alluvial bottoms of the larger streams, tongues of flood-plain forest extend far west. In the eastern part of this transition area, the presence of bald cypress (*Taxodium distichum*), sweet gum (*Liquidambar Styraciflua*), river birch (*Betula nigra*), black gum (*Nyssa sylvatica*) and sycamore emphasize the relationship to the forest of the alluvial lands of the Mississippi embayment. Westward, the number of species decreases. Even more striking is the change in form and stature which is apparent in going westward from the forest through the savannah areas.

Toward the north, where savannah forests are no longer a prominent feature of the upland, the transition area is one where valley slopes are wooded and uplands are occupied by prairie. The character of the forest vegetation varies with slope exposure and position on the slope. On sheltered northerly slopes, small areas of mesophytic woods (sugar maple, basswood, red oak) are western outposts of more eastern vegetation types. Their herbaceous layer contains species characteristic of mesophytic forests over a wide area, as *Orchis spectabilis, Asarum canadense, Claytonia virginica, Dentaria laciniata, Dicentra Cucullaria, Sanguinaria canadensis, Hybanthus concolor, Viola eriocarpa, Circaea quadrisulcata* var. *canadensis, Aralia racemosa, Hydrophyllum appendiculatum, Phlox divaricata* (var. *laphamii*), and *Solidago latifolia* (Palmer and Steyermark, 1935; Gates, 1940). Such forest vegetation reaches northeastern Kansas and is identical with that found in comparable sites in the Dissected Till Plains (early Pleistocene age) in southeastern Nebraska and southern Iowa.

Narrow extensions of deciduous woodland are seen far west in the Prairie region. In the Sand Hills of Nebraska, eastern and western arboreal species meet. There, western yellow pine (*Pinus ponderosa* var. *scopulorum*) sometimes occupies the tops of hills on whose lower slopes are small areas of deciduous woodland. The most conspicuous trees of the deciduous woodlands of the Sand Hills are *Tilia americana, Ostrya virginiana, Fraxinus pennsylvanica* var. *lanceolata,* and *Juniperus virginiana.* These form the "linden–cedar–ironwood–ash association" described by Pool (1914). Other

Near the western limits of its range, bur oak becomes shrubby. Buffalo County, Nebraska. (Courtesy, U. S. Forest Service.)

trees commonly associated with these dominant species are box-elder, hackberry, Kentucky coffee-tree, black walnut, cottonwood (*Populus Sargentii,* from the west), bur oak, willow (*Salix amygdaloides, S. interior*), elm (*Ulmus americana, U. fulva,* and *U. Thomasi*). The trees are smaller than farther east; in fact, many are dwarfed and almost shrubby. The undergrowth plants of these western outposts of eastern deciduous forest are not abundant; a few species of more mesophytic forests eastward are seen, as *Sambucus canadensis, Evonymus atropurpeus, Rhus radicans, Parthenocissus quinquefolia, Smilax herbacea, Pilea pumila, Phryma Leptostachya, Osmorhiza longistylis,* and *Cystopteris fragilis.*

NORTHERN DIVISION: AREA NORTH OF THE GLACIAL BOUNDARY

Interglacial and postglacial migrations from the Interior Highlands—the center and area of best development and greatest diversity of oak–hickory forest—resulted in the spread of oak–hickory forest over large areas of glaciated land in the Central Lowland. Here it frequently alternates with prairie, for in the Central Lowland prairie vegetation extends far eastward.

Not all of the land north of the glacial boundary onto which oak–hickory forest has spread is of the same age. That immediately adjacent to the Ozark Plateau on the north is of early Pleistocene (Kansan) age. It has been available for vegetational occupancy for perhaps 400,000 years. That adjacent to the northeast in Illinois is younger, although still old (Illinoian) drift; it has been available for about 200,000 years. Oak–hickory communities also occur frequently in the Driftless area, but this area is included in the Maple–Basswood region. In contrast to these older areas, the area of eastward extension of oak–hickory forest in northeastern Illinois, Indiana, and Ohio, and the area of northward extension on the western border of forest from northern Iowa to Canada, are on young (Wisconsin) drift where much less time has been available for vegetational development —perhaps a maximum of 30,000 years at the southern border, and 20,000 years or even less in the remainder of the area.

Two sections are recognized in the Northern Division of the Oak–Hickory region, the **Mississippi Valley section,** extending northward from the Ozarkian area, and the **Prairie Peninsula section,** extending toward the northeast.

MISSISSIPPI VALLEY SECTION

This section extends from the Missouri River northward, for the most part to the west of the Mississippi River, but occupying also the Mississippi River hills (approximately the northern half of the Mississippi Border area of Vestal, 1931) as far north as the Driftless area. In southern Minnesota, it swings northwestwardly, forming the western margin of forest northward into Canada. Inclusions of oak–hickory in the Driftless section of the Maple–Basswood region and northward in the St. Croix Valley could be interpreted as an interrupted lobe of the Oak–Hickory region. A considerable part of this area is generally assigned to the Prairie Formation. Deciduous forest is largely confined to stream valleys and protected ravine slopes; prairie occupies the intervening land.

The area of old (Kansan) drift to the north of the Ozark upland is the section known to physiographers as the "Dissected Till Plains." Its eastern boundary is near the Mississippi River from St. Louis northward to a point west of the western terminus of the Illinois-Wisconsin boundary.[3] The "Western Young Drift section" of physiographers (including areas covered by drift of Iowan and later substages of the Wisconsin) adjoins the Dissected Till Plains on the north. Its lobate southern margin reaches southward to Des Moines in central Iowa. This is an area of youthful topography and low relief, whose surface has been changed but little since the retreat of glacial ice. It contrasts strongly in topography with the Dissected Till Plains

[3] A narrow band of Kansan and older drift which borders the Driftless area on the west is included in the "Driftless section" of physiographers, because it is topographically similar to that section and contrasts with the young drift on the west (Fenneman, 1938). This interpretation is followed here, and the narrow strip covered by old drift in northeastern Iowa and southeastern Minnesota is included in the Driftless section of the Maple–Basswood region, as it is vegetationally similar to that area.

whose surface is submaturely to maturely dissected, with a relief of about 100 to 300 feet.

The relative areas occupied by forest and prairie have changed from time to time with climatic shifts. Drier periods, such as the post-Wisconsin xerothermic period, favored the extension of prairie at the expense of forest; more humid periods favored forest expansion, and westward extension along streams. Northward migrations, first from the unglaciated Ozark Upland onto the Kansan drift area, much later onto the young drift area, resulted in the spread of oak–hickory forest northward from the Ozark center. Many of its species, as *Quercus Shumardii* var. *Schneckii, Pinus echinata, Bumelia lanuginosa* and a large number of shrubs and herbs, although generally distributed in the Ozarks, rarely extend northward beyond the Ozarks (Palmer and Steyermark, 1935). Some, as *Quercus stellata* and *Q. marilandica,* extend a considerable distance over the old drift, although dropping out before its limits are reached. Others have pushed far upstream in the Prairie region and northward into the Young Drift area. Notable among these is *Quercus macrocarpa,* which reaches to and beyond the limits of deciduous woodland, into southeastern Wyoming. Migrations from the east have enriched the forest flora of the valley woodlands of the Prairie region.

The deeply loess-covered hills, bordering the Mississippi River, support a variety of vegetation types ranging from dry grassland to comparatively mesic forests on ravine slopes. In these, white oak is abundant, and sugar maple and basswood (*Tilia neglecta*) are occasionally present. The common trees of the adjacent bottomland forests are silver maple, elm, willow, ash, pin oak, and river birch, of which only the last is a southern species.

A number of distinct forest communities have developed in the valleys of the Prairie region west of the Mississippi River. These differ in composition and mesophytism; they sometimes form broken bands at different elevations on the slopes, their distribution being related to soil moisture. Some phase of oak–hickory forest occupies the drier slopes and hilltops. This is in contact with the prairie of the upland, or is bordered by a shrub community which intervenes between woodland and prairie. The most mesophytic forest communities occupy protected ravine slopes, northerly slopes, or lower slopes where there is a permanent supply of ground water. Sugar maple and basswood are the dominant trees of these mesophytic communities in the eastern part of the area. Westward, sugar maple drops out (the limit of its geographic range is reached) and basswood alone may be dominant, or share dominance with red oak. Such valley woodland communities in central Iowa have been discussed by Aikman and Smelser (1938), in eastern Nebraska by Pool, Weaver, and Jean (1918), and by Aikman (1927).

The maple–basswood (or maple–linden) community of north slopes in central Iowa is made up largely of the two dominants which compose about 90 per cent of the stand in both larger and smaller size classes (Table 35, column 1). The oak–hickory community of southwest slopes is

Table 35. Composition of woodland communities in Iowa: *1,* Maple–basswood; *2,* Oak–hickory; *3,* Transition; *4,* Bur oak–hickory. (Data for *1, 2,* and *3* from Aikman and Smelser, 1938, for North Woods tract, Story and Boone Counties; data for *4* from Conard, 1918, for area near Grinnell.)

	1	*2*	*3*	*4*
Number of trees*	84	104	104	70
Tilia americana	35.7	. .	32.7	10.0
Acer saccharum nigrum	57.1	5.8	9.6	. .
Ulmus fulva	7.1	. .	32.7	. .
Quercus borealis maxima	. .	17.3	5.8	. .
Quercus alba	. .	71.1	. .	. .
Fraxinus americana	. .	5.8	. .	. .
Carya ovata	. .	. .	3.8	15.7
Carya cordiformis	. .	. .	3.8	. .
Ulmus americana	. .	. .	11.5	. .
Quercus macrocarpa	. .	. .	. .	37.1
Ulmus sp.	. .	. .	. .	24.3
Acer Negundo	. .	. .	. .	10.0
Prunus serotina	. .	. .	. .	2.8

* Trees approximately 10 inches and over, d.b.h.

dominated by white oak and red oak in the larger size classes (Table 35, column 2), while hickories constitute about one-third of the 2 to 8 inch d.b.h. class. In a transitional community on flood plains, basswood, red and white elm, and sugar maple are the principal species (Table 35, column 3). Bur oak is an important constituent of many of the upland communities (Table 35, column 4). Disturbance does not greatly modify the forest composition, as most of the species sprout freely from stumps, thus regenerating a forest whose composition is similar to that of the original stand. Two important timber cover types are recognized, (1) a white oak type in which white oak predominates and black oak and hickories are the principal associates, and (2) a white oak–black oak–red oak type on the better, more moist and well-drained soils (Geneaux and Kuenzel, 1939). Occasional boreal relic communities in Iowa, in which balsam fir and yew are conspicuous, contrast strongly with the prairie and prevailing types of forest vegetation (Conard, 1932, 1938).

The valley woodlands of eastern Nebraska are representative of the western outposts of deciduous forest in the old glaciated area. The principal forest communities recognized by Aikman (1927) are the bur oak–bitternut hickory, the black oak–shellbark hickory, and the red oak–basswood. The first of these occupies the more xerophytic slopes and hilltops along the

Missouri River and its tributaries. In addition to the dominant species (*Quercus macrocarpa* and *Carya cordiformis*), a number of other trees are present, including *Q. Muhlenbergii* and trees characteristic of more mesophytic communities. The black oak–shellbark hickory community is intermediate in mesophytism; it is more limited in distribution than the other two communities. In addition to the dominants (*Q. velutina* and *C. ovata*), the more important species are *Q. Muhlenbergii, Gymnocladus dioicus, Tilia americana,* and *Q. borealis* var. *maxima*. The red oak–basswood community occupies "the best protected slopes and ravines along the rivers in the eastern part of the state and in the southeast all of the lower slopes and the north exposures of higher ones" (Aikman, 1927). Hophornbeam (*Ostrya virginiana*) is a usual component of this community. Associated species are few, and generally limited to the southeastern part of the state. The limit of range of red oak is reached before that of basswood; hence upstream along the Missouri and Niobrara Rivers, the community has but one dominant, *Tilia americana,* which extends westward into the Sand Hills of north-central Nebraska (p. 178).

Because forest composition here at the western limits of forest is in part dependent on the ranges of species, equivalent communities differ from place to place. This is reflected by the different community names applied by different authors (Pool, Weaver and Jean, 1918; Weaver, Hanson and Aikman, 1925; Aikman, 1927). The last of these authors considered forest communities of the whole of eastern Nebraska, the others of limited areas. Composition of forest communities near Peru, in southeastern Nebraska (as derived from transects given by Pool, Weaver and Jean) is given in Table 36.

Table 36. Valley woodlands of eastern Nebraska: *1, 2,* and *3,* Oak and oak–hickory communities of upper slopes; *4,* Red oak community; *5* and *6,* Basswood community of lower slopes. (Data from transects by Pool, Weaver, and Jean, 1918.)

	1	*2*	*3*	*4*	*5*	*6*
Number of trees*	23	18	18	25	46	25
Quercus macrocarpa	26	22	11	..	..	..
Quercus Muhlenbergii	18	..	22	16	3.2	..
Quercus velutina	26	66	22	..	..	..
Quercus borealis maxima	..	11	11	76	6.5	8
Carya ovata	30	..	33	8	..	..
Carya cordiformis	..	..	..	..	..	4
Ulmus fulva	..	..	..	..	15.2	12
Fraxinus lanceolata	..	..	..	..	8.7	4
Juglans nigra	..	..	..	..	..	12
Tilia americana	..	..	..	..	67.4	60

* As trees are smaller near their western limits, stems 6 inches and over, d.b.h., are included.

The change in forest composition from east to west, as one by one the tree species reach the limits of their ranges, is best illustrated by examples. In eastern Iowa, *Acer saccharum* is a dominant in the maple–basswood community; in central Iowa its place is taken by *A. saccharum* var. *nigrum,* and farther west, in eastern Nebraska, maple is replaced by red oak in the most mesophytic communities. In the oak–hickory communities, the number of species of oaks and hickories decreases westward. For example, *Quercus alba,* abundant eastward, drops out before eastern Nebraska is reached. *Q. macrocarpa* extends farther than any other oak, at its western limits being a low shrub, often less than a meter in height (Pound and Clements, 1900). *Carya ovata* and *C. cordiformis* extend farther west than other hickories. Of the more mesophytic species, *Tilia americana* is remarkable in its westward extent.[4]

From south to north, there is also a change in woodland communities. At the western limits of deciduous woodland in the north, bur oak is the only oak, and hickories are not present. Elm, ash, and box elder near the streams, and bur oak savannah and scrub between valley woodlands and prairie are the deciduous forest communities of this northwestward extension. Far north, paper birch and aspen, deciduous outposts from the Hemlock–White Pine–Northern Hardwoods region occur. Aspen is abundant in the Turtle Mountains of northern North Dakota, where it may dominate over large areas (Bergman, 1912).

The deciduous forest belt of Minnesota is a northwestward extension from the Mississippi Valley. A part of it, known as the "Big Woods," belongs to the Maple–Basswood Forest region (Chapter 11). Oak forest and oak savannah surround the Big Woods and extend, thence, northwestwardly as a very narrow band separating the conifer forest (Minnesota section of the Hemlock–White Pine–Northern Hardwoods region) from the prairie. This band, and the interrupted forest belt west of the Big Woods is bur oak savannah (Daubenmire, 1936; Ewing, 1924).

Where the prairie-forest border extends northward into Canada,[5] elements of three forest formations and of prairie meet. Here, white elm, boxelder, basswood, and bur oak from the Deciduous Forest, red pine, white pine, and white cedar from the Great Lakes-St. Lawrence forest to the east,

[4] The ability of *T. americana* (*T. glabra*) to grow in the area of the grassland climax, although in situations with abundant soil moisture, emphasizes its difference, ecologically, from the several species of the Mixed Mesophytic Forest region. Failure to distinguish the somewhat glabrous species as *T. neglecta* and *T. floridana* from *T. americana* has resulted in statements concerning the presence of the latter species in regions where it does not occur. *T. americana* is northern and western, only rarely occurring in the Mixed Mesophytic Forest region (northern part) and Western Mesophytic Forest region (northern and western parts).

[5] In area distinguished by Halliday (1937) as the Aspen–Oak section of the Boreal Forest region.

and balsam poplar, black spruce, and white spruce from the Boreal Forest emphasize the transitional character of the area.

Separated areas of oak forest referable to the Mississippi Valley section of the Oak–Hickory region lie to the northeast of the Mississippi River and east of the St. Croix River in Wisconsin. These are often mixed oak woods made up of *Quercus ellipsoidalis, Q. alba, Q. macrocarpa, Q. coccinea, Q. borealis* var. *maxima,* and *Ostrya virginiana,* with occasional *Tilia americana,* and, northward, *Pinus Strobus.* White oak is frequently the dominant species; southward, shellbark hickory is usually associated with the white oak. The common hazel (*Corylus americana*) is abundant in the white oak woods; *Anemonella thalictroides* and *Uvularia sessilifolia* are characteristic herbaceous plants (Butters and Rosendahl, 1924). In the bands of oak forest bordering prairie areas, *Quercus macrocarpa* is the dominant species (Weaver and Thiel, 1917). The mixed oak forest of the upper Mississippi Valley is closely related to the oak forests of the Driftless area, from which it probably migrated. It is frequently interrupted by prairie. Ability to occupy a variety of soils, both sandy and basic, suggests climatic control. However, the occasional basswood and sugar maple, and the greater frequency of these species on ravine slopes indicate the transitional character of these areas.

PRAIRIE PENINSULA SECTION

The "Prairie Peninsula" as mapped by Transeau (1935) extends eastward from the Mississippi River across Illinois and western Indiana, with outliers as far east as central Ohio, and as far north as southern Wisconsin and southern Michigan. The name is used because of the frequent occurrence of prairie in a peninsula-like projection from the prairies of the Mississippi Valley.[6]

Forests of this area are prevailingly oak–hickory, hence it is considered as a part of the Oak–Hickory Forest region. However, the oak–hickory forest area here referred to as the Prairie Peninsula section is somewhat more extensive than the Prairie Peninsula proper, although it approximates it in form. Its boundaries are indefinite as disjunct areas of oak–hickory forest occur within the adjacent forest regions. It extends from the northern border of the Ozark Hills of southern Illinois (where it adjoins the Western Mesophytic Forest region), northward into southern Wisconsin, between the Driftless area on the west and the northwestern arm of the Beech–Maple Forest region which borders Lake Michigan. Eastward in Indiana and southern Michigan it is more interrupted. In southwestern Indiana, an arbitrary boundary between it and the Western Mesophytic Forest region

[6] Prairie areas are shown on the Forest Regions map by stippling.

is drawn at the glacial boundary (margin of Illinoian drift).[7] Within the Tipton Till Plain area of Wisconsin drift in Indiana, oak–hickory, beech–maple,[8] and prairie communities occur in frequent juxtaposition. Across this tension zone an irregular arbitrary boundary between the Oak–Hickory and Beech–Maple regions extends to northeastern Indiana and southern Michigan. This approximates the boundary of "upland oak forest" in Indiana as mapped by Gordon (1936). (See p. 305, boundary of Beech–Maple Forest region.)

The occupancy by oak–hickory forest of a considerable area within and at the borders of the Prairie Peninsula is an event which took place, in part during, in part subsequent to the post-Wisconsin xerothermic period. Pollen profiles of bogs in the Prairie Peninsula indicate the early entrance of oaks in this area and a time of maximum occurrence in a warm dry period (the "xerothermic period," period IV) preceding the present (see Chapters 14 and 17). Climatic succession, coupled with progress in the erosion cycle on the young drift plains, tends to replace prairie by oak–hickory forest, and eastward, oak–hickory forest by more mesophytic types. Within the Prairie Peninsula proper, climatic limitations (especially low precipitation-evaporation ratios, low relative humidity in summer, and high proportion of growing season precipitation) preclude development of mesophytic forests, except in local situations, and result in permanence of prairie and oak–hickory communities.

In the eastern part of the Prairie Peninsula section, local factors are determinative, for this is a tension zone between oak–hickory and beech–maple communities or between oak–hickory and mixed mesophytic communities.

Even to the east of the boundary of the Prairie Peninsula section, areas of oak forest are of sufficient extent to be shown on small-scale maps (Chapman, 1944). Thus, over several counties in the Wisconsin till plains of south-central Ohio, the original forest type was oak. In hilly areas of central Indiana, oak–hickory forest occupies the drier slopes (Friesner and Potzger, 1937; Potzger, 1935; Potzger and Friesner, 1940a). In areas of low relief, there is sometimes a relation between degree of dissection (or organization of a drainage pattern) and advance in hydrarch succession. This has been demonstrated by comparing adjacent areas where drainage has reached different stages of development. The prairie and oak–hickory communities occupy areas without organized drainage; while adjacent areas with well developed drainage are occupied by beech–maple forest (Sears,

[7] South of the glacial margin in the Western Mesophytic Forest region, many areas of oak–hickory forest occur (Chapter 5). In some of these there is no indication of change, the oak–hickory community is climax (see Mauntel's Woods, p. 145 and Table 26).

[8] Developmental stages, such as the mixed hydro-mesophytic forests of moist flats and depressions are included.

1926). The presence of low morainal ridges, usually sandy or gravelly, on the otherwise gently undulating till plain usually results in alternation of oak or oak–hickory communities (on the ridges), and beech–maple or transitional and developmental communities (on the till plain).

In the northern lobe of the Prairie Peninsula in southwestern Michigan, prairies occur on glacial outwash or on coarse valley filling. The original areas were separated from one another; although nearly treeless, bur oak trees were scattered on the prairies. White oak, or white oak and black oak occupied many of the sandy ridges. The original extent of the prairies and oak openings can now be judged only from differences in soil, as agricultural utilization has obliterated most of the natural vegetation. The soil of the prairies was dark in color, due to its high humus content, thus contrasting with the lighter more yellowish forest soil. In plowed fields, circular yellowish patches surrounded by dark prairie soil may indicate the former position of trees (Veatch, 1927; Wing, 1937).

Similarly, the oak lobe extending northward in Wisconsin is a mosaic of prairies, oak openings, and denser oak woods interspersed here and there with small tamarack swamps and marshy stream borders. The prairies are now almost entirely in cultivation, or are invaded by dense young mixed oak stands. Their former presence is often indicated only by the presence of a few characteristic prairie plants (Gould, 1941). An increase in forest area, at the expense of prairie, followed the early settlement of the country as a result of decrease in the frequency of prairie fires. A few of the original oak openings remain where large bur oaks in open formation, some of the trees 250 years old, dominate the landscape (Stout, 1946). Sandy and gravelly morainal ridges are occupied by mixed oak or oak–hickory woods in which prairie plants are conspicuous.

In the broader western part of the Prairie Peninsula section (Illinois and northwestern Indiana) climatic control is more complete. Oak–hickory forest, although widely distributed, is generally limited to slopes of shallow and ill-defined ravines or of low morainal ridges. Prairie occupied the gently rolling to flat intervening areas, where expanses of black prairie soil in the cultivated fields, and the occasional prairie plant along fence-rows are all that remain to indicate former occupancy by prairie. The frequent alternation of prairie and oak woods is now indicated best by differences in color of soil. Prairie soils are black; woodland soils are brownish or yellowish, the darker soils those onto which forest invasion of prairie was comparatively recent. Forest invasion of prairie has progressed most rapidly along ravines where erosion breaks the prairie sod and facilitates forest invasion. Thus, boundaries between prairie and forest soils are not always the boundaries between prairie and forest communities; the forest may have advanced onto prairie soil. Where such occupancy is long continued, the soil becomes modified to a yellow-gray silt loam (McDougall, 1919, 1925;

An ungrazed oak–hickory woodland in Illinois. (Courtesy, U. S. Forest Service.)

Woodard, 1925). Differences in forest composition are correlated with length of the period of occupancy, showing successional development which may take place as forest advances onto the prairie upland. "High content of soil organic matter, dominance in an earlier stage of bur oak, elm, and yellow oak, and complete *absence* of white oak, are attributed to a history of development from prairie upland only a few hundred years ago" (Marberry, Mees and Vestal, 1936).[9] Different stages of development sometimes occur side by side in a single prairie grove (Gleason, 1912); isolated groves may differ greatly in specific content (Gleason, 1913).

Throughout much of the Prairie Peninsula section, the more pronounced ravine slopes are occupied by mixed forest communities in which sugar maple may be a dominant species; beech in the east,[10] and red oak and basswood in the west are codominant with the maple. Southward, additional

[9] This statement was made concerning "University Woods" in Champaign County, Illinois, where the present forest is a highly mixed type.

[10] Beech reaches the western limits of its range (in the Prairie Peninsula) in the eastern tier of counties along the southern half of the eastern border of Illinois; it is absent from the prairie region of northwestern Indiana (see map of distribution of *Fagus grandifolia,* Transeau, 1935).

species are often abundant. Thus these ravine slope forests (called post-climax communities) approximate the climax types of surrounding regions —beech–maple to the east and northeast, maple–basswood to the north and northwest, and mixed mesophytic of the Western Mesophytic region to the south.

The forests of low terraces and well-drained bottomlands are composed of an admixture of mesophytic slope forest species and flood-plain species, together with a number of other species among which walnut, ash, elm, coffee-nut, hackberry, bur oak, and honey locust are often conspicuous. In low flood-plain and streamside communities, silver maple, white elm, and cottonwood are most abundant, often composing about three-fourths of the stand. Sycamore is generally present, except in the north, in all areas not subject to prolonged overflow. The number of species in the bottomland forests is generally greater in the south than in the north, for some of the southern species, as pecan and river birch, extend northward into this section (Telford, 1926; Fuller and Strausbaugh, 1919; McDougall, 1925).

A number of upland forest communities, in general related to soil and topography, may be distinguished. These usually contain a high percentage of oaks.

A large part of the Illinoian drift area of southern Illinois is relatively undissected till plain. Here local factors determine the forest types. The fine-textured soil with impervious subsoil, with consequent poor drainage and aeration, is wet in spring but dry and hard in summer. Much of the area is occupied by wet prairie in which swamp species prevail. Pin oak is a conspicuous tree of wet areas, where it occurs singly or in groves, and accompanied by shingle oak (*Quercus imbricaria*), river birch, and honey locust. In less wet but equally poor soil areas, post oak is dominant. In fact, this whole area is mapped as post oak upland, as this species is its most widespread and characteristic tree (Telford, 1926). On the poorest soils, post oak occurs in pure stands, or may be accompanied by blackjack oak. On slopes where streams cut through the flats, an oak–hickory or mixed hardwood type of forest prevails (Telford, 1926; Vestal, 1931).

Forests are less extensive on the Wisconsin drift, where stream dissection has only begun, than on the old drift of western Illinois. Locally within the Wisconsin drift area, more or less extensive sand deposits introduce distinct vegetational features. The largest of these are the Kankakee sand plains, an area of some 3000 square miles partly in Illinois and partly in Indiana. Low ridges in the sand plains are forested; black oak (of small stature) is usually the dominant tree. In open stands, prairie species are abundant in the undergrowth. In more mesophytic situations, often on the sides of the ridges or between the ridges, bur oak, or bur oak and white oak dominate. An admixture of additional species, such as red oak, walnut, hackberry, wild cherry in the successionally most advanced communities gives a mixed aspect to the forest (Gleason, 1910). The lowlands of the sand plains are occupied by swamp and wet prairie communities.

The majority of the upland forest soils are intermediate in character between the extremes of the post oak upland and the sand deposits. On these, white oak and black oak are the dominant species in many areas (Table 37, area 1). In more mesophytic situations, red oak and sometimes

Table 37. Composition of upland forest types of the Prairie Peninsula section: *1,* Grazed oak–hickory woods in Warren County, Indiana; *2,* A "virgin stand" of "upland mixed hardwoods" in McLean County, Illinois. (Data for *2* from Telford, 1926.)

	1		*2**
	Canopy	*Understory*	
Number of trees	100	42	81
Quercus alba	58	23.8	21.0
Quercus velutina	16	. .	6.1
Quercus imbricaria	5	2.4	. .
Quercus borealis maxima	4	2.4	1.2
Quercus macrocarpa	2	. .	. .
Carya ovata	6	19.0	9.0
Carya cordiformis	2	9.5	
Fraxinus americana	1	2.4	. .
Ulmus americana	1	2.4	12.3
Ulmus fulva	1	4.9	
Tilia americana	2	. .	. .
Juglans nigra	2	21.4	. .
Prunus serotina	. .	4.9	1.2
Ostrya virginiana	. .	4.9	. .
Acer saccharum	. .	2.4	30.9

* Canopy and understory not distinguished; the high percentage of *Acer saccharum* may be due to understory trees; percentage based on number of trees per acre.

sugar maple accompany the white oak (Table 37, area 2). The most mesophytic upland forest type is a mixed forest in which sugar maple, red oak, white ash, and sometimes basswood are predominant species. The more mesophytic forests, if long undisturbed, have a mull humus and a gray-brown forest soil. The undergrowth contains a variety of widespread forest mesophytes, as *Botrychium virginianum, Cystopteris fragilis, Uvularia grandiflora, Smilacina racemosa, Hepatica acutiloba, Actaea pachypoda, Sanguinaria canadensis, Hydrophyllum appendiculatum, Phlox divaricata, Eupatorium rugosum,* and *Solidago latifolia.*

The composition of a series of forest communities of increasing mesophytism, from a white oak ridge type (where oaks comprise over 90 per cent of the canopy) to mesophytic flat and north slope forests (where sugar maple is abundant) is shown in Table 38.

Few, if any, areas of oak–hickory forest remain in undisturbed condition;

Table 38. Composition of a series of forest communities of increasing mesophytism in Morton Arboretum, Lisle, DuPage County, Illinois; *1,* A white oak ridge; *2,* Slope; *3,* A lower flat; *4,* North slope.

	1	*2*	*3*	*4*
Number of canopy trees	85	89	121	88
Quercus alba	60.0	30.3	8.3	8.0
Quercus borealis maxima	18.8	31.5	28.9	22.7
Quercus macrocarpa	10.6	. .	5.0	5.7
Quercus ellipsoidalis	5.9	. .	. .	. .
Acer saccharum	. .	13.5	18.2	26.1
Fraxinus americana	. .	16.8	11.6	20.4
Tilia americana	. .	4.5	6.6	12.5
Juglans nigra	2.3	1.1	5.0	3.4
Ulmus fulva	1.2	1.1	8.3	1.1
Ulmus americana	. .	. .	7.4	. .
Carya cordiformis	1.2	. .	. .	. .
Prunus serotina	. .	1.1	.8	. .

grazing is general, and some cutting has been done in almost every area. Second-growth stands of post oak (in the post oak upland area), of scrubby black oak and black oak–hickory on sandy soils, and of white oak–black oak or white oak–black oak–hickory on clayey or intermediate soil types, are the usual representatives of the oak–hickory upland forest of the Prairie Peninsula section. These are similar in composition to old-growth types.

CHAPTER 7

The Oak–Chestnut Forest Region

A great area trending in a northeast-southwest direction from southern New England and the Hudson River Valley to northern Georgia is included in the Oak–Chestnut Forest region.[1] Here oaks and (formerly) chestnut are so abundant in most situations as to characterize the region. Various oak, oak–chestnut, and oak–chestnut–tuliptree communities occupy the climax sites. Mixed mesophytic communities, if they occur at all, are generally confined to coves and lower ravine slopes.

The name, oak–chestnut, is used for this region although the Oak–Chestnut association, which characterized it, no longer exists in unmodified form. It is now so changed by the death of chestnut that its original composition can be determined only in the areas most recently invaded by blight, and there only because dead chestnut is still standing. The name is retained because it is impossible as yet to predict the final outcome of the partial secondary successions everywhere in progress.[2] Opening up of the canopy releases some of the undergrowth trees, and may temporarily favor the least tolerant of these. The replacement of chestnut by oaks seems to be taking place in many sites; in some, tuliptree has increased; in others, more mesic and shade-tolerant species appear to be favored. Future changes in canopy composition may result in changes in the humus layer (because of differences in decomposition products of different species of trees) and later, in the character of the undergrowth. Only after several generations of forest could an equilibrium be reached, and the dominants of the new climax be determined.

The Oak–Chestnut Forest region covers all of the Blue Ridge Province of physiographers, and all of the Ridge and Valley Province except its northern end in the Hudson-Champlain Valley and its southern end from southern Tennessee southward. It extends onto the Piedmont Plateau in northern

[1] This is by no means coextensive with the "chestnut–chestnut oak–yellow poplar forest" of Shantz and Zon (1924) which includes not only the Oak–Chestnut region as here defined, but also the Mixed Mesophytic and Western Mesophytic Forest regions.

[2] Where chestnut was an important constituent of second-growth sprout hardwoods, natural replacement is resulting in stands in which oaks are dominant (Korstian and Stickel, 1927).

Throughout the Oak–Chestnut Forest region, the chestnut trees are dead or almost dead, killed by a fungus disease. Where chestnut alone was dominant, all canopy shade is gone; only trees intolerant of shade can grow, and slowly re-establish a forest, but not a climax forest. (Courtesy, National Park Service.)

Virginia and over all of the Piedmont of Maryland, southeastern Pennsylvania, and northern New Jersey. From the Potomac River northward, it includes a strip of Coastal Plain. It covers a small area of the New England Province in southern New England and all of the Reading Prong Highlands of Pennsylvania and New Jersey (considered as part of the New England Province).

Along its western border, this region is in contact with the Mixed Mesophytic Forest region. The boundary between these two regions (approximately at the Cumberland and Allegheny fronts) is fairly sharp except in its middle part where a transition belt is evident. Here the vegetation boundary is east of the physiographic boundary (p. 75). This is where ridges of the Ridge and Valley Province are crowded near the plateau escarpment. Northward, the Oak–Chestnut region is in contact with the Hemlock–White Pine–Northern Hardwoods region whose boundary is near the southern and eastern limits of the Glaciated Allegheny Plateau in northern Pennsylvania and southeastern New York. In southern New England the vegetation boundary bears no relation to physiographic boundaries. On the south and east, the Oak–Chestnut region is in contact with the Oak–Pine region of the Piedmont Plateau. This boundary between Oak–Chestnut and Oak–Pine crosses the southern part of the Great Valley (Ridge and Valley Province), then swings southward around the Southern Appalachians (Blue Ridge Province), where it follows the inner border of the Piedmont at the foot of the Blue Ridge Mountains. Northward in Virginia, it begins to diverge somewhat from the Blue Ridge, thus including in the Oak–Chestnut region the inner Piedmont of Virginia, which in places is almost mountainous. From the Potomac River northward, the boundary extends across the Coastal Plain of Maryland, Delaware, and New Jersey, reaching the Atlantic Coast at the projecting end of the Navesink Highlands, south of New York Bay. This part of the boundary across the Coastal Plain follows approximately the outer margin of Cretaceous sediments.

On the whole, the Oak–Chestnut region is mountainous; only toward the northeast does it extend over nonmountainous terrain. Everywhere the occurrence of oak–chestnut communities seems intimately related to slopes; only rarely do they occupy flats. Where there are extensive flat or nearly flat areas, as on the Harrisburg peneplain in the broad valleys of the Ridge and Valley Province, oak–chestnut communities do not occur; white oak forests are there the rule. Where streams have cut below the Harrisburg peneplain producing slopes of a younger erosion cycle, oak–chestnut forest is not found; such situations may be occupied by mixed mesophytic communities.

Oak–chestnut communities are not limited to this region; outliers occur in the Mixed Mesophytic Forest and Western Mesophytic Forest regions to the west, very locally in the Beech–Maple region (on the plains of Lake

Ontario) and in the Northern Appalachian Highlands Division of the Hemlock–White Pine–Northern Hardwoods region.

Five sections are distinguished in the Oak–Chestnut Forest region. The area of greatest diversity of vegetation, of finest forests, is the **Southern Appalachians.** The **Northern Blue Ridge** section extends north from Roanoke, Virginia, into southern Pennsylvania. The **Ridge and Valley section,** extending from Pennsylvania to southern Tennessee is a third section. The **Piedmont section** extends from Virginia northeastward almost to the Delaware River. The **Glaciated section** includes that part of the region north of the glacial boundary, in northern New Jersey, New York, and southern New England, and a small area of unglaciated land south of the boundary in New Jersey and Pennsylvania.

SOUTHERN APPALACHIANS

The term, Southern Appalachians, is commonly used for the southern section of the Blue Ridge Province.[3] It includes the mountain mass south of Roanoke Gap and is composed of the Blue Ridge (in the limited sense) along the southeast border, the Unaka Range, including the Great Smoky Mountains, along the northwest, and a number of more or less transverse ranges and peaks in the interior (as the Black Mountains, Grandfather Mountain, Nantahala Mountains, etc.). Elevations range from about 1300 or 1500 feet at the margin of the Piedmont in North Carolina, and 1500 feet in stream valleys of the western slope, to 6711 feet at the summit of Mt. Mitchell near the eastern border of the section and 6680 feet on Clingman's Dome in the Great Smoky Mountains. The great relief—over 5000 feet—naturally results in zonal distribution of forest vegetation.

This is an area of great antiquity, and one whose mountains have not been base-leveled. The most extensive of all peneplains of the East—the Schooley peneplain—is but indistinctly recorded in a "subsummit peneplain" in parts of this mountainous area. Here mountains rise more than 2500 feet above this oldest peneplain level. This continuously mountainous area was a refugium where many plants, unable to find suitable habitats on the vast peneplain, persisted during the period of peneplanation. Many old endemics, relics of the Tertiary flora, of which a considerable number have their closest relatives in eastern Asia, are found in these mountains.[4] Later

[3] It is not synonymous with the "Southern Appalachians" of foresters, which may refer to the entire Appalachian Highland south of Pennsylvania (Society of American Foresters, 1926; Frothingham and Stuart, 1931).

[4] Cain (1943) discusses the "Tertiary character of the cove hardwood forests of the Great Smoky Mountains National Park." Asa Gray, in 1842, in his "Notes of a botanical excursion to the mountains of North Carolina" mentioned many of these rare plants of the Southern Appalachians. Later, in 1878, and again in 1884, he emphasized the Tertiary character of the flora and pointed out its Asiatic relationship (see Sargent, 1889).

View in the Pisgah National Forest, North Carolina. The paler trees on the mountain slopes are chestnut in bloom; the ascendant white staminate catkins are seen in the immediate foreground. (Courtesy, U. S. Forest Service.)

peneplains, of which the Harrisburg is most important, are represented only by incipient peneplains. The lowland of the French Broad River at an elevation of 2100 to 2200 feet is referable to the Harrisburg peneplain; its vegetation—prevailingly oak and oak–pine—resembles that of the southern part of the Great Valley.

The forests of the Southern Appalachians, especially of the Great Smoky Mountains, are outstanding in wealth of species, and in size attained by the forest trees.[5] Due to the diversity of topography and range of altitude, there are great differences in forest vegetation in different parts of the section.

Various classifications of the forests of this section have been suggested. Frothingham and Stuart (1931) group the forest types of the Southern Appalachians into three classes—"moist slope and cove, dry slope and ridge, spruce . . . forests"—and state that about one-third of the mountain forest area belongs to the first class, nearly two-thirds to the second class,

[5] In "Big Trees of the Great Smokies" by S. Glidden Baldwin, M.D. (from *Southern Lumberman,* issue of December 15, 1948), the circumference breast high of a number of trees is given. Among them are tuliptree, 23 ft. 8½ in.; silverbell, 11 ft. 9 in.; cucumber magnolia, 18 ft. 3½ in.; black cherry, 12 ft. 4 in.; white ash 15 ft. 5 in.; beech 11 ft. 6 in.; hemlock, 19 ft. 9 in.; and yellow or sweet buckeye, 15 ft. 11 in.

and smaller areas to the spruce type.[6] The first class includes the cove forest (mixed mesophytic), lower moist-slope forest (in which chestnut, together with several oaks and hickories, forms the bulk of the forest), and upper moist-slope forest (a "northern hardwoods" type in which yellow birch, sugar maple, and beech are important species). The "dry slope and ridge forest," chiefly on southerly and westerly exposures, is made up largely of oaks and pines together with an admixture of other drought-resistant trees.

An earlier classification of forest types (Society of American Foresters, Committee of the Southern Appalachian Section, 1926) is the most usable classification presented. Three principal groups, each with a number of forest types, are recognized: (1) Northern forest, mostly at high altitudes, including spruce–fir, hemlock–yellow birch, and northern hardwood; (2) moist slope and cove forest, including buckeye–basswood, northern red oak, cove hardwood, yellow poplar, cove hemlock, white pine, chestnut, and white oak; (3) dry slope and ridge forest, including chestnut oak, black oak–scarlet oak, and pitch pine–mountain pine. The second and third of these groups include the forests of moderate elevations.

We shall consider first, the forest types of moderate elevations, taking up the "northern forest" types in their altitudinal sequence. The principal forest types represented at moderate elevations are chestnut or oak–chestnut, mixed mesophytic or cove hardwood, and hemlock or hemlock–hardwood in mesic and submesic situations; oak and oak–pine on dry slopes and ridges and on certain low-elevation situations. Higher elevations are occupied by birch–beech–maple or "northern hardwood" forest, with or without hemlock, and spruce–fir forest.

Forests of Moderate Elevations

CHESTNUT OR OAK–CHESTNUT COMMUNITIES

Because of the elimination of chestnut by blight, the oak–chestnut forest, the most characteristic community of the region, no longer occurs in its original condition. Few data are available giving its composition other than in general terms. Data given here (in Tables 40, 42, 43, 44) are based on published records and on areas in which dead chestnut was still standing at the time of the study. Extensive use of chestnut in the tanning industry, and clean cutting for pulpwood and charcoal over great areas of the mountains, earlier decimated the forests.

Oak–chestnut communities originally covered most of the mountain slopes and much of the rolling upland of the subsummit peneplain (Schooley peneplain). Thus the range of elevation was from about 1300

[6] As "Southern Appalachians" is used by these authors in a more extended sense (an administrative area of foresters), the proportions of the several forest classes will not be correct for the Southern Appalachians proper.

feet to about 4500 feet or even higher on southern exposures. In many areas, chestnut was the dominant tree, often occurring in almost pure stands. The Society of American Foresters (1926, 1932, 1940) refers to this as a chestnut type and records the term "oak–chestnut" as a synonym. Some of the associated species as there listed occur in one, some in another phase of the oak–chestnut community. The number of tree species in typical oak–chestnut communities is not large, although the number of tree species occurring in the Southern Appalachian section of the Oak–Chestnut Forest region probably exceeds that in any other section or region of the Deciduous Forest. This is because complex mixed mesophytic communities occupying the coves greatly add to the number of species in the region.[7] Many variations in composition of the oak–chestnut forest are seen, at least some of which are related to soil moisture and slope exposure. Some are transitional communities related to the mixed mesophytic communities; some are transitional to higher altitude forests.

Almost throughout, the oak–chestnut forest is characterized by an abundance of ericaceous shrubs in the understory. Particularly striking, are the masses of flame Azalea (*Rhododendron calendulaceum*) which are familiar to every visitor to the Southern Appalachians. Only in the best situations—where the oak–chestnut forest contains some of the species of the mixed mesophytic forest of the coves—is the ericaceous layer absent; there, an herbaceous layer, resembling that of mixed mesophytic communities, is seen.

At the lower elevations the oak–chestnut forest on the best sites (northerly or moist) was a dense forest of very large trees, mostly chestnut and tuliptree (commonly called yellow poplar in the Southern Appalachians). Tuliptree decreases and oaks increase in abundance upward. Hemlock may be abundant in the canopy layer, or, absent in the canopy, may be a dominant in the young growth replacing the dead chestnut. Accessory species include some of the "cove hardwoods." The admixture of cove hardwoods, species of the mixed mesophytic forest, suggests a relationship of this community of the Oak–Chestnut region to the Mixed Mesophytic association. In undergrowth (shrub and herb layers) this community resembles the mixed mesophytic communities and is unlike the oak–chestnut communities of the Mixed Mesophytic Forest region. *Rhododendron* and *Kalmia* may be present (particularly if hemlock is a constituent of the canopy), but the aspect is largely controlled by the deciduous species, as *Hamamelis virginiana, Stewartia ovata, Ilex montana, Cornus alternifolia, Lindera Benzoin, Parthenocissus quinquefolia,* and the distinctly mesophytic herbaceous layer with an abundance of mull humus

---

[7] This will explain the statement made by Weaver and Clements (1938) that "it is in this association that the number of species of trees reaches a maximum." Actually, it is the Mixed Mesophytic association that has the greatest number of tree species.

species. This more mesophytic phase of the lower elevation oak–chestnut forest is well represented by parts of the Joyce Kilmer Forest (near Robbinsville, North Carolina, in the Nantahala National Forest, Table 43, area 2) and in the Great Smoky Mountains on ravine slopes adjacent to cove forests (Table 40, area 1).

Scarcely less mesophytic as to site, is the forest in which oaks (*Q. montana, Q. borealis* var. *maxima, Q. alba, Q. velutina,* and *Q. coccinea*) are the principal associates of chestnut. Such oak–chestnut communities may contain a few cove hardwoods; they usually have an ericaceous shrub layer. In the drier situations, the forest was less dense, with smaller and sometimes crooked trees. At higher elevations, red oak is often a dominant species; chestnut oak also is abundant; white oak occurs as a dominant with chestnut on some exposed southerly slopes. Yellow birch, red maple, and occasional spruce (*Picea rubens*) are indicative of the transition to higher altitude forest types.

The more typical oak–chestnut forest (in which chestnut and oak prevail), is widespread and often interrupted only by the cove forests of deep concavities on the slopes. This is the "chestnut type" or "chestnut slope type" of foresters (Reed, 1905; Society of American Foresters, 1926, 1932, 1940). Chestnut may have been the dominant species, in which case the present forest community is almost entirely secondary due to the killing of most of the canopy by blight. What the future forest will be is still problematic—sometimes one, sometimes another of the usual understory species predominating. The chestnut and oak–chestnut forests usually have a well-developed heath layer in which *Rhododendron, Azalea, Kalmia, Vaccinium, Leucothoe, Menziesia,* etc. may be abundant. Forests along the Appalachian Trail near Wayah Gap in the Nantahala Mountains, on the slopes of Grandfather Mountain (Table 42, type 2), and of Mt. Mitchell, are representative, as are also many parts of the east slope of the Blue Ridge. The similarity of this, the most extensive forest type of the Southern Appalachians, to the higher elevation summit oak–chestnut communities in the Cumberland Mountains section of the Mixed Mesophytic Forest should be noted.

MIXED MESOPHYTIC COMMUNITIES OR COVE HARDWOODS

Cove forests of the Southern Appalachians are typical mixed mesophytic communities. They have long been referred to as mixed forests (Harshberger, 1903), cove forest (Frothingham and Stuart, 1931; Cain, 1931), cove hardwood forest (Cain, 1943), or unnamed, but with principal trees listed, and clearly recognized as distinct from the prevailing chestnut or oak–chestnut communities (Ashe, 1897).

Within a region of such high rainfall as the Southern Appalachians (approximately 70 inches at moderate elevations, more at higher elevations) the limitation of mixed mesophytic communities to coves and lower north slopes is in striking contrast to their more general distribution in the Cumberland Mountains. In the Southern Appalachians, they appear to be "postclimax." It is probable that this spatial limitation in a region often visited by botanists and ecologists is in part the cause of the widespread lack of recognition of a Mixed Mesophytic association covering a large area. The relationship of oak–chestnut and mixed mesophytic communities is clearly indicated in the Southern Appalachians, although the derivation of the former from the ancestral mixed forest is perhaps not so evident here as where spatial relations are reversed, as in the Cumberland Mountains.

No sharp boundary can be drawn between the richer chestnut, chestnut–tuliptree, or oak–chestnut communities of mesic slopes, and the mixed mesophytic communities of coves. Typical representatives of the forests of the two situations are, however, quite unlike—in canopy composition, in woody undergrowth, and in herbaceous layer.

The finest cove hardwood forests are in the Great Smoky Mountains. Some of these are made up almost entirely of deciduous species; in others, hemlock is present or even a dominant (Tables 39, 40). This is the same condition as was seen in the Cumberland Mountains, in the Mixed Mesophytic Forest region (p. 52). The composition of the cove hardwood forest varies with altitude, and with the topographic situation. At higher

Table 39. Mixed mesophytic communities or cove hardwoods in the Great Smoky Mountains: *1,* West-facing cove on slopes of Alum Cave Creek at about 3000–3500 feet; *2,* Slopes of north-facing ravine on Mt. Mingus at about 3500–4000 feet; *3,* North-facing cove on slopes of Thunderhead Mountain at 4500–5000 feet.

	1	2	*3*
Number of trees	117	166	73
Tilia spp.	24.8	18.1	. .
Acer saccharum	22.2	4.8	4.1
Quercus borealis maxima	14.5	4.8	. .
Aesculus octandra	10.3	16.9	23.3
Halesia monticola	10.3	6.6	9.6
Tsuga canadensis	2.5	18.7	1.4
Betula lutea	. .	2.4	32.9
Betula lenta	.8	8.4	. .
Fagus grandifolia	. .	2.4	17.8
Prunus serotina	. .	. .	8.2
Castanea dentata	4.3	2.4	. .
Fraxinus americana	3.4	5.4	. .
Liriodendron tulipifera	1.7	3.6	. .
Carya cordiformis	3.4	1.2	. .
Acer rubrum	. .	3.6	1.4
Magnolia Fraseri	1.7	.6	. .
Amelanchier laevis	. .	. .	1.4

levels it grades into the beech–birch–maple of northerly slopes.[8] In stream bottoms and ravines, the proportion of hemlock increases. Such hemlock–hardwoods communities have been referred to as "hemlock bottom" (Reed, 1905), "cove hemlock" (Soc. Amer. Foresters, 1926), "cove hemlock–beech" (J. H. Davis, 1930), and "Tsugion" (Cain, 1943).

The cove hardwoods community or mixed mesophytic community of the coves is a forest made up of some 25 or 30 tree species (not all in any one cove), of which six or eight may be dominants in varying proportions in different stands. The most abundant species—those attaining dominance in some stands—are sweet buckeye, basswood (*Tilia heterophylla*), sugar maple, silverbell (*Halesia monticola*), tuliptree, beech, yellow birch, hemlock, and occasionally chestnut.[9] Other canopy species commonly present are white ash (including Biltmore ash), wild cherry, cucumber magnolia, red maple, red oak, and bitternut hickory; of smaller trees, *Magnolia Fraseri, Amelanchier laevis,* and *Acer pensylvanicum* are frequent. Only in the addition of the silverbell tree (*Halesia monticola*) does the canopy of the Southern Appalachian cove hardwood community differ from typical mixed mesophytic communities of the Cumberlands. This is the only one of the trees of the Tertiary forest which becomes a dominant in the Southern Appalachians and is poorly represented and of small stature in the Cumberlands. In size of canopy trees, the mixed mesophytic communities of the Smokies and of the Cumberlands are essentially similar—only a few species are known to attain somewhat larger size in this more southern section.

As in the Mixed Mesophytic Forest region, a number of segregates are distinguishable. Because of the large number of canopy species and relatively small area occupied by mixed mesophytic communities (in proportion to the prevailing chestnut communities), separated stands often appear to differ from one another. Although some are referable to more or less clearly defined segregates, all may well be thought of as mixed mesophytic communities. Cain (1943) distinguishes seven segregates—buckeye–basswood, sugar maple–silverbell, yellow birch, beech, hemlock–beech, hemlock–tuliptree, and hemlock—in the "cove hardwood complex" of the Greenbrier region of the Great Smoky Mountains. As is evident in Tables 39 and 40, additional segregates might be recognized.

In those communities which are dominantly deciduous, a large number of

---

[8] In the Southern Appalachians, and especially along the major river systems, the lower part of the beech population is essentially the "white beech." At somewhat higher elevations, there is an obvious genetic blending of white and red beeches. At higher elevations where there is often an abrupt increase in the amount of beech, the beech is the "red beech," which, on the eastern slope of the Southern Appalachians, reaches its best development at about 3000 to 4000 feet. The beech which occurs just below the spruce–fir forest is the "gray beech" (fide W. H. Camp, 1948).

[9] In the moist cove forests the dead chestnut trees are rapidly becoming unrecognizable, while those of drier sites are still standing.

Cove forest in the valley of Alum Cave Creek, Great Smoky Mountains National Park. Leucothoë (in the foreground) and Rhododendron here form a dense shrub layer near a stream.

Table 40. Forest communities of Ramsey Fork of Greenbrier, Great Smoky Mountains: *1*, Chestnut slopes; *2, 3, 4,* and *5,* Mixed mesophytic or cove hardwoods communities at increasing elevations. Area *5* extends approximately to the lower limits of spruce.

	1	2	*3*	*4*	*5*
Number of trees	158	76	62	162	161
Castanea dentata	30.5	. .	. .	. .	. .
Tsuga canadensis	19.0	5.3	6.4	14.8	14.3
Liriodendron tulipifera	12.7	13.2	11.3	. .	. .
Acer rubrum	13.3	21.0	1.6	2.5	.6
Quercus montana	7.6	. .	. .	. .	. .
Nyssa sylvatica	5.7	. .	. .	. .	. .
Oxydendrum arboreum	1.3	. .	. .	. .	. .
Robinia Pseudoacacia	.6	. .	. .	. .	. .
Tilia spp.	. .	13.2	14.5	17.9	19.9
Halesia monticola	. .	17.1	12.9	2.5	5.0
Aesculus octandra	. .	1.3	14.5	14.8	16.1
Prunus serotina	. .	. .	3.2	17.3	13.0
Acer saccharum	. .	. .	8.1	14.2	15.5
Betula lenta	4.4	3.9	8.1	3.7	. .
Betula lutea	. .	7.9	4.8	3.7	11.2
Fraxinus americana	.6	6.5	4.8	2.5	1.2
Fagus grandifolia	1.9	2.6	9.7	4.3	3.1
Magnolia Fraseri	. .	5.3	. .	. .	. .
Magnolia acuminata	. .	2.6	. .	. .	. .
Quercus borealis maxima	2.5	. .	. .	.6	. .

deciduous small tree and shrub species occur in the undergrowth. These, except for a few species more abundant at higher elevations, are the same as in the mixed mesophytic forest of the Cumberlands. The prevalence of deciduous species in the arboreal layers in large measure determines the nature of the humus layer and hence of the herbaceous growth. The luxuriant herbaceous layer, containing a variety of ferns and mull geophytes (as *Trillium, Dentaria, Erythronium, Dicentra, Disporum,* etc.), is characteristic of all typical prevailingly deciduous mixed mesophytic communities. Cain (1943) lists, with constancy and frequency per cents, the vernal and aestival flora of the cove hardwoods of the Smokies.

Cove hardwood forests are not confined to the Great Smoky Mountains. Comparable communities are seen on lower north slopes in the Nantahala River Gorge where the rich herbaceous growth is strikingly different from that which prevails over most of the slopes. In the Joyce Kilmer Forest, a memorial tract in the Nantahala National Forest, the hemlock cove type is well represented, as is also a prevailingly deciduous type suggesting by its large amount of chestnut, a transition to the prevailing chestnut and chestnut–tuliptree communities of that area (Table 43). Cove hardwoods occur locally in valleys in the Blue Ridge, for example along the Pacolet River above Tryon, where the vernal flora is particularly rich. In the Black

Mountains, Davis (1930) distinguishes a "cove climax association" in which *Castanea, Liriodendron, Tsuga, Quercus borealis* [*maxima*], and *Acer rubrum* are most frequent. He states that there are over 40 species of trees in this community—only about two-thirds of this number in any one

The hemlock–mixed mesophytic community of broader flats in ravines of the Joyce Kilmer Memorial Forest is similar to ravine forests of the Great Smokies. *Rhododendron maximum* in bloom. Nantahala National Forest, North Carolina. (Courtesy, U. S. Forest Service.)

cove. Related to the abundance of chestnut, oak, and hemlock, these cove forests contain an abundance of heath shrubs. In this respect they are not typical mixed mesophytic communities. The cove forest of the Black Mountains is more or less intermediate or transitional between the mixed mesophytic cove forest and the rich chestnut slope forest, displaying characteristics of each type.

In those communities in which hemlock is a dominant species, *Rhododendron maximum* is usually dominant in the undergrowth. Some hemlock forests, however, do not have a Rhododendron layer. In Ravenel's Woods, near Highlands, N.C., both types of hemlock forest are well represented (Oosting and Billings, 1939). The mor humus layer which normally results from the dominance of hemlock and of Rhododendron is not favorable to the development of a luxuriant herbaceous layer. Such plants as *Mitchella repens, Viola rotundifolia, Trillium undulatum, Galax aphylla, Medeola virginiana, Goodyera pubescens, Dryopteris spinulosa intermedia,* and *Dennstaedtia punctilobula* are the usual herbaceous species of the hemlock communities. A larger proportion of these species, and more individuals of each occur where Rhododendron is absent or scattered.

The hemlock community commonly occupies the floors of ravines, broad valley flats at high elevations (around 4000 feet, more or less), and parts of the high-level plateau which is such a prominent feature of the southeastern part of the Southern Appalachians.[10] Such areas are sometimes referred to as "spruce flats" (a local usage in the Great Smokies), or hemlock bottom (Reed, 1905). The community appears to be climax in nature; it is reproducing itself (with perhaps minor oscillations in abundance of its canopy species). It is not distinct from the more typical mixed mesophytic communities having a lower percentage of hemlock.

It should be apparent that the Southern Appalachians are closely related to, if not a part of, the Mixed Mesophytic Forest region. Their wide separation from that region—by the Great Valley of the Ridge and Valley Province—is one reason for not including them in the Mixed Mesophytic Forest region. The prevalence of chestnut and oak–chestnut communities on mesic slopes, and the essential continuity of such communities northward into the Northern Blue Ridge where true mixed mesophytic communities are all but absent, justifies the exclusion of this area from the Mixed Mesophytic region even though its cove hardwoods are splendid examples of the Mixed Mesophytic association.

OAK AND OAK–PINE COMMUNITIES

The sharp contrast in appearance between the prevailing oak–chestnut communities of slopes and the oak and oak–pine of some of the low inter-

---

[10] This is the subsummit peneplain, the Southern Appalachian representative of the Schooley peneplain.

montane plateaus is a vegetational feature of the Southern Appalachians. The Asheville basin (which represents the Harrisburg peneplain) is occupied by a mixture of oaks and pine, in which the southern red oak (*Q. falcata*) is prominent. Post oak and blackjack oak (*Q. stellata, Q. marilandica*) are also present. Some of the low hills are occupied chiefly by white oak, or sometimes by a white oak–black oak community in which a *Vaccinium* layer may be well developed. Some of the most mesic situations are occupied by white pine or combinations of white pine and white oak. Comparable communities occupy other low plateau situations. Davis (1930) refers to the vegetation of the Asheville basin as the "plateau type."

This vegetation, so unlike that of the surrounding mountains, resembles that of the Great Valley to the west. It appears to be more than coincidence that these incipient areas of the Harrisburg peneplain should be vegetationally similar to the more extended peneplain areas in the Ridge and Valley section.

On the lower slopes of the Blue Ridge adjacent to the Piedmont, and on outlying spurs of the Blue Ridge which rise above the Piedmont, oaks, hickories, and pines occur with chestnut. The oaks generally predominate—although in some areas about one-half of the trees are pines.

Ashe (1897) distinguishes three divisions of the forest of the lower mountains: (1) table mountain pine, on the east and south slopes of the Blue Ridge; (2) shortleaf and pitch pine, in the basin of the French Broad, on river hills of the Swannanoa, and on other low hills below 2800 feet in elevation; (3) white pine, occurring in isolated small stands along the south and east slopes and crest of the Blue Ridge and on low hills westward, usually between 2800 and 3800 feet. The "table mountain pine" type also occurs at low elevations along the west border of the Southern Appalachian section. It is called "pine heath" by Cain (1931), who states that "its most abundant species . . . are *Pinus echinata* and *P. pungens,* which occur in pure or mixed stands. Other associates of these are *P. rigida, P. virginiana, Robinia Pseudoacacia, R. hispida, Acer rubrum, Nyssa sylvatica,* and *Castanea dentata.*" Ericaceous shrubs are abundant; hence, the name "pine heath." This pine community on the south slopes of hills may be sharply delimited from the chestnut woods of the north slopes.

Forests of the Higher Elevations

At higher elevations the regional forest gives way to more northern types of forest. On south slopes the chestnut forest may extend to about 5000 feet or even higher. On northerly slopes, it gives way at about 4500 feet to a "northern hardwood" forest. This in turn grades into the spruce–fir of the highest forest belt. In ravines which are occupied by cove forests, the transition is more gradual, due to shifting of dominance of some of the cove species, and increasing prominence of cherry, yellow birch, and buckeye, with the appearance of scattered spruce sometimes as low as 4500 feet.

Two principal forest types are recognized on the upper mountain slopes: (1) a "northern hardwood" type which has also been variously referred to as "northern hardwood climax association" (Davis, 1930), "sugar maple slope" (Reed, 1905), "beech–maple community" (Brown, 1941), and (2) the spruce–fir forest. Transitions, of course, occur not only between these two types, but also between each of these and lower elevation forest types.

"NORTHERN HARDWOOD" FOREST TYPE[11]

In the northern hardwood forest type, the most important trees are sugar maple, yellow birch, beech, and buckeye. White ash, wild black cherry, silverbell, cucumber, chestnut, and occasionally basswood, are accessory species which may be present. That this is distinct from the Hemlock–White Pine–Northern Hardwoods Forest is apparent in its generic as well as specific composition. Buckeye and silverbell are representatives of distinctly southern genera; the species of *Tilia* which occur are southern—*T. americana* of the North is not a constituent of this forest. At its best, this is a dense forest of rather large trees.

At higher elevations dwarfing becomes pronounced and the stand more open. These open and dwarfed or gnarly hardwood stands which occur on some of the summits and higher slopes, and often adjacent to "balds," are sometimes referred to as "orchards": the "subalpine orchard association" in the Black Mountains (Davis, 1930); the "beech orchard," "chestnut orchard," and "orchard woodland" in the Great Smokies (Cain, 1931).

Among the shrubs of the northern hardwood forest belt, *Viburnum alnifolium, Acer pensylvanicum,* and *Acer spicatum* suggest the northern relationship of this community. The herbaceous layer, too, contains a number of northern or mountain species, as *Maianthemum canadense, Oxalis montana, Circaea alpina, Houstonia serpyllifolia,* and *Aster acuminatus,* as well as many of the mesophytes of lower elevation forests. The herbaceous layer is luxuriant in the lower part of the northern hardwood belt; in the "orchards," a large admixture of grass and sedge gives it a characteristic aspect. In early spring, *Erythronium americanum, Claytonia virginica, Phacelia fimbriata,* and *Houstonia serpyllifolia* may carpet the ground. Where chestnut or oak prevails in the "orchard woodland," as on the slopes of Wayah Bald in the Nantahala Mountains, heath shrubs, chiefly Azalea, are abundant.

Variations in composition of the northern hardwood forest are due in part to differences in slope exposure and altitude, and in part to proximity of other forest types. The greatest variation is seen in the Great Smokies, where some high altitude forests resemble cove forests, except for the great

[11] This name is used in deference to foresters' usage. It resembles the northern hardwoods only in the abundance of beech, sugar maple, yellow birch, and certain of the shrubs.

Oaks at the higher elevations are irregular in form and smaller in stature than in lower-elevation forests, and the forest less dense. Nantahala National Forest, North Carolina. (Courtesy, U. S. Forest Service.)

increase in the amount of yellow birch, and might be classified as cove forests; others are almost pure stands of one or another of the usual dominants with only a small admixture of other species. The beech–maple forest type is represented on Roan Mountain where these two dominants comprise 85 per cent of the individuals 10 inches and over, d.b.h., and over 90 per cent of those 4 inches and over, d.b.h. (see Table 41, data from D. M. Brown, 1941). A somewhat more complex type is represented on the slopes of Grandfather Mountain by the "Sugar Maple Slope" type of Reed (1905) where beech, sugar maple, basswood, yellow birch, and hemlock make up about 75 per cent of the individuals 5 inches and over, d.b.h., while hemlock, yellow birch, and sugar maple prevail among the largest individuals (Table 42, type 3). In the "northern hardwood" forest type in the Black Mountains, the "dominant and most constant trees are *Acer saccharum, Fagus grandifolia, Aesculus octandra, Tilia heterophylla michauxii, Betula lutea, Fraxinus americana, Prunus serotina,* and locally, *Halesia carolina*"[12] (J. H. Davis, 1930).

[12] *Halesia monticola* is not always separated from *H. carolina.* The former species, listed in the cove forests of the Great Smoky Mountains (p. 201), is the large-flowered Blue Ridge and Appalachian Highland species.

Table 41. Forest communities of the higher slopes of Roan Mountain: *1,* Beech–maple community; *2,* Spruce–fir community. (Data from D. M. Brown, 1941.)

	1		*2*	
Inches d.b.h.	4–9	10 & +	4–9	10 & +
Number of trees	181	75	182	124
Fagus grandifolia	81.2	74.7	..	.8
Acer saccharum	13.8	10.7	..	..
Betula lutea	.6	8.0	4.4	21.0
Betula lenta	1.1	..	..	..
Betula nigra	.6	1.3	..	..
Aesculus octandra	1.6	1.3	..	.8
Amelanchier laevis	.6	4.0	..	..
Acer spicatum	.6	..	2.2	.8
Sorbus americana	..	..	1.1	..
Abies Fraseri	..	..	61.5	19.4
Picea rubens	..	..	30.8	57.3

SPRUCE–FIR FOREST TYPE

The spruce–fir forest type of the Southern Appalachians, although not a deciduous forest type, nevertheless occurs in the midst of a region dominantly deciduous. It is characterized by the dominance of red spruce, *Picea rubens,* and southern balsam or Fraser fir, *Abies Fraseri.* One of its dominants, red spruce, is a characteristic species of the Northern Appalachian Highland Division of the Hemlock–White Pine–Northern Hardwoods region; the other, Fraser fir, is an endemic of the Southern Appalachians. This type is thus distinct from any northern conifer community.

Spruce enters at a lower elevation than fir, usually at about 4500 feet. It is present in small numbers in the higher parts of the deciduous communities with which the spruce–fir forest is in contact, or as a codominant with yellow birch in a yellow birch–red spruce forest type (which includes also a number of species of the north slope and cove forests).[13] Higher, spruce and fir may be codominant, or either one may dominate, constituting 80 per cent or more of the stand (Cain, 1935a).

The red spruce forest occurs at a lower elevation than the fir forest, usually around 5000 to 5500 feet. Red spruce attains a much larger size than does Fraser fir, occasionally reaching a height of 130 or 140 feet, and trunk diameter of 4 feet (Cain, 1935a; J. H. Davis, 1930). Hence, in the red spruce forest the superior layer is made up entirely or almost entirely of spruce, while the small tree layer may be dominantly fir. In many places the spruce forest is interrupted by rocky spurs on which *Rhododendron catawbiense* forms dense thickets. During the bloom season of Rhododen-

[13] The yellow birch–red spruce type of the Society of American Foresters (1932, 1940).

The high-elevation red spruce forest with an undergrowth of purple Rhododendron (*Rhododendron catawbiense*). Middle Creek, Black Mountains of North Carolina. (Courtesy, U. S. Forest Service.)

dron these streaks of color alternating with the dark green of the spruce forest are a magnificent spectacle. Within the spruce forest, Rhododendron is more scattered, and accompanied by other shrubs, as *Viburnum alnifolium, Rubus canadensis, Vaccinium pallidum,* and *V. erythrocarpum.*

The Fraser fir forest occupies some of the higher summits and northerly slopes, usually above 6000 feet. Only one arborescent layer is distinguishable. The interior aspect of the fir forest is entirely different from that of any other community of the Southern Appalachians, for mosses and liverworts form a dense cover over the soil and the innumerable fallen trees. *Oxalis montana* and *Lycopodium lucidulum* are frequent in the moss mats. In both the spruce and fir forest, the spinulose shield-fern is abundant.

The contrast in composition of the red spruce and Fraser fir forests is well illustrated by Cain's tables (illustrating areas in the Great Smoky Mountains) which show 88.9 per cent of the basal area to be red spruce in the spruce type and 90.7 per cent to be fir in the fir type. On Roan Mountain, the larger trees (10 inches d.b.h. and over) of the spruce–fir community as described by Brown (1941) are *Picea rubens* (57.3 per

A forest of Fraser fir occupies some of the higher summits of the Southern Appalachians, usually above 6000 feet. The luxuriant carpet of mosses and liverworts, the ferns, and the mountain oxalis give a distinctive aspect to this forest. Mt. LeConte, Great Smoky Mountains National Park.

cent), *Abies Fraseri* (19.4 per cent), *Betula lutea* (21 per cent), and *Acer spicatum* (0.9 per cent). In this community, spruce constitutes only 30.8 per cent of the 4 to 9 inches d.b.h. class, while fir constitutes 61.5 per cent (Table 41). On Grandfather Mountain, the spruce–fir forest (called "mountain type" by Reed, 1905) covers the upper slopes (Table 42, type 4). Here Carolina hemlock (*Tsuga caroliniana*) is a constituent species, often localized on rocky slopes.

Balds

The occasional treeless mountain tops of this area are known as "balds." Some are grassy—the grassy balds; some are more or less densely covered with a variety of heath shrubs—the heath balds. Adjacent to either type, the forest communities are composed of more or less dwarfed and twisted or gnarly trees. The "orchards" (p. 207), sometimes abruptly, sometimes gradually, merge into the grassy balds.

Grassy balds, as the name implies, are covered by a more or less continuous stand of grasses and grass-like plants, dotted with cushions of *Polytrichum,* patches of ferns and other herbaceous plants, and occasional shrubs. Grassy balds suggest by their appearance some of the subalpine slopes of the Rocky Mountains.

Heath balds are famous for their exceeding beauty. Some are covered by a dense stand of shrubs, among which *Rhododendron catawbiense, R. maximum, R. carolinianum, R.* (*Azalea*) *calendulaceum* (and others), *Kalmia latifolia, Vaccinium* spp., *Gaylussacia baccata, Aronia melanocarpa, Prunus pensylvanica,* and *Viburnum cassinoides* are conspicuous. Smaller shrubs and ground heaths include *Leiophyllum Lyoni, Gaultheria procumbens, Epigaea repens,* and *Galax aphylla.* On some of the balds, the shrubs are grouped in patches sometimes of one species, sometimes of several, while open grassy stretches intervene. There is no more beautiful sight than such a mountain top when the purple Rhododendron (*R. catawbiense*) and Azaleas are in bloom. "Rhododendron Gardens" on Roan Mountain, the heath balds of the Smokies (such as Myrtle Top on Mt. Le Conte, Andrews Bald, Brushy Mountain, etc.), and of the Black Mountains, and Wayah Bald in the Nantahala Mountains, are well-known examples.[14]

The problems of succession on the balds, of cause and of permanence of balds, are beyond the scope of a work on the Deciduous Forest. Certainly in some instances there is evidence of tree invasion in the balds; in others there is an abrupt boundary between forest and bald, and little or no evidence of forest invasion. Opinions differ as to cause or permanency of balds. Brown (1941) discusses evidence of succession in the balds of

[14] Brown (1941); Cain (1930, 1931); Davis (1930); Camp (1931); Wells (1936, 1937).

The Bald on Roan Mountain, North Carolina, with purple Rhododendron in bloom. (Courtesy, U. S. Forest Service.)

Roan Mountain, and, after giving a historical summary of views bearing upon bald succession, discusses the problem of origin of balds. To this and the references there cited, the reader is referred for further consideration of balds.

Topographic and Altitudinal Sequences

The consideration of forest types of the Southern Appalachians as more or less distinct entities does not sufficiently depict the relations of these communities to topography and to altitude, and the contacts and intergradations between communities which are observed in the field. By selecting a number of situations where different communities are in juxtaposition, where community differences may to some extent at least be correlated with changes in topography or of altitude, a more concrete picture of the vegetation may be gained. Also, those places commonly visited may be made to fit into the broad picture of forest types.

GREAT SMOKY MOUNTAINS

As many of the communities have reached their best development in the Great Smoky Mountains, this area will be considered first. Secondary, as well as primary forest types are represented, for only about 40 per cent of the total area of the Great Smoky Mountains National Park is occupied by virgin forest.

Almost regardless of slope exposure and steepness, chestnut forest prevails at moderate elevations. All of the large chestnut trees have died in relatively recent years; the virgin chestnut forest is but a ghost forest. Secondary successions, the assumption of dominance by some one or several of the usual understory species will in time result in the development of mature communities. Which of these may ultimately be climax for such sites only several forest generations can determine. On some of the lower ridges, a pine–heath community occupies the dry south slopes, giving way abruptly to the prevailing chestnut. Locally, at moderate elevations, and not always on southerly slopes, the pine–heath community interrupts the prevailing deciduous forest. Contacts may be abrupt and sometimes appear to be related to substratum rather than slope exposure. At the lowest elevations and in some of the outlying valleys where a red soil is seen, oak forest has developed. This suggests a correlation with the prevailing oak communities of the zone of the red and yellow soils (lateritic soils) of the South.

In valleys with a gray-brown forest soil and on flats at low elevations (1500 feet ±), the original forest was very mesophytic—with hemlock, beech, buckeye, sugar maple, sweet gum, magnolias (*M. tripetala, M. Fraseri*), and thickets of Rhododendron and Leucothoë. Sometimes, the whole valley was filled with giant tuliptrees, "avenues of tulip like the big trees of California" (accounts of older inhabitants). These communities of the low elevation valleys are all but gone. Above 1500 feet, only small areas of "flats" have developed along the ravines. Some of these are locally known as "spruce flats." In these areas, hemlock is the dominant species, with tuliptree, beech, red and sugar maple, yellow birch, and silverbell.

Where the chestnut-clad slopes are cut by ravines, the chestnut forest is interrupted by narrow bands of hemlock, birch and tulip near the stream if the valley is open to the south or shallow and rather dry, or by areas of "cove hardwoods" where more protected ravines or concavities indent the slope. These mixed mesophytic communities (cove hardwoods) are limited in extent and often give way rather abruptly to the prevailing chestnut slope forest.

Examples of the mixed mesophytic communities of coves are given in Table 39 (p. 200). The first of these (area 1, a west-facing cove in the slopes of Alum Cave Creek at about 3000 to 3500 feet elevation) is the most typical mixed mesophytic community. Here five species (basswood, sugar maple, red oak, buckeye, and silverbell) make up 82 per cent of the canopy. This community is the ecologic equivalent of the sugar maple–basswood–buckeye community of the Cumberland Mountains (cf. Table 2). In the second area (area 2, on the north slopes of Mt. Mingus, at an elevation of 3500 to 4000 feet) hemlock, basswood, and buckeye together make up about 54 per cent of the canopy. A high elevation cove forest (area 3, a north-facing cove on the slopes of Thunderhead Mountain at an elevation of 4500 to 5000 feet) illustrates the increased abundance of beech (red beech), yellow birch, and buckeye at high elevations. These species here constitute 74 per cent of the canopy.

A sequence of ravine slope communities in Ramsey Fork of Greenbrier, from lower to higher elevations, is illustrated by Table 40 (p. 203), where area 1 is a chestnut–hemlock–tuliptree community of lower southerly slopes, and areas 2, 3, and 4 are successively higher mixed mesophytic communities on more or less northerly slopes along the stream. Here the increased importance of cherry at higher elevations (about 3700 feet) is well shown. This is comparable to the higher slopes of Black Mountain in the Cumberlands, where cherry increases in amount. The mixed mesophytic communities illustrated by Tables 39 and 40 have the typical luxuriant herbaceous layer of true mixed mesophytic communities everywhere. The rich vernal flora is quite distinct from that of adjacent communities in which chestnut is (or was) a dominant species, and is essentially like that of the Cumberland Mountains. In the "spruce flats" and other areas where hemlock is abundant, Rhododendron usually is dominant in the shrub layer, and the herbaceous layer is sparse.

High elevation deciduous forests of beech, birch, and buckeye; mixed deciduous–coniferous communities (transitions from the deciduous to the spruce–fir); and spruce–fir communities occupy the higher slopes, except where interrupted by "orchards" and "balds."

ROAN MOUNTAIN

Roan Mountain is situated to the northeast of the Great Smoky Mountains, and constitutes one of the very short mountain masses or cross ranges which extends southward from the Unaka Range. Its slopes are drained by tributaries of the Tennessee River, near which the elevation is about 2500 feet. From this general level, the mountain mass rises to an elevation of 6285 feet at Roan High Knob, the highest point on the mountain.

The lower slopes of the mountain, to an elevation of 3500 or 4000 feet, were, for the most part originally covered by oak–chestnut forest. That cove forests (mixed mesophytic communities) were represented is indicated by Harshberger's list of trees—*Castanea, Liriodendron, Acer, Tilia, Aesculus,* and *Fagus* (1903, 1911). The higher slopes are occupied by beech–maple and spruce–fir communities, while balds—grassy balds, alder balds, and Rhododendron balds—occur on the broad ridges above 5500 feet. These communities have been considered in detail by D. M. Brown (1941), from whose tables of density, data for Table 41 were taken. Except for the small individuals (less than 4 inches d.b.h.), beech far exceeds maple in the "beech–maple community," constituting about 75 per cent of the larger individuals. In the spruce–fir community, spruce is dominant among the larger trees, while fir is dominant in the 4 to 9 inch d.b.h. class and all smaller size classes.

BLACK MOUNTAINS

In and near the Black Mountains, a short range which culminates in Mt. Mitchell, the highest peak in the Southern Appalachians (6711 feet), the vegetational sequence may be traced from the Piedmont Plateau on the east or from the Asheville Basin on the southwest, upward through the middle elevation forests to the "northern hardwood" and conifer forests (Davis, 1930). The vegetation of the Asheville Basin and of the lower outlying valleys in the Great Smoky Mountains area is similar—oak forests, or oak and pine. In the Asheville area, southern red oak (*Q. falcata*) is abundant. The Piedmont forest is similar to that of interior basins, except for the abundance of yellow pine (*P. echinata*).

Upward from these basal forests, the several types of what Davis calls "Appalachian Forest Formation" are seen. As in the Smokies, the prevailing forest of slopes (the "mesic slope association" of Davis) is one in which chestnut is dominant, and accompanied by a number of species of oak. On drier slopes, pines (*P. rigida* and *P. pungens,* or lower, *P. echinata* and *P. virginiana*) are dominant, or codominant with oaks and chestnut.

Cove forests occupy the coves, the best situations. Davis distinguishes a "cove climax association," a "cove hemlock–beech association," and a "high cove association." The first of these is the richest forest, and originally was an extensive forest type. This cove forest is unlike the cove mixed mesophytic forests of the Smokies. In it, chestnut, tuliptree, hemlock, red oak, and red maple are the most frequent species. It thus is similar to the forest community illustrated in Table 40, area 1, a less rich and less mesophytic community than the superb cove hardwoods of the Great Smokies.

The communities of higher slopes of the Black Mountains, as in the Great Smokies, include a "northern hardwood" forest type, communities transitional to the spruce–fir forest, the "orchard" and "savannah" types of dwarfed deciduous trees, spruce–fir forest, and both grassy and heath balds.

GRANDFATHER MOUNTAIN (LINVILLE AREA)

The Grandfather Mountain region of the Southern Appalachians is of interest because here a residual mountain mass rises above the high plateau which represents the Schooley peneplain. Data on the original forest cover of this area were given by Reed (1905), who studied the vegetation of 16,000 acres. He distinguished four forest types—hemlock bottoms, chestnut slope, sugar maple slope, and mountain (Table 42). Elevations range from 3800 feet on the Linville River near Linville (on the high plateau) to 5964 feet on Grandfather Mountain. On the high plateau, valleys are broad with extensive nearly level bottomlands (peneplain features). Here are the "hemlock bottoms," where about 65 per cent of the large trees are hemlock.

Yellow birch is second to hemlock in number and size. The subordinate position of beech, both as to number and size, is evident from the table. The undergrowth in the "hemlock bottoms" is made up almost exclusively of Rhododendron, which forms almost impenetrable thickets.

Table 42. Forest types of the Linville or Grandfather Mountain area of North Carolina: *1,* "Hemlock bottom"; *2,* "Chestnut slope"; *3,* "Sugar maple slope"; *4,* "Mountain type." Data adapted from tables given by Reed, 1905.*

	1		*2*		*3*		*4*	
Inches d.b.h.	5–19	20–54	5–19	20–44	5–19	20–57	5–19	20–48
Number of trees per acre	56.6	24.9	86.7	15.3	89.5	23.3	110.9	19.6
Hemlock	25.3	64.6	3.0	7.5	4.7	28.1	.8	16.8
Yellow birch	40.1	19.4	6.8	3.8	12.2	10.7	10.3	22.9
Red maple	4.7	4.5	13.3	6.3	2.6	4.4	..	..
Chestnut	4.8	2.8	45.4	47.4	6.1	9.6	..	..
Red oak	..	..	4.3	15.3	..	..	..	..
White oak	..	..	2.7	7.5	..	..	..	..
Chestnut oak	..	..	6.3	5.7	..	..	..	..
Sugar maple	..	..	2.1	1.3	14.5	18.2	1.0	3.6
Basswood	..	..	..	..	13.6	9.1	..	..
Beech	14.7	.7	4.4	.1	31.4	.6	12.4	.8
Cucumber tree	..	..	4.0	1.7	1.9	2.4	..	..
Yellow buckeye	..	..	..	..	6.2	6.1	2.8	1.9
White ash	..	..	..	..	3.3	6.4	..	..
Fraser fir	..	..	..	..	..	..	14.6	.4
Red spruce	3.7	1.0	..	..	..	..	53.7	48.2
Other species	6.8	6.8	7.8	3.3	3.5	4.3	4.4	5.3

* In the original tables, data given represent a survey of 604 acres, or 3.8 per cent of the whole tract. The tables include diameter and number of trees per acre, based on an average of 121 acres for the hemlock bottom, 236 acres for the chestnut slope, 115 acres for the sugar maple slope, and 24 acres for the mountain type. Common names are used here as in the original tables.

The slopes rising above the broad flats of the old peneplain are covered either by "chestnut slope" or by "sugar maple slope" forests. The chestnut slope type (which covered 41 per cent of the area of the entire tract studied) clothes the southeast, south, southwest and west exposures, extending from the hemlock bottoms to the tops of some of the lower mountains of the area, and to 4600 feet on the slopes of Grandfather Mountain where it gives way to the "mountain type" of forest. The canopy of the chestnut forest of these high elevations (everywhere higher than any examples mentioned in the Great Smokies) is almost half chestnut; red oak is second in abundance. (This chestnut and red oak dominance is fairly general in the higher elevations of the chestnut or oak–chestnut forest type.) The "sugar maple slope" type (covering 10 per cent of the tract) occupies the northwest slopes of

Grandfather Mountain, and occurs in a few other isolated patches. It is a "northern hardwood" forest in which hemlock, yellow birch, and sugar maple are most abundant among the larger trees. It includes many of the cove hardwoods, some of which reached very large size in this community. The "mountain type" of forest covers the upper slopes of Grandfather Mountain and the summit of Grandmother Mountain, approximately 18 per cent of the total area of the tract. It is a forest in which the typical trees are red spruce, Fraser fir, and Carolina hemlock. Much of the forest of this tract is now badly cut-over, but it is still possible to distinguish the several types characterized by Reed. The chestnut slope forest is now entirely gone; many places have been modified by fire, and a heath bald aspect brought about through the killing of trees.

NANTAHALA MOUNTAINS AND SURROUNDING AREA

The southern end of the Southern Appalachians is deeply dissected by the Little Tennessee River and its tributaries. Elevations range from less than 2000 feet along the Nantahala and Little Tennessee Rivers to about 5500 feet in the Nantahala Mountains. The Nantahala River for some distance flows through a gorge whose steep slopes in a few spots support mixed mesophytic forest with rich herbaceous undergrowth. Such a community on a very steep north-facing slope, with beech, sugar and black maple, buckeye, tuliptree, and silverbell, contrasts strongly with the oak woods which prevail over much of the surrounding slope. The herbaceous growth, too, is strikingly different from that more generally seen. The many ferns, including *Athyrium thelypteroides, A. pycnocarpon, Phegopteris hexagonoptera, Adiantum pedatum,* and *Cystopteris fragilis,* emphasize the luxuriance of the herbaceous layer which includes such pronounced mesophytes as *Trillium erectum, Uvularia grandiflora, Disporum lanuginosum, Caulophyllum thalictroides, Astilbe biternata, Dicentra canadensis, Tiarella cordifolia, Viola canadensis, Phacelia bipinnatifida,* and *Hydrophyllum canadense*. No ericaceous shrubs are present, instead are *Lindera, Asimina, Rubus odoratus, Cornus alternifolia, Aristolochia durior*—plants of the mixed mesophytic forest. The contrast between this community and the vegetation seen on most of the mesic slopes in this area (cf. Joyce Kilmer Forest) suggests some cause other than slope exposure for the outstanding richness. The underlying rock here is calcareous, which produces a richer soil than is usual in the Blue Ridge Province. Oosting (1941) very briefly refers to "rich hardwoods on thin alkaline soil" in the Nantahala Gorge.

Not far to the west of the Nantahala Gorge, and within the Nantahala National Forest, a forest tract known as the Joyce Kilmer Memorial Forest contains a number of representative lower slope and ravine communities of superior quality "with individual trees of hemlock 70 inches in diameter (almost 6 feet through) and 130 feet tall, and yellow poplar . . . which

Table 43. Joyce Kilmer Memorial Forest, Nantahala National Forest, North Carolina: *1,* North slope chestnut community; *2,* Transition between chestnut and mixed mesophytic communities on a low ridge sloping north; *3,* Mixed mesophytic or cove hardwoods community of ravine slopes; *4,* Hemlock–mixed mesophytic community of ravine bottoms.

	1	*2*	*3*	*4*
Number of trees	40	91	142	151
Castanea dentata	82.5	50.5	16.9	2.0
Liriodendron tulipifera	5.0	15.4	16.2	15.9
Tilia spp.	2.5	2.2	21.8	16.6
Fagus grandifolia	. .	2.2	12.7	13.9
Halesia monticola	. .	6.6	12.7	5.3
Acer saccharum	2.5	2.2	10.6	1.3
Tsuga canadensis	. .	. .	3.5	35.1
Magnolia acuminata	5.0	6.6	.7	2.0
Magnolia Fraseri	2.5	. .	1.4	.6
Quercus montana	. .	6.6	. .	. .
Betula lenta	. .	4.4	1.4	4.0
Betula lutea	. .	. .	. .	.6
Fraxinus americana	. .	3.3	1.4	1.3
Carya sp.	. .	. .	.7	. .
Aesculus octandra	. .	. .	. .	.6
Ilex opaca	. .	. .	. .	.6

are 80 inches in diameter and 170 feet tall"[15] (Table 43). This tract, which lies in the Little Santeetlah Creek drainage, is partly enclosed by two spurs extending eastward from the Unaka Range. The north slope forest here is dominantly chestnut, and is in strong contrast to the area just considered on the Nantahala River. These north slope communities, with their admixture of cove hardwood species, have a woody and herbaceous undergrowth which is more closely related to that of the mixed mesophytic forest than to the oak–chestnut forest. Heaths are absent, and soft-leaved woody and herbaceous plants prevail. Different areas in the forest illustrate varying tendencies in the replacement of dead chestnut, a phenomenon which has been referred to a number of times. For example, on a north slope (area 1 of Table 43) where chestnut formed over 80 per cent of the canopy, hemlock is dominant in the reproduction. On a low ridge sloping north (area 2), where the forest is only half chestnut, the undergrowth contains many of the cove hardwoods which may replace dead chestnut and convert this part of the forest into a cove hardwoods or mixed mesophytic community similar to that now occupying the most mesic ravine slopes (area 3). In spring, the continuous ground cover of ferns and mull humus herbs emphasizes the mixed mesophytic character of these ravine slope forests. A hemlock–mixed mesophytic community which is similar to ravine forests

[15] Mimeographed booklet of the U. S. Forest Service on the Joyce Kilmer Memorial Forest.

in the Great Smokies occupies the broader flats in ravines (area 4). This differs from the "hemlock bottom" of the Grandfather Mountain area, which is at a much higher elevation, in the large number of tuliptrees and small number of yellow birch. South slopes are occupied by oak–chestnut communities in which ericaceous plants are present.

On higher slopes in the Nantahala Mountains, chestnut and oaks, particularly red oak and chestnut oak predominate. The ericaceous shrub layer over large areas is dominated by flame Azalea which in June transforms the forest into a vast floral display. These mountains do not have "northern types" of forest near their summits; the trees of the oak–chestnut forest become reduced in size and wind-twisted, and are more widely spaced near the mountain tops. Wayah Bald, a heath bald on which Azalea is dominant, is the culminating peak (5335 feet).

THE BLUE RIDGE

The Blue Ridge, the bordering range of the Southern Appalachians on the south and southeast, rises abruptly from the Piedmont Plateau. Viewed from the east or south it is an imposing mountain range; from the west, it rises but little above the interior plateau, although its cernuous crest is marked by a number of more or less prominent summits.

The forest of the outward slope merges with that of the Piedmont; chestnut, oaks, hickories, and pines are the dominant trees. Above the Piedmont transition, the forest of the slopes is (or was), in most places, dominantly chestnut, with an ericaceous layer. Locally in valleys which indent the slopes, if the soil is suitable, a greater variety of trees grow. An outstanding example is seen along the slopes of the Pacolet River above Tryon, where mixed mesophytic communities occur. An area of Porter loam, derived from deeply weathered Whiteside Granite, is particularly rich in plant species (Peattie, 1928).

The chestnut–red oak type of upper slope forests (in which there is some chestnut oak) is well illustrated on many of the higher slopes (for example, on Standing Indian Mountain). Cove forests of small extent interrupt the oak–chestnut forest.

On the plateau which is in many places the summit of the Blue Ridge, land utilization has been extensive. However, southward, a considerable proportion of the land is in forest, although few primary stands remain. Ravenel's Woods, also known as the Primeval Forest, near Highlands, North Carolina, is an excellent example of the high plateau forest (elevation 4400 feet). In it three communities are distinguishable: (1) an oak–chestnut community of slopes (somewhat disturbed); (2) a hemlock–mixed mesophytic community, in which hemlock, red maple, silverbell, chestnut, and birch (*Betula lenta*) compose about 80 per cent of the canopy; and (3) a hemlock–Rhododendron community, in which over half of the canopy trees are hemlock, and Rhododendron forms a continuous layer (Table 44). Within the hemlock forest (2 and 3 above), Oosting and Billings (1939) distinguish two communities which they designate as Rhododendron and Polycodium types. The hemlock–Rhododendron community of Ravenel's Woods is very similar to the "hemlock bottoms" type distinguished by Reed (1905) in the Linville area (cf. Table 42).

Table 44. Communities of Ravenel's Woods, the "Primeval Forest" near Highlands, North Carolina: *1,* Oak–chestnut community of gentle southwest slope; *2,* Hemlock–mixed mesophytic community of flats removed from creeks; *3,* Hemlock–Rhododendron community near to streams.

	1	*2*	*3*
Number of trees	38	96	50
Castanea dentata	34.2	11.5	4.0
Quercus montana	34.2	. .	. .
Quercus alba	10.5	. .	. .
Quercus borealis maxima	7.9	5.2	2.0
Acer rubrum	7.9	16.6	4.0
Betula lenta	5.3	10.4	16.0
Tsuga canadensis	. .	29.2	56.0
Halesia monticola	. .	13.5	4.0
Liriodendron tulipifera	. .	4.2	6.0
Prunus serotina	. .	. .	4.0
Pinus Strobus	. .	. .	4.0
Magnolia Fraseri	. .	7.3	. .
Magnolia acuminata	. .	1.0	. .
Amelanchier laevis	. .	1.0	. .

Northward, where elevations are lower, land utilization has all but obliterated original vegetation. All evidence points to the prevalence of oak–chestnut communities.

NORTHERN BLUE RIDGE

This section is coextensive with the physiographic division of the same name. It extends from the Roanoke River Gap in southern Virginia northward into southern Pennsylvania. North of the Potomac River, where the Blue Ridge of Virginia (Blue Ridge proper) ends in Elk Mountain, the adjacent ridges to the east—the Blue Ridge of Maryland and Catoctin Ridge—are included in the Northern Blue Ridge section. Throughout its extent, the section is narrow, generally only a single mountain range, sometimes expanding to 12 or 14 miles in width.

The forests of the Northern Blue Ridge lack the luxuriance and variety which are distinctive features of the Southern Appalachian section. This is in part related to less favorable climate, in part to a less varied topography and lesser range of altitude, and in part to the physiographic history of the area. The highest elevations reached are on Stony Man (4010 feet) and Hawks Bill (4049 feet), both in the Shenandoah National Park. Unlike the Southern Appalachian area, the Northern Blue Ridge was greatly reduced at the time of development of the Schooley peneplain; only isolated knobs and low swells rose above the general level. Because of the ready accessibility from the Piedmont on the east and the Great Valley on the west, the entire area has been subjected to lumbering. Much of the forest cover is very young. The best of the Northern Blue Ridge is included in the Shenandoah National Park, an area of 183,311 acres.

Oak–chestnut communities prevail throughout all of the Northern Blue Ridge. Some of the most mesophytic situations in valleys at the base of the mountains contain more mesophytic communities in which beech, hemlock, and sometimes white pine may be present. Tuliptree is abundant in the secondary communities of the lower slopes. In the most mesophytic coves of higher elevations (3000 feet more or less), sugar maple, basswood, and red oak are the usual dominants, while chestnut, sweet birch (*Betula lenta*), and ash are common species. The herbaceous layer of such forest communities contains a variety of species including *Asarum canadense, Cimicifuga racemosa, Sanguinaria canadensis, Aruncus dioicus* (*A. sylvester*), *Viola canadensis, V. eriocarpa, Aster divaricatus,* and *Solidago*

Oak woods of the Northern Blue Ridge, where black snakeroot or black cohosh (*Cimicifuga racemosa*) is one of the abundant herbaceous plants. Forest of white and red oaks, hickories, and ash; Catoctin Range, Frederick County, Maryland. (Courtesy, U. S. Forest Service.)

latifolia which are common to most mesophytic communities. In such areas, the humus layer is mull. In less mesophytic areas where the organic material is imperfectly incorporated, ericaceous shrubs are common and none of the herbaceous mesophytes mentioned above occur. On rocky or windswept slopes, chestnut oak dominates; table mountain pine (*P. pungens*) is not uncommon. Local rock-slide communities, in which *Betula lenta* is dominant, are similar to those of ridges in the Ridge and Valley section (p. 235) where they are more frequent and better developed. On mountain tops which have been subjected to lumbering and fire, bear oak (*Quercus ilicifolia*) forms a dense cover.

At the northern end of the section, summit elevations are low and flat plateau-like tops prevail. Sandy soils and early destruction of the forest cover, together with repeated burning, have resulted in great areas of bear oak and sweet fern (*Comptonia peregrina*). The general slopes originally supported oak–chestnut forest, interrupted on ravine slopes by mixed forests, or in sheltered ravines, by hemlock or hemlock–white pine communities. Caledonia State Park in southern Pennsylvania is situated in one of these mesophytic valleys where hemlock, white pine, tuliptree, and white oak are most abundant, and Rhododendron forms dense thickets. White oak and red oak, together with pine and hemlock, form a transition community on the lower slopes below the dry oak–chestnut upper slopes.

In the Maryland portion of the section, Blodgett (1910) distinguishes a number of physiographic divisions, listing some of the commoner or characteristic species of each. The greater abundance of beech in valleys of the foothills of the Catoctin than on the Piedmont to the east is emphasized. Local communities here are mixed forests best referred to the mixed mesophytic type. Some contain hemlock, some are entirely deciduous. The Middleton Valley (between the Catoctin and Blue Ridge of Maryland) is reduced to the Harrisburg peneplain level (as are valleys of the Ridge and Valley Province). Locally on ravine slopes cut in this valley, mixed mesophytic communities have developed.

The Shenandoah National Park, whose range of elevation is from 600 to 4049 feet, may be considered as representative of the Blue Ridge of Virginia. The forests of the lower slopes are all secondary. Red cedar is common where limestones of the adjacent valley to the west outcrop on the lower mountain slopes. Black locust and sassafras occupy clearings and old fields on the slopes, and tuliptree the coves. On the Blue Ridge foothills along the Piedmont Plateau, only the occasional old white oak, black oak, and tuliptree give any clue to the nature of the original forest cover. Dogwood evidently was abundant in the oak forest.

On the general slopes, the prevalence of oak–chestnut forest is apparent everywhere. Concavities low on the slopes contain tuliptrees, while high on the slopes birch is abundant, or combinations of sugar maple, basswood, and birch along with the ever-present red oak. Forest variations along the upper slopes and crests are related to slope exposure, steepness, and concavity or convexity of slope. Mesophytic red oak–sugar maple–basswood communities, or groups of hemlock in northerly concavities alternate with red oak–chestnut communities whose undergrowth contains Azalea, mountain laurel, and blueberries, or with oak–chestnut communities with a continuous heath layer. If the crest is at a low elevation (2500 feet or less) tuliptrees are present in the north slope coves. On windswept slopes and knobs, chestnut oak is abundant, and pines dominate locally. Trees are widely scattered on rocky crests and rock slides, where also there are a few distinctive herbaceous plants (as *Sedum telephioides* and *Potentilla tridentata*). Only at the highest elevations (slopes of Hawks Bill Mountain) are spruce and fir present. The fir is considered by some to be a distinct species, *Abies intermedia* (Fulling, 1936), by others to be a variety of the northern balsam fir, *Abies balsamea* var. *phanerolepis* (Fosberg and Walker, 1941). It is not the Southern Appalachian endemic species, *Abies Fraseri.*

White Oak Canyon, on the east slope of the Blue Ridge in the Shenandoah National Park, is the most extensive area of undisturbed forest in the Park. Here variations in forest composition through a vertical range of about 1500 feet, and over slopes of varying steepness and direction may be seen. Such variations in places are quite abrupt; communities in which sugar maple, basswood, and red oak, or sugar maple, basswood, red oak, and hemlock are dominant (or comprise over 50 per cent of the canopy) may be adjacent to ones in which chestnut oak is dominant, or chestnut, red oak, and chestnut oak. In the forest of the canyon as a whole, oaks and chestnut comprise about 50 per cent of the canopy of 17 species. Red oak and chestnut, which are most abundant, comprise about 30 per cent of the canopy trees; sugar maple, hemlock, chestnut oak, white oak, and basswood (abundance in order named) comprise about 50 per cent of the canopy. All of the variations or communities recognizable may be grouped into two principal types, one in which sugar maple is present, and another in which sugar maple is absent (Table 45). In the first of these, sugar maple forms about 16 per cent of the canopy; all oaks (red, chestnut, and white) and chestnut taken together, form about 40 per cent of the canopy; all 17 canopy species are present. In the second group, oaks and chestnut comprise about 88 per cent of the canopy, with chestnut, white oak, chestnut oak, and red oak the trees of greatest abundance; neither sugar maple nor basswood are present in any of these, and the canopy contains only eight species. Even the most mesophytic communities do not approach the cove hardwoods of the Southern Appalachians in luxuriance. None has the aspect of a mixed mesophytic community.

Table 45. Percentage composition of forest types of White Oak Canyon, Shenandoah National Park: *1*, The forest as a whole; *2*, Those communities in which sugar maple is present; *3*, The oak–chestnut communities, *i.e.*, communities from which sugar maple and basswood are absent.

	1	*2*	*3*
Number of canopy trees	719	564	155
Castanea dentata	15.2	9.8	34.8
Quercus borealis maxima	15.2	15.2	14.8
Acer saccharum	12.5	15.9	. .
Tsuga canadensis	12.1	13.5	7.1
Quercus montana	11.1	9.0	18.1
Quercus alba	9.2	6.2	20.0
Tilia neglecta	7.4	9.4	. .
Betula lenta	4.9	5.7	1.9
Liriodendron tulipifera	3.3	4.2	. .
Carya spp.	2.5	3.2	. .
Acer rubrum	2.5	2.5	2.6
Fraxinus americana	1.5	1.8	.6
Ulmus fulva	1.1	1.4	. .
Robinia Pseudoacacia	.8	1.1	. .
Juglans cinerea	.4	.5	. .
Nyssa sylvatica	.3	.4	. .
Juglans nigra	.1	.2	. .

RIDGE AND VALLEY SECTION

This section of the Oak–Chestnut Forest region includes a large part of the Ridge and Valley Province of physiographers—nearly all of the Middle section and about half of the Southern section.[16] It extends from southern Tennessee northeastward to eastern Pennsylvania, almost to the glacial boundary near the Delaware River. It lies between the Appalachian Plateaus to the west and the Blue Ridge to the east, extending northeastward beyond the terminus of the Blue Ridge. Both boundaries are rather clearly defined through most of their length; only that part of the boundary which approximates the Virginia-West Virginia line is indefinite. The section averages about 40 miles in width in Tennessee, widening northward to about 65 miles in northern Virginia and reaches a maximum width of 80 miles in central Pennsylvania.

Along the western boundary (northwestern side), this section of the Oak–Chestnut Forest is in contact with the Mixed Mesophytic Forest region northward into Pennsylvania. Along its northern boundary in Pennsylvania it is in contact with the Northern Appalachian Highland Division of the Hem-

[16] The Oak–Pine region extends northward into Tennessee in the Ridge and Valley Province.

lock–White Pine–Northern Hardwoods region. Along the western boundary in Virginia and West Virginia, there is a pronounced transitional belt between the Oak–Chestnut and Mixed Mesophytic regions. In Pennsylvania some features of the mixed mesophytic forest, some of the oak–chestnut, and some of the northern hardwoods are seen along a narrow belt following more or less the margin of the Appalachian Plateaus.

The broad topographic features of the section are suggested by its name. In general, the area is a lowland above which rise rather level-crested longitudinal ridges, sometimes crowded, sometimes widely spaced. These features are the result of certain geologic and physiographic processes which have influenced vegetation as well as determining topography. "The larger steps in the process were: (1) General peneplaning, (2) upwarping, (3) reduction of the weaker rocks to plains at lower levels, (4) further uplift and dissection" (Fenneman, 1938, p. 197). The strata were intensely folded at the close of the Paleozoic era—hence the longitudinal arrangement of features, for different rocks erode at different rates. The whole area was reduced to the level of the Schooley peneplain; present level-crested ridges represent this peneplain. After upwarping, erosion reduced all areas of weaker rocks to lower levels than that retained by more resistant strata; thus the "valleys" of the section were produced. These represent the Harrisburg peneplain. Well known examples are the Great Valley, the Shenandoah Valley (a part of the Great Valley), the Nittany Valley, and the Hagerstown Valley. Subsequent erosion (after post-Harrisburg uplift) has resulted in the trenching of many of the valleys (as at Natural Bridge, Virginia) and the more or less mature dissection of some parts of the Harrisburg peneplain.[17]

Vegetationally, these processes are important for they have produced widely different habitats: the steep-sided ridges on resistant sandstones; the flat or rolling lowland on shales and limestones; and the ravine slopes and bluffs of the latest erosion cycle along streams cut into the lowland. Previous vegetation, too, was profoundly affected by these far-reaching physiographic changes (see Chapter 17).

In brief, the forests of the ridges are oak–chestnut except in sheltered ravines or coves, where mixed mesophytic communities or hemlock communities are seen; those of the valley floors (Harrisburg peneplain), whether flat or rolling, are prevailingly oak with white oak the most characteristic species; those of ravine slopes along streams which indent the Harrisburg peneplain are beech or mixed mesophytic.

Differences between the southern and northern parts of the section are due in part to latitude and the ranges of species involved, in part to transitional features related to adjacent regions.

---

[17] Some will recognize additional peneplains or will question the reference of all parts of the "valleys" to the Harrisburg peneplain. These questions are discussed by Fenneman, 1938.

Almost anywhere in the Ridge and Valley section, on the steep slopes of the ridges where the soil is shallow and stony, forests of this aspect are seen. Chestnut oak, pine, and heath shrubs prevail. (Courtesy, U. S. Forest Service.)

Forests of the Ridges or Mountain Ranges

The ridges have been formed where stronger rocks (sandstone and conglomerate) have resisted erosion. These often extend unbroken for many miles, with slopes but slightly indented by streams; in places they are cut by gaps. They may decline in elevation toward their ends, or may bifurcate, or be joined by other ridges.[18] The sandstones which form their crests are often steeply inclined or almost vertical. The slopes in many places are vast rock slides—tumbled masses of lichen-covered sandstone blocks—or steep more or less barren shale slopes; in others they are relatively gentle forest-clad slopes. Very few of the ranges attain elevations sufficient to result in well-defined altitudinal variations in vegetation.

The forests of the ridges are easily accessible from the wide and fertile valleys; this resulted in early cutting and in repeated culling. In many places cutting for charcoal long since destroyed the forest. Almost everywhere, injury due to forest fires has been severe. Except in relatively in-

[18] These features are intimately related to geologic structure which results in various types of mountains.

accessible spots, and in parts of the closely placed ranges near the middle of the western border, few good areas of forest remain. Some that do are exceptional in type rather than representative. Oaks dominate in most stands of young secondary forest, and generally represent the former oak–chestnut forest. That a variety of types did exist is still evident from differences in secondary communities and from remnants of old-growth forests.

The forest of the slopes of the ridges is far from uniform throughout the section. Many communities have developed in response to unlike conditions of different areas. In addition to differences related to northern or southern location (the section is nearly 700 miles in length), and to the usual factors of slope exposure and steepness, are those brought about by the nature of the rocks and dip of strata (influencing soil moisture and development of soil mantle). Only local studies can take all of these into consideration;[19] each ridge is more or less a unit and generalizations for the section as a whole or even for parts of the section will often not be correct for local areas.

MIDDLE AREA

Along the middle part of the western border (from the vicinity of Spruce Knob, West Virginia, southward for about 100 miles) is a mountainous region of closely placed ranges. This is a transition area with features of the Mixed Mesophytic Forest region to the west and of the Oak–Chestnut Forest region to the east. Here, although the physiographic province boundary follows Back Allegheny Mountain (west of the Greenbrier River), the forest region boundary is best placed at the east base of Allegheny Mountain. To the west of this boundary buckeye and white basswood are conspicuous constituents of the forest. Eastward, many of the lower slope and valley communities could be considered mixed mesophytic communities, although these most characteristic species are absent or rare. Mixed forests with hemlock, beech, tulip, red maple, red oak, white oak, and sometimes sugar maple, basswood, and white pine occupy the most mesic habitats in the ravines, while oaks prevail on the slopes often with tuliptree and, in places, white pine. The limitation of mixed mesophytic communities to the most mesic sites and the prevalence of oaks in ravine slope forests in situations where mixed mesophytic communities would prevail farther west, serve to identify this area as part of the Oak–Chestnut region. White oak is usually the dominant tree in mesic slope forests, with tuliptree an important constituent. Chestnut was present but in this part of the deciduous forest it was infected early and all trees died so long ago that its presence is now indicated only by rotted stumps and occasional sprouts. On drier slopes, chestnut oak is dominant—in what once was a chestnut–chestnut oak community. Pines are common on the cliffs.

---

[19] As an example of a local area with distinctive features, see Allard, 1946.

View in the mountainous western border of the middle area of the Ridge and Valley section. Much of the forest is made up of young second-growth oak stands; here and there, areas of older forest are indicated by the larger and more irregular crowns. George Washington National Forest, Virginia. (Courtesy, U. S. Forest Service.)

In a representative area in the Wilson Creek drainage in George Washington National Forest (south of Warm Springs, Virginia) the mesophytic forest of the valley flat contains beech, hemlock, tuliptree, red oak, white oak, red maple, sugar maple, basswood, hickory, sour gum, and walnut; beech is locally dominant; in higher areas sugar maple may be dominant. On very steep northerly slopes, hemlock is abundant, with chestnut oak, sweet birch, hickory, cucumber tree, red oak, and an admixture of more mesophytic species. The shrubby growth contains some ericaceous species —as Kalmia, Azalea, Rhododendron—and some soft-leaved species—as *Hamamelis, Viburnum acerifolium, Acer spicatum,* etc. Dry rocky southerly slopes support a stand of gnarly chestnut oak, with an undergrowth dominantly ericaceous. A more extensive forest type—white oak–tuliptree–black oak—occupies the more gentle slopes (Table 46, area 1). Dogwood is abundant in the understory. The poor herbaceous and shrub layer is one rather typical of oak woods, with *Vaccinium, Azalea,* and species of *Desmodium* and *Lespedeza* abundant. Higher on the slopes, chestnut oak and black oak increase (Table 46, area 2); a poorer chestnut oak type occupies the uppermost slopes, except for cliff margins where pines prevail.

In the little Laurel Run Natural Area in the George Washington National Forest, a primary stand of white pine and hemlock occupies the coves, while the adjacent slopes and ridges are clothed in oak forest in which chestnut was an original constituent.

Virgin forest along the Wilson Creek road; George Washington National Forest, Virginia. (Courtesy, U. S. Forest Service.)

The forests of the generally separated ridges of the eastern half or two-thirds of the middle area of the Ridge and Valley section from southern Pennsylvania across Maryland, West Virginia and northern Virginia are for the most part much disturbed and secondary. Blodgett (1910) mentions some of the features of North Mountain and Tonoloway Ridge in Maryland. The West Virginia and Virginia ranges are comparable vegetationally to the Wilson Creek area already described, but with less indication of mixed

Table 46. Composition of oak forests of slopes in Wilson Creek drainage, George Washington National Forest, Virginia: *1,* White oak–tuliptree–black oak type on gentle southeast slopes; *2,* Chestnut oak–black oak–tuliptree type of higher slopes.

	1	*2*
Number of trees	181	41
Quercus alba	53.0	9.7
Liriodendron tulipifera	18.8	17.1
Quercus velutina	12.1	17.1
Quercus montana	8.3	34.1
Carya spp.	3.8	14.6
Castanea dentata (stumps)	1.1	4.9
Nyssa sylvatica	1.6	. .
Quercus borealis maxima	1.1	. .
Robinia Pseudoacacia	. .	2.4

mesophytic communities in ravines of the lower slopes. Beech is essentially absent; sugar maple is occasionally localized in the most favorable sites very near the base of the mountains. White oak appears to have been the dominant species on average slopes, and chestnut oak on upper rocky slopes. Bear oak communities are frequent on ridge crests. The dominance, in the ground layer of the secondary slope forests, of a typical oak-forest herbaceous vegetation, indicates the prevalence, in the original cover, of oak forest over a considerable proportion of the area.

Distinctive vegetational features have developed in the Massanutten complex where the mountain slopes are of shale (Allard, 1946). Open shale barrens with expanses of bare shale and open straggly or park-like woodland of oaks (mostly *Q. montana* and *Q. borealis* var. *maxima*) or of oaks and pine (*P. virginiana*) with a few other trees are most characteristic. These suggest the "oak barrens" of the Knobs Border area of the Mixed Mesophytic Forest region (p. 120). Where some soil has accumulated—on the tops of shale knobs or on the lower slopes—closed forest communities develop.

In the secondary communities which cover most of the middle area (in the transition belt as well as on the more separated ranges), little or no indication is seen of mixed mesophytic communities. Groves of white pine occur on northerly slopes; scrub pine is abundant on shaly south slopes; young tuliptrees occupy some of the coves; oaks prevail over most of the slopes.

SOUTHERN VIRGINIA AND TENNESSEE

Farther south, the transition belt between the Oak–Chestnut and the Mixed Mesophytic region is narrow. Mixed mesophytic communities occur more frequently in ravines of ranges near the Appalachian Plateau than in those of ranges separated by broad expanses of valley (Harrisburg peneplain). In the area adjacent to the Cumberland Mountains there is some evidence of segregation into communities comparable to the association-segregates of the Cumberland Mountains—beech–maple–basswood–buckeye, white oak–beech–maple, white oak–chestnut, and higher slope chestnut communities.

White oak is less abundant than in the middle area. Chestnut is (or rather was) the dominant tree of mesic slope forests, as is true in the Southern Appalachians. However, the chestnut slopes were interrupted by mixed mesophytic coves. Hemlock communities occur locally.

A number of forest communities are rather well illustrated on the slopes of Salt Pond Mountain near Mountain Lake, Virginia. The prevailing slope forest was dominantly chestnut (before the chestnut was killed). Red oak and white oak were the principal associates, forming about 14 to 45 per cent of the canopy (Table 47). As red oak was second in abundance

Table 47. Oak–chestnut forest communities on the slopes of Salt Pond Mountain, Mountain Lake, Virginia. (Data secured in 1932.)

Number of trees	117	50
Castanea dentata	84.6	56.0
Quercus borealis maxima	11.1	22.0
Quercus alba	2.6	22.0
Acer rubrum	.8	. .
Magnolia acuminata	.8	. .

to chestnut in the reproduction, the future forest may be one dominated by red oak, for growth will be rapid with release of the young individuals already present. (Red oak is dominant on some of the higher slopes.) White oak, red maple, sweet birch, cucumber tree, service berry, striped maple, mountain holly (*Ilex montana*), and witch-hazel are other species of the small tree and tall shrub layer. Species of *Azalea* and *Vaccinium* are present although the undergrowth is not dominantly ericaceous.

Hemlock forest in which Rhododendron is abundant occupies sheltered coves and slopes bordering the lake. This is essentially the same as the hemlock–Rhododendron community of Ravenel's Woods in the Southern Appalachians (cf. Table 44, area 3). The northern species which are present (as *Oxalis montana, Trillium undulatum, Clintonia borealis,* and *Betula lutea*) are equally a part of mountain forests southward.

On windswept ridge crests, dwarfing is sometimes conspicuous. Almost any of the species of the slope forest may be present in the summit scrub. Not only is stature greatly reduced, but leaf size is also only a fraction of normal. Bear oak (*Q. ilicifolia*) is dominant in some places, birch (*B. lenta*) in others. White oak, red oak, sour gum, sassafras, and chestnut—fruiting at heights of 5 to 10 feet—are common. The large number of shrub and herbaceous species is in contrast to the condition in secondary bear oak communities on much disturbed and frequently burned summits (p. 235).

A beech–white oak or white oak–beech community of lower mountain slopes of southerly exposure is indicated by poor remnants and scattered trees on Clinch Mountain and Copper Ridge. Scattered areas of beech woods in which there are occasional buckeyes suggest mixed mesophytic communities and the possibility that early culling removed the more desirable species.

The southernmost part of the section (vicinity of Knoxville and southward) lacks the mountain ridges so characteristic of all the area northward. Its low ridges and knobs are a result of dissection of the valley floor, hence their vegetation will be considered with that of valleys of the section.

PENNSYLVANIA

The Ridge and Valley section reaches its greatest width in Pennsylvania. The eastern (or southeastern) one-fourth to one-third of the section is valley —the Great Valley—while the wider western (or northwestern) strip contains a large number of ridges and narrower valleys.

Most of the ridges are covered with secondary forests; in only a few places are there remnants of primary stands. As farther south, oaks prevail in most of these secondary communities. Hemlock and white pine are, however, much more frequent than in the secondary communities of ridges farther south, often forming dense pure stands. Sweet birch is dominant on many rocky upper slopes. A number of more or less distinct oak communities are seen; red oak, white oak, chestnut oak, and scarlet oak are the principal dominants in the several oak communities. Chestnut was a constituent of most of the oak communities, both primary and secondary, but in what proportion it is now impossible to determine, for this area is near to the original center of infection by chestnut blight, and by 1930, 100 per cent of the trees were infected and 51 to 100 per cent dead (Frothingham and Stuart, 1931, map, p. 50, prepared by Office of Forest Pathology, Bureau of Plant Industry). Beech is almost absent even in the more mesophytic communities, although it is a constituent of forests of ravine slopes cut below the Harrisburg peneplain.

Where the valleys (of the Harrisburg cycle) between the ridges are narrow, or where the old valley floor has been considerably dissected, the slopes of ravines of the latest erosion cycle may not always be readily distinguished from the mountain slopes. The forests of these lowest slopes are essentially mixed mesophytic communities. They may contain a large proportion of hemlock and white pine, or may be prevailingly or entirely deciduous, with beech, basswood, sugar maple, tuliptree, ash, red maple, and red oak among the canopy species. In the abundance of hemlock and white pine, in the absence of buckeye, and in the lesser number of canopy species usually present, these mixed mesophytic communities are more or less transitional to the northern forests of the Appalachian Plateau to the north (Allegheny section of the Hemlock–White Pine–Northern Hardwoods).

The most mesic community of the mountain ranges of this part of the section occupies the higher mountain valleys; it is a hemlock or hemlock–white pine community in which Rhododendron is abundant or dominant in the undergrowth. This is represented by numerous secondary and disturbed stands, as well as by a few isolated primary stands among which the Alan Seeger and Detweiler Run State Forest Monuments are outsanding (p. 236). This is not the prevailing forest type, nor is there any indication that it is the regional climax. White pine formerly was abundant on some of the mountain slopes, and there is evidence that a mixed oak–chestnut–white pine community was more or less widespread especially on northerly slopes. Some such areas are now occupied by poor secondary stands of oaks, red maple, birch, and tuliptree. Hemlock, too, is more or less scattered through the mixed oak forests of northerly slopes. The admixture of white pine and hemlock is more noticeable in the northern part of the section and suggests the transition to the Hemlock–White Pine–Northern Hardwoods of the Appalachian Plateau adjacent to the north.

Comparable to the red oak–sugar maple–basswood community of the Northern Blue Ridge (p. 224) are the red oak–birch and red oak–basswood communities such as are seen in the mountains north of Chambersburg. These vary with exposure and position on the slopes; birch is more abundant on the uppermost slopes, becoming dominant in some situations. *Acer spicatum* and *Acer pensylvanicum* are common species of the undergrowth. Lower, birch drops out, red oak increases, and white oak enters, together with a variety of species including hickories, ash, and tuliptree.

In the district of the Seven Mountains (southeast of State College, Pa.), a broad belt of mountain ranges on closely folded strata where the very resistant Silurian sandstones are responsible for the ridges, an interesting variety of mountain communities may be seen. This area will be used to illustrate the variety of communities possible in a limited area; not all of these will be seen on all the mountains, nor are all of the communities of the Pennsylvania ridges present here. Most of the forest communities are secondary, although a few primary areas remain on dry upper slopes and rock slides where chestnut oak or birch are dominant, and in mountain valleys where, in the Alan Seeger and Detweiler Run State Forest Monuments, fine areas of hemlock, white pine, and oaks are preserved.

The general impression is one of oak slopes interrupted here and there by groves of pine and by hemlock or red maple ravines. A variety of oak communities include mixed oak, white oak–chestnut oak, chestnut oak–scarlet oak–black oak, scarlet oak or scarlet oak–pitch pine, chestnut oak, and bear oak. Those in which white oak or scarlet oak are abundant are usually at lower elevations; dogwood is abundant in the understory in white oak communities. Chestnut was a constituent of all these communities.

The scarlet oak or scarlet oak–pine communities occupy the driest slopes, often with westerly or southerly exposure. Heaths (*Kalmia, Vaccinium, Gaylussacia, Gaultheria,* and *Epigaea*) are common, and sweet fern (*Comptonia peregrina*) a frequent associate. In denser stands, especially those of sprout origin, the undergrowth is very sparse.

The white oak communities are generally on less steep or northerly slopes. White oak may be dominant, or one of the species in mixed oak communities in which white oak, chestnut oak, black oak, red oak, scarlet oak, and red maple are associated. These may or may not have an ericaceous layer; *Hamamelis virginiana, Viburnum acerifolium,* and *Corylus cornuta* occur frequently. Many of the younger secondary or sprout forests have almost no undergrowth.

Typical chestnut oak communities occupy exposed and rocky mountain slopes, and rock slides more or less soil-covered. Some of these are primary stands; the low-branching, broad-crowned trees of such sites were not desirable to the lumberman. Chestnut oak is the dominant species; chestnut, as indicated by old stumps, was second in abundance. Among other species of this chestnut oak community are sweet birch, red oak, red maple, sour gum, and white pine. The young tree layer is composed of the same species (with chestnut oak dominant), and *Acer pensylvanicum* almost always present. The continuity of the ericaceous layer (made up chiefly of *Kalmia, Vaccinium,* and *Gaylussacia*) is conspicuous in all except the denser stands, and the paucity of species pronounced. The secondary chestnut oak community (of similar sites) is marked by the dominance of uniform-sized chestnut oaks; its shrub layer is much less continuous because of the more even and closed canopy of the secondary stand. Some mixed oak stands in which chestnut oak is the dominant species occupy less exposed sites. These usually have a thick leaf litter (chestnut oak leaves decompose very slowly) and very poor shrub layer; herbaceous plants are almost lacking.

A dwarf tree community (about 10 feet in height) in which bear oak is dominant occupies some of the sandstone ridge-tops. With the bear oak are dwarf chestnut oak, chestnut, sassafras, witch-hazel, pitch pine, and a scattering of smaller ericaceous shrubs. This suggests the dwarf forests of ridge-tops in southern Virginia (p. 233), but the contrast in paucity of shrub and herb species indicates its secondary character.

The sweet birch rock-slide community occupies the most tumbled rocky slopes where no soil has accumulated. There is no shrub layer, although there may be a few scattered individuals of *Kalmia, Acer pensylvanicum,* and *Acer spicatum.* The birch pioneers in such situations, and retains dominance to maturity. Chestnut oak enters in the most favorable spots. This, together with the fact that the chestnut oak community frequently occupies partially soil-covered rock slides indicates the gradual replacement of the birch community by chestnut oak.

Pine communities are represented by small groups of old trees—white pine and pitch pine—and by secondary scrub pine (*P. virginiana*) communities with which pitch pine and pignut hickory are associated. Except for occasional smaller individuals there is no undergrowth other than a few herbaceous plants (as *Danthonia spicata, Antennaria plantaginifolia, Potentilla simplex*), cushion mosses, and Cladonia.

More or less flat-bottomed mountain ravines are occupied by red maple or hemlock communities. The red maple community usually has an admixture of hemlock and sometimes of white pine. With the red maple are red oak, tuliptree, basswood, sweet birch, ash, and hickory. Where hemlock is present, Rhododendron is abundant in the understory.

The hemlock forest type of ravines is represented by dense second-growth stands of almost pure hemlock, hemlock–white pine, hemlock–red maple, or hemlock–Rhododendron communities. The best of these occupy broad flat-bottomed ravines in which the stream is but slightly entrenched. In primary stands white oak grows with the hemlock, while on slightly higher bordering slopes it is the dominant species. The Alan Seeger State Forest Monument well illustrates these two primary forest communities (Table 48). The hemlock type occupies the lower flats; hemlock, white oak, and red maple are the most abundant canopy species, together composing

Table 48. Composition of two communities of the Alan Seeger State Forest Monument, Seven Mountains, Pennsylvania: *1,* Hemlock–Rhododendron community of the lower ravine flats; *2,* Oak forest of slightly higher parts of the ravine flats.

	1		*2*	
	Canopy	*Second layer*	*Canopy*	*Second layer*
Number of trees	83	61	33	30
Tsuga canadensis	37.3	29.5	. .	13.3
Quercus alba	16.9	4.9	39.4	70.0
Quercus borealis	6.0	11.5	27.3	6.6
Acer rubrum	14.5	8.2	9.1	3.3
Pinus Strobus	7.2	9.8	15.1	3.3
Carya glabra	6.0	. .	. .	. .
Nyssa sylvatica	6.0	1.6	3.0	. .
Betula lenta	3.6	14.7	. .	. .
Magnolia acuminata	1.2	13.1	. .	. .
Liriodendron tulipifera	1.2	3.3	. .	. .
Fraxinus americana	. .	1.6	3.0	. .
Prunus serotina	. .	1.6	. .	. .
Acer saccharum	. .	. .	3.0	. .
Quercus montana	. .	. .	. .	3.3

about 70 per cent of t[illegible]odendron is abundant; the ground vegetation is sparse an[illegible]s but few species—*Maianthemum canadense, Cypripedium acau*[illegible]*olygala pauciflora, Viola blanda, Mitchella repens,* and *Dryopteris noveboracensis.* On slightly higher parts of the ravine flat, white oak is dominant; red oak and white pine are the principal associates, these three species comprising about 80 per cent of the canopy. The understory is open; Rhododendron is lacking. A mixed evergreen-deciduous community of hemlock, white pine, white oak, tuliptree, and chestnut (represented by old dead trunks) occupies the northerly slopes rising above the ravine flat.

The Detweiler Run State Forest Monument is a ravine slope forest somewhat comparable to, but much more extensive than the slope forest above the hemlock flats of Alan Seeger. It is a mixed forest of hemlock, white pine and deciduous species which, on the northerly slope, extends almost to the ridge crest. On the southerly slope it gives way to a chestnut oak rock-slide community.

A relic bog known as Bear Meadows State Forest Monument is an unusual feature of the Ridge and Valley section (p. 481). In appearance and floristics it resembles the bogs on the Appalachian Plateau in the Hemlock–White Pine–Northern Hardwoods region. Extensive sedge communities (increased by damming of the bog outlet), bog shrub communities (in which *Nemopanthus mucronata* and *Chamaedaphne calyculata* are conspicuous), bog forest border (in which are spruce, fir, hemlock, white pine, red maple, sour gum, etc.), hemlock–Rhododendron zone, and mixed oak forest on bordering slopes demonstrate the developmental stages in this bog succession which lead to the establishment of the physiographic climax of ravine flats and the regional oak forest. Very few such bog areas are known in this section; Blodgett (1910) mentions bog pockets on North Mountain in Maryland; boggy flats occur in the vicinity of Mountain Lake on Salt Pond Mountain, Virginia.

Forest Communities of the Valley Floors (Harrisburg Peneplain Level)

Throughout the entire north-south extent of the Ridge and Valley section, the dominance of white oak in the forest communities of the valley floor is a unifying character. Although almost all of the area is in cultivation, this feature is obvious in the numerous scattered old white oaks near towns, farmhouses, and school buildings, in the occasional white oak groves, and in the rare stands of little-disturbed forest.

The valley floors of the Harrisburg peneplain level cannot be thought of as plains, although some fairly extensive level areas are seen. Much of the land is slightly rolling; some, especially near the larger streams and toward the south end of the section, is more or less maturely dissected, but with low relief. Many low ridges or swells rise above the general level of the valley floors. In spite of these topographic variations, the vegetational uniformity is pronounced.

While white oak is the dominant species, and frequently the only one remaining, there is ample evidence to indicate that few, if any, pure stands actually occurred in the original forest cover. Frequent accompanying species include tuliptree, especially on the low swells, hickories, a number of oaks (red, black, and at the southern end of the section, Spanish oak), and white pine (in the northern part of the section). In the better stands, white oak is represented in all size classes, indicating the climax nature of the valley white oak forest type. These white oak woods differ greatly from those of the mountain slopes in the non-ericaceous character of the undergrowth.

Where low shaly ridges rise slightly above the valley floors (ridges of a type which cannot be included with the mountain range ridges of the section), black oak is generally more abundant, and the forest community may be considered as white oak–black oak–hickory. It may contain scarlet oak, chestnut oak, tuliptree, and a number of other species, thus suggesting the mixed oak forest of the mountain slopes. In the north, white pine, and sometimes pitch pine and scrub pine; in the south blackjack and Spanish oaks, and southern pines are constituents of the woodland, the latter abundant in young secondary stands.

The white oak community of the level or slightly rolling valley floor is particularly characteristic of this topographic division of the Ridge and Valley section. Although the universal presence of this community is often indicated only by old white oaks, the better stands contain a variety of species in addition to the dominant white oak. More or less representative areas in the northern part of the section (where white pine is an infrequent constituent) are seen in and near State College, Pennsylvania. Beneath the canopy, which is made up of white oak with black, red, and scarlet oaks, occasional white pine and hickory, are young trees of the same species, together with a few additional species—dogwood, wild cherry, *Sassafras, Crataegus,* and *Amelanchier.* The shrubby undergrowth includes *Cornus racemosa, Ceanothus americanus, Sambucus canadensis, Rubus occidentalis, Rosa* sp., *Vitis, Celastrus scandens, Parthenocissus quinquefolia,* and *Rhus radicans,* most of which are somewhat indicative of a limestone substratum, as are also some of the herbaceous species. Among the plants common in the herb layer are *Geranium maculatum, Uvularia perfoliata, Smilacina racemosa, Desmodium* spp., *Amphicarpa bracteata, Aster* spp., and *Solidago caesia.*

A splendid example of the white oak forest was seen south of Berryville, Virginia, in the Shenandoah Valley (Table 49). While only about 63 per cent of the canopy trees in the sample are white oak, no other trees are as

Table 49. Composition of a white oak forest in the Shenandoah Valley (Harrisburg peneplain) near Berryville, Virginia.

	Canopy	*Second layer*	*Small tree layer*
Number of trees	110	44	
Quercus alba	62.7	40.9	x
Carya tomentosa	18.2	36.4	x
Quercus velutina	8.2	. .	x
Carya cordiformis	5.4	6.8	x
Quercus borealis maxima	2.7	. .	x
Juglans nigra	1.8	4.5	x
Quercus Muhlenbergii	.9	. .	. .
Nyssa sylvatica	. .	4.5	x
Morus rubra	. .	2.3	x
Fraxinus americana	. .	2.3	x
Ulmus fulva	. .	2.3	x
Cornus florida	. .	. .	XX
Celtis occidentalis	. .	. .	x
Celtis pumila georgiana	. .	. .	x
Cercis canadensis	. .	. .	X
Crataegus sp.	. .	. .	x
Prunus americana	. .	. .	x
Prunus serotina	. .	. .	x
Sassafras albidum	. .	. .	x
Viburnum prunifolium	. .	. .	x

x, species present in understory; X and XX, subdominant species.

large as the white oaks. In appearance, this is a white oak forest. White oak is likewise the most abundant species of the middle-sized trees, and is well represented in the small tree layer. All evidence points to the continued dominance of white oak. The largest trees, 3 feet and over d.b.h., are widely spaced and others not so large are well scattered or sometimes grouped. The small tree layer is dense, with *Cornus florida* the dominant species. The shrubby and herbaceous layers are similar to those of the preceding areas, but a larger number of species is present because of the lack of disturbance.

The dominance of white oak on the Harrisburg peneplain in the Ridge and Valley section is not without parallel in other forest regions. It should be recalled that white oak communities were noted on the tongues of Lexington (Harrisburg) peneplain extending into the Knobs Border area of the Mixed Mesophytic Forest region, and in a large area of the Low Hills Belt adjacent to the Teays drainage from northeastern Kentucky to the Pittsburgh area (pp. 94, 121). It will be seen that white oak is a dominant in large areas of the Piedmont Plateau—Piedmont peneplain of Harrisburg age (p. 264). The problem involved in the apparent correlation of white oak forest and Harrisburg peneplain will be discussed later (Chapter 17).

Forests of Slopes Along Streams Indenting the Harrisburg Peneplain

In many places streams have entrenched their valleys below the valley floors (Harrisburg peneplain) of the Ridge and Valley section. Along these rejuvenated streams are slopes produced in a later erosion cycle. The soils are distinct from the old soils of the valley floor. Forest communities on these slopes are totally unlike those of the valleys whether level or rolling, and unlike those of the mountain slopes.

Mixed mesophytic forest communities occupy the slopes of valleys cut into the Harrisburg peneplain. They occur locally throughout the entire north-south extent of the section. Beech is an almost universal constituent of these mixed mesophytic communities, although often present in small numbers. In the Ridge and Valley section it is largely confined to such situations—except in the transition belt near the western border of the section, and in the ravines of the southern ranges. *Tilia heterophylla* and *T. neglecta, Acer saccharum, Liriodendron, Quercus borealis* var. *maxima, Quercus alba, Fraxinus americana,* and *Juglans nigra* are frequently present. These mixed mesophytic communities are usually small in extent and often fragmentary. Only rarely are areas seen in which any considerable proportion of the twenty odd canopy species recorded in the fragments occur. The gray-brown (melanized) forest soil and rich mull humus so characteristic of the Mixed Mesophytic region have developed here. The herbaceous layer is luxuriant and contains many of the species common in the mixed mesophytic forest. It contrasts strongly with that of the white oak forests of valley floors and of the oak–chestnut communities of the mountain slopes. No ericaceous species are present in herb or shrub layers. Ferns and spring geophytes are abundant.

One of the best areas of mixed mesophytic forest in this topographic division of the section is seen in the gorge of Cedar Creek at Natural Bridge, Virginia (Table 50). Seventeen canopy species, of which three (basswood, sugar maple, and tuliptree) compose a little over 50 per cent of

Table 50. Composition of the canopy of a mixed mesophytic forest community in the gorge of Cedar Creek at Natural Bridge, Virginia (sample of 182 canopy trees).

Tilia heterophylla	22.0	*Quercus Muhlenbergii*	2.2
Acer saccharum	17.6	*Fagus grandifolia*	1.6
Liriodendron tulipifera	13.2	*Ulmus fulva*	1.6
Quercus borealis maxima	9.3	*Juglans nigra*	1.6
Tsuga canadensis	8.8	*Pinus Strobus*	1.1
Fraxinus americana	7.1	*Nyssa sylvatica*	.5
Carya spp.	7.1	*Juglans cinerea*	.5
Quercus alba	2.7	*Thuja occidentalis*	.5
Magnolia acuminata	2.2		

the stand, are present. One (arbor vitae) is an accidental, being generally localized on limestone outcrops. Another phase of mesophytic forest, one in which hemlock is dominant and beech second in abundance, is confined to small areas of the ravine flat. It is quite comparable to the hemlock–mixed mesophytic communities of ravines in the Mixed Mesophytic Forest region. As there, ground vegetation is sparse and contains a few evergreen species (*Mitchella repens* and *Asarum virginicum*). Where the valley is wider and southerly slopes thus exposed to the sun, a xeric community of red cedar and oaks with a few of the species of the mixed mesophytic forest (sugar maple, tulip, and red oak) is seen.

Local Edaphic Communities

On many limestone outcrops red cedar is abundant, often forming definite red cedar communities in which oaks, hophornbeam, and redbud are accessory species. One such area is referred to by Cain (1931) as juniper woodland. One was mentioned above in the Natural Bridge area. On the lowest slopes of the Blue Ridge along the Shenandoah Valley—where limestones of the Great Valley are tilted up against the older crystalline rocks of the Blue Ridge—red cedar is abundant. Almost anywhere that limestone outcrops it may be expected.

Open red cedar communities (if primary) always contain patches of xeric herbs. Species commonly occurring include *Asclepias tuberosa, A. verticillata, Monarda fistulosa, Salvia lyrata, Aster oblongifolius, Kuhnia eupatorioides, Agave virginica,* and a few tall grasses, especially *Andropogon scoparius.* Scattered shrubs and stunted trees (*Rhamnus caroliniana, Rhus aromatica, Cercis canadensis, Ostrya virginiana, Juniperus virginiana*) indicate the ultimate though exceedingly slow replacement by forest.

Résumé

A brief résumé of the outstanding features of the forest vegetation of the Ridge and Valley section will emphasize the prevalence of oak (originally oak–chestnut) communities on the mountain slopes, and mesophytic hemlock, hemlock–white oak, or hemlock–white pine–oak physiographic climax communities in the mountain valleys; the dominance of white oak forest on the valley floors (Harrisburg peneplain); and the local but widespread occurrence of mixed mesophytic communities on the ravine slopes formed in the latest erosion cycle. The great difference in climax communities on the three topographic divisions of the section raises a question as to the nature of the regional climax, if indeed, a region of such diverse topography can have a regional climax. If the most mesophytic community is to be considered the regional climax, then mixed mesophytic forest should be thought of as the climax. That it is nowhere reached on the mountain slopes may be due in part to the inhibiting effects of soil as related to underlying rock and in part to physiographic history. The same may be said of valleys, except that the soil is an old soil of an upraised peneplain, rather than one being derived from the generally siliceous rocks of the mountain ranges. Here also, physiographic history seems to have had a profound effect upon the nature of present-day vegetation. On the ravine slopes formed in the latest erosion cycle, the vegetation is developing in response to present forces both topographic and climatic. Although very limited in extent, it seems logical to assume that mixed mesophytic forest is the potential climax of the area; that by its development the extent of the mixed forest, greatly restricted at one or more times during the Tertiary and early Pleistocene, may expand eastward into what we now know as the Oak–Chestnut region. The outliers are forerunners in a development which would take thousands of years to complete, for it must await the development of land surfaces and of soils no longer related to the Harrisburg cycle. That these mixed mesophytic communities are not relics of a former more extensive mixed mesophytic region (mixed forest of the Tertiary) seems certain because of their limitation to surfaces produced in the last (or present) erosion cycle. These problems are discussed in Part III.

PIEDMONT SECTION

The Piedmont section extends from southern Virginia, where it is a narrow belt adjacent to the Blue Ridge, northward, widening in northern Virginia, and north of the Potomac River including the entire width of the Piedmont Province and an adjacent strip of Coastal Plain. The Piedmont section terminates at the north, some distance south of the glacial boundary[20] and approximately at the northern boundary of the Piedmont Upland of southeastern Pennsylvania.

Physiographically, the narrow belt in Virginia is Inner Piedmont, where "abundant monadnocks are confined to a fairly definite belt, generally 15 to 20 miles wide, at the foot of the Blue Ridge. This belt differs geologically from the smoother lands to the east" (Fenneman, 1938, p. 139). It also differs vegetationally from the land to the east, agreeing rather with the Blue Ridge in some of its forest characters.

The strip of Coastal Plain in Maryland, Delaware, and New Jersey which is included in the Piedmont section coincides approximately with the area of Cretaceous sediments. It is the Tension Zone of Hollick (1899) and the middle district of Stone (1908, 1911). Its outward (or seaward) boundary is, however, more irregular than can be shown on a small scale map. It is in part related to Pleistocene terraces, the older of which are occupied by forest vegetation more closely related to that of the Piedmont Plateau than to that of the Coastal Plain. Soil differences between the

[20] The boundary is a vegetational one; for explanation of its location see p. 248.

younger outer part of the Coastal Plain and the older inner (and generally higher) parts help to emphasize the vegetational boundary. This part of the Piedmont section in New Jersey which is in the Coastal Plain Province of physiographers is often referred to by the local inhabitants as "Piedmont." It is so referred to by Harshberger (1911, p. 461).

Very little primary forest vegetation remains in the Piedmont section. Occasional groups of old trees are doubtless remnants of primary stands. Some areas have been continuously forested although modified since the time of the early settlements by cutting for timber and for firewood, by grazing, and by fire, as well as by the killing of chestnut by disease. Other areas were once cleared and their forests have started anew. The variety of secondary communities is, therefore, great and because of the long period of human utilization and modification, these differences frequently indicate little as to original differences in the forest cover. Only intensive local studies which take into consideration the nature and period of utilization as well as the local environmental factors can bring the maze of communities into a semblance of order. Very little literature is available for any part of the Piedmont section.

Virginia

In general, the usual Piedmont woodland is composed predominantly of oaks—chestnut, white, black, red, post, and blackjack—with white oak more frequent toward ravines, and post oak and blackjack oak most abundant in wide interstream areas. Dogwood is usually prominent. Among the great number of forest communities distinguishable in the hilly Inner Piedmont of Virginia may be mentioned white oak–black oak, white oak–black oak–tulip, chestnut oak, and chestnut oak in combination with other oaks, pine and pine–oak communities. These latter are suggestive of the Pine–Oak region adjacent to the southeast. A more or less representative and fairly old example of the white oak–black oak community contains white and black oaks as dominants, and tuliptree, red maple, and sour gum as accessory species of the canopy; and in the understory dogwood, black oak, white oak, tuliptree, hickory (*C. tomentosa*), red maple, sour gum, and wild cherry. The shrub layer is made up of heaths (*Azalea, Vaccinium, Gaylussacia*), *Castanea pumila, Viburnum prunifolium, Ilex verticillata,* and *Lindera Benzoin.* The herbaceous growth is characteristically that of oak woods, including species of *Desmodium* and *Lespedeza, Cassia nictitans, Amphicarpa, Baptisia tinctoria, Viola pedata, Gerardia, Hieracium,* and *Solidago caesia.* In some stands, the tuliptree is much more abundant, becoming codominant with the oaks. White oak, black oak, scarlet oak, sour gum, and hickory are combined in another variant of the white oak–black oak community. Younger second-growth stands of more uniform-size trees are frequent, with various combinations of species; for example, scarlet, chestnut, white, and black oaks, sour gum, red maple, and dogwood, with a low shrub layer of *Vaccinium* and *Gaylussacia;* scarlet, black, white, post, shingle, and southern red oaks, scrub and pitch pines, sour gum, tuliptree, hickory; black, white, scarlet, post, blackjack, and willow oaks, sassafras, hickory, tuliptree, red maple, and dogwood.

Old-field communities of *Andropogon* and *Sorghastrum* with scattered pines, or of black locust, sassafras, and persimmon are frequent. Local moist areas on valley flats are quite different, with pin oak, red maple, willow oak, swamp white oak, and sweet gum. These are more frequent in the Piedmont Lowland.[21] River birch (*Betula nigra*) is common along streams.

One of the most mountainous parts of the Inner Piedmont is the Bull Run-Pond Mountain area in northern Virginia. In this area (discussed by Allard and Leonard, 1943), the original forest of the

> highlands was Oak–Chestnut in composition, the chestnut oak, *Quercus montana,* being the major dominant, with the chestnut, *Castanea dentata,* formerly occupying a high place in its composition, together with red oak, *Quercus rubra* [*Q. borealis* var. *maxima*], black gum, *Nyssa sylvatica,* tulip poplar, *Liriodendron tulipifera,* and hickory (*Carya*).

Due to elimination of chestnut, this is changing to an oak–hickory forest. Chestnut oak sometimes forms almost pure stands on the ridges. Table mountain or knob cone pine (*Pinus pungens*) occurs on high windswept ridges; white pine is local; pitch pine is an invader in pastures but is less frequent than scrub pine. Red cedar, also, is a frequent invader of old pastures. The most mesophytic forest community of the area occupies the deeper and moister soils of lower slopes. It is made up of white, chestnut, red and black oaks, sour gum, hickory, tuliptree, white ash, and some beech. This is the nearest approach to a mixed mesophytic community (which it is called by Allard and Leonard) in the Inner Piedmont of Virginia; it differs markedly from that association in the large number of oaks and in the presence of black oak; only the beech suggests the mixed mesophytic.

North of the Potomac River

The Piedmont section north of the Potomac River lacks the hilly or mountainous Inner Piedmont belt seen adjacent to the Blue Ridge in Virginia. Instead, adjacent to the Blue Ridge, there is an area known as the Piedmont Lowland (Lancaster-Frederick Lowland), which is now almost entirely utilized. Nothing remains of primary vegetation; scattered groves of old white oaks are evidence of the former prevalence of this species.

Scattered old trees and secondary woodlands (long modified by human disturbance) indicate something as to the nature of the forest cover in the

[21] Physiographers distinguish the Piedmont Upland and Piedmont Lowland, the latter extending southward from Pennsylvania in a tongue to the east of the hilly Inner Piedmont.

hilly and rolling Piedmont Upland of Maryland. Early in this century (before the chestnut blight eliminated chestnut) chestnut was abundant. Shreve (1910a) states that

> the commonest tree of the Lower Midland [Piedmont Upland west of Parr's Ridge] today is the Chestnut, but it is altogether likely that its ability to send up suckers as compared with the slow growth of oak seedlings, or even oak suckers, is responsible for its predominance.

In a tract which he considered to be near the average in composition, 35 per cent of the trees were chestnut, 30 per cent oaks (white and black), and 28 per cent hickories (mockernut, pignut, and bitternut). In some woodlands, chestnut was uncommon, and chestnut oak the leading species. With the elimination of chestnut, the oaks have assumed greater prominence. Tuliptree, which was occasionally present in the oak–chestnut woodland, has now taken a position of codominance with the oaks. All of the oak (or oak–tuliptree) woods are secondary, although some are fairly old stands. The similarity in composition of stands of different ages indicates that all represent the same successional stage. The tulip in all stands may in part be replacing dead chestnut, and if so, its percentage may be expected to decrease in later stages because of closing of canopy. Most of these secondary Piedmont woodlands have not had time to develop to a stage in which more tolerant species take part. However, small beech trees may be found in many of them, suggesting successional advance. On ravine slopes where a deep and rich soil has developed, beech is more abundant, and beech, tuliptree, and white oak often the prevailing species. The richest and most varied herbaceous growth, of a character contrasting strongly with that of the oak–chestnut (or oak) woodlands, is seen in these slope forests. Such communities are the Piedmont representatives of mixed mesophytic forest. They are particularly well developed along the Fall Line, where beech is more abundant than elsewhere.

The Piedmont vegetation extends out onto the Coastal Plain in the northern counties of both the Eastern and Western Shore districts of Maryland, where the upland forests are deciduous, with oak, chestnut, and hickories predominating. Thus they contrast with the evergreen upland forests of the more southern counties of these districts, where pines (loblolly and scrub) predominate, and chestnut is rare and confined to gravelly soils (Chrysler, 1910; Shreve, 1910b). These evergreen upland forests belong in the Oak–Pine region. The boundary between the two forest regions (Oak–Chestnut and Oak–Pine) is very irregular, for it is in part correlated with the boundary between two of the Pleistocene terraces (the Wicomico, which is higher, and the Talbot, which is lower). Shreve (1910b) has shown that on the Eastern Shore the dominant vegetation of the Talbot is related to that farther south, in tidewater Virginia and North

Carolina, while that of the Wicomico is related to the vegetation of the Piedmont Plateau.

The forests of ravine slopes of the Wicomico terrace are dominated by chestnut and chestnut oak where the soil is gravelly, while on loam soils a community of mixed mesophytic relationship is seen. In this the principal trees are beech, white oak, Spanish oak, sycamore, pignut hickory, red mulberry, wild black cherry, hackberry, and holly (Shreve, 1910b). Zon (1904), working in the Western Shore district, distinguished the forests of the tops of higher hills and gentle slopes where scrub pine prevails; of the more abrupt ravine slopes (middle slopes) where chestnut and other hardwoods predominate; and of the lower slopes where beech and yellow poplar (tuliptree) assume prominence, especially in the smaller size classes. Where the slopes are gentle these types overlap and a chestnut–pine type may develop (Table 51). Comparable types are distinguished in the "Coastal Plain hardwood region" of Delaware (Sterrett, 1908).

In the rolling Piedmont Upland of southeastern Pennsylvania, beech is more abundant than on the Piedmont farther south. It is seen in company with a variety of other species on many of the ravine slopes, although

Table 51. Composition of forest types (secondary) of southern Maryland (Western Shore district) as distinguished by Zon, 1904: *1,* Forest of middle slopes (average of seven acres); *2,* Forest of lower slopes (average of three acres); *3,* Forest of gentle slopes of broad ravines (average of six acres). (Data from Zon's Tables I, II, III.)

	1		*2*		*3*	
Inches d.b.h	12	2–11	12	2–11	12	2–11
Trees per acre	48.7	133.1	29.2	181.8	52.1	135.5
Chestnut	81.2	46.6	43.0	6.5	73.4	38.1
White oak	5.6	11.6	1.3	3.3	6.7	7.6
Beech	3.5	8.9	16.3	32.3	. .	. .
Red oak	5.6	8.0	8.9	4.7	2.6	2.0
Red maple	.9	6.9	1.3	4.7	. .	. .
Hickory	.3	6.8	2.5	18.6	.3	2.2
Sweet gum	.9	4.2	1.3	1.4	. .	.3
Yellow poplar	.3	1.8	16.3	18.0	. .	. .
Black gum	.3	1.7	5.1	3.7	. .	. .
Scrub pine	.9	1.3	. .	1.0	17.0	47.3
Persimmon	.3	.5	. .	. .	. .	. .
Witchhazel	. .	.4	. .	. .	. .	. .
Red cedar	.3	.3	. .	. .	. .	. .
Black cherry	. .	.3	. .	. .	. .	. .
Black walnut	. .	.2	2.5	1.0	. .	. .
Mulberry	. .	. .	. .	3.1	. .	. .
Sycamore	. .	. .	1.3	.6	. .	. .
White elm	. .	. .	. .	.4	. .	. .
Other species	. .	.3	. .	.6	. .	2.5

more commonly on northerly exposures. Harshberger in 1904 called attention to what he termed "mixed deciduous forest formation" in southeastern Pennsylvania, saying that the original forest was a mesophytic one and listing a large number of trees of which it was composed (see p. 40). He did not distinguish distinct communities. Sterrett (1908) also lists a number of species in the ravine forests of the Piedmont of northern Delaware. Gordon, in 1941, considering a part of Chester County (in southeastern Pennsylvania) states that forest of the "Chestnut Oak–Chestnut–Yellow Poplar type" is most extensive. In this, dogwood is most abundant in the understory. "A shrub layer of heaths, especially of dwarf blueberries and Pinxter (Pink Azalea), occasionally Mountain Kalmia, Aromatic Wintergreen and Spotted Pipsissewa show the extremely acid condition of the soil in the drier portions of the forest." Here, as in the Piedmont of Maryland, "there is unmistakable evidence of a mixed mesophytic cover" which occupied the ravine slopes, and bordered the water courses. The trees listed by Gordon in a typical stand south of West Chester are:

> Yellow Poplar, White Oak, Red Oak, Beech, Chestnut, Pignut Hickory, Shagbark Hickory, Red Maple, White Ash, Wild Black Cherry, American Hornbeam, and Flowering Dogwood. Red Elm, Black Walnut, Butternut, and Bitternut Hickory are also to be found in such habitats. Considerable more Beech occurs on north-facing slopes; more White Oak, Red Oak and Hickory are to be found on the south-facing slopes.

Something of this slope contrast (on very gentle slopes) may be seen in woodland of the Westtown School, where beech is very rare in an oak–hickory–tuliptree community on the south slopes, and abundant (moderate sized trees) in the north slope woodland which originally contained a considerable proportion of chestnut. On steeper slopes, the contrast is much greater. Such slope contrasts on Gulph Creek in the South Valley Hills were described by Harshberger (1919). The north-slope woods is a mixed mesophytic community with beech, tuliptree, red, white and black oaks, basswood, and chestnut, and a rich herbaceous layer, while the south slope is a dry oak–chestnut type with post, chestnut, blackjack, and black oaks, sassafras, chestnut, sour gum, and dogwood, and an ericaceous shrub layer.

The greater frequency of mixed mesophytic communities in southeastern Pennsylvania (northern part of Piedmont section) than farther south seems to be related to the distribution of forest types in pre-Pleistocene time (see Chapter 17).

Hemlock communities of small extent occupy some of the steeper north slopes along streams crossing the Piedmont of Maryland and Pennsylvania (Shreve, 1910a; Harshberger, 1904). These may be groves of hemlock, or hemlock in combination with other species, as beech, tuliptree, butternut, and red maple. Hemlock occurs in one locality in a slope forest of the Maryland Coastal Plain (Shreve, 1910b, p. 122).

Areas known as "serpentine barrens" occur locally in the Piedmont of Maryland and Pennsylvania where the prevailing granitic soils are interrupted by outcrops of serpentine. The appearance and composition of the vegetation change abruptly. Instead of woodland (or bluegrass pasture and farms) there is a grassland dominated by *Andropogon scoparius, Bouteloua curtipendula,* and *Sorghastrum nutans.* Although these are prairie dominants, the accompanying forbs are, for the most part, not plants of the prairie region. *Phlox subulata* is common and conspicuous.[22] The soil of these grasslands is a dark, almost black, magnesium rendzina resembling the rendzina soils of some prairies. It is in strong contrast with the yellow-brown and gray-brown soils of the region. Scattered red cedars, and groves of post oak or blackjack oak border the serpentine grassland. In the woodland of adjacent slopes black and white oaks are dominant. Succession, immeasurably slow, and dependent on the leaching of magnesium, results in a sequence of communities leading to the establishment of the regional vegetation.

GLACIATED SECTION

The southern boundary of the Glaciated section of the Oak–Chestnut region is the southern boundary of the area known to foresters as "central hardwoods" or as "sprout hardwoods," because of the abundance of reproduction by this method (Map, p. 249). This section thus extends a short distance south of the glacial boundary[23] in New Jersey and eastern Pennsylvania. It extends northward in the Hudson River Valley to beyond Albany, and eastward across southern New England.

Except on the west, the limits of this section bear little relation to physiographic boundaries. It includes parts of four physiographic provinces: Coastal Plain (Long Island and extreme southeastern Massachusetts and the older Cretaceous surface in New Jersey); Piedmont (extreme eastern Pennsylvania and northern New Jersey); Ridge and Valley Province (the Hudson-Champlain section from Pennsylvania to north of the Mohawk River); New England Province (southern part of New England Upland, including New Jersey and Hudson Highlands, and the southern end of Seaboard Lowland in Massachusetts and Rhode Island). The one unifying feature in the physiographic history of the greater part of the section is glaciation. Most of the area was covered by ice of the Wisconsin stage; smaller areas to the south of the Wisconsin terminal moraine in Penn-

---

[22] For more information concerning the serpentine barrens, see Shreve (1910a); Pennell (1910, 1912); Wherry (1932); and Gordon (1941).

[23] For map showing glacial boundary, see p. 498.

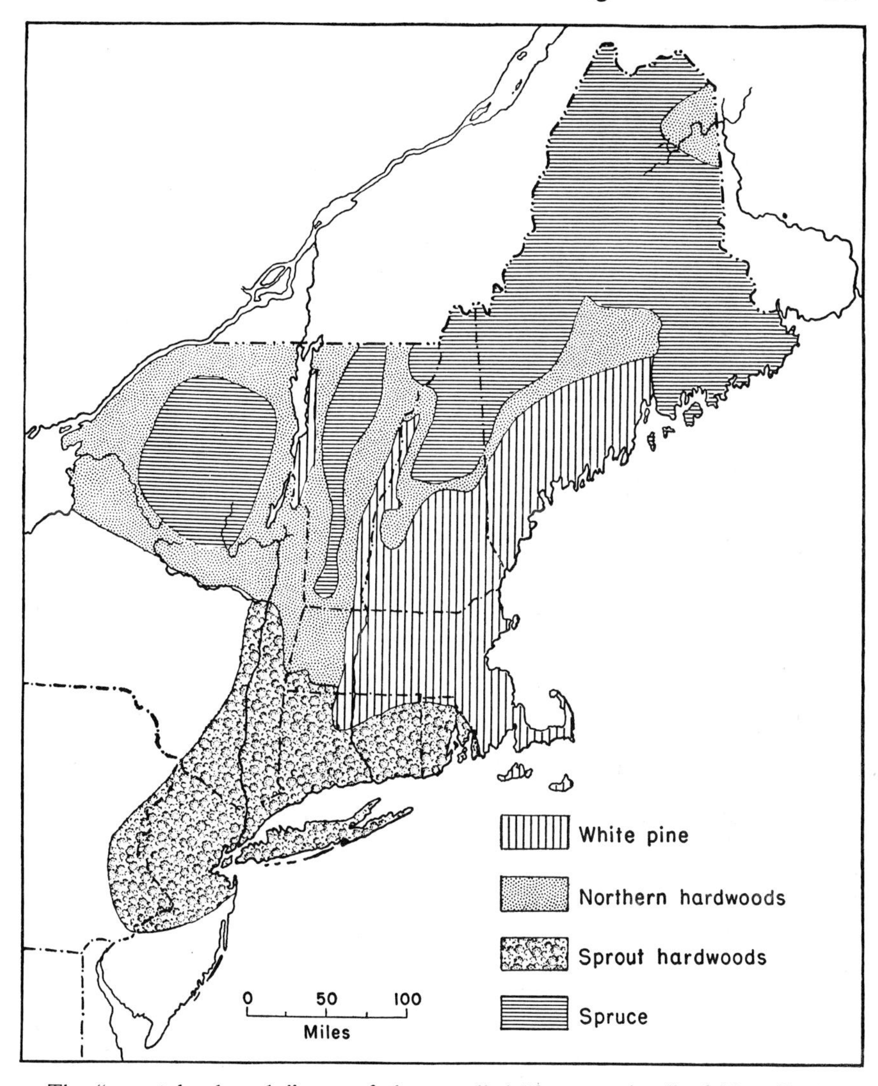

The "sprout hardwoods," one of the so-called "forest regions" of New England, characterize the Glaciated section of the Oak–Chestnut Forest region. (Map of "Forest Regions of New England," after Hawley and Hawes, 1912.)

sylvania and New Jersey were reached by Illinoian or by Jerseyan glaciation. The southern boundary of this vegetation section is not far to the south of the southern limits of these early ice advances.

Topography varies greatly in the Glaciated section. Although modified by glaciation, pronounced initial differences are due to geologic features. More or less rugged upland areas, as in the granitic districts, along the trap ridges and longitudinal ridges of the Ridge and Valley Province, contrast

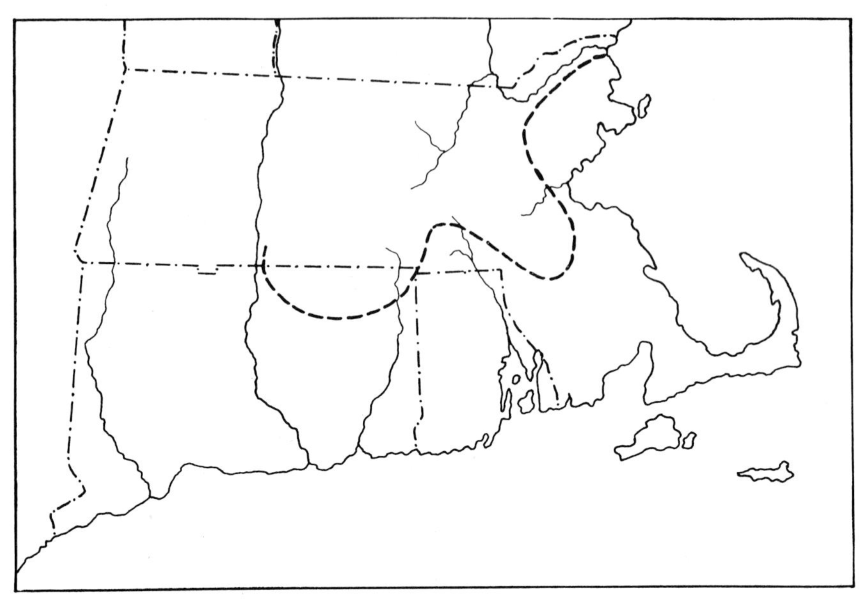

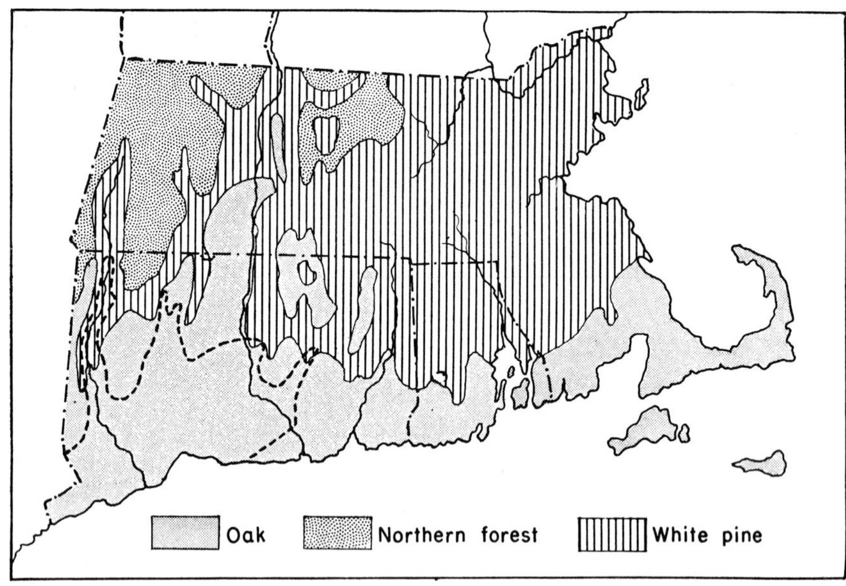

The boundary between more northern and more southern forest types in New England is indefinite, and hence maps differ. Above, the heavy broken line separates the red cedar–gray birch and the white pine old-field communities (after Raup, 1940). Below, an oak type is distinguished from more northern types; south of the broken line, mixed mesophytic communities occur (after Bromley, 1935).

strongly with the smoother lowlands. The smooth outwash plains of Long Island, the sand plains of Connecticut and southeastern Massachusetts are quite different from the usually rolling lowlands. Local features in the vegetation of this section are in large part related to topography; no general account can include the many variations.

The vegetational contact along the west and north is with the Hemlock–White Pine–Northern Hardwoods region. The transition is fairly abrupt to the west of the Hudson River where the contact is along the margin of the Appalachian Plateau. Across southern New England the boundary is indefinite. This is shown by the lack of accordance of maps. Hawley and Hawes (Map, p. 249) include all but a small part of northern Connecticut and Rhode Island and none of Massachusetts (except the southwestern corner) in their "sprout hardwoods" region, which is characterized by the abundance of chestnut and of oaks (red, white, black, scarlet, and chestnut). Bromley (Map, p. 250) places the boundary between the "oak region" and the "white pine region" farther south, excluding northwestern and northeastern Connecticut from the oak region (which extends north in the Connecticut Valley well into Massachusetts) and the northern half or more of Rhode Island, and including southeastern Massachusetts. Lunt (1938) also recognizes an "oak region of Connecticut and Rhode Island." Nichols' statements (1913, 1914) imply that he considered all of Connecticut except the northwest part to belong to the area in which a mixed forest predominantly of chestnut and oaks is climax. His map (1935) showing the boundary between the Deciduous Forest and the Hemlock–White Pine–Northern Hardwoods Forest places over half of Connecticut in the latter region. It is, however, too generalized and on too small a scale to be taken as indicating the exact position of this boundary. Raup (Map, p. 250) draws a boundary (transition) between red cedar–gray birch and white pine old-field communities across northeastern Connecticut and eastern Massachusetts and suggests that this boundary "may be a reflection of fundamental differences in growing conditions which were effective in the primeval forests of this region." All are in agreement in distinguishing between a more southern forest type in which oaks and (formerly) chestnut predominate, and a more northern type or types in which white pine is more or less abundant, and hemlock, beech, sugar maple, and yellow birch common.

Sprout Forests

The forests of the Glaciated section of the Oak–Chestnut region are prevailingly secondary communities of sprout origin. In appearance these are usually even-age stands, in which some 25 or 30 species may be represented. The trees are mainly of sprout origin, the sprouts having originated after clean-cutting; occasional trees from seed are included in the stands

(Schwarz, 1907; Frothingham, 1912). The ability to sprout appears to be retained indefinitely; hence, sprout woods may be one, two, or even three times removed from the old-growth stand from which they originated. In most parts of the deciduous forest, the ability to sprout is displayed only by young trees and by a few species. The sprout habit is, however, characteristic of chestnut everywhere, although here it is more prolific than elsewhere. Nichols (1913) mentions a chestnut stump with 375 coppice shoots; only a small number of these could grow to any size. Schwarz (1907) gives the average number of chestnut sprouts in one clump in one-year-old stands as 60. Secondary sprout stands are common in the Hemlock–White Pine–Northern Hardwoods region (pp. 338, 397). A sprout forest continues, in general, the composition of the original stand, although the more prolific sprouters become relatively more abundant. Seedlings have little opportunity for development in sprout stands because of the rapid growth of sprouts. Secondary communities of seedling origin develop in old-field succession.

Primary Forests

A number of unlike primary forest types (other than local, edaphic, and developmental communities) occur in the Glaciated section: oak or oak–hickory, oak–chestnut, mixed mesophytic, and hemlock–hardwoods. The pitch pine or pine–oak forest type of the sand plains is an edaphic community originally fairly extensive. The forest communities of undrained depressions, swamps and bogs and those of alluvial flats are developmental communities due to local edaphic conditions. Some of these, because of the persistence of the habitat, may sometimes be thought of as physiographic types (Bromley, 1935).

OAK OR OAK–HICKORY FOREST TYPE

The oak or oak–hickory forest type is one in which white oak is the most characteristic and usually the most abundant species. This is a somewhat xeric forest type with (in remnant stands) the trees rather widely spaced. The old forest[24] in Stonington, Connecticut, on a low hill bordering the Sound, was representative of this type; this was described by Nichols (1913) and later discussed by Raup (1941). "The character trees are white oak and black oak (*Quercus velutina*), especially the former, associated with which are shagbark hickory and red maple" (Nichols, 1913). Nichols expressed some doubt as to the primeval nature of this stand; Raup found (from a study of ring counts) that the forest had been "seriously damaged in 1815, presumably by the hurricane of that year." There is evidence, however, that it had not changed much in its composition, that it had been, prior to 1815, similar in composition to the recently destroyed stand, and that the type would have continued into the next generation.

[24] This forest was largely destroyed by the hurricane of 1938.

Although the species associated with white oak are not the same in all examples of this forest type, nor are the undergrowth constituents always the same, a white oak or oak–hickory type may be recognized at intervals throughout the section. A community (somewhat comparable to that in Stonington) was described on Montauk (at the east end of Long Island), where old, but low and gnarled oaks dominate (Taylor, 1923). Nichols (1914) refers to an oak–hickory forest type in Connecticut, saying that it is "the most familiar type of woodland encountered along the trap ridges near New Haven, and in many sites . . . may represent the ultimate formation." Bromley (1935) mentions examples near Stamford, Connecticut. The communities thus far mentioned typify the so-called "oak region." The soils here are "in a transition zone between the gray-brown podzolic soils farther south and west, and the podzols of the north" (Lunt, 1938). The area is within the "Brown Podzolic Soils" zone as mapped by Kellogg (Map, p. 26). Raup (1938) distinguishes a "white oak–hickory association" in the Black Rock Forest in the Hudson Highlands which is "characteristically restricted to the northeast sides of the crowns of the hills, and to fine soils which are relatively stable." It also occurs in an intermediate position between a more xeric black oak ridge crest community and a more mesic ravine community (New Jersey Highlands near Morristown). Old white oaks on the Triassic Lowland of New Jersey (Piedmont Lowland) and in the Hudson Valley suggest the white oak community of the valley floors of the Ridge and Valley section (p. 238).

A type of oak–hickory forest which occurs frequently on the very gently rolling to flattish surface of the Piedmont Lowland is related to high water table.[25] Such hydro-mesophytic species as pin oak, swamp white oak, sweet gum, and red maple occupy the poorly drained spots. White oak is common on slightly better-drained sites, and may be associated with other oaks and hickories (*Quercus borealis* var. *maxima, Q. velutina, Carya ovata, C. tomentosa*), and in places with beech. The aspect is similar to that of Illinoian till plain forests of the Western Mesophytic Forest region (p. 133). Variations in composition related to depth of the water table, and dissimilarities of canopy and understory suggest the possibility of further development if drainage improves.

OAK–CHESTNUT FOREST TYPE

Chestnut oak, black oak, and red oak communities, or combinations of these, represent the former oak–chestnut forest type. In the sprout hardwood stands, chestnut increased in relative abundance to such an extent that, at the beginning of this century, it formed about 50 per cent of the standing timber of Connecticut (Nichols, 1913). Since then it has been so

[25] An excellent example is Mettler's Woods in Franklin Township, Somerset County, New Jersey, west of New Brunswick.

nearly eliminated that it can no longer be referred to as a dominant and its name does not enter into community names. The chestnut oak communities are essentially the equivalent of the chestnut–chestnut oak communities still recognizable farther to the southwest. The red oak communities are comparable to the red oak–chestnut communities of other sections of the Oak–Chestnut region. Oak–chestnut communities have recently become, or are changing to oak and oak–hickory in most instances. As elsewhere, the chestnut oak community occupies the steeper slopes and more or less stabilized rock slides of more rugged parts of the section (Raup, 1938) and morainal hills in the Coastal Plain part (Conard, 1935; Cain, 1936a). Nichols (1913) mentions a chestnut oak or chestnut oak–pignut community as a stage in upland succession, destined to be replaced by oak–hickory. Some of the earlier descriptions of stands now referable to the chestnut oak type emphasize the importance of chestnut. Harshberger (1910) said of the forest on the Navesink Highland of New Jersey (Cretaceous belt of the Coastal Plain where an iron hardpan is responsible for the elevation) that the principal trees, in order of relative abundance, were chestnut, chestnut oak, black oak, white oak, scarlet oak, and red oak; with these were pignut hickory and, rarely, beech and tuliptree. Now, chestnut is gone and the oaks dominant. The ericaceous layer (of *Kalmia, Vaccinium, Gaylussacia,* and *Azalea*), however, readily distinguishes such oak forests from the oak or oak–hickory type just considered.

Many recent descriptions ignore chestnut entirely in listing principal trees. Thus, Tryon (1930) and Raup (1938) give the principal trees of the chestnut oak type as chestnut oak (62.37 per cent), red oak (26.83 per cent), white oak (5.67 per cent), hard maple (4.22 per cent), and red maple (0.41 per cent). Dogwood is a dominant in the second layer of some of the chestnut oak communities (especially the more mesophytic variants) and is absent from others. *Kalmia* and other ericaceous shrubs (*Vaccinium, Gaylussacia,* and *Azalea*) often form a dense undergrowth. A forest on Long Island, in which the principal trees are chestnut oak, black oak, chestnut, red maple, sweet birch, and beech, is referred to by Transeau (1908) as the dominant forest of the region.

Black oak communities (except on the sand plains) are generally localized on ridge crests. Black oak may be abundant in the chestnut oak communities (Cain, 1936a).

The red oak community is the most mesophytic of the oak–chestnut communities and often contains the largest number of species in its superior layer. Chestnut was originally one of the principal species. Other oaks and hickories are usual constituents, and the tuliptree is a frequent associate. Dogwood is abundant in the understory; heath shrubs, although present, are (except *Kalmia*) subordinate to other species. Nichols considered a forest of this type to be climax, characterizing it as follows:

Along the trap ridges the climax forest of this region is best developed on the lower slopes, where there is a relatively constant supply of ground water, available throughout the growing season. By far the most abundant and most characteristic tree here is the chestnut (*Castanea dentata*). With it are commonly associated *Liriodendron tulipifera* and some of the more mesophytic trees of the preceding stage, e.g., *Quercus rubra* [*Q. borealis* var. *maxima*], *Quercus alba* and *Acer rubrum,* while scattered through the forest, sometimes abundant locally, are other mesophytic trees, such as *Acer saccharum, Fagus grandifolia, Fraxinus americana, Prunus serotina, Tilia americana,* and *Tsuga canadensis.* As species of secondary importance may be mentioned *Carpinus caroliniana, Cornus florida, Ostrya virginiana* and *Sassafras variifolium.*

In the Hudson Highlands, this type was referred to by Tryon (1930) as "hardwood slope type," and by Raup (1938) as "red oak association." There it occupies the lower slopes of hills, especially of northerly exposures. The admixture of such species as beech, sugar maple, basswood and hemlock suggests the transition to mixed mesophytic and hemlock–hardwoods types.

MIXED MESOPHYTIC AND HEMLOCK–HARDWOODS TYPES

Mixed mesophytic forest communities, although not always recognized as such, occur frequently in the Glaciated section, particularly in that part which extends over the New England Upland of Connecticut and New Jersey (New Jersey Highlands). In northwestern New Jersey (within the glaciated part of the Ridge and Valley Province) mixed mesophytic communities appear to be absent, although the richer forests in sheltered ravines (as on the lower slopes of Kittatinny Mountain) are suggestive of mixed mesophytic. Beech, however, is generally absent, in which respect these communities differ from the most mesophytic communities of New England and the Hudson and New Jersey Highlands.

In the New England Upland (including the Hudson and New Jersey Highlands), a number of variants are seen. Hemlock is present in some, and absent in other areas. Beech is usually, but not always, one of the more abundant species. No line of demarcation can be drawn between the richer oak forests in which white oak is a dominant and the mixed mesophytic type, nor between the more mesophytic phases of the red oak type and the mixed mesophytic type. A community such as was described by Nichols as climax, if seen today after the chestnut is gone, and some of the suppressed understory trees released, would doubtless resemble, if not be identical with communities now interpreted as mixed mesophytic. Certainly in some mesophytic stands beech appears to be taking the place of chestnut. This is in part due to the rapid growth of root shoots, a mode of reproduction frequently displayed by beech, especially northward. Bromley considers that the mixed mesophytic forest type is the climatic climax of the region, and maps it as occurring over approximately the southwestern two-thirds of Connecticut (Map, p. 250). He states that it is composed principally of oak, chestnut, hickory, sugar maple, beech, tuliptree, sweet birch, and a varying admixture of hemlock, and that some twenty or more tree species occur in this community. In addition to the above, the following may be seen: red, white, and chestnut oaks, shellbark, pignut, and mockernut hickories, red maple, yellow birch, white ash, butternut, basswood, and black cherry. This is similar to the climax type characterized by Nichols. Raup (1938) groups the more mesophytic communities of ravines and lower north slopes of Black Rock Forest (Hudson Highlands) as "cove forests," distinguishing mixed hardwood, hemlock–mixed hardwood, and beech–maple types. Beech is a principal species only in the last of these, which is a beech–sugar maple–yellow birch community and thus is allied to the northern forest. The mixed hardwood and hemlock–hardwood types are comparable to Bromley's mixed mesophytic type. Lutz (1928) considers the hemlock–hardwoods forest to be climax for southern New England. The development of such communities is related to post-Pleistocene migrations (see Chapter 17).

Old-Field Succession

Communities of old-field successions and those developing after severe disturbance of the former forest growth (by fire, cutting, grazing) occupy a greater proportion of the area of the Glaciated section than do remnants of old-growth stands. The red cedar–gray birch community is the characteristic old-field community of this section (Map, p. 250). The two species may be codominant, or either one may occur in pure stands. This community may gradually be replaced by a hardwood community and this in turn by a hemlock–hardwood community (Lutz, 1928).

Developmental and Edaphic Communities

A great variety of developmental and edaphic communities related to the unlike topographic and soil areas of the section occur. Those of Connecticut were discussed by Nichols in a series of papers (1913–20), and are more or less representative of the section as a whole. Communities of uplands, lowlands (including lakes and bogs), stream valleys, and seacoast areas are included. In the large number of ponds and lakes, and of swamps and bogs with their attendant vegetation, this section of the Oak–Chestnut region differs from all regions previously discussed, and resembles the Beech–Maple and Hemlock–White Pine–Northern Hardwoods regions. Only its coastal swamps are similar to those farther south (in the Oak–Pine and Southeastern Evergreen regions).

The southern white cedar (*Chamaecyparis thyoides*) is the dominant tree of certain of these swamps or bogs. These occur from Massachusetts southward, and are well represented in Connecticut and on Long Island (Nichols, 1915; Bromley, 1935; Taylor, 1916). Southern white cedar also occurs in bogs on Kittatinny Mountain where it is associated with the northern black spruce (*Picea mariana*) and sometimes with larch (*Larix laricina*).

Swamp forest (of deciduous species) occupies many of the glacially formed depressions. In these, red maple, white elm, pin oak, swamp white oak, sour gum, and sometimes sweet gum are the principal species. These occur in mixed or almost pure stands, and sometimes in combination with the southern white cedar (Nichols, 1915; Cain and Penfound, 1938; Bromley, 1935; Taylor, 1923).

Pitch pine is the dominant tree on sand plains of Connecticut, Rhode Island, and southeastern Massachusetts (Britton, 1903; Nichols, 1914; Bromley, 1935; Olmsted, 1937; and mentioned in many early historical and travel records), and in the Hudson-Mohawk pine barrens between Albany and Schenectady (Bray, 1930). Many places in the sand plains are open with very sparse vegetation, or are occupied by various herbaceous communities, and sometimes by a more or less dense stand of *Andropogon scoparius*. *Andropogon* communities occupy the finer soils of the outwash plain on Long Island—Montauk Downs and Hempstead Plains (Taylor, 1923; Blizzard, 1931; Cain, Nelson and McLean, 1937). These are all communities which may persist indefinitely—edaphic climaxes. There is evidence, however, of replacement by oaks, but some question as to whether development can, in all cases, progress to this stage.

A few rather unusual types of woodland (or groves) occur in the coastal belt of this section. A maple–holly forest covers the oldest dunes on Sandy Hook. This is a wood of strikingly mesophytic aspect, made up of large trees of holly (*Ilex opaca*) and red maple, together with occasional red cedar, wild black cherry, and hackberry, and a tangled understory overrun with lianes (Chrysler, 1930). It is interpreted as a sand dune or salt spray climax leading from Ammophila→shrubs→red cedar→holly. A forest community in which holly is a constituent also occurs near the east end of Long Island; here beech and red maple are the principal species (Taylor, 1923). Persimmon (*Diospyros virginiana*) forms a grove at Lighthouse Point near New Haven (Hawley, 1909; Nichols, 1913). This is a northern outpost for this species. On the Elizabeth Islands of Massachusetts (southwest from Woods Hole), pure stands of beech occur. "These trees grow nowhere very tall, averaging perhaps, 30 to 40 feet, and their low, flat, leafy crowns meet overhead, forming a thick roof through which a subdued light filters." Usually, however, a number of other species (white and black oaks, hophornbeam, sassafras, red maple, sour gum, and witch-hazel) occur with the beech (Fogg, 1930).

Floristic Relations

The consideration of the Glaciated section cannot be closed without some mention of its floristic relations. It is an area of overlap of northern and southern species, of dovetailing of northern and southern forest communities (Nichols, 1913). The southern element is represented by such trees as *Diospyros virginiana, Chamaecyparis thyoides,* and *Ilex opaca,* all of which are abundant in the local communities of which they are a part. This southern element also includes *Liquidambar Styraciflua* of the deciduous swamps, and, less pronouncedly southern, the dominant white and chestnut oaks. It is indicative of relatively recent northward migrations (see Chapter 17). The northern element is best represented in the bogs which harbor many relic species. The area is a meeting ground of trees of the central deciduous and northern forests, and to some extent, of the shrubby and herbaceous species of these forests.

CHAPTER 8

The Oak–Pine Forest Region

The characteristic appearance of this region, brought about by the dominance of oaks and pines over much of the area, is responsible for the name "Oak–Pine." However, except on poorer soils and in drier sites, pines are more or less temporary and are ultimately replaced by deciduous species. As there are no virgin forests remaining, the nature of the climax forest must be deduced from successional trends, the oldest of undisturbed secondary stands, and the few old trees which antedate settlement. Oaks (white, black, post, red, and southern red) and hickories (white and pignut) are most common and widespread. On this basis, the region under consideration might well be termed the "Eastern Oak–Hickory Forest," or considered as a great eastern lobe of the Oak–Hickory region (p. 163). However, pines do persist in areas less suitable to deciduous species, and they were prominent in the original forest cover. Sourwood and sweet gum are almost constant associates of the dominant oaks and hickories. As these trees are not a part of oak–hickory communities of the Oak–Hickory Forest region (as defined in this book), they further serve to distinguish the climax communities of the Oak–Pine region from those of the Oak–Hickory region. Many of the species of the central deciduous forest are abundant here; a considerable number more characteristic of the south Atlantic and Gulf Coastal Plain occur in favorable situations. This region is transitional between the central deciduous forest and the prevailingly evergreen forests of the Southeast. The name "Oak–Pine" emphasizes this transition. The abundance of loblolly pine (*Pinus Taeda*) affords a strong contrast with the central deciduous forest; the infrequence or localization of longleaf pine (*Pinus palustris*) contrasts with the Southeastern Evergreen Forest region.

The region as here defined extends as a broad band from southern New Jersey southward to Georgia and thence westward to the alluvial lands of the Mississippi basin. Here it is interrupted but appears again west of the Mississippi where it occupies a considerable area in southern Arkansas, northwestern Louisiana and adjacent Texas. Its inner margin on the Atlantic slope is in contact with the Oak–Chestnut region; on the Gulf slope east of the Mississippi, with the Oak–Chestnut, Mixed Mesophytic, and Western

Mesophytic Forest regions; on the Gulf slope west of the Mississippi, with the Oak–Hickory region.

In New Jersey, Delaware, and Maryland, it covers the Coastal Plain, except its innermost (Cretaceous) border (p. 242). In Virginia it includes the Coastal Plain north of the James River, and much of the Piedmont.

The abundance of pine will in general serve to distinguish the vegetation of the Oak–Pine region from that of the adjacent Oak–Chestnut region where the boundary (or rather transition) between the two crosses the Piedmont in Virginia. This boundary in Virginia is related to soils. On that part of the Piedmont which is occupied by the Oak–Chestnut forest (p. 242), the soils belong to the Chester series of the gray-brown forest soils, while in the adjacent Oak–Pine region the soils belong to the Cecil series of the red and yellow, or lateritic, soils of the South. The area of overlap of the Oak–Pine region onto the Coastal Plain corresponds fairly well with the area of Sassafras soils.[1]

Through the Carolinas and adjacent Georgia (its most typical part) the Oak–Pine region is more or less coextensive with the Piedmont, extending over but a small part of the Coastal Plain in northern North Carolina. In Georgia, a lobe of the Oak–Pine region extends northward from the Piedmont into the Great Valley of northwestern Georgia and adjacent Tennessee. In Alabama and Mississippi, the regional boundaries cross, rather than approximate, physiographic boundaries, thus including parts of the Piedmont, Great Valley, Appalachian Plateau, and Coastal Plain. Furthermore, the boundaries here on the Gulf slope are so indistinct that they must be placed more or less arbitrarily on a basis of the occurrence or abundance of species more properly belonging to the more northern Mixed Mesophytic and Western Mesophytic regions or to the more southern Gulf Coast vegetation. Their position is further complicated by the belting of the Gulf Coastal Plain. That part of the region west of the Mississippi is on the Coastal Plain, and is limited on the north by that physiographic boundary.

Most of the Oak–Pine Forest region has a rolling topography where gentle slopes prevail. Some of it is flat, particularly parts on the north Atlantic Coastal Plain; some of it is rough and hilly, especially near the Blue Ridge or in line with the southern extension of mountainous provinces in the South. Almost everywhere it is indented by streams whose valleys have more abrupt slopes than the average upland slopes.

Two sections are recognized: The **Atlantic Slope** section, whose streams drain toward the Atlantic Ocean, and the **Gulf Slope** section, whose streams drain to the Gulf of Mexico. The division between the two is about midway across Georgia. The Gulf Slope section is interrupted by the loess bluffs and alluvial plain of the Mississippi River.

[1] For extent, location, and characteristics of these soils, see maps in Marbut, 1935.

A shortleaf pine and hardwood stand in the Lee Experimental Forest, Buckingham County, Virginia. The forest type most frequently seen in the Oak–Pine Forest region is one made up of pines and oaks, or of pines with a layer of hardwood reproduction. Younger stands are dominantly pine; in older areas deciduous species have largely replaced the pines. (Courtesy, U. S. Forest Service.)

ATLANTIC SLOPE

The features of this section are best exemplified on the Piedmont Plateau, of which the Piedmont of North Carolina is most representative. This is an ancient upland, underlain by crystalline and metamorphic rocks,[2] and long subjected to erosional processes. The topography is typically rolling. The general upland represents the Piedmont (Harrisburg) peneplain; monadnocks, such as Pilot Mountain (p. 267) rise above the general level, and stream valleys of a later erosion cycle are cut below it. The soils (except those of the northern end of the section from the Chesapeake Bay district to New Jersey) belong to the red and yellow or lateritic soil group.

From the days of the early settlements, land utilization with consequent destruction of forests, subsequent abandonment of fields, and soil erosion have profoundly influenced vegetation. Blocks of even-age pines (*Pinus Taeda, P. echinata, and P. virginiana*) in pure or mixed stands, uneven-age hardwood forests in which oaks and hickories prevail, and mixed forests of oaks and pines are scattered about on the upland. Bottomland forests are usually of small extent[3] and composed of deciduous species among which the flood-plain trees are prominent. Ravine bluffs or steep banks rising above the bottomlands (slopes below the general upland level of the Piedmont), although areally unimportant, are of particular interest because of their mixed hardwood community in which beech is a significant species. Throughout the Piedmont, differences of soil and topography are reflected in differences in forest composition.

Climatic differences within the area of the Atlantic Slope section, related to latitude and altitude, result in differences between forest composition in the North and South. Some of the species of the more southern forests become less abundant or drop out entirely before the northern limits of the section are reached. The dominant and characteristic loblolly pine is one of these. Some of the coastal plain species, more or less frequent in the lowlands near the eastern margin of the Piedmont, rarely, or never occur near the mountain border. The Piedmont and Coastal Plain parts of the section differ in relative amounts of bottomland and upland swamp. In the Piedmont, the strips of bottomland are usually narrow and not swampy; upland swamps are rare for few level and poorly drained areas remain in the uplands. In the Coastal Plain, the streams of the Chesapeake Bay district are estuarine near their mouths, coastal marshes and alluvial swamps are prominent; many flat and undissected interstream areas remain (because of the more youthful topography) and hence upland swamps are frequent.

The Piedmont

Virgin forest, as it was before the local appearance of the white man, is today completely lacking in the region. Only small isolated stands of relatively great age (200 to 300 years) and relatively little disturbance remain as apparent

[2] Small areas of Triassic sediments do not modify the general picture.

[3] Some exceptions will be found where the Oak–Pine region extends onto the embayed section of the Coastal Plain.

relicts of what the original extensive forest of the uplands must have been like. Such stands are few since little of the area has been free from disturbance by man at some time. . . . In the early days of settlement, large tracts of forest were cleared for cultivation. Land was cheap and little thought was given to agricultural practices that would maintain soil productivity. The heavy winter rains soon washed away the best of the topsoil from unterraced fields and, after several years of cultivation without fertilizing, they became relatively unproductive. It became common practice to abandon such fields and clear new land. This procedure has continued to the present day, although lands now being cleared have usually been cultivated before, in some instances probably more than once (Oosting, 1942).

These practices have resulted in a patchwork of fields, second-growth forest communities of various ages, and culled hardwood stands of varying composition. Only the forests of the few situations unfavorable for cultivation, such as ridges, knolls, and stream bluffs, and the occasional small tract on the better sites which has escaped clean cutting for cultivation remain, although even these have been selectively cut for generations. These remnants must be used in reconstructing the picture of original vegetation, and in determining forest potentialities.

In the original forest cover three more or less parallel but irregular and broken belts—an "eastern pine belt," a belt of "deciduous forests," and a "western pine belt"—were distinguishable (Ashe, 1897). These three belts are indistinctly indicated on soil maps as: (1) a belt where the Cecil sandy loam has developed over a considerable area; (2) a belt with large areas of Georgeville and associated soils (the deciduous belt); and (3) the inner belt of Cecil clay loam and Cecil sandy loam (Marbut, 1935).

Today, in crossing the Piedmont, one passes through pine woods, deciduous woods, mixed woods, old fields in various stages of successional development, and farm lands scattered over the rolling upland.[4] Young pine stands appear almost irrespective of soils, as pines are almost invariably the first invaders of abandoned fields.

For the most part the three species of pine occur as even-aged stands, either pure or mixed, which become established, characteristically, on abandoned fields from 1 to 10 years after cultivation ceases. Even-aged stands of pine over 80 years old are infrequent, and stands under 40 years of age predominate (Korstian and Coile, 1938).

Where the soils are sandy and pines more or less abundant in older mixed stands, such herbs as sensitive brier (*Schrankia*) and birdfoot violet (*Viola pedata*) are plentiful. Open old-field communities—where the pines are young and widely spaced—contain an abundant more or less showy

[4] Duke Forest, an area of approximately 5000 acres of rolling Piedmont land, well illustrates the majority of existing forest types of the Piedmont (see Korstian and Maughan, 1935).

herbaceous flora mingled with tall grasses (especially *Andropogon*) and sedges.

With the occupancy of abandoned fields by pines, the potentially deciduous forest lands may now be pine-covered, whereas originally the pines were confined to hill tops. This has tended to obscure the three forest belts originally distinguishable.

Several types of upland deciduous forest may be recognized (Table 52). In the most widespread type, generally occupying the better soils, white

Table 52. Upland (climatic climax) forest communities of the Piedmont of North Carolina: *A,* White Oak; *B,* Post Oak. (Data from Oosting, 1942, Tables 16, 19.)

	A. White Oak		*B. Post Oak*	
	Overstory	*Understory*	*Overstory*	*Understory*
Quercus alba	43.9	4.1	21.4	7.3
Carya spp.	31.7	19.5	8.3	13.6
Quercus stellata	8.5	.2	45.2	6.5
Quercus velutina	4.9	2.2	. .	.6
Quercus coccinea	3.7	.4	3.5	1.3
*Quercus rubra**	2.4	.6	7.1	.6
Quercus borealis maxima	1.2	.6	2.3	.2
Quercus marilandica	. .	. .	1.1	3.0
Nyssa sylvatica	1.2	6.1	. .	4.9
Diospyros virginiana	1.2	.2	. .	.2
Pinus Taeda	1.2	1.6	. .	.2
Pinus echinata	. .	. .	10.7	6.7
Cornus florida	. .	30.9	. .	23.0
Oxydendrum arboreum	. .	24.0	. .	6.7
Acer rubrum	. .	5.3	. .	9.7
Juniperus virginiana	. .	2.4	. .	8.8
Liriodendron tulipifera	. .	.2	. .	.2
Amelanchier canadensis	. .	.2	. .	. .
Liquidambar Styraciflua	. .	.2	. .	2.8
Prunus serotina	. .	.4	. .	. .
Ilex decidua	. .	.4	. .	. .
Morus rubra	. .	.4	. .	. .
Fraxinus americana	. .	. .	. .	2.6
Ulmus spp.	. .	. .	. .	.9
Cercis canadensis	. .	. .	. .	.2
Chionanthus virginicus	. .	. .	. .	.2

* *Quercus falcata.* Nomenclature as in Oosting, 1942.

oak is the most abundant species (locally the dominant species) and hickories and other oaks (black, post, red, southern red, scarlet) are well represented, together with a scattering of other species.[5] A white oak–post

[5] This has been referred to as the white oak–black oak–red oak type (Coile, 1933; Korstian and Maughan, 1935; Oosting, 1942), or more briefly, as the white oak type (Oosting, 1942).

oak (or post oak) type, in which post oak is more abundant than white oak occupies somewhat drier situations and poorer soils than the former type. Both of these may be thought of as oak–hickory forest, and it seems "reasonable to recognize the two forest types as variants of the oak–hickory climax which are determined by site" (Oosting, 1942). A post oak–blackjack oak type occupies the poorest situations where there is a relatively impermeable subsoil or where erosion has been extreme. It is seen most frequently in the eastern belt, but is everywhere rather limited in extent.[6]

The shrub and herbaceous growth of the oak woods is sparse as compared with the more centrally located forest regions. It is made up of species which are of widespread occurrence in oak woods, not only here, but in the more central and western forest regions, and other species more southern in distribution. Among these may be mentioned *Vaccinium arboreum, V. vacillans, Viburnum rufidulum, V. prunifolium, V. acerifolium, Ceanothus americanus, Callicarpa americana, Chionanthus virginicus, Chimaphila maculata* (a species present in a large percentage of woods), *Desmodium* spp., *Lespedeza* spp., *Clitoria mariana, Cassia nictitans, Ascyrum hypericoides, Coreopsis major, Sericocarpus asteroides,* and *S. linifolius.*[7] Occasionally the younger second-growth oak stands have an almost continuous heath layer made up of *Kalmia, Gaylussacia* and *Vaccinium.* Such communities are conspicuous on some of the flatter sandy upland areas toward the northern end of the Piedmont district in Virginia.

Bands of bottomland forest along the larger streams interrupt the upland vegetation. River birch, black willow, cottonwood, sycamore, and sweet gum are the most frequent streamside trees. A mixed bottomland hardwood community develops where flats are old enough and wide enough. Sweet gum, willow oak (*Quercus Phellos*), white and winged elms, red maple, tuliptree, ash, and southward, water oak (*Quercus nigra*) and sugarberry (*Celtis laevigata*) are common species of the bottomland forest. Oosting (1942) describes such a community on the Eno River, north of Durham, North Carolina (Table 53). Fifteen species are represented in the overstory; of these, willow oak, sweet gum, swamp red oak, and white oak are most abundant, making up 66 per cent. Ten of the overstory species and nine additional species are represented in the understory; of these, *Acer floridanum* is most abundant.[8] A number of shrubs and woody climbers are present; poison ivy and cross-vine are abundant; and the evergreen *Gelsemium sempervirens* is frequent.

Where slopes of northerly exposure rise more or less abruptly above

[6] Oosting (1942) considers this type to be preclimax when it develops on undisturbed soils, and subclimax when it develops on eroded soils.

[7] This list includes species noted in a number of widely separated areas on the Piedmont of North and South Carolina.

[8] Oosting regards this community as postclimax.

Table 53. Lowland (postclimax) forest of the Piedmont of North Carolina. (Data from Oosting, 1942, Table 22.)

	Overstory	*Understory*
Quercus Phellos	26.8	4.6
Liquidambar Styraciflua	18.3	9.2
Quercus rubra pagodaefolia	15.5	1.7
Quercus alba	7.0	1.3
Pinus Taeda	8.4	. .
Carya ovata	4.2	9.2
Quercus lyrata	4.2	. .
Nyssa sylvatica	2.8	.4
Carya carolinae-septentrionalis	2.8	. .
Acer rubrum	2.8	6.3
Acer floridanum	1.4	23.5
Celtis occidentalis	1.4	.8
Quercus velutina	1.4	. .
Quercus stellata	1.4	. .
Ulmus americana	1.4	2.5
Cornus florida	. .	16.0
Ilex decidua	. .	6.7
Ulmus alata	. .	6.3
Carpinus caroliniana	. .	3.4
Juniperus virginiana	. .	2.9
Morus rubra	. .	2.1
Fraxinus americana	. .	1.7
*Quercus rubra**	. .	.8
Carya glabra	. .	.4

* *Quercus falcata.* Nomenclature as in Oosting, 1942.

the bottomland, a forest community unlike any of the upland communities is seen. This community contains a variety of mesophytic species among which beech is almost universally present, and particularly characteristic. White oak and tuliptree are common associates; red maple, walnut, black cherry, and occasionally sugar maple (toward the north and near the mountains) occur, as do also a few of the trees of the upland oak forests and of the bottomland forest. The area occupied by this community in the original forest cover was not large; mere fragments of the community remain today. In composition and in appearance this slope community suggests the mixed mesophytic forest, in fact it may be considered as a Piedmont representative of that forest association. It develops only on slopes being formed in the present erosion cycle, not on slopes of the rolling upland of the Piedmont or Harrisburg peneplain (see Chapter 17).

The developmental stages in the reforestation of abandoned fields on the Piedmont and the successional relations of the more mature communities have received considerable attention from investigators in that area (Crafton and Wells, 1934; Korstian and Maughan, 1935; Billings, 1938; Coile, 1940; Oosting, 1942). That "competition between individuals of the forest vegetation for soil moisture is a highly significant factor in the growth, development, and reproduction of forests in the Piedmont plateau" was demonstrated by Korstian and Coile (1938) by using trenched plots. In general, the developmental sequence of communities is herbaceous →pine invasion→pine forest→hardwood invasion (establishing a deciduous understory)→mixed pine–oak-→oak or oak–hickory.

Communities which do not belong to the prevailing upland forest types or their secondary developmental stages, or to lowland and streamside types are but rarely seen. Local variations, of course, occur, but even local communities departing widely from the more or less monotonous and ubiquitous types are rare. Along the Fall Line, exposures of granite occur where rock successions are displayed with their moss and lichen mats and their rock or shallow soil herbs, some of which are local in occurrence or endemic on rock outcrops of the Southeast. Among these are *Portulaca Smallii, Talinum teretifolium, Arenaria brevifolia,* and *Diamorpha cymosa* (Oosting and Anderson, 1939).

Where pronounced monadnocks rise above the Piedmont—such as Pilot Mountain, an isolated 2413 foot peak rising 1500 feet above its surroundings—the forest vegetation, except of the lower slopes, resembles that of the Oak–Chestnut region rather than that of the Oak–Pine region. The Blue Ridge relationship is emphasized by the abundance of chestnut oak and the presence of pitch pine (Williams and Oosting, 1944). Such mountains are vegetationally, as geologically, outliers of the Blue Ridge.

The Coastal Plain or Embayed District

North of the James River, the Oak–Pine region extends across the Coastal Plain of Virginia; north of the Potomac, it is confined to but is not coextensive with the Coastal Plain. There, the inner part of the Coastal Plain is occupied by the Oak–Chestnut forest (p. 194). The inclusion of the Coastal Plain of Virginia north of the James River in the Oak–Pine region and south of that river in the Southeastern Evergreen Forest region, is based upon floristic differences between the two areas. The latter contains a large number of characteristic species of the South Atlantic or South Atlantic and Gulf Coastal Plain, as *Sarracenia flava, Drosera capillaris, Aletris aurea,* and many more (Fernald, 1937). Furthermore, the presence in this area of longleaf pine (*Pinus palustris*), the most abundant and characteristic tree of the Southeastern Evergreen region, and of live oak (*Quercus virginiana*) allies it with the Southeast. North of the James River, these species are lacking. It should not be understood that typical coastal plain species are lacking farther north; many are found in New Jersey and in isolated stations northward. The boundary at the James River coincides with a soil boundary: to the south are soils of the Norfolk series of the red and yellow soils group; to the north are soils of the Sassafras series of the gray-brown or podzolic forest soils.

This district of the Atlantic Slope section differs topographically from the Piedmont because of its more youthful topography and drowned valleys. Flat interstream areas are frequent, and hence upland swamps are found; alluvial swamps and marshes border the streams. Relief is slight, and sloping areas are of relatively small extent. These topographic features with their accompanying vegetation are particularly well developed in Maryland. In New Jersey, larger areas are uninterrupted by streams or have small stream valleys but slightly below the general level. This is the location of the famous Pine Barrens of New Jersey, which must be considered apart from other forest communities of the district. Between the Pine Barrens and the Piedmont and along the Delaware River is a strip of Coastal Plain with more relief, which is occupied by deciduous forest and has been included in the Oak–Chestnut region (p. 242).

CHESAPEAKE BAY AREA

Upland and lowland forests are represented in the Chesapeake Bay area —Eastern and Western Shores of Maryland (Shreve, 1910b; Chrysler, 1910). The upland forests of drained or sloping areas resemble those of the Piedmont farther south. Loblolly pine, in pure stands or mixed with Virginia pine or with deciduous species, is the most abundant tree in secondary forests developed on abandoned lands. On flat clay uplands, where drainage is poor, sweet gum, willow oak, pin oak, and sour gum mingle with the usually more abundant white oak. Upland swamp forest communities, which are poorly represented or absent in the Piedmont, occur on both clay and sandy soils. Those on the former are not sharply delimited from mixed pine–deciduous communities of the drier clay uplands; the deciduous species are more abundant and the scrub pine absent. Upland swamp forests in sandy soils may be prevailingly coniferous or prevailingly broad-leaved. The latter are developmentally older. The trees usually present are willow oak, white oak, sweet gum, red maple, water oak, cow oak, black gum, sweet bay (*Magnolia virginiana*), holly, and dogwood. Throughout the sandy soil areas of the uplands, cutting and clearing have greatly increased the amount of pine. Sterrett (1908) states that originally there was very little pine in the "mixed pine and hardwood region" of southern Delaware, and that the same types of forest prevailed as in the hardwood belt of the Coastal Plain to the north.

Lowland forests vary in physiognomy and composition in relation to degree of overflow and permanence of abundant water. Two principal types, hardwood and cypress swamp, may be distinguished. The lowland hardwood forest communities frequently contain an admixture of loblolly pine, and sometimes of southern white cedar. (Cedar swamps are well developed in the Pine Barrens.) Sweet gum is almost everywhere the most abundant species; red maple, willow oak, pin oak, and sour gum are the usual associates, and in areas seldom reached by high water, tuliptree and beech may occur. This community is the equivalent of the bottomland forest of the Piedmont (p. 265). The cypress swamps, or "river swamps" (Shreve, 1910b) occupy lands adjacent to but upstream from the brackish marshes. Cypress swamps are here at their northern limit. They are northern outliers of a community more typical of the Southeast. The Pocomoke River Swamp, toward the southern end of the Eastern Shore of Maryland, is separated from the well-known Dismal Swamp of North Carolina and Virginia by Chesapeake Bay. It "appears to be an isolated northern extension of the larger development on the mainland" (Beaven and Oosting, 1939). The forest is dominated by bald cypress (*Taxodium distichum*), swamp black gum or southern tupelo (*Nyssa sylvatica* var. *biflora*), and red maple. Nearly pure stands of white cedar occur near the upland borders of the cypress swamp where the soil is peaty; in less peaty areas, the lowland hardwood forest forms a transition to the upland forest. Here are oaks, chiefly water oak, willow oak, cow oak, and white oak, with which are associated tuliptree, river birch, beech, and occasional trees of the upland forest (Beaven and Oosting, 1939).

PINE BARRENS OF NEW JERSEY

The Pine Barrens occupy the greater part of the Coastal Plain of New Jersey exclusive of its inner, western, and southeastern borders—the middle, coast, and Cape May districts of Stone (1908) and Lutz (1934)—which are occupied by deciduous forest communities. The area appears as an extensive plain clothed with forests of pitch pine and scrubby oaks, which are interrupted here and there by expanses of shrubby oaks and pines—the "Plains"—and by bogs and swamps with their cranberries and white cedars, or sedgy marshy savannahs.

Topographically, this is an area of low relief through which flow sluggish streams in shallow valleys. The soils are sandy, in places gravelly, and generally strongly podzolized. It is separated from the coast by a band of salt marsh and coastal strand and dunes,[9] behind which is a broken strip of oak woods four to five miles in width.

Vermeule, in 1888, described the Pine Barrens as follows:

> Casting the eye over this plain from any point of vantage, such as one of the small round hills which rise from its surface here and there at wide intervals, we see an unbroken extent of the dark-green pine forest as far as the limit of

[9] The whole area, under the name "Northern Pine Barren–Strand District" was described briefly by Harshberger in his Phytogeographic Survey of North America (1911); the several pine barren communities and associated cedar swamp, transitional, savannah, and other communities more fully in the "Vegetation of the New Jersey Pine Barrens" (1916). Lutz (1934) emphasized causal factors, particularly in relation to the "pitch pine plains." Harshberger (1911, 1916), Taylor (1912), and Lutz (1934, 1934a) discuss causes for the existence of the Barrens. To these sources the reader is referred for more detailed information concerning this interesting area.

vision, stretching away in long, gentle swells, as level as the ocean itself. So level is it in places that the greater height of the timber in the swamps gives the appearance of a ridge on what is really the lowest ground of the plain, and the uninitiated may be deceived thereby. It is often impossible for even a practiced eye to judge which way the ground descends, and in passing along the long, straight roads cut through the timber, it frequently seems that one is always at the bottom of a hollow, the ground appearing to rise in both directions, whereas it is either level or sloping all one way (Vermeule, in Geol. Surv. N. J., 1900).

The most characteristic forest community of the Pine Barrens is a more or less open stand of pitch pine (*Pinus rigida*), with which are associated occasional yellow or shortleaf pine (*P. echinata*). In rare instances, shortleaf pine is the dominant species. Young pines are frequent in the understory, although in many places oaks (*Quercus marilandica, Q. stellata, Q. velutina, Q. coccinea,* and occasionally, *Q. montana* and *Q. falcata*) form a lower layer, or occasionally attain canopy size. The most abundant shrubs are the scrub oaks (*Q. ilicifolia* and *Q. prinoides*) and the heath shrubs (*Kalmia latifolia, K. angustifolia, Gaylussacia baccata, G. frondosa, G. dumosa, Vaccinium vacillans,* and *Lyonia mariana*). Where the shrub layer is open, a few ground shrubs (as *Leiophyllum buxifolium, Arctostaphylos Uva-ursi, Hudsonia tomentosa,* and *H. ericoides*) appear.

Mixed pine and oak woods occur frequently in the Barrens, and small dominantly deciduous stands are not uncommon. Along the margins of the Barrens, the pine woods gives way to oak. An oak strip a few miles in width intervenes between the Barrens and the coast.

The Pine Barrens are interrupted by the so-called "Plains," which "are characterized by low tree growth, i.e., less than 11 feet in height" (Lutz, 1934). The scrub oak (*Q. ilicifolia*) is abundant as are also a considerable number of other shrubs. In fact, "the greater part of the lesser vegetation is shrubby." The apparently dwarf trees are sprouts from repeatedly burned bases. In fact, "the Plains owe their existence to repeated fires." Communities transitional between the Plains and the Pine Barrens occur. The broom crowberry (*Corema Conradii*) is a characteristic ground shrub of the "Plains"; here, also, the pyxie (*Pyxidanthera barbulata*) is particularly abundant.

Strips of swamp, in which the southern white cedar (*Chamaecyparis thyoides*) is dominant, occupy estuaries indenting the Barrens. These cedar swamps are separated from the pine communities by low but sometimes fairly abrupt slopes heavily clothed with shrubs and small deciduous trees. The early lumbering of white cedar (100 to 200 or more years ago) resulted in an increase of deciduous swamp species. The cedar swamps are floristically interesting because of the open bog inclusions with both northern and southern bog species, and particularly because of the occurrence here of the rare and local curly grass fern (*Schizaea pusilla*). Southern white cedar swamps are characteristic of the Atlantic and Gulf Coastal Plain, and occur frequently in the Southeastern Evergreen Forest region. The northern outposts of this swamp forest type (in the Glaciated section of the Oak–Chestnut region) have already been mentioned (p. 256).

Floristically, the sandy Coastal Plain of New Jersey should be considered as an integral part of the Coastal Plain to the south. However, although its swamp forest communities are outliers of swamp forest communities of the Southeastern Evergreen region, its upland forests belong with the Oak–Pine region. Its dominant conifer, pitch pine, is not a constituent of the Southeastern Evergreen Forest region.

GULF SLOPE[10]

As the Oak–Pine region on the Gulf slope occupies parts of several physiographic provinces (Piedmont, Ridge and Valley, Appalachian Plateau and Coastal Plain), it displays considerable topographic and soil, and hence vegetational, diversity. Its boundaries, both on the north and south are so indistinct that they can be but arbitrarily drawn. More than any other part of the Oak–Pine region, this is a transition belt where the ranges of trees of the central hardwood forest and of the evergreen forest of the Southeast overlap.

Longleaf pine, elsewhere confined to the Southeastern Evergreen Forest region, here is (or was) abundant in places on the Appalachian Plateau (Warrior Coal Basin), in the Coosa Valley (southern part of the Ridge and Valley Province), on the Alabama representative of the Blue Ridge (parts of it locally called Talladego or Rebecca Mountains) where it ascends to 2000 feet, and on the Piedmont where it is a prominent constituent of the forest of nearly every county. Other species more characteristic of the Coastal Plain, among which are *Magnolia virginiana* and *Persea palustris* (wet ravines in the Blue Ridge), likewise extend north into this section.[11] Similarly species more characteristic of the Appalachian Upland extend far south on the Gulf slope. *Quercus montana* and *Castanea dentata* range south to beyond the Fall Line in Alabama; the latter is said to have been abundant in southeastern Alabama (Southern Red Hills) many years ago. Such overlapping ranges serve to emphasize the overlapping forest types. Loblolly pine forests extend well north of the limits of the Oak–Pine region into the Western Mesophytic Forest region. Mixed mesophytic forest communities extend southward across the Oak–Pine region

[10] Much of the material for the part of the section east of the Mississippi has been taken from Harper (1943), Mohr (1901), and Lowe (1921); for the part west of the Mississippi from Foster (1912), Tharp (1926), and Turner (1937).

[11] Probably the most surprising examples of coastal plain species in the Appalachian Upland are *Croomia pauciflora* (Harper, 1942a) and *Sabal minor* (Harper, 1938), the former in the northeastern part of Alabama, the latter nearer the Fall Line.

along valley slopes and in ravines. These examples of overlap will explain the ill-defined regional boundaries.

In Georgia, the inner boundary along the foot of the Blue Ridge is fairly well defined. In northwest Georgia, it swings northward to include the Great Valley in southern Tennessee; thence southwestward along the southeast-facing escarpment of the Appalachian Plateau, which it follows almost to Birmingham, Alabama. From here westward, there is little or no relation to physiographic areas. The general direction is northwestward in Alabama, then southwestward across Mississippi. West of the Mississippi River the small area of Oak–Pine forest is limited at the north by the Coastal Plain–Ouachita Mountain boundary as far as western Arkansas; thence it extends southward across Texas to the Brazos River. The outer boundary of the section follows approximately along the Fall Line until, in Alabama, that turns northward; from thence the boundary, highly indefinite and crossing belts of the Coastal Plain, extends in a southwestwardly direction to the Loess Hills near the mouth of the Yazoo River. West of the Mississippi River, the Oak–Pine region includes portions of southwestern Arkansas and northwestern Louisiana, and sends a lobe into eastern Texas.

Georgia is by location partly on the Atlantic, and partly on the Gulf slope. The greater part of the Oak–Pine region of that state is on the Piedmont.

Alabama occupies a central position in the Gulf Slope section. Here, the vegetational features are highly diversified, for parts of several distinct physiographic areas are included in the Oak–Pine region. Harper, in his "Forests of Alabama" (1943), has mapped and characterized fifteen "regions" in the state, several with subdivisions, 20 divisions in all (Map, p. 274). Mohr (1901) delimited fourteen "floral areas." Of the areas defined by Harper, the following are included here in the Oak–Pine region: the "Basin Region" (a subdivision of the "Coal Region"), the Coosa Valley (which is the southwestern end of the Great Valley), the Blue Ridge, the Piedmont, and a large part of the Shortleaf Pine Belt. These correspond in general to the following of Mohr's subdivisions: the Warrior basin and upper Cahaba Valley, the Coosa basin, the mountains of metamorphic rock (which includes the Blue Ridge and Piedmont), and the Shortleaf Pine Belt, on the Coastal Plain.

The Oak–Pine region in Mississippi extends as a broad band across the belted coastal plain, including parts of the Tennessee River Hills, the northeast Prairie Belt, the Flatwoods, and the North Central Plateau—areas distinguished by Lowe (1921).

The diversity of vegetation in the Gulf Slope section of the Oak–Pine region is apparent from detailed maps; however, it is not possible to distinguish each of these areas in a broad treatment. Five subdivisions are made: the Piedmont of Georgia and Alabama, the Great Valley (or Appalachian Valley), the Warrior and Cahaba basins, the Coastal Plain area in Alabama and Mississippi, and the Texas–Louisiana–Arkansas area.

Piedmont

No original forest remains in the Oak–Pine region in Georgia. The ubiquitous oak and oak–pine woods, frequently interrupted by peach orchards on the Atlanta plateau, are similar to those already treated in the Piedmont district of the Atlantic Slope section of which this is a continuation. Higher elevations and deeper dissection on the Dahlenega Plateau of northern Georgia result in some similarity with the adjacent Blue Ridge. The occurrence of longleaf pine more or less locally on the Piedmont of western Georgia allies this area with the Gulf slope.

The Piedmont of Alabama (including the Blue Ridge)[12] may be considered as the most representative part of the Oak–Pine region of the Gulf slope. Except for the abundance of longleaf pine and the occasional other species of the Southeastern Evergreen region, this area is similar to the Piedmont district of the Atlantic slope. A part of the area is mountainous—Cheaha (or Chehawhaw) Mountain reaches an elevation of 2400 feet—and here the chestnut oak (*Q. montana*) and chestnut[13] of the Appalachian Upland are in contact with longleaf pine forests. Longleaf pine reaches an elevation of 1900 or 2000 feet on the slopes of Cheaha, and covers the tops of the ridges on some of the lesser mountains.[14] That there was a fine development of longleaf pine forest in this area is clearly shown by Mohr's (1897) statement concerning forests in Clay County:

> There I found the foothills and narrow valleys between them, at an elevation of from 1400 to 1500 feet, covered with a truly magnificent forest of *Pinus palustris,* yielding to the acre as much merchantable timber as the best class of pine lands in the coast pine belt from Alabama to Texas.

This fine type of longleaf pine forest was somewhat exceptional; most of it is composed of less stately trees. "At its vertical limit of distribution

---

[12] Harper (1943) distinguishes a Blue Ridge region in Alabama.

[13] Chestnut began to die out here before the blight was introduced to America. Mohr, in 1901, wrote "The chestnut, originally one of the most frequent trees of these forests, is at present rarely found in perfection. The older trees mostly show signs of decay, and the seedlings, as well as the coppice growth proceeding from the stump, are more or less stunted. It is asserted by the old settlers that this tree is dying out all over the mountain region, where at the beginning of the second half of the century it was still found abundant and in perfection." Reed (1905a) said concerning a forest tract of 30,000 acres in Coosa County: "The chestnut was formerly a common tree throughout the longleaf pine land, but during the last fifty years it has been dying out. At present the dead stubs are often found scattered among the pines, but sound living trees are extremely scarce. Only one perfect specimen was seen during the two months spent upon the tract."

[14] On Weogufka Mountain, "on the southern exposures the pure longleaf forest extends up the slopes and covers the tops of the ridges" (Reed, 1905a).

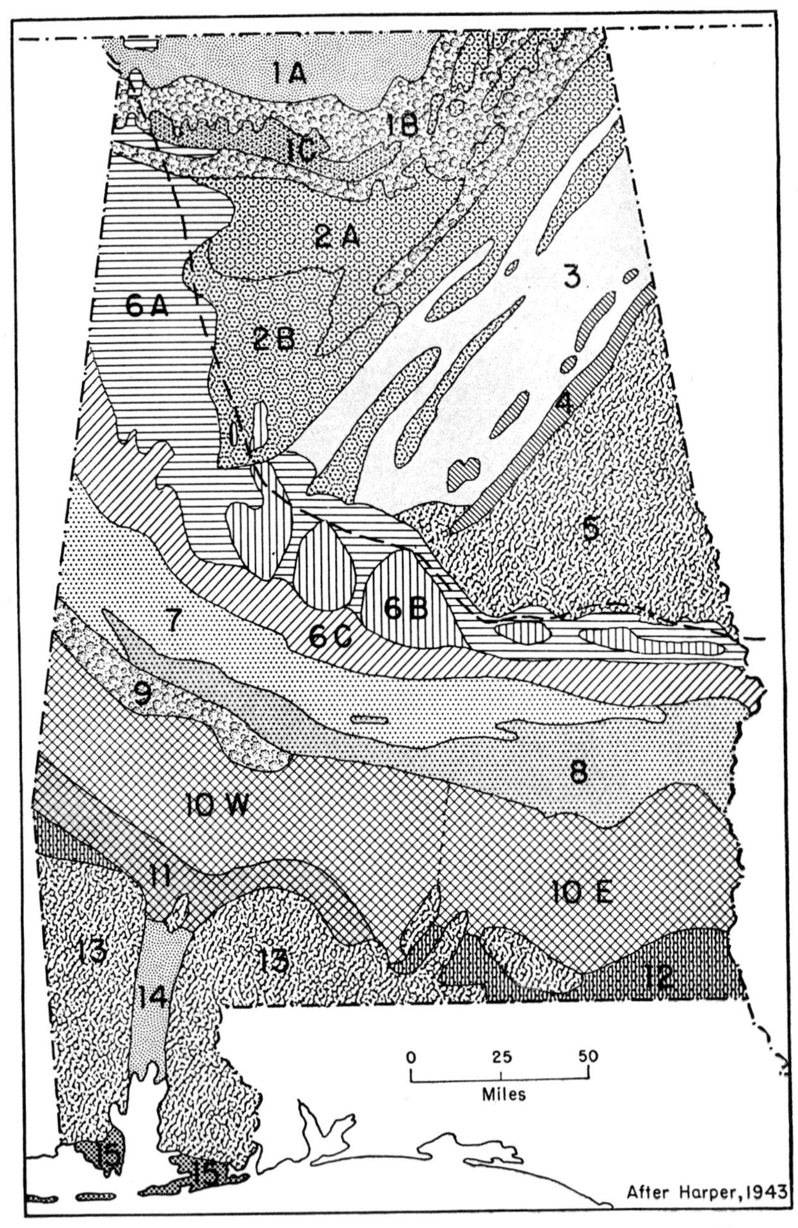

Map of Alabama, showing the "regions" distinguished by Harper. This map illustrates the control of geologic features in the distribution of vegetation. Toward the northeast, bands in general lie in a southwest-northeast direction in accordance with the structural features of the Appalachian Highlands. South and west of the Fall Line (heavy broken line), a concentric banding is evident which reflects differences in lithologic character of the coastal plain sediments, here in a pattern known as a "belted coastal plain." (*Continued on facing page*)

[on Cheaha] the pine is suddenly replaced by mountain oak [*Q. montana*], chestnut, and pignut hickory" (Mohr, 1901). Over much of the Piedmont upland, dry oak and pine woods prevail. The most abundant species are the pines (*P. Taeda, P. palustris, P. echinata*), the oaks (*Q. falcata, Q. stellata, Q. velutina,* and in the driest situations, *Q. marilandica*), and the hickories (*C. tomentosa, C. glabra*) with dogwood abundant in the understory.

Ravine slopes are occupied by a more mesophytic forest in which may be found beech and other hardwoods—tuliptree, white ash, maple,[15] white oak, holly, redbud, bigleaf magnolia, and *Halesia carolina*. This is a Gulf slope representative of the Mixed Mesophytic Forest association.

The lowland forest is a mixed hardwoods community made up of lower slope mesophytes and of trees of alluvial bottoms, together with a considerable admixture of pine. It is similar in character to that of the Piedmont of the Atlantic slope (p. 265). The broad-leaved trees listed by Reed (1905a) as composing the forest which covers the "creek land" (i.e., which "covers the bottoms of the valleys and sometimes reaches a short distance up the lower slopes") are: oaks (white, cow, post, chestnut, black, Spanish, red, willow, water), three or four species of hickory, beech, ash, tuliptree, sour gum, sweet gum, two or three species of magnolia [probably *M. virginiana, M. macrophylla,* and *M. tripetala* as these are listed for the district by Harper], dogwood, sourwood, red and sugar maple, blue beech, a species of elm [probably *Ulmus alata*], basswood, holly, hornbeam (*Ostrya*), and river birch. These greatly predominate over the pines.

Great Valley

The southern end of the Ridge and Valley Province from below Knoxville southwestward across northwestern Georgia and northeastern Alabama to

[15] Mohr lists *Acer Rugelii;* Harper lists *Acer rubrum* and *Acer leucoderme.*

(*Continuation of legend for illustration on facing page*)

The regions distinguished by Harper are:

- *1.* Tennessee Valley
  - *1A.* Chert Belt
  - *1B.* Tennessee Valley proper
  - *1C.* Little Mountain
- *2.* Coal Region
  - *2A.* Plateau
  - *2B.* Basin
- *3.* Coosa Valley
- *4.* Blue Ridge and outliers
- *5.* Piedmont
- *6.* Central Pine Belt
  - *6A.* Shortleaf Pine Belt
  - *6B.* Longleaf Pine Hills
  - *6C.* Eutaw Belt
- *7.* Black Belt
- *8.* Blue Marl Region
- *9.* Post Oak Flatwoods
- *10.* Southern Red Hills
  - *10W.* Western Division
  - *10E.* Eastern Division
- *11.* Lime Hills
- *12.* Lime-sink Region
- *13.* Southwestern Pine Hills
- *14.* Mobile Delta
- *15.* Coast Strip

the Fall Line is included in the Oak–Pine region. The transition from the oak communities of the Ridge and Valley section of the Oak–Chestnut region to the oak–pine communities of this section is gradual. The fertile soil of the valley lands is almost entirely under cultivation and little remains to indicate the nature of the original forest. An increase southward in the amount of loblolly pine on the hills and ridges, and, near the border of Alabama, the presence of longleaf pine distinguish the southern end of the Great (or Appalachian) Valley—the Coosa Valley region of Harper (1943).

The longitudinal ridges of chert and sandstone which rise above the valley have poor thin soils, and are occupied by forests comparable to those of the mountains of the Piedmont (p. 273). In Cherokee and Etowah counties, Alabama, areas of sandy and gravelly soils are occupied by longleaf pine flatwoods "resembling some near the coast even to the extent of containing pitcher-plants" (Harper, 1943). In the low flatwoods, sour gum, sweet gum, and loblolly pine are common trees.

Warrior and Cahaba Basins

The drainage basins of the Warrior and upper Cahaba rivers in Alabama have developed in a region of younger rocks of the Coal Measures. This is a moderately hilly area, most rugged along its southern borders. The Warrior Basin is the southern end of the Cumberland Plateau. Elevations range from about 100 to 800 feet, considerably lower than in the Warrior Tableland[16] to the north. "There are dry oak and pine forests, now including considerable second growth, on the ridges or uplands, dense thickets of spruce (or cliff) pine on the brows of bluffs, beech and various other hardwoods on steep rich slopes, and still others along streams." (Harper, 1943). The forests of ravine slopes, with their beech, tuliptree, basswood, white oak, white ash, wild cherry, cucumber tree, red oak, sugar maple, bigleaf magnolia, holly, redbud, and yellowwood, are southward extensions of mixed mesophytic forest. As in other parts of the Oak–Pine region of the Gulf slope, there is a pronounced admixture of southern species—*Magnolia virginiana* along creeks, *Quercus laurifolia* on sandy river banks, *Quercus Durandii* on limestone hills, and *Pinus palustris* in the poorest soils. The latter covers isolated areas of sandy soil near the northern limits of the Warrior Basin.

Floristically, this area is of particular interest because of the occurrence of certain endemic species. *Neviusia alabamensis* grows on shaded sandstone cliffs of the Warrior River; *Croton alabamensis,* an evergreen shrub of tropical affinities, grows in rocky woods in the Tuscaloosa and Little Cahaba valleys (Mohr, 1901; Harper, 1922).

[16] This area, although transitional in character, was included in the Mixed Mesophytic Forest region (p. 113).

Coastal Plain Area in Alabama and Mississippi

In the vicinity of Tuscaloosa, Alabama, the outer boundary of the Oak–Pine region diverges from the Fall Line which there turns northward. The remaining area of the Gulf Slope section of the Oak–Pine region is on the Coastal Plain. Because this is a belted coastal plain (i.e., with more or less parallel bands differing from one another in geologic and topographic features), its vegetation is arranged in more or less distinct belts which extend in a general southeast-northwest direction (Map, p. 274).

The Shortleaf Pine Belt is the inner (northeasterly) of these belts, and extends from Tuscaloosa northwestwardly.[17] It is an area of rounded hills and broad often swampy valleys. The upland forest of this area is "distinguished by the frequency of the shortleaf pine [*P. echinata*] among the hardwood trees, mostly upland oaks and hickories, the pine having originally constituted about one-half of the tree growth" (Mohr, 1901). The loblolly pine is abundant in second-growth stands.

A second belt of the Coastal Plain, the Black Belt or Northeast Prairie Belt, is characterized by the occurrence of originally treeless areas occupied by prairie vegetation. This belt has already been referred to in the Western Mesophytic Forest region into which it extends in northern Mississippi and southern Tennessee (p. 158). It is a prominent belt of the Coastal Plain of Alabama, where it will again be mentioned in the Southeastern Evergreen Forest region (p. 289). The prairie soil, derived from the Cretaceous limestone, is black when wet—hence the name Black Belt; it supports typical prairie communities. The belt is further characterized by the groves of trees, chiefly oaks, on the "higher reddish soil areas which dot the prairie surface like islands." Stream bottoms within this belt were originally occupied by dense hardwood forests.

Only the southern end of Pontotoc Ridge, a third coastal plain belt, extends into the Oak–Pine region. The vegetation of this belt belongs with the region to the north.

The Flatwoods Belt—a narrow band only three to fifteen miles wide—is a rather flat area of heavy soils, waterlogged during rainy periods, dry and hard at other times; yellow and loblolly pines and blackjack, post, and Spanish oaks are the principal trees. Southeastwardly, this belt extends nearly half-way across Alabama where, because of its many southern species, it must be included in the Southeastern Evergreen Forest region.

The North Central Plateau, to which about one-half of the area of the Oak–Pine region in Mississippi belongs, is a maturely eroded plain of about 400 to 600 feet in elevation trenched by many streams with broad

---

[17] It includes that part of the Central Shortleaf Pine Belt of Harper (1943) to the northwest of Tuscaloosa, which is approximately the Shortleaf Pine Belt of Mohr (1901), and the southern end of the Tennessee River Hills of Lowe (1921).

valleys. The uplands are well drained; their forests a mixture of pines and oaks with hickories and a few other species, or pure stands of pine. Yellow pine was more abundant in the original forest; loblolly pine is everywhere the most abundant tree in second-growth stands. Westward, as the belt of loess hills is approached, pines are less abundant and the number of hardwood species greater. The loess hills here belong in the Western Mesophytic Forest region; the mixed mesophytic communities of these hills extend southward, gradually changing in character with the admixture of broad-leaved evergreen species of the Southeastern Evergreen Forest (pp. 161, 301).

Texas–Louisiana–Arkansas Area—The Coastal Plain West of the Mississippi River

Except at its southern end in Texas, the land occupied by oak–pine forest is rolling to hilly; most of it is maturely dissected. The larger streams of this area flow in broad alluvial valleys whose hardwood forests contrast in appearance and species with the pine–oak and oak–hickory communities of the upland.

The upland forests of this area have much in common with those of the Oak–Hickory region which is adjacent on the north. The transition from the Oak–Hickory to the Oak–Pine region is indicated by the strong admixture of pine in the Ouachita Mountains (p. 173), and by the occurrence of oak–hickory communities on the Coastal Plain of southern Arkansas in the Oak–Pine region. An area of overlap rather than a boundary lies between these two regions. The oak–hickory forest of the best soils is made up of oaks (*Q. alba, Q. Prinus, Q. falcata, Q. Shumardii*), hickories (*C. ovata, C. tomentosa, C. cordiformis*), and a number of other species. A black oak–hickory type resembling that in the mountains to the north except for the presence of sweet gum and Shumard red oak, occurs in less superior soils, while a post oak–blackjack oak type on the poorest soils is essentially like that in the mountains (Turner, 1937). The admixture of pines in the several oak and oak–hickory types is not confined to the Coastal Plain. However, the abundance of loblolly pine on the Coastal Plain is in contrast to its infrequency in the mountains, and thus may serve to characterize the Oak–Pine region.

Through all except the southern part of this vegetation area in Texas, yellow pine was originally the more abundant pine. Thus the northwestern part of Louisiana was referred to as "shortleaf pine uplands" by Foster (1912). Tharp (1926) distinguished two subdivisions in his "pine–oak forest" of eastern Texas, a northern in which yellow pine is dominant, and a southern in which loblolly pine is dominant. Three oaks (*Q. stellata, Q. marilandica,* and *Q. falcata*) are generally to be found with the pines, the two former always as understory species, the last sometimes attaining the dominant layer. Sweet gum, also, may be associated with the pines. Toward the southern margin of this area, the admixture of longleaf pine indicates a transition to the Southeastern Evergreen Forest.

The bottomland forests of the Texas–Louisiana–Arkansas area of the Oak–Pine Forest region are essentially tongues of the hardwood forest of the alluvial plain of the Mississippi River or of the streams of the longleaf pine belt. They contain a large number of species including trees of wide distribution in alluvial soils almost throughout the deciduous forest, trees of moderately southern range (as *Taxodium distichum, Quercus lyrata, Q. Phellos, Q. Prinus,* and *Celtis laevigata*), trees of more circumscribed, generally southwesterly range (as *Ulmus crassifolia, Carya Buckleyi*), and evergreen trees of decidedly southern range (as *Magnolia grandiflora, Persea borbonia,* and *Quercus virginiana*).

CHAPTER 9

The Southeastern Evergreen Forest Region

The most prominent vegetational feature of much of this region is the preponderance of evergreen trees. The longleaf pine forests of the sandy uplands—differing in character and associates on soils of different texture and moisture content—dominate the landscape of much of the Atlantic and Gulf Coastal Plain. The maritime live oak forests, draped with Spanish moss (*Tillandsia usneoides*), also emphasize the evergreen aspect of these southeastern forests. The evergreen *Magnolia grandiflora* is a prominent tree of the hammocks of northern Florida and adjacent Georgia and Alabama, and of mesophytic forest communities of ravine slopes and low bluffs in southern Alabama, Mississippi, and Louisiana. There is scarcely a forest community, upland or lowland, without some admixture of coniferous or broad-leaved evergreen trees and shrubs. The question may well arise as to why this region should be included with the Deciduous Forest Formation.

Communities in which deciduous trees predominate do occur; and deciduous species may form the subdominant layer in pine communities. Successional trends, and the status of the several communities (developmental or climax), together with the controlling or modifying effects of topography, soil, and fire, point unmistakably to the potentially greater importance of broad-leaved deciduous trees.

Although areally dominant, the longleaf pine forest (the community largely responsible for the name "southeastern evergreen forest" so frequently used) is but an edaphic climax so modified and stabilized by recurring fires that it is considered a fire subclimax. (H. H. Chapman, 1932; Garren, 1943, and the many references there cited.) The live oak forest of the maritime belt is interpreted by Wells (1939, 1942) and by Wells and Shunk (1938) as a salt spray climax. The admixture of broad-leaved evergreens almost throughout the area is related to the low latitude (and consequently long growing season and mild winters), and is indicative of a transition to the subtropical forest of southern Florida. In Florida, where one after another of the deciduous species reaches the southern limits of its range, the transition to the Subtropical Evergreen Forest is most pronounced. Southern Florida is in no sense a part of the deciduous forest; warm temperate and tropical species prevail (J. H. Davis, 1943).

The deciduous (or prevailingly deciduous) communities, and the role of deciduous species in succession will be emphasized because of their genetic relation to the deciduous forest as a whole.

The Southeastern Evergreen Forest region is confined to the Coastal Plain, and except at the north and in the Gulf States, is essentially co-extensive with that physiographic area (see boundaries of Oak–Pine region, pp. 267, 271). It extends from the James River in Virginia almost to the limit of forest in Texas; peninsular Florida is not included. In the Mississippi embayment, bottomland forests essentially like those near the coast extend northward on the Mississippi alluvial plain to the southern tip of Illinois, sending tongues into the adjacent Western Mesophytic and Oak–Hickory Forest regions (pp. 150, 159, 174, 178).

The Coastal Plain is a relatively young land area. Its oldest rocks, along the inner border, are Cretaceous in age; most of the area is underlain by Tertiary sediments. That all of the Coastal Plain has not been a land surface continuously since its initial emergence from the sea is indicated by a series of terraces, successively younger toward the coast, now generally considered to be Pleistocene in age. On the Atlantic Coastal Plain, the series of terraces occupy the greater part of the area; the older ones are much dissected, the younger still undissected and their surfaces comparatively smooth.[1] On much of the Gulf Coastal Plain, where local relief is frequently greater, the older terraces are poorly, if at all, represented, or their places are taken by stream terraces. In Louisiana, Holland (1944) distinguishes two major physiographic areas, an older, which he terms the Tertiary Uplands, and a younger, the Quaternary Lowlands, comprising four Pleistocene terraces and several Recent alluvial surfaces. The terraces may be in part fluvial, and in part marine in origin. "The Pleistocene terraces have been to some extent correlated in age with the several glacial invasions" (Fenneman, 1938). During the periods of glaciation a great volume of water was locked up in the ice sheet and the sea in proportion withdrawn from the land.

Intermittent rise and fall of the sea relative to the land, the deposition of new material over an older surface, the emergence of a bare coastal strip, all profoundly affect vegetation and the possibilities of migration. The relatively recent development of a coastal plain vegetation, particularly that of the outer (younger) terraces, must be kept in mind in any interpretation of its features (see Chapter 17). The greater amount of overlap of ranges of Appalachian and southern (Coastal Plain) species on the Gulf than on the Atlantic slope may well be related to differing extent of Pleistocene terraces.

The Southeastern Evergreen Forest is floristically distinct from all other

[1] Maps of these terraces, their names, and tentative age, may be found in Fenneman (1938); also see Chapter 16.

forest regions of temperate North America and contrasts strongly with the adjacent ancient Appalachian Upland. Except in the deciduous forest communities and in parts of the older inner belt of the Coastal Plain, a large proportion of its species are essentially limited to the Coastal Plain. Fernald (1931) emphasized this distinctness as follows:

> Descending to the broad outer Coastal Plain of Atlantic North America, the region of siliceous and acid peaty soils of Tertiary or later origin, we come to a change in the flora so striking as immediately to challenge attention; for, whereas the distinctive non-endemic genera of the southern Appalachian region are those shared with temperate eastern and central Asia or with southern and eastern Europe, the Old World elements in the indigenous Coastal Plain flora of Atlantic North America are prevailingly the groups with wide tropical and subtropical range, the families, tribes, genera, and sections shared by the Americas with tropical and subtropical Africa, tropical and subtropical Asia, and Australia but for the most part unknown in temperate Europe and largely unrepresented in temperate Asia.

On a floristic basis, the Southeastern Evergreen Forest region might well be excluded from the Deciduous Forest. Nevertheless, there are distinct floristic relationships with the ancient Appalachian Upland. The flora of the hardwood communities of the Coastal Plain has evidently migrated from the Appalachian area. Thus, in the mesophytic hardwood forests of slopes and ravines, a number of wide-ranging species of the deciduous forest may be seen, as *Parthenocissus quinquefolia, Hydrangea arborescens, Trillium sessile, Laportea canadensis, Aristolochia Serpentaria, Polygonum virginianum, Actaea pachypoda, Impatiens pallida, Eupatorium rugosum, Adiantum pedatum, Phegopteris hexagonoptera,* and *Polystichum acrostichoides.* With these, are generally associated a few more southern forms. This flora of the mesophytic hardwood forests contrasts strongly with the more typical coastal plain and southern flora of the region. Some of the so-called "coastal plain species" are derived from closely related Appalachian species, or are represented on the Appalachian Upland by isolated relic occurrences. The majority are more southern in their affinities.

The vegetation of this region is in part warm temperate-subtropical, in part distinctively coastal plain, and in part temperate deciduous. It is made up of a variety of widely different forest communities—coniferous, mixed coniferous and hardwood, deciduous hardwood, and mixed deciduous and broad-leaved evergreen hardwood—interrupted here and there by swamps, bogs, and prairies. The large number of unlike communities is related to the diverse environmental conditions of the region. Differences in age of the land surface, and hence in degree of dissection; differences in soil materials (coastal plain deposits) and hence in topography, soil texture, and soil moisture relations; differences in fire frequency and hence in fire effects; and finally, differences in land utilization and hence in secondary communities even on adjacent and environmentally like tracts, all contribute to the multitude of communities of the region.

Interstream areas are flatter and more poorly drained on the younger (outer) terraces, stream bottoms are broader, and slopes limited to the immediate vicinity of bottoms. Here the evergreen shrub-bogs ("bays" or "pocosins"), the cypress swamps and cedar bogs, the savannahs and flatwoods are extensively developed on the flat uplands; cypress and bottomland hardwood forests of the lowlands are unexcelled. Sand hills—old dunes associated with ancient shore lines—are favorable for the development of xeric pine communities. Coastal features—dunes, beaches, tidal estuaries, and lagoons—have their own characteristics favoring distinct vegetation types.

On the Gulf Coastal Plain, the greater diversity in the nature and degree of consolidation of sediments (which have resulted in the development of a typical belted coastal plain) is reflected in bands and patches of unlike vegetation (Map, p. 274). It is also evident on detailed soil maps where the contrast between the Atlantic and Gulf Coastal Plains is marked. Such terms as "red lands," "red hills," "black belt," and "lime-sink region," have reference to soil (and topographic) areas related to underlying calcareous deposits. Their vegetation differs from the more widespread types. On the whole, the red soils, of which the Orangeburg is representative, are generally derived from calcareous deposits and are more fertile than the prevailing widespread sandy soils of the upland and support a higher proportion of hardwood trees. The "Black Belt" gets its name from the dark color of its soils, clay soils with a high calcium carbonate content. Here the original cover was prairie and open deciduous forest. Only a detailed study of the vegetation of the Coastal Plain can take cognizance of these many striking variations. (For details see Hilgard, 1860; Mohr, 1901; Harper, 1906, 1913, 1914, 1928, 1943, and other papers; Wells, 1928, 1942.)

The impression gained in passing from the Oak–Pine region southward across the Gulf Coastal Plain is one of pronounced and sudden changes in vegetation. In places, the woodlands are almost entirely deciduous, and only the abundance of Spanish moss (*Tillandsia*) gives the impression of southern forest. Through the Black Belt, showy flowers (white *Oenothera, Phlox, Verbena,* etc.) are abundant. Everywhere, slight changes in level result in differences in communities: oak and hickory groves, or pine woods on the hills; swampy forest with sweet gum, water and willow oaks, sugarberry, and, perhaps, beech and tuliptree in low flats, always with a dense and tangled undergrowth of shrubs and vines, among which sweet bay magnolia is often conspicuous. On rounded hills, "hammocks," the evergreen magnolia, or beech and magnolia are sometimes dominant, while in lower situations they are mingled with a variety of other species. In the deep swamps, cypress dominates. However, there may be only a few feet difference in elevation between deep swamps and broad-leaved woodlands. In the southern belt of the Coastal Plain (Southwestern Pine Hills, Lower Coast Pine Belt, or Longleaf Pine Belt), the dominance of longleaf pine gives an aspect quite unlike that of the other belts. Here is a grassy ground-cover and abundance of characteristic flowers and ferns—chain fern, *Zygadenus,* lupine, *Baptisia,* and, most conspicuous of all, the white and red pitcher plants (*Sarracenia Drummondii*). Pine and sweet bay flats, with their dense tangle of shrubs and lianes, interrupt the longleaf pine woods.

The coniferous evergreen forest communities exceed all others in area occupied. Of these, the longleaf pine forests (of which there are several types) are the most characteristic and unifying vegetation feature of this southeastern region. These are essentially the same in composition, except for minor species, through the 1200 miles extent of this region. Equally similar in composition throughout, but more circumscribed in area, are the bottomland forests. Because of this vegetational unity, no sections are distinguished. Floristic divisions, based on characteristic endemic species, have long been recognized (Harshberger, 1911). As this region is in part transitional to another forest formation—the Subtropical Broad-leaved Evergreen Forest—and hence not a typical part of the Deciduous Forest Formation, its communities are characterized briefly. For more detailed information on the plant communities of the Coastal Plain, the environment, and the effects of fire, the reader is referred to the literature on that area. The extensive bibliographies in papers by Wells (1942) and by Garren (1943) should be consulted.

Pine and Pine–Oak Forest Communities

A number of more or less unlike communities dominated by pines are usually distinguished. Prominent among these are the longleaf pine forests of sand hills and flat sandy stretches, the two-storied forests of pines and oaks, the longleaf–slash pine forests of the southern border of the area, and the loblolly or loblolly–longleaf pine forests of moist soils. Throughout the northern end of the region, and along its northern border on the Gulf Coastal Plain, loblolly pine is (and was originally) more abundant than longleaf pine. Southward, it was more generally confined to moist soil in shallow depressions or along the borders of the so-called "bays." As a result of disturbance it has increased greatly so that in many places where longleaf pine originally prevailed, loblolly (or old-field) pine is now the dominant species.

LONGLEAF PINE FORESTS

The most xeric situations of the Coastal Plain are sand hills. These form a belt known as the "Sand Hills" along the inner border of the Coastal Plain of the Carolinas and adjacent Georgia. Elsewhere in areas of deep sand deposits similar hills occur. A broad rolling sandy belt extends across southern Georgia, western Florida and southern Alabama into Mississippi. Longleaf pine alone of the pines is able to grow in this habitat. On the driest ridge-tops it may be stunted, but generally is a tree of stately proportions.

The more open xeric forests are two-storied communities, with a layer of scrubby oaks—chiefly turkey or fork-leaved blackjack, *Quercus laevis* (*Catesbaei*), and perhaps some upland willow oak or bluejack, *Quercus cinerea,* and blackjack, *Quercus marilandica.* Bare areas of white sand are invaded by a few herbaceous xerophytes, and later by wire-grass (*Aristida stricta*). In the better areas where disturbance has been less severe, and where moisture conditions are not too extreme, there are shrubs and a variety of attractive flowers which make the sand hills quite colorful in season.[2] The longleaf pine subclimax of the sand hills has been very generally replaced by scrubby oak communities as a result of cutting and frequent burning.

In the more southern sand hills or pine hills belt of Alabama

> on the hills and broad swells of the table-lands the long-leaf pine reigns supreme. The high forest is almost bare of undergrowth and its monotony is frequently unbroken for long distances, no other trees or shrubs appearing among the tall trunks of the pine, which spread their gnarled limbs at a height of from 40 to 65 feet above the ground. It is only in the accidental openings of the forest that a second growth of the predominating species takes possession of the ground, which if interfered with by human agency, is replaced by black jack and Spanish oaks, not rarely accompanied by dogwood (*Cornus florida*) and the glandular summer haw (*Crataegus elliptica*) . . . On the sterile ridges deeply covered with the mantle of loose white sands . . . the long-leaf pine becomes stunted and is more or less replaced by the barren or turkey oak and blue jack, trees rather below medium size, often dwarfed and scrubby (Mohr, 1901).

These southern pine hills forests are carpeted with grasses of many species, and with abundant herbaceous plants.[3]

There are all gradations between the pure longleaf pine forests, the two-storied pine–oak forests, and the scrubby oak communities. There is evidence that with protection from fire, successional development might ultimately lead to the establishment of an oak–hickory forest. Fire, if not too frequent, is considered to be a factor of great importance in maintaining the subclimax longleaf pine forests. Longleaf pine is capable of growing in a

---

[2] Wells and Shunk (1931) emphasize the pioneer and early successional stages of vegetation on the coarser sands of the North Carolina Coastal Plain. Coker (1911) describes the sand hills of South Carolina near Hartsville, and Harper (1906), of the Altamaha grit region of Georgia. In these papers can be found lists of herbaceous and woody plants. Mohr (1901) and Harper (1906) give lists of the herbaceous plants of the "rolling pine uplands" or "dry pine barrens."

[3] The West Florida Pine Hills (Harper, 1914) are a part of the broad southern belt of longleaf pine forest.

wide variety of habitats, from the driest to permanently moist. It does not, however, occupy land subject to inundation. In moist and wet soils, its dominance is shared with other species. The forests of moist or wet flats are frequently referred to as "flatwoods."

PINE FLATS, SAVANNAHS, AND BAYS

A hardpan layer is responsible for the poor drainage of much of the flat pineland. In other places, topography and the lack of stream dissection are responsible. Except in the wettest parts, longleaf pine may maintain dominancy among the tall trees; loblolly pine (*P. Taeda*) is frequently associated with it, or becomes the dominant tree. Near the Gulf Coast and in northern Florida, Cuban pine or slash pine (*P. caribaea*) enters this community. In the wettest areas, pond pine (*P. rigida* var. *serotina*) may be the only coniferous tree. Between the denser mixed pine stands of moist flats, and the open pine and evergreen shrub community (bay), or pine and grass–sedge community (savannah), there are all gradations. The "bays" and the "savannahs" are upland bogs in which there is a "well defined humus-infiltrated soil layer always present." This varies in color (gray, gray-black, or black) and in thickness (7 to 40 cm.), and overlies a sandy soil (Wells and Shunk, 1928).

The "bays," or shrub-bogs, are a dense tangle of evergreen shrubs, lianes, and small trees, among which are interspersed some deciduous species; this dense growth is usually overtopped by a scattering of loblolly and pond pines. In floristic composition and in prevalent growth form, these "bays" are a part of the Subtropical Evergreen rather than the Deciduous Forest Formation.

The "savannahs" or upland grass–sedge bogs are dominated by grasses and sedges among which are numerous showy flowers. The savannah in bloom is a beautiful community. Not the least of its showiness is due to the bright yellow trumpet-shaped leaves of the yellow pitcher plant (*Sarracenia flava*) which sometimes covers large areas. The local endemic, Venus' fly-trap (*Dionaea muscipula*), occurs in savannahs and openings in the shrubby bogs of wet pine flats. The popularly used term, savannah, refers alike to grass–sedge areas with scattered trees, and to treeless areas. The latter are sometimes called prairies, but differ both in habitat and floristics from the true prairies of the Gulf slope. The savannahs, or grass–sedge communities "are mostly of recent origin in the historical period due to destruction of an earlier shrub-bog vegetation under higher fire incidence of the white-man era" (Wells, 1942).

LOBLOLLY PINE AND PINE–HARDWOODS FORESTS

The loblolly pine (or pine–oak) communities of the northern end of the region and of its northern border in the Gulf States are transitional be-

Longleaf pine forest in the "flatwoods" belt of southeastern Georgia, a northward extension of the East Florida flatwoods. Saw-palmetto (*Serenoa repens*), inkberry (*Ilex glabra*), and grasses are the principal ground-cover plants. (Courtesy, U. S. Forest Service.)

tween this region and the adjacent Oak–Pine region. In historic time, loblolly pine has increased greatly, due to its rapid invasion of abandoned fields. In many parts of the Southeastern Evergreen region in which loblolly pine was originally fairly local, it is now more abundant than longleaf pine. The loblolly pine–hardwoods community also has increased its extent at the expense of shortleaf (yellow) pine and of longleaf pine (Pessin, 1933; Soc. Amer. Foresters, 1940). In Virginia, "pine barrens" in which loblolly pine is the dominant species are comparable to the longleaf pine barrens farther south. In both areas, *Quercus laevis* (*Catesbaei*) is regularly associated with the pines, and characteristic dry pine barren species, as *Pyxidanthera barbulata* and *Arenaria caroliniana,* are found (Fernald, 1937).

The second-growth loblolly pine–hardwoods community is now an important cover type in the Coastal Plain, occupying moist and fertile soils. In addition to pine, it may contain a large number of hardwood species

A fine stand of loblolly pine, with which oaks and other hardwood species are mingled. Francis Marion National Forest, South Carolina. (Courtesy, U. S. Forest Service.)

including oaks (white, willow, southern red, swamp red, post, swamp chestnut, water), hickory, sweet gum, sour gum, red maple, ash, holly, etc. (Garver, et al., 1931). Yaupon (*Ilex vomitoria*) is a common shrub of the southern loblolly pine forests, sometimes forming a more or less continuous layer, sometimes scattered, or intermingled with the deciduous hardwoods of the understory.

SLASH PINE FORESTS

Slash pine (*Pinus caribaea*), although a constituent of southern longleaf pine forests, is more abundant in a narrow belt along the Gulf Coast and southernmost Atlantic Coast, and in peninsular Florida. In this coastal belt it may occur in pure stands, or associated with other pines and oaks, principally evergreen species. Such forests are well developed on the coastal islands, sometimes in situations where the trees become partly buried by wind-blown sand (Pessin and Burleigh, 1941; Harper, 1943).

Slash pine is also a constituent of hydric communities,[4] where it is associated with pond cypress, *Taxodium ascendens,* or swamp tupelo, *Nyssa sylvatica* var. *biflora* (Penfound and Watkins, 1937).

PRAIRIES AND ASSOCIATED FOREST COMMUNITIES

On soils derived from calcareous strata in the Gulf Coastal Plain, the vegetation contrasts with that of the generally noncalcareous belts on either side. Among the trees, deciduous species prevail except on calcareous rock outcrops where red cedar (*Juniperus virginiana*) is abundant, or on the highest ridges capped with siliceous deposits where pines mingle with the deciduous trees. Natural openings occupied by prairie vegetation occurred in the original forest cover, and a few areas of typical prairie vegetation still remain.

Two prairie belts have been distinguished, the "Black Belt" which extends across Alabama, curving northward in Mississippi (where it passes out of the region), and the "Jackson Prairie" which crosses southern Mississippi. A third prairie area in southern Louisiana, to the south of the longleaf pine region, occupies the most recent Pleistocene terrace. Unlike the treeless savannahs of the Atlantic Coastal Plain, the grass and herb communities of these prairie belts are made up principally of species of the interior. Their forest communities, too, are for the most part outliers of deciduous forest: post oak on some of the flat, poorly drained hard soils and mixed hardwoods in the deeper and fertile soils of the lowlands (Mohr, 1901; Lowe, 1921; Harper, 1943).

[4] Harper (1943) refers the slash pine of wet soil to *Pinus Elliottii;* Little (1944) lists *P. Elliottii* as a synonym of *P. caribaea.*

Swamp Forests of the Uplands

As these are similar to the more extensive swamp forests which border streams in the region, they are considered but briefly here.[5] The typical *Nyssa–Taxodium* swamp forest of shallow lakes and ponds is not unlike the comparable bottomland type (p. 291). Swamp black gum or swamp tupelo (*Nyssa sylvatica* var. *biflora*), pond cypress (*Taxodium ascendens*) and pond pine (*Pinus rigida* var. *serotina*) are characteristic trees, all growing where water stands for long periods, but is subject to considerable fluctuation in depth. These are accompanied by a number of other trees, some evergreen, some deciduous. Slash pine (*P. caribaea*)[6] is associated with gum and cypress on the Gulf Coast. In the deep swamps there is very little shrubby or herbaceous vegetation; near the margins this increases in amount. Between the upland swamp forest and the typical "bay," there is no line of demarcation. Under the influence of fire, trees may disappear and a shrub-bog result. Upland swamps may grade into river swamps.

Southern white cedar (*Chamaecyparis thyoides*) forms almost pure stands in some upland depressions. Wells (1942) considers its establishment to be related to fire "which occurred at a time of subhigh water when the raw surface peat is not destroyed." Because this is a community growing on peaty accumulations, it is to be regarded as a bog rather than a swamp. "Without fire . . . old white cedar stands give way to a subclimax forest of red bay (*Persea*), ti-ti (*Cyrilla*), and sweet bay (*Magnolia virginiana*) trees, of which the red bay is dominant."

Bottomland Forests

The forests of alluvial land in the Southeastern Evergreen Forest region occupy a much larger proportion of the total land area than do flood-plain forests in any other region. Because bottomlands are often broad and hence their forests a conspicuous part of the vegetation, and because these forests are economically very important in the region, bottomland forests are considered apart from the other vegetation. Alluviation in these valleys, which slowly tends to raise some part of the bottomland above flood levels, and trenching of older deposits have resulted in the formation of "second bottoms," of low alluvial "ridges" in interstream areas in the bottoms, and of natural levees. These are most distinct in the Mississippi bottoms. The growth of natural levees near the streams decreases the possibility of drain-

---

[5] More information concerning these communities may be found in the following references: Mohr (1901); Penfound and Watkins (1937); Harper (1906, 1914); Hilgard (1860); Wells (1942); Buell and Cain (1943); Korstian (1924); Korstian and Brush (1931).

[6] See footnote 4, p. 289.

age in interstream bottomland areas, and often results in extensive interstream bottomland swamps and bayous.

The streams crossing the Coastal Plain are sluggish and generally carry a considerable amount of silt. In many cases, their mouths are obstructed by spits and bars, tending to increase the extent of swamp. Each of these sluggish streams is bordered by a band of bottomland forest. Near the coast, great swamp areas extend over the flat interstream uplands along the sluggish tributaries. Between upland swamp and river swamp there is here no line of demarcation, although there is a difference in the character of the sediment.

The great alluvial plain of the Mississippi River in its southern part interrupts the continuity of pine uplands which extend beyond it into Louisiana and Texas. The forest of the southern part of the alluvial plain of the Mississippi in Louisiana and Mississippi differs but little from its northern part in Tennessee, Missouri, Illinois, and Kentucky, or even from the extensions of this forest which follow tributaries upstream—along the Wabash in Indiana, and along the Tradewater and Green rivers in the alluviated basin of Kentucky. All of the Mississippi alluvial plain, therefore, should be included in the Southeastern Evergreen Forest region.

Three subdivisions of bottomland forest may be recognized: swamp forest (sometimes referred to as deep swamps, true swamps or sloughs), hardwood bottoms (or glade bottoms), and ridge bottoms (or cane ridges). Where levees border the stream, these separate the bottomland forest on the landward side of the levee, from stream margin communities (bands of black willow, hackberry, pecan, poplar, and sycamore) on the streamward slope. The swamp forest occupies land on which water stands throughout the year except during extreme droughts; its principal trees are cypress and tupelo gum. The hardwood bottoms are subject to frequent overflow, and are usually covered with water through the late winter and spring; the forest is made up of a large number of species. The ridge bottoms, elevated but a few feet above the general level of the bottoms, are covered by water only during floods; water drains off quickly instead of standing for months as in the hardwood bottoms. This forest contains some of the species of the hardwood bottoms, and a larger number of oaks and hickories. These subdivisions are recognizable in the Mississippi bottoms and along streams of the Coastal Plain; bottomlands throughout the region are very similar. Ridges are less often seen on the Atlantic slope than in the Mississippi alluvial plain, and overflow is more frequent.

SWAMP FORESTS (CYPRESS–TUPELO)

The most extensive swamps of this region are the Dismal Swamp (Kearney, 1901), the Pamlico Swamp, the Santee River swamp, the Okefenokee Swamp, the numerous bayou and slough swamps of the Mississippi,

A cypress swamp in North Carolina. (Courtesy, U. S. Forest Service.)

especially those of the Delta, and the swamps of the St. Francis River basin and Reelfoot Lake in the area of the New Madrid earthquake of 1811.

The swamp forests along streams in the low flat Coastal Plain border may merge with upland swamp forests. The line of demarcation is usually at the limit of alluviation. Alluvial swamps and nonalluvial swamps differ in soil or substratum, and to a considerable extent in species. Pond cypress (*Taxodium ascendens*), swamp tupelo (*Nyssa sylvatica* var. *biflora*), and pond pine (*Pinus rigida* var. *serotina*) are more characteristic of the latter which occur frequently in acid pinelands; bald cypress (*Taxodium distichum*) and water tupelo (*Nyssa aquatica*) are characteristic of the former. Upland ponds and swamps or bays may be the heads of "branches" or "branch swamps" which in turn lead to creek and river swamps. In the creek swamps along streams flowing from the pineland swamps, the swamp tupelo may be the dominant species, or there associated with bald cypress (Hall and Penfound, 1939a). Any large inland swamp area, as the Okefenokee Swamp, is in part upland (pond cypress) swamp, and in part river (bald cypress) swamp (Wright and Wright, 1932). The swamps of the Mississippi are all alluvial or bottomland swamps.

The principal trees of the river swamps and sloughs are bald cypress and water tupelo; with these are occasional other trees, as silver maple (*Acer saccharinum*), red maple (*Acer rubrum* and var. *Drummondii*), water and pumpkin ash (*Fraxinus caroliniana, F. tomentosa*), and, in the neighborhood of the pinelands, swamp tupelo. Pecan is a frequent constituent in the Mississippi bottoms.[7] Water-elm (*Planera aquatica*) forms thickets in the swamp forests. The cypress in these situations develops a wide flaring base and numerous knees; the water tupelo develops a swollen convex base and arching roots. These species may grow in pure or mixed stands, and always where water stands all or much of the year. Except in areas of very shallow water, there is little undergrowth in dense stands. The swollen bases of the water tupelo, however, afford a favorable habitat for mosses and liverworts, which grow profusely just above the water line, and for the gray polypody, *Polypodium polypodioides;* decaying logs and stumps also form footholds for various swamp herbs (Coulter, 1904; Hall and Penfound, 1939; 1943).

Reproduction in these swamp forests is dependent upon fluctuation in water level. Seeds of the dominant species cannot germinate under water. Adequate oxygen supply is available only when the water surface is lowered to the ground level (Demaree, 1932; Shunk, 1939). The development of enlarged bases and knees seems also to be related to gas exchange (Kurz and Demaree, 1934).

Near the borders of the deeper swamps, many other species of trees are associated with the cypress and tupelo, or replace them, resulting in the growth of a mixed hardwood stand of bottomland trees.

HARDWOOD BOTTOMS

The most extensive bottomland forests are those referred to as glade bottoms or hardwood bottoms. Although flooded for a considerable period, the surface through much of the year is dry. The hardwood bottoms forest of the Mississippi Valley consists of the following species, arranged in order of abundance: sweet (red) gum, red maple, swamp chestnut (cow) oak, swamp red oak, shingle oak, overcup oak, willow oak, elm, sassafras, hackberry (*Celtis laevigata*), papaw, dogwood, and Carpinus (Chittenden, 1905a). In the Mississippi Delta area, water oak and live oak are the principal species, along with sweet gum, green and pumpkin ash, red maple (*Acer rubrum* var. *Drummondii*), and tupelo. The southern aspect of this forest is emphasized by the abundant palmetto (*Sabal minor*). Live oak becomes more abundant toward the coast, and is the dominant tree near the bayou levees. "The forest on the hardwood bottoms of the Congaree River, in South Carolina, consists chiefly of red gum, cottonwood (*Populus heterophylla*), white ash, elm, sycamore, hackberry, some few oaks, and red and silver maples" (Chittenden, 1905a).

The forest of the hardwood bottoms is usually very dense, and in it, trees reach very large size, many over three feet in diameter. Small trees and

[7] Cain (1935) gives statistical data concerning a cypress swamp at Hovey Lake, in southwestern Indiana, almost at the upstream limits of this community.

Bottomland forest, with large sweet gum, in the Yazoo basin of Mississippi. The abundant lianes are a feature of bottomland forests. (Courtesy, U. S. Forest Service.)

shrubs are frequent and lianes luxuriant. In the southern part of the region, at least some of these are evergreens. Dense patches of cane, and tangles of Smilax often prevent tree reproduction.

RIDGE BOTTOMS

What are locally known as ridges occur in the bottomlands. These, although elevated but a few feet above the surrounding bottoms and almost imperceptible to the eye, are free of water most of the year; their soil is better drained and better aerated. Here are found some of the finest bottomland forests. Sweet gum is, as on the lower land, the predominant tree; with it are oaks (including white oak), shagbark hickory, and pecan. In addition, most of the species of the wetter bottoms are found. In the highest and best drained areas in the delta plains of the Yazoo, occasional beech and magnolia trees occur with the bottomland trees (Lowe, 1921). These ridges are (or were originally) covered by dense stands of cane.

SUCCESSION

The communities of river bottoms are developmentally related. Swamp and river-border communities (the latter of willow, cottonwood, river birch, silver maple, boxelder, hackberry, green ash, pecan, sycamore) are, as deposition proceeds, replaced by mixed bottomland forest—the sweet gum–red maple–swamp oak forest—of the hardwood or glade bottoms type. The stages in development differ from place to place, as indeed do the constituent species of the several communities. Black willow is a frequent, sometimes dominant, stream border species. River birch, cottonwood, silver maple, boxelder, green ash, hackberry or sugarberry (*Celtis laevigata*), pecan, and sycamore are usual constituents of marginal communities. In the delta of the Mississippi, the usual early stages of succession are black willow→hackberry–pecan→poplar–sycamore (fide Penfound). Excessive local deposition, resulting in the formation of natural levees and of the low cernuous ridges between bayous, permits the entrance of less hydric species, and results in the establishment of the sweet gum–white oak–hickory forest, or ridge bottoms type. The frequent appearance of mesophytes in the best drained sites points toward the establishment of a mesophytic forest containing some of the species of the slope forests.

BOTTOMLAND FOREST COMPOSITION[8]

Most data on the composition of bottomland forests do not separate the communities of the several bottomland habitats. All hardwoods are considered together, thus developmental sequences and correlations with site

[8] Additional material on bottomland forests may be found in the literature cited, and in Harper, 1906, 1914, 1943; Mohr, 1901; Tharp, 1926; Turner, 1937; Hazard, 1933; C. S. Chapman, 1905.

factors are obscured. However, as bottomland forests differ greatly in specific content from all other deciduous forest communities, a few examples of percentage composition are given (Tables 54, 55). These are selected

Table 54. Composition of bottomland hardwood forests in southern Illinois at the northern end of Mississippi alluvial plain: *1,* Maple–elm–pin oak–sweet gum forest in the Cache bottomland; *2,* Sweet gum–swamp white oak–pin oak on better drained parts of Cache bottomland; *3,* White oak–hickory–sweet gum forest in the Wabash bottoms. (Data from Telford, 1926.)

	1	*2*	*3*
Acer saccharinum	27.3	. .	. .
Ulmus americana	27.3	4.8	6.8
Quercus palustris	13.0	12.9	. .
Quercus bicolor	. .	25.8	. .
Quercus alba	1.3	. .	29.2
Liquidambar Styraciflua	11.7	37.1	19.2
Carya spp.	. .	9.7	21.1
Nyssa sp.	9.0	. .	6.2
Fraxinus sp.	7.8	1.6	4.3
Quercus falcata pagodaefolia	2.6	1.6	. .
Quercus velutina	. .	. .	8.1
Quercus Prinus	. .	1.6	. .
Platanus occidentalis	. .	1.6	. .
Populus deltoides	. .	1.6	. .
Morus rubra	. .	1.6	. .
Sassafras albidum	. .	. .	2.5
Juglans nigra	. .	. .	1.9
Gleditsia triacanthos	. .	. .	.6

so as to illustrate forests of widely separated areas and different predominant species.

Forest types of Table 54 are selected to illustrate decreasing hydrophytism: a maple–elm–pin oak–sweet gum forest in the Cache bottomlands of southern Illinois, a type common between the sloughs and well-drained benches; a sweet gum–swamp white oak–pin oak forest of better drained parts of the Cache bottomland; a white oak–hickory–sweet gum forest in the Wabash bottoms of southern Illinois, a type occupying "ridge bottoms" and well drained benches. Examples of hardwood bottomland forests of South Carolina, Missouri, and Tennessee are given in Table 55. In these, glade and ridge bottoms are not separated.

Hardwood Forests of the Uplands and Slopes

To the casual observer traversing the Atlantic Coastal Plain, its vegetation appears to be made up almost entirely of pine forests of one or another type (and the related savannah and bog-shrub communities) on the broad and poorly dissected interstream areas, alternating with broad strips of bottomland forest. On the Gulf Coastal Plain, due to the greater diversity of soils which the belted character of this plain introduces, mixed pine–hard-

Table 55. Composition of bottomland hardwood forest: *1,* "Virgin hardwood bottomland, South Carolina" (average of 37 acres); *2,* "Virgin hardwood bottomland, Missouri" (average of 76 acres); *3,* "Hardwood type," Dyer County, Tenn. (based on nine sample acres). (Data for *1* and *2* from Chittenden, 1905a; for *3,* from Clark, 1908; communities not distinguished; common names as in published tables.)

	1, S. C.	*2, Mo.*	*3, Tenn.*
Inches d.b.h.	18–50	18–49	18–46
Number per acre	20.11	24.89	13.64
Red gum	44.4	36.0	35.1
Cottonwood	20.3	. .	. .
Ash[1]	15.2	1.0	1.5
Maple[2]	3.4	22.5	4.8
Elm	2.3	11.3	2.8
White oak	. .	10.4	16.3
Hickory	. .	.7	14.8
Black gum	. .	7.3	11.4
Black oak	. .	9.2	. .
Red oak[3]	7.5	. .	4.1
Willow oak	. .	. .	6.5
Sycamore	4.1	. .	6.5
Hackberry	2.2	. .	. .
Other species	.6	1.6	.7

[1] Listed as white ash in South Carolina and Missouri.
[2] Listed as red maple in Missouri.
[3] Probably *Quercus falcata* or its var. *pagodaefolia.*

wood and hardwood communities are more frequently encountered on the uplands. Even here, the interruption of the more xeric upland types by broad strips of bottomland is pronounced. Cutting of pines has in some instances changed mixed communities into deciduous communities, hastening perhaps, the establishment of oak and oak–hickory forest, a successional tendency which is suggested by undergrowth and accessory species.

OAK–HICKORY FOREST

Indications from developmental trends point to the possibility of the establishment of an oak–hickory climax over much of the upland, if and when the slow accumulation of humus in the sandy soils makes conditions favorable for the dominance of deciduous trees. Through much of the area the inhibiting effects of fires tend to maintain the pine subclimax. Climatically, the region is capable of supporting deciduous forest on its uplands comparable to that on the adjacent Piedmont.

MESOPHYTIC MIXED HARDWOODS

Occasionally, on the low but abrupt slopes between upland and bottomland, groups of mesophytic deciduous trees are seen which belong neither to

(Oosting, 1942).

the pine→oak–hickory sere of the uplands, nor to the bottomland forest. The best indicator species is beech.[9] Its associates differ from slope to slope and from one geographic area to another. Beech, white oak, red maple, and sweet gum were noted in several such situations in South Carolina. Harper (1943) frequently mentions beech and a number of other hardwoods (as tulip, sweet gum, white oak, bigleaf magnolia, holly, cucumber magnolia, and rarely, basswood) as occurring on ravine slopes and bluffs of Alabama, in "rich woods," or occasionally on "second bottoms." In the pinelands of Texas, groves of beech occur on slopes (Tharp, 1926). In Louisiana, "Yellow poplar was once abundant in the well-drained coves of the slope between the valley (Red River) and the sandy pine region adjoining" (Foster, 1912). Beech and magnolia (*M. grandiflora*) are also present.

This ravine slope forest is a mesophytic mixed hardwood community in which beech is the most universally present and the most abundant species, comprising approximately one-third of the canopy layer. Data on composition are taken from Akerman (1925) in Virginia, and from the writers' observations in a forest near Wilmington, North Carolina (Table 56).

Akerman describes the habitat as follows:

> In Surry County [Virginia] the land usually slopes abruptly from the uplands to the stream beds and the slopes are indented by many ravines where the smaller water courses head back into the uplands. These steep slopes and ravines are covered with hardwood stands. The composition of these stands may be illustrated by giving counts made in the northern part of the county, on an acre basis: Beech 55, yellow poplar 45, bitternut hickory 35, elm 15, sycamore 5, sweet gum 5, white oak 5, red maple 5, black walnut 5, butternut 5 (Table 56, area 1).

A North Carolina example of this forest type at Washington Creek north of Wilmington occupies slopes of ravines cutting into the Wicomico terrace (Table 56, area 2). The soil is a medium fine sand blackened above with humus. Beech comprises a little over one-third of the canopy in a stand with 13 canopy species. Oaks (mostly white oak) and hickories (mostly white) each comprise about one-fifth of the canopy. Florida sugar maple (*Acer floridanum*), tuliptree, sweet gum, holly, walnut, and white ash are other trees of this area. Additional species are represented in the undergrowth in which Florida maple is most abundant. Shrubs are plentiful and include a number of evergreen southern species, thus emphasizing the subtropical relationship. Among the woody species are *Castanea pumila,*

[9] This is "white beech," which is "typically a tree of the southern and southeastern coastal plain. It is not unusual to find it around the cypress–tupelo swamps. . . . It is common along many of the southern rivers and forms quite good stands up to about 500 feet elevation in Virginia (as along the bluffs of the James River). In the Mississippi drainage it comes northward toward the glacial boundary; along the Atlantic coast it is plentiful northward into Maryland" (W. H. Camp, in litt., 1948).

Table 56. Composition of mesophytic mixed hardwoods communities of ravine slopes: *1*, Surry County, Virginia (data from Akerman, 1925; size classes not stated); *2*, Slopes of Washington Creek, Pender County, North Carolina (sample of 164 canopy trees).

	1	*2*	
		Canopy	*Under*
Fagus grandifolia	30.6	36.6	x
Liriodendron tulipifera	25.0	6.1	x
Carya cordiformis	19.4	. .	. .
Carya tomentosa	. .	15.8	x
Quercus alba	2.8	14.0	x
Acer floridanum	. .	9.1	X
Ulmus alata	8.3	. .	. .
Liquidambar Styraciflua	2.8	5.5	. .
Quercus velutina	. .	4.3	x
Carya pallida	. .	3.1	. .
Platanus occidentalis	2.8	. .	. .
Acer rubrum	2.8	. .	. .
Juglans nigra	2.8	1.2	x
Juglans cinerea	2.8	. .	. .
Ilex opaca	. .	1.8	x
Fraxinus americana	. .	1.2	x
Quercus falcata	. .	.6	. .
Quercus nigra	. .	.6	. .
Cornus florida	. .	. .	x
Oxydendrum arboreum	. .	. .	x
Carpinus caroliniana	. .	. .	x
Morus rubra	. .	. .	x
Nyssa sylvatica	. .	. .	x
Symplocos tinctoria	. .	. .	x
Aralia spinosa	. .	. .	x

x, species present in understory; X, subdominant species.

Asimina parviflora, Hamamelis virginiana, Ilex glabra, Evonymus americanus, Vaccinium arboreum, Gelsemium sempervirens, and *Callicarpa americana.* The herbaceous layer is made up of species of the central deciduous forest; characteristically southern or coastal plain plants are lacking. This situation regarding woodland herbs was noted in Virginia by Fernald (1937) who pointed out that such habitats have "an unexpected number of species characteristic of the richer woodlands . . . of the interior."

That this is a forest community more mesic than the oak–hickory of the uplands, yet distinct from that of the alluvial bottomlands, is at once apparent. It is essentially the same as the community of comparable situations in the Oak–Pine region (p. 265). Here on the Coastal Plain it is occupying slopes cut in the Pleistocene terraces. At least a part of its species have migrated from the older land area of the interior; a part are of southern origin; and at least one, *Acer floridanum,* is a coastal plain derivative of an upland species. If the community of mature topography, the most mesic community, is to be regarded as climax, then the climax of the northern half of the region (the Gulf strip is different) is a mixed deciduous forest having certain resemblances to the Mixed Mesophytic association.[10] It certainly is in large part derived from that major association, and suggests the climatic possibility of further encroachment of mixed mesophytic forest. Such encroachment would be impossible as long as the present sandy soils and immature topography persist.

BEECH–MAGNOLIA FOREST

Farther south, the mingling of broad-leaved evergreens with deciduous trees results in another type of mesophytic hardwood forest—the beech–magnolia forest—in which the deciduous beech and evergreen magnolia (*M. grandiflora*) are the principal species. This forest community occupies dissected river bluffs and their associated ravines (habitats similar to those of the mesophytic mixed hardwood community just considered) and "hammocks." Its soils are generally fertile, often slightly calcareous.

The beech–magnolia forest is well represented in the "red lands" and "hammock regions" of northern Florida, southern Georgia, and southern Alabama; and in Mississippi and Louisiana, on ravine slopes in the Prairie and Montgomery terraces, and in the "Cane Hills" or loess bluffs of the Mississippi River.

The composition of the beech–magnolia forest varies in different parts of its range. Among other trees frequently associated with the two principal species are *Ilex opaca, Quercus alba, Q. nigra, Q. laurifolia, Q. Prinus, Acer floridanum, Ostrya virginiana,* and *Cercis canadensis.* Other less frequent species include *Quercus Shumardii, Q. Muhlenbergii, Halesia carolina, H. diptera, Morus rubra, Fraxinus americana, Juglans nigra, Carya tomentosa,* and *Tilia* sp. Southward additional broad-leaved evergreens enter, among which are *Osmanthus americanus, Persea borbonia,* and *Prunus caroliniana,* gradually changing the appearance of the community by the prevalence of the evergreen growth form. Shrubs also are partly deciduous, partly evergreen species—species of the deciduous forest, species of the broad-leaved evergreen forest. Where evergreen species prevail in upper layers, the ground vegetation is poor, and includes a few saprophytes and root parasites; *Mitchella repens* is frequent. However, delicate herbs of more northern mesophytic forests are not lacking.

Specific mention must be made of the Apalachicola River bluffs because

---

[10] Wells (1928, 1942) and Wells and Shunk (1928) refer to a beech–maple climax in this area. The community bears little resemblance to the Beech–Maple climax, containing, in all its woody layers, species which never enter the Beech–Maple region. This mixed forest of slopes of the Coastal Plain was derived from the same ancestral forest as was the Beech–Maple, and by migrations which must have begun soon after the uplift of the Schooley peneplain, but which have only recently reached the younger land areas (see Chapters 17, 18).

of their outstanding examples of the beech–magnolia forest, of the occurrence there of a large assortment of herbaceous mesophytes, and particularly because of the occurrence there of two endemic trees, *Torreya taxifolia* and *Taxus floridana,* and the somewhat less local *Croomia pauciflora.*[11] In this area and the nearby "Marianna Red Lands," herbaceous mesophytes are abundant. *Sanguinaria, Hepatica, Dentaria, Anemonella, Trillium* (*Hugeri* and *lanceolatum*), *Polygonatum biflorum, Arisaema Dracontium, Phlox divaricata, Stellaria pubera, Scrophularia marilandica, Solidago latifolia,* and the ferns, *Polystichum acrostichoides* and *Phegopteris hexagonoptera,* suggest more northern forests. Some of these are not known farther south. *Pachysandra procumbens,* a characteristic and abundant species in the Western Mesophytic Forest region, occurs as a disjunct in rich woods of the Marianna Red Lands and also farther west in the loess hills of Mississippi (Harper, 1914; C. A. Brown, 1938).

A belt of hilly and strongly dissected land extends along the eastern border of the alluvial plain of the Mississippi River. This is variously known as the Loess Hills, Bluff region, and Cane Hills. The last name has reference to the former abundance of cane (*Arundinaria gigantea*) in the forests of this area. The material[12] of which the bluffs are composed is calcareous, hence the soil of this belt is fertile. The forests are composed almost entirely of hardwoods; pines were lacking in the original growth but have entered in old fields and clearings, and red cedar appears on denuded slopes (Dunston, 1913).

Because of the north-south extent of the loess hills, the composition of the forest changes somewhat from north to south, as more southern species enter, and more northern species drop out. The transition from the mixed mesophytic communities on the bluffs of the Western Mesophytic Forest region (p. 159) to the beech–magnolia community of the Southeastern Evergreen Forest region is gradual. In fact, the forest of the southern end of the loess bluffs is a mixed mesophytic community modified chiefly by the presence of the evergreen magnolia as one of the dominants. This results in a pronounced difference in physiognomy, which is further emphasized by the presence of various evergreen undergrowth shrubs, and by the drapery of Spanish moss (*Tillandsia*).

The beech–magnolia forest of the loess hills is made up of a large number of species, not all of which occur in any one forest stand. Among these, in addition to beech and evergreen magnolia, are tuliptree and sweet gum (which in places are dominant), oaks (*Q. alba, Q. Shumardii, Q. Prinus, Q. Durandii, Q. velutina, Q. nigra*), hickories (*C. tomentosa, C.*

---

[11] See "A pilgrimage to Torreya" by Asa Gray, 1875, in "Scientific papers of Asa Gray" (Sargent, 1889).

[12] The nature of this material is open to question. Although most frequently referred to as loess, it is not everywhere so interpreted.

Table 57. Composition of canopy and understory of beech–magnolia forest of Louisiana: *1,* Near north shore of Lake Ponchartrain along Bayou Cane; *2,* On ravine slopes cut in Montgomery Terrace north of Baton Rouge. (Both areas grazed and probably modified by cutting.)

	1		*2*	
Number of trees in canopy (C) and understory (U)	77 (C)	62 (U)	47 (C)	17 (U)
*Magnolia grandiflora**	33.8	8.1	14.9	17.6
Fagus grandifolia	6.5	1.6	51.1	17.6
Quercus nigra	2.6	3.2	10.6	11.7
Quercus Prinus	18.2	8.1	4.2	5.9
Pinus Taeda	13.0	32.3	. .	. .
Liquidambar Styraciflua	10.4	11.3	. .	. .
Liriodendron tulipifera	. .	. .	6.4	5.9
Tilia sp.	. .	. .	6.4	. .
Quercus falcata pagodaefolia	6.5	1.6	. .	. .
Quercus Shumardii	. .	. .	2.1	11.7
Quercus alba	2.6	. .	. .	. .
*Quercus virginiana**	1.3	. .	. .	. .
Carya spp.	1.3	3.2	2.1	. .
Morus rubra	1.3	6.4	. .	. .
Ulmus alata	1.3	. .	. .	. .
*Ilex opaca**	1.3	1.6	. .	. .
Ostrya virginiana	. .	. .	2.1	11.7
Carpinus caroliniana	. .	17.7	. .	11.7
Cornus florida	. .	1.6	. .	. .
*Persea borbonia**	. .	1.6	. .	. .
*Prunus caroliniana**	. .	. .	. .	5.9
Nyssa sylvatica	. .	1.6	. .	. .

Area *1* illustrates influence of adjacent pinelands and swamp; total relief, 15 feet. Area *2* is situated in a section where pines are absent and where dissection is more pronounced; additional species in adjacent ravines are *Fraxinus americana* and *Prunus serotina.* Broad-leaved evergreens (*) are represented in both layers of both communities.

myristicaeformis), basswood (*Tilia neglecta*), cucumber tree (*Magnolia acuminata*), walnut and butternut (*Juglans nigra, J. cinerea*), elms (*Ulmus americana, U. fulva*), hackberry (*Celtis laevigata*), and a number of smaller trees as *Ilex opaca, Carpinus caroliniana, Ostrya virginiana, Morus rubra, Castanea pumila,* and *Prunus caroliniana.* Cane, originally abundant but now almost everywhere destroyed by cattle,[13] sometimes reached a height of 20 feet. The beech–magnolia forest, during the blooming period of the magnolia, is one of the most beautiful of the hardwood forests.

The successional relations of this forest community, and its relationship to other mesophytic forest communities are of interest because of their

[13] Hilgard (1860) said that "Although at present, the cane on the hills has mostly been destroyed by cattle, it may sometimes be seen in protected spots, covering them from the foot to the summit—as was the case generally before the country was settled."

bearing on the distribution and migration of the major associations of the deciduous forest (Chapter 17).

The beech–magnolia forest, with its Florida maple and variety of other mesophytes, is regarded by Gano (1917) as the climax of forest development on the clay hammock lands of the uplands of northern Florida. Harper (1906) says that the boundary between sand hills and hammocks in the Altamaha grit region of southern Georgia is never sharp, and that hammock vegetation may be expanding, if the situation is free from fire. On the loess hills, it appears to be climax. Everywhere except along the Mississippi River it is isolated from other mesophytic forest communities.

Its principal species vary in relative abundance in different areas. Either magnolia or beech is most abundant (the former has been removed more frequently). Cutting and grazing have greatly modified this forest type. However, two examples are given (Table 57) illustrating difference in dominance of its two characteristic species. In the Tunica Hills of Louisiana (a strongly dissected area with steep-sided ravines and considerable relief) the forest more nearly resembles the mixed mesophytic forest, containing some of its species in all layers.

Southward in Florida, the Beech–Magnolia climax is replaced by an evergreen climax forest in which magnolia is the principal tree, "with live oak, pignut, laurel oak, holly, and various understory trees, shrubs, and saw-palmetto" (Kurz, 1942). The magnolia climax develops on ancient dunes well back from the coast. The successional stages leading to the establishment of this climax type were pointed out by Kurz. This is a climax community of the Broad-leaved Evergreen Forest Formation.

EVERGREEN OAK FOREST

Evergreen oak forests, in which live oak(*Quercus virginiana*) is usually the predominant species occur in a narrow belt along the South Atlantic and Gulf coasts—in about the same general area as slash pine forests. Other evergreen oaks (*Q. virginiana* var. *geminata,* and *Q. myrtifolia*) and slash pine (*Pinus caribaea*) may be constituents of the coastal forest. "Of the magnificent groves which once lined the shores of the Gulf and its numerous inlets, but few remain" (Mohr, 1901). Like the slash pine forests (p. 289), the live oak and oak–pine communities often occupy dunes on coastal islands where successional development may lead to the replacement of pine and pine–oak by live oak (Penfound and O'Neill, 1934). On shell mounds along the coast, red cedar (*Juniperus virginiana*) may be associated with the live oaks (Harper, 1943). The species on the coastal dunes are tolerant of salt spray (Oosting, 1945; Oosting and Billings, 1942). Wells (1939, 1942) considers the live oak forest to be a "salt spray climax."

Live oak is a tree of the coast. On alluvial soils somewhat more removed from the coast, evergreen oak forests may have an admixture of deciduous species, and sometimes of evergreen magnolia (Penfound and Howard, 1940). In Louisiana, "live oaks, draped with Spanish moss, are the dominant trees of the coastal ridges, but give way as we proceed inland to a mixed forest consisting chiefly of live oak, water oak, red or sweet gum, elm, hackberry and magnolia, often with an undergrowth of dwarf palmetto or switch cane, or both" (Viosca, 1933). Live oak appears to be the final stage in succession on abandoned fields in the vicinity of New Orleans (Bonck and Penfound, 1945).

The evergreen oak forest, with drapery of Spanish moss, is characteristic of the coastal strip of the South Atlantic and Gulf coasts. (Courtesy, U. S. Forest Service.)

CHAPTER 10

The Beech–Maple Forest Region

This region is characterized by the development of a climax in which beech (*Fagus grandifolia*) and sugar maple (*Acer saccharum*) are the dominant trees of the canopy. Although often referred to as the Maple–Beech Forest (Weaver and Clements, 1938), beech is usually the most abundant canopy tree, while sugar maple dominates in the understory.[1] The Beech–Maple Forest region occupies much of the Till Plains of Ohio and Indiana, the western end of the Glaciated Allegheny Plateau in northern Ohio and western Pennsylvania, the southern part of the Great Lake section of physiographers, almost surrounding lakes Erie and Ontario, and extends across the southern half of the Lower Peninsula of Michigan (there indented by a lobe of the Oak–Hickory Forest region) and into southeastern Wisconsin, where it begins to merge with other types. In southern Ontario, it is the area designated by Halliday (1937) as the Niagara section of the Deciduous Forest region, often mapped as the Canadian hardwoods belt.

The southern boundary of the region, from northeastern Ohio to western Indiana follows the southern limit of Wisconsin drift. Along most of this boundary, the Beech–Maple Forest region is in contact with the Mixed Mesophytic Forest or its western ecotone, the Western Mesophytic Forest region. Where this boundary crosses strongly dissected areas it is less distinct than elsewhere; a transition from the more southern forest is then evident. The western boundary is irregular and in places indistinct; it is indented by lobes of the Oak–Hickory Forest, which extend northeastwardly across Indiana and into southern Michigan.[2] The northern boundary in Michigan, shown on the map as a fairly smooth line, is more or less arbitrary and lies within a transition zone. Here, northern pines enter (Map, p. 306), and climax dominance shifts from beech and maple to hemlock and northern hardwoods. However, to the very northern limits of the Hemlock–White

[1] Maple is dominant in the canopy in some of the lower-lying parts of beech–maple woods in this region; it is often the dominant in beech–maple communities in the Hemlock–White Pine–Northern Hardwoods Forest region.

[2] These boundaries are in part based upon the writer's field observations, in part on statements in the literature, and in part on maps by A. G. Chapman (1944), Gordon (1932, 1936, 1937a), Nichols (1935), Transeau (1935), and Veatch (1932).

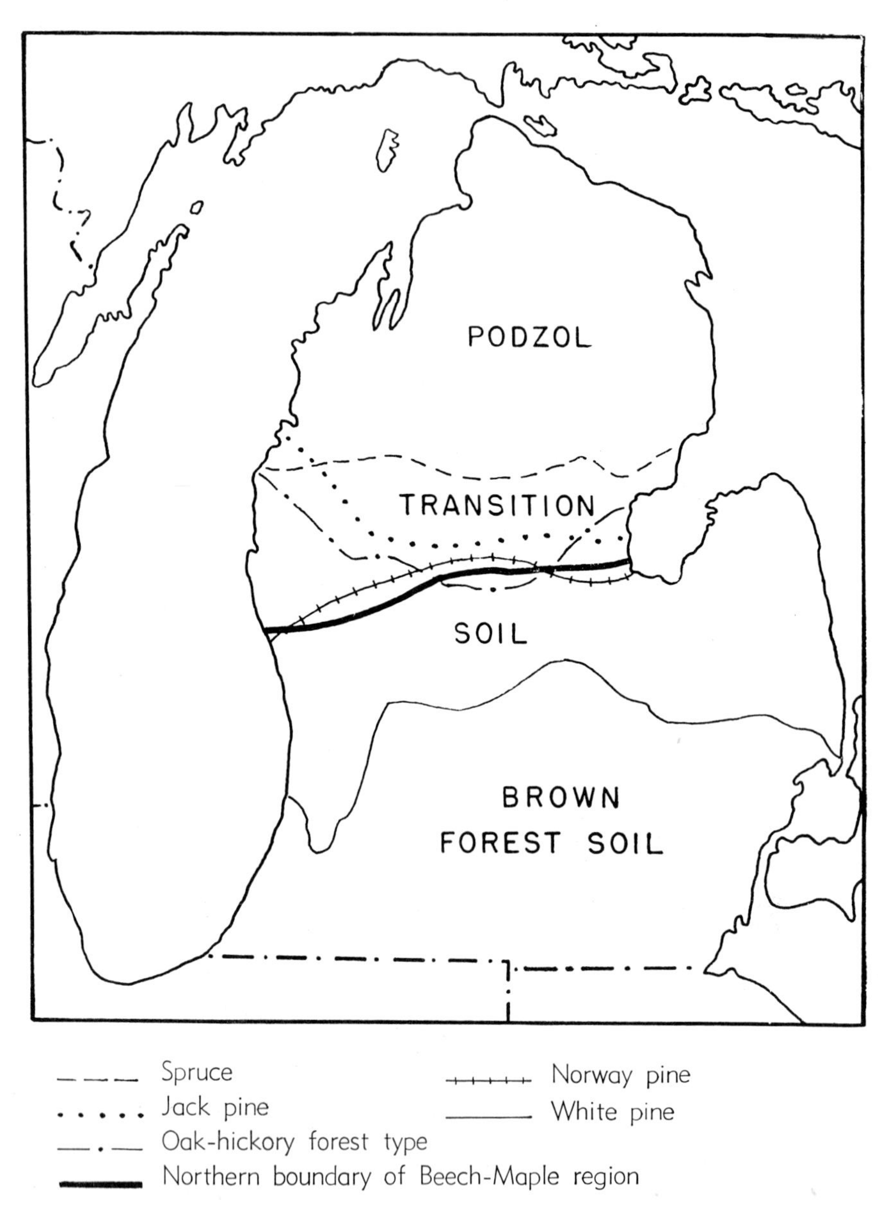

Map of Lower Michigan showing boundary between the Beech–Maple and Hemlock–White Pine–Northern Hardwoods regions, and its relation to tree ranges and soils. This boundary lies in a transition where the ranges of northern and southern trees overlap, and where the soils are transitional between the podzols and the gray-brown soils of the deciduous forest. The limit of oak–hickory communities (extensions from the Michigan lobe of the Prairie Peninsula section) is also reached in this transition zone. (After Veatch, 1932, with regional boundary added.)

Pine–Northern Hardwoods region and in its mountainous eastern division, beech–maple communities (or maple, or maple–basswood) alternate more or less frequently with communities of the type usually recognized as regional climax.

The Beech–Maple region is entirely within the area covered by the last or Wisconsin ice sheet. This youthful land surface contrasts with the great unglaciated territory of middle and lower latitudes of the United States, and with the older drift areas (Kansan and Illinoian) of the Mississippi, lower Ohio and Missouri River valleys—territory occupied by the forest regions thus far considered. In all this younger glaciated area, all vegetation was destroyed by the ice. All true soil, also, was destroyed, either removed by glacial scouring or deeply buried beneath the mantle of drift.

After recession of the ice, the weathering of parent material from which soil would be formed began. Invasion by plants moving in from the unaffected area south of the Wisconsin glacial boundary began. The amount and nature of this invasion was in part controlled by the character of the mineral surface or substratum—the bare area onto which plant propagules came. Under the combined influence of weathering and a vegetation cover (at first sparse, later continuous) soil development proceeded. Invasion was in part controlled by climate, gradually changing as the ice receded farther and farther northward and finally disappeared. The nature of the substratum influenced the establishment of species, the climate exerted a selective influence on type of plant, hence, type of vegetation. Waves of migration, so to speak, progressed northward from the unglaciated land to the south. How fast and how far migration could take place depends in part upon the species involved—its migration capacity, and the physiologic requirements of the growing plant. At a single jump, a species can move no farther than its seed may be carried by natural means. Establishment of that seed in the new site will depend on the suitability of environment. A shade plant must be preceded by shade-producing species. A plant requiring organic material in the soil will fail to establish itself on the raw glacial drift; it must await the incorporation into the soil material of decaying organic material from plants which have preceded. Reproduction and hence the opportunity to increase its kind in the new location, and to move farther cannot take place until a lapse of time, how much time depending on the length of the juvenile period of the particular species involved.

At every place upon the drift mantle, revegetation involved the slow process of plant succession beginning on a primary bare area—an area unaffected by previous occupancy by plants. The rapidity of that succession was in part affected by the availability of suitable invaders. It was in part affected by the contour of the land. Its end stage or climax was determined by climate.

The cool moist climate which followed the recession of the ice from its more southern borders favored the establishment of a coniferous forest—a spruce–fir forest such as now occupies a broad band north of the Great Lakes. That spruce–fir climax could not prevail even in the most suitable topographic situations until the slow processes of plant succession and of soil development, progressing hand in hand and interdependent, had paved the way. That such a climax did at one time prevail is shown by the pollen records of bogs. Hand-in-hand with the development of a spruce–fir climax, soil development must have resulted in great areas of podzol. Deterioration of the podzol and gradual change to a podzolic soil must have accompanied the infiltration of hardwood species.

Further amelioration of temperature permitted the entrance of more southern species. Decrease in humidity—the advent of a drier climate—was unfavorable to the earlier moist climate entrants and dominants. As they died out, their places were taken by plants moving in from the west. Later, increasing humidity and continued lessening of glacial cold permitted the readvancement northward of southern mesophytes. The evidences of these changes are found in the fossil pollen record and in the pattern of existing vegetation. Further disentanglement of the postglacial waves of migration will be left to the final discussion of vegetational history (Chapter 17). At present, let us keep uppermost the thought of the time element, of the difference in time available for the development of soil and of vegetation to the south and to the north of the Wisconsin glacial boundary.

Some 20,000 or 30,000 years, more at the southern margin, less in the north, have been available for migration into and occupancy of the Wisconsin drift area. On the Illinoian drift plains which from Ohio to the Mississippi River lie to the south of the Wisconsin border, some 200,000 years have been available. The unglaciated land to the south (exclusive of the Coastal Plain) has been continuously occupied by plants for millions of years. Time is an important factor in explaining some of the contrasts, both vegetational and soil, between adjacent regions.

The vegetation of the younger glaciated area, where only about 30,000 years have elapsed since the melting of the last ice cap, can be understood only if this time factor is considered. The vegetation of this young area has changed and is changing, not alone as a result of development of plant successions, not alone as a result of changing climates of post-Pleistocene times, but also as a result of slow migrations of plants which have not had time enough to reach their climatic limits. The regional soil boundary, between the melanized soils to the south and the podzolic soils to the north, lies approximately at the Wisconsin glacial boundary. Gradual melanization of the podzolic soils, which takes place most rapidly on moist but well drained slopes, will shift the soil boundary northward, just as slow migration of plants is resulting in a transition zone along the southern border of the younger drift.

Species composition of communities—of developmental and climax stages—is subject to the slow changes induced by entering migrants capable of taking a place in seral stages or in climax stages. Boundaries between "vegetation regions" do not exist as such in the younger glaciated area. Boundaries must be chosen arbitrarily at the approximate borders of the temporary dominance of particular climax types in the great area involved. A dynamic approach to the vegetation of this great area as a whole, an approach such as is frequently used in developmental studies in limited areas, will enable us to relate and to differentiate the several generally recognized forest regions of the area, to understand the pattern of interpenetrating climaxes.[3]

Because of the youth of the land surface, developmental stages of plant successions occupy large areas, sometimes almost obscuring the regional characters. This is in marked contrast with the regions to the south, where developmental communities (except those of secondary successions) are almost lacking. Some of the developmental stages perpetuate stages of preceding climatic successions. For example, communities of bogs are relics, dating back to an early post-Pleistocene northward migration of vegetation of the Northern Conifer Forest. Relic oak–hickory communities, climax in character, owe their development to a somewhat later post-Pleistocene dry period. Sandy or gravelly morainal ridges are usually occupied by oak–hickory forest, situations too dry for beech–maple forest. Prairie communities (both dry and wet prairie) and prairie species in the drier black oak–white oak forests are relics of a post-Pleistocene dry period. Some developmental stages, because of the dominance in them of white oak, may simulate oak–hickory forest; others, because of the number of tree species, may simulate mixed mesophytic forest (Sampson, 1930). Phases of swamp forest originally covered such large areas in the Beech–Maple Forest region that they may be shown on generalized state vegetation maps (Gordon, 1936; A. G. Chapman, 1944). The "Great Black Swamp" on the lake plain of northwestern Ohio was one of the most extensive of these swamp forest areas; in northeastern Ohio, also, extensive swampy tracts occur. In such areas, the climatic climax, beech–maple, could develop only on knolls and low ridges, or in areas where stream dissection has progressed far enough to produce good drainage (Shanks, 1939).

The Beech–Maple Forest region is so uniform throughout its extent that no sections need be recognized. The transitional belt along the southern border is not sufficiently distinct to warrant its recognition as a section. The local variations related to stage in the succession and to topography afford variety as far as forest type is concerned.[4] As similar variations and similar

[3] The term, "interpenetrating climaxes," was first used by Dansereau (1944) who called attention to this feature in the forests of Quebec.

[4] The many unlike communities occurring in a single county in southeastern Michigan are described by Bingham, 1945.

successional development occur almost throughout, except in marginal transition areas, these do not detract from the uniformity of the region as a whole.

Forest Areas

An examination of a number of forest areas, geographically scattered over the area of the Beech–Maple Forest region, will demonstrate the nature of that forest as it occurs in different parts of the region and on unlike topographic situations.

FORESTS OF THE MARGINAL BELT

In the southwestern part of the region, at Turkey Run State Park, Indiana, a number of forest communities including climax and developmental stages are well preserved (Table 58). This is a strongly dissected area in which the sandstone bedrock is exposed. Erosional, not glacial topography largely controls the forest sites. The area is located about ten miles within the border of the Early Wisconsin drift. Its vegetation is transitional in character. Small areas, slightly modified by cutting, and referable to beech–maple, do occur (Esten, 1932). Local hemlock communities occupy some of the steeper slopes and plateau margins (Daubenmire, 1930; Friesner and Potzger, 1932, 1932a). Communities which may be considered as attenuated mixed mesophytic occupy parts of the dissected upland (beech, tuliptree, sugar maple, 71 per cent) and mesic slopes (beech, tuliptree, sugar maple and basswood, 62 per cent). On southerly and westerly slopes, an oak forest (white oak and red oak, 82 per cent) suggests the oak forests of the Western Mesophytic Forest region and the mesophytic oak forests of the Oak–Hickory region adjacent to the west and southwest. Succession on alluvial flats leads from the marginal sycamore community to sycamore–walnut–elm,[5] then to sugar maple–walnut–elm, and finally to a community closely approaching the mixed mesophytic forest of the slopes. Developmental tendencies in the area strongly indicate that beech–maple is but a stage in the climatic succession leading to the establishment of a mixed mesophytic climax in the area—but of mixed mesophytic lacking its two most characteristic species, whose range limits do not extend this far. It will be noted that *Tilia americana* (not *T. heterophylla*) is one of the canopy species in this mixed mesophytic community. This suggests the transitional position of this forest, as *T. americana* is a dominant in the northern maple–basswood forest. *Aesculus* is represented by *A. glabra* in developmental communities.

A forest area in southwestern Ohio within 20 miles of the southern limit

[5] In the abundance of walnut in developmental stages of the flood-plain succession, this area resembles the western part of the deciduous forest. In the Mixed Mesophytic Forest region, this species is regularly a constituent of the climax.

Table 58. Percentage composition of forest communities of Turkey Run State Park, Indiana.

A. Upland and slopes: *1,* Rolling and dissected upland; *2,* Mesic slopes; *3,* Southerly and westerly slopes.

B. Developmental series (*4→5→6*) on alluvial flats leading toward establishment of a mixed mesophytic forest; note resemblance of community illustrated by *6* to that illustrated by *2.*

	A			*B*		
	1	*2*	*3*	*4*	*5*	*6*
Number of canopy trees	238	82	60	28	64	108
Fagus grandifolia	33.2	20.7	5.0	. .	1.5	16.6
Liriodendron tulipifera	20.6	14.6	1.7	. .	1.5	4.6
Acer saccharum	17.2	13.4	. .	3.6	39.1	13.9
Tilia americana	. .	13.4	. .	. .	. .	1.8
Tsuga canadensis	. .	8.5	5.0	. .	. .	. .
Quercus alba	7.5	1.2	63.3	. .	. .	6.5
Quercus borealis maxima	5.0	2.4	18.3	. .	. .	1.8
Carya sp.	3.4	3.7	5.0	. .	. .	3.7
Carya ovata	2.9	2.4	1.7	. .	1.5	.9
Fraxinus americana	3.4	2.4	. .	. .	3.1	6.5
Prunus serotina	.4	2.4	. .	. .	. .	. .
Nyssa sylvatica	3.8	. .	. .	. .	. .	. .
Sassafras albidum	1.7	. .	. .	. .	. .	. .
Ulmus fulva	.8	3.7	. .	. .	3.1	. .
Gymnocladus dioicus	. .	1.2	. .	. .	1.5	. .
Aesculus glabra	. .	. .	. .	. .	6.3	10.2
Juglans cinerea	. .	4.9	. .	. .	6.3	.9
Morus rubra	. .	. .	. .	. .	1.5	. .
Quercus macrocarpa	. .	. .	. .	. .	1.5	. .
Fraxinus quadrangulata	. .	. .	. .	3.6	. .	.9
Celtis occidentalis	. .	2.4	. .	3.6	4.7	11.1
Ulmus americana	. .	1.2	. .	14.3	10.9	9.3
Juglans nigra	. .	1.2	. .	28.5	15.6	10.2
Platanus occidentalis	. .	. .	. .	46.4	1.5	.9

of the Early Wisconsin ice sheet is selected to illustrate the beech–maple forest of rolling and mildly dissected land close to the southern boundary of this forest region (Table 59). Here, in the forest as a whole, the two dominants together comprise between 80 and 90 per cent of the canopy which contains 15 species. As demonstrated by a transect across this forest, it is made up of three more or less distinct parts differing in relative abundance of the two dominants, and in the presence of a third dominant, tuliptree, in one part. These differences, on unlike topographic divisions of a single area, may explain the varying composition of small isolated woodlots in the Beech–Maple region. Sugar maple is most abundant in the understory (in part perhaps because of an interval of grazing); however, the representation of accessory species of the canopy in the undergrowth is such as to suggest that they will be more abundant in the canopy of the ensuing forest

Table 59. Canopy composition of Hueston's woods, north of Oxford, Ohio; based upon a transect extending (*1*) across the rolling and mildly dissected upland, (*2*) down the slopes, to (*3*) an alluvial terrace; area as a whole in column *W*.

	1	*2*	*3*	*W*
Number of canopy trees	203	25	38	266
Fagus grandifolia	72.4	36.0	. .	58.6
Acer saccharum	7.4	64.0	89.6	24.4
Liriodendron tulipifera[1]	10.3	. .	. .	7.9
Fraxinus americana	4.9	. .	. .	3.7
Nyssa sylvatica	2.0	. .	. .	1.5
Ulmus fulva	1.5	. .	. .	1.1
Ulmus americana	. .	. .	5.3	.7
Prunus serotina	.5	. .	. .	.4
Celtis occidentalis	.5	. .	. .	.4
Quercus alba	. .	. .	2.6	.4
Carya ovata	. .	. .	2.6	.4
Ostrya virginiana	.5	. .	. .	.4
Gleditsia triacanthos[2]	x	. .	. .	x
Carya cordiformis[2]	x	. .	. .	x
Aesculus glabra[2]	. .	x	. .	x

[1] Very large trees removed many years ago.
[2] Presence in the area, but beyond limits of transect, is noted by x.

than in the present one. The contrast between this beech–maple forest, and the mixed mesopyhtic forest illustrated by Table 21 is significant when it is realized that these two areas are but 35 miles apart, the one on the younger glaciated area, the other on the older (Illinoian) drift. There is no climatic boundary between the two areas.

Table 60. Composition of forest near Canfield, Mahoning County, Ohio (sample of 142 canopy trees; 78 second-layer trees).

	Canopy	*Second layer*		*Canopy*	*Second layer*
Fagus grandifolia	40.1	53.8	*Fraxinus americana*	1.4	1.3
Acer saccharum	27.5	26.9	*Liriodendron tulipifera*	.7	3.8
*Quercus alba**	14.8	2.6	*Magnolia acuminata*	.7	3.8
Acer rubrum	7.7	1.3	*Carya cordiformis*	. .	1.3
Nyssa sylvatica	5.6	3.8	*Ostrya virginiana*	. .	1.3
Quercus borealis maxima	1.4	. .			

* Includes standing trees and stumps left from recent cutting.

White oak is a frequent constituent of beech–maple forests. Table 60 gives canopy and second-layer composition of such a forest near Canfield, Mahoning County, Ohio on the Glaciated Allegheny Plateau. The mor humus layer overlying a pale A-horizon contrasts with the mull humus of the two areas previously considered. The increased dominance of beech and sugar maple in the second layer points to the continuance of the beech–maple type. The presence of *Maianthemum* and of *Dryopteris spinulosa* var. *intermedia* in the herbaceous layer is strongly suggestive of northern forests. This area, although less than 20 miles within the glacial boundary, is on the Late Wisconsin drift. There has been less time for soil development than on the two areas on Early Wisconsin drift already considered. But even in this area, there is some indication of change; on well-drained but moist slopes mull humus replaces the mor of the rolling area, and overlies a gray-brown forest soil. However, the forest of such sites is referable to the Beech–Maple association.

Where moraines of strong expression border the drift area, or extend inward more or less at right angles to the margin (interlobate moraines), the topography is favorable to other forest types. Where there is considerable relief which has been increased by stream dissection, tongues of mixed oak forest or of mixed mesophytic forest (and its developmental stages) have penetrated into the younger glaciated area. This is illustrated by Chapman's map (1944) of Ohio forests which shows a lobe of mixed forest extending northward from the Unglaciated Allegheny Plateau toward the Cuyahoga Valley along an interlobate moraine.

In less hilly morainal areas, swamp forests occupy the depressions and usually some type of oak forest occupies the swells. Although it is entirely possible that occupancy of morainal swells of the eastern part of the region by oak dates back to the post-Wisconsin xerothermic period, there is usually evidence of development toward a forest at least approaching the beech–maple type. The Graber woods, northeast of Wooster, Ohio, is an excellent example; it is, viewed in its entirety, a virgin white oak forest (Table 61). However, swells, flats, and depressions differ from one another in soil and humus type, and in composition of the occupying forest community, both of which are related to stage in plant succession. A red maple–white elm community occupies the wetter depressions; flats (with mor humus layer) are occupied by red maple–white oak, with an admixture of red oak, ash, beech, and occasional other species. These communities are phases of the swamp forest more extensively developed on flatter parts of the drift and on the lake plains. The presence of beech suggests progress toward a mesophytic forest type; the absence of sugar maple is related to poor drainage and aeration. On low swells white oak predominates in the canopy; beech and sugar maple are abundant in the 3 to 4 inch d.b.h. and smaller size classes; here is a somewhat matted humus varying to fairly good mull. Occasional white oak areas lack beech and red maple. Locally, beech is well represented among canopy species and predominates in the second layer, while sugar maple and beech dominate in smaller size classes; in such communities a good mull has developed. The developmental trend in depressions and on swells is

White oak community in Graber's Woods, Wayne County, Ohio. Small sugar maples and beeches are indicative of possible future development toward a type more in accord with the present regional climax. (Photo by O. D. Diller, Associate State Forester, Ohio.)

Table 61. Percentage composition of canopy (sample of 208 trees) and second layer (sample of 131 trees) of a white oak forest occupying a morainal area; the Graber woods, Wayne County, Ohio. Five communities are distinguishable in the forest: red maple–white elm in depressions, red maple–white oak on flats, white oak with an admixture of mesophytic and hydro-mesophytic species occupying low swells, white oak, and a community in which beech is prominent. Although these communities are not separated in the table, the number of communities in which each species occurs (presence) is indicated by *P*.

	Canopy		*Second layer*	
	Per Cent	*P*	*Per Cent*	*P*
Quercus alba	45.7	4	10.7	2
Acer rubrum	19.2	4	8.4	5
Fraxinus americana	7.7	5	14.5	3
Quercus borealis maxima	7.7	4	6.8	3
Fagus grandifolia	6.7	3	32.8	3
Acer saccharum	3.8	3	3.1	2
Ulmus americana	2.9	2	6.1	3
Carya ovata	2.9	2	2.3	3
Quercus bicolor	1.4	1	..	..
Carya cordiformis	.5	1	3.8	2
Nyssa sylvatica	.5	1	4.6	3
Liriodendron tulipifera	.5	1	..	..
Quercus palustris	.5	1	..	..
Prunus serotina	..	..	2.3	3
Ostrya virginiana	..	..	4.6	2

toward the establishment of a white oak–beech forest with evidence (from undergrowth) of the ultimate establishment of beech–maple. The development is similar to that seen on the undissected Illinoian till plain where a comparable succession leads to the establishment of a beech physiographic climax (p. 133).

An area of beech–maple forest (in the North Chagrin Reservation of the Cleveland Metropolitan Park System) studied by Williams (1936) is of particular interest because of the transitional nature of some of its communities. This is located on the margin of the Glaciated Allegheny Plateau, along the west side of the Chagrin River. The strongly dissected river bluff with its ravines and flat-topped spurs extending out from the adjacent tableland affords a variety of habitats. The relatively level land is occupied by the regional beech–maple forest. Williams interprets the forest of ravine slopes and adjacent margin as a "beech–hemlock–oak–chestnut mictium." He suggests the replacement of the hemlock–beech areas by beech–maple, and considers that beech in the mictium "represents an entering wedge of the climatic climax." In the light of community relations existing in all of the dissected areas of the marginal part of the younger glaciated area, it seems better to recognize, not a "beech–hemlock–oak–chestnut mictium" but a mixed mesophytic forest community which is entering the area in climatic succession. It is occupying the slopes and immediate margins of the adjacent spurs, not the spur tops where most of the beech–hemlock is located. As dissection increases, as it ultimately must, this mixed mesophytic community may increase in extent. The hemlock or beech–hemlock community may be crowded out as the contour changes, although it is probable that a small percentage of hemlock may continue in the slope forest. The areal dominance of the beech–maple climax is evident, and it may be expected to retain this dominance under present climatic conditions as long as flat to rolling upland remains. A number of northern shrubs and herbs in the forest area as a whole (*Sambucus pubens, Lonicera canadensis, Maianthemum canadense, Streptopus amplexifolius, Habenaria orbiculata, Aralia nudicaulis,* and *Galium boreale*) show a relation to the northern forest.

The distribution of different forest types in relation to topography in this general section of the Beech–Maple Forest region has been well shown by Shanks (1942) in his study of Trumbull County, Ohio. Beech–maple occupies nearly half of the area, and swamp forests and upland oak forests most of the remaining land. He points out that the mixed mesophytic communities are restricted to slopes with "a good supply of water and good internal drainage (soil aeration)." Chestnut and chestnut oak communities, transition types, and relic communities comprised but a small part of the original forest cover.

FORESTS OF THE TILL PLAINS AND LAKE DISTRICT

The areas thus far considered are in more or less marginal positions in the Beech–Maple Forest region, or occupy dissected areas more advanced in the erosion cycle than is usual in the younger glaciated area. Few good forest tracts remain on the Till Plains or in the Lake District, except in poorly drained places unfit for agriculture, or on dry sandy beach ridges, neither of which sites support the climax vegetation. The evidence from isolated woodlots (usually grazed) and from scattered remnants of the primeval forest is that beech–maple forest originally occupied all the more favorable sites.

The Till Plains and Lake District for the most part are characterized by low relief where minor differences in elevation (a foot or two), resulting in differences in soil moisture and aeration, determine the forest type. The more compact poorly aerated and poorly drained soils of the slightly lower places support forests in which hydro-mesophytic trees (as *Quercus bicolor, Ulmus americana*) may comprise about half the canopy. The better aerated slightly higher places are occupied by the climax dominants (*Acer saccharum, A. saccharum* var. *nigrum, Fagus grandifolia,* and sometimes *Tilia americana*).

Friesner and Ek (1944) have described such an area on the Tipton Till

Table 62. Percentage composition of canopy (based on trees over 10 inches d.b.h.) of Shenk's woods, Howard County, Indiana. Area divided into: *1*, Swamp forest, or "Quercus–Ulmus community"; *2*, Transition; *3*, Maple–beech, or "Acer–Quercus–Fagus community." (Data from tables by Friesner and Ek, 1944.)

	1	*2*	*3*
Number of trees	55	49	56
Ulmus americana	20.0	6.1	. .
Quercus bicolor	18.1	8.2	. .
Quercus alba	. .	12.2	. .
Quercus borealis maxima	10.1	4.1	. .
Acer saccharum	6.7	10.0	73.2
Fagus grandifolia	1.7	10.2	9.0
Tilia americana	1.7	14.3	3.6
Ulmus fulva	10.1	8.2	7.1
Quercus macrocarpa	3.4	4.1	1.8
Fraxinus americana	3.4	6.1	1.8
Fraxinus quadrangulata	1.7	2.0	. .
Carya cordiformis	1.7	2.0	. .
Carya ovata	1.7	4.1	. .
Platanus occidentalis	3.4	2.0	. .
Quercus Shumardii	. .	6.1	1.8
Aesculus glabra	1.7	. .	1.8
Acer saccharinum	5.0	. .	. .
Quercus Muhlenbergii	3.4	. .	. .
Celtis occidentalis	1.7	. .	. .

Plain of Indiana (Shenk's woods, Howard County) which is partly swamp forest and partly beech–maple forest (Table 62). The authors recognize a Quercus–Ulmus community and an Acer–Quercus–Fagus community, the names reflecting dominance as determined by percentage of basal area of stems 3 inches and over d.b.h. In the more typical parts of the former community, which may be called swamp forest (area 1 of Table 62) the hydro-mesophytic species comprise about 45 per cent of the trees over 10 inches d.b.h., and 60 per cent of the crown cover; while in the latter, 73 per cent of the trees over 10 inches d.b.h. are sugar maple, and the hydro-mesophytic species are lacking (area 3 of Table 62). Certain communities of the Till Plains forests (such as area 2 of Table 62) may, because of the large number of canopy species and lack of pronounced dominants, simulate mixed mesophytic communities. They are, however, developmental communities of the transitional type described by Sampson (1930, 1930a), and will ultimately be replaced by the regional climax.

A forest tract in the southern part of the Lake District (Berkey woods, Kosciusko County, Indiana) is treated statistically by Potzger and Friesner (1943). This area is flat to very slightly rolling, and has no streams, therefore is poorly drained in some parts. The dominant species are *Acer saccharum* (41.7 per cent), *Fagus grandifolia* (19.4 per cent) and *Ulmus*

Thomasi (15.5 per cent), which make up 76.6 per cent of the trees over 10 inches d.b.h. as determined from tables given by Potzger and Friesner, or 28.1, 46.6, and 13.7 per cent respectively of the canopy trees of the less wet part of the woods (writer's field notes). Potzger and Friesner interpret this forest community as "typical mixed mesophytic," although the poor drainage and large amount of elm are strongly suggestive of a late stage of the elm–ash–maple phase of the swamp forest of the region. Neither the aspect of this forest nor the habitat (poor drainage and aeration) agree with characteristics of the true mixed mesophytic. Rather this is one of the transitional communities which, only because of the large number of tree species, simulates a mixed mesophytic community; it is an advanced developmental stage of the swamp forest succession. Good representation of the three dominant species of the canopy in all size classes indicates a continuance of this type of forest until better drainage and aeration, which will come with mature topography, permit further development and the final establishment of a climax beech–maple forest. In the same general area, dry black oak–white oak woods (with some hickory) occupy deep gravelly soil of morainal ridges and sand plains. Well-drained but moist slopes are occupied by the beech–maple forest type; in such a site near Goshen, Indiana, 98 per cent of the canopy trees are sugar maple and beech. The diverse communities of this section are in part due to differences in topography and soil, in part to the location in the tension zone between the Beech–Maple and Oak–Hickory regions.

In all probability, the forest tract which the largest number of students of forest ecology have looked at as an example of the Beech–Maple association is Warren's Woods, near Three Oaks, in Berrien County, Michigan. Here is an excellent example of the mesophytic climax toward which all successions in the region bordering the south end of Lake Michigan are developing. Nowhere in the Chicago region (Cowles, 1901) was there a mature example of the Beech–Maple climax; therefore, this tract across the lake was the objective of many a field trip. Fortunately preserved as a state reservation, this forest still remains as an outstanding example of the beech–maple forest type (Table 63). The only ecologic publication concerning the area is a statistical study by Cain (1935b) in which data are based on twenty-five quadrats each 10 by 10 meters. To all appearances, this is a virgin forest tract; however, Cain states that selective cutting "took out about 10 per cent of the timber 45 to 50 years ago. The scattered trees removed were basswood, tulip poplar, white oak and black cherry." The area is flat to slightly rolling, with pronounced though low slopes toward the Galien River, which borders the virgin forest on the east and south. A beech–maple forest, in which these trees now form over 90 per cent of the canopy, occupies the general flat upland. In a more sloping part of the forest, the dominance of beech and maple is less complete (52 per cent of

Table 63. Percentage composition of canopy of Warren's Woods, near Three Oaks, Berrien County, Michigan: *1,* Forest of general level; *2,* Forest of lower and sloping parts.

	1 (232 trees)	*2* (79 trees)
Fagus grandifolia	62.1 } 93.1	19.0 } 51.9
*Acer saccharum**	31.0 }	32.9 }
Ulmus americana	3.4	7.6
Tilia americana	1.3	16.4
Liriodendron tulipifera	.9	..
Quercus borealis maxima	.4	2.5
Carya ovata	.4	..
Platanus occidentalis	.4	3.8
Celtis occidentalis	..	7.6
Fraxinus americana	..	6.3
Prunus serotina	..	2.5
Juglans nigra	..	1.3

* Some *Acer rubrum* may be included.

the canopy) and basswood is more abundant (16 per cent). Reproduction indicates the continuance of the present forest type.

Not all forest areas of southern Michigan are beech–maple. Some, as the Fred Russ Forest south of Decatur, are beech–maple in part, and in part predominately white oak, but showing definite developmental trends toward beech–maple. These white oak areas contain a high proportion of middle-size maple, and smaller beech. Thus the succession is white oak→an increase in sugar maple→beech–maple (Table 64), a trend comparable to that already noted in morainal areas of the marginal belt in Ohio (p. 313).

That part of the Beech–Maple region to the north of Lake Erie in extreme

Table 64. Canopy composition of the Fred Russ Forest (property of Michigan State College) south of Decatur, Michigan, in which two distinct communities are recognizable: *1,* Beech–maple; *2,* White oak or white oak–maple.

	1	*2*
Number of canopy trees	100	85*
Fagus grandifolia	48	2.3
Acer saccharum	23	11.8
Quercus alba	1	57.6
Quercus borealis maxima	1	8.2
Ulmus americana	8	..
Juglans nigra	8	8.2
Liriodendron tulipifera	4	4.7
Tilia americana	3	4.7
Prunus serotina	3	..
Fraxinus americana	1	1.2
Nyssa sylvatica	..	1.2

* In area *2* there are a large number of sugar maples not quite of canopy size and many smaller beech trees, indicating development toward the type illustrated by area *1.*

Views in the white oak–maple (left) and beech–maple (right) communities of the Fred Russ Forest near Decatur, Michigan. The maple, usually of smaller size than the white oaks, and the small beech trees of the white oak–maple community are indicative of a trend toward beech–maple forest. (Photos by C. I. Arnold, Russ Forest.)

southern Ontario and along the south side of Lake Ontario in western New York contains a large variety of species, at least a few of which, notably chestnut and cucumber magnolia, are almost limited to this part of the region. A considerable number of deciduous forest species are known in Canada only in this area, and reach the northern limits of their Canadian range here (Macoun, 1894). However, such species "occur as scattered individuals or groups, either on specialized sites or within the characteristic association of the Section. This association . . . consists primarily of beech and sugar maple, together with basswood, red maple and (northern) red, white, and bur oak" (Halliday, 1937).

Transitions and Inclusions

The vegetation of the Beech–Maple Forest region reflects, in its transitions and inclusions of communities referable to other climaxes, the influence of changing climates of postglacial time.

Near its southern boundary (which is the southern boundary of Wisconsin ice), continuing migrations from the south are resulting in an infiltration of southern species into the beech–maple climax and an increase in the number of canopy species, thus bringing about the development (in the topographically most mature situations) of communities suggesting mixed mesophytic (cf. Table 58, community 2). Apparently the climatic succession is not yet completed.

That part of the Beech–Maple region in New York State displays some affinity to the Oak–Chestnut region, from which it is separated by the northern part of the Allegheny Plateau (occupied by Hemlock–White Pine–Northern Hardwoods Forest). In fact, Bray (1930) maps the vegetation of the glacial lake plain bordering Lake Ontario on the south as similar to that of the Hudson River Valley, and of the Susquehanna Valley. The latter, once serving, via the Finger Lakes district, as an outlet of glacial meltwater, may have served as a migratory route over which southern species reached this eastern end of the Beech–Maple region, and from it, the Niagara district of southern Ontario. This migration probably took place during the post-Wisconsin xerothermic period.

Well within the region, the southern influence is not pronounced; instead, remnants of a former more northern vegetation persist. Bogs, infrequent in the southern part of the region, become increasingly numerous northward. Bog communities contain a high proportion of northern species. Northern conifers still persist in most of the bogs, generally dominating the earlier forest stages. These northern coniferous bog border communities are in sharp contrast to the prevailing deciduous forest vegetation. Such relic communities are further evidence of a great postglacial climatic succession (see Chapters 14, 15, 17). Bogs, occupying the deeper depressions where peaty soils have accumulated, and swamp or hydro-mesophytic forests on poorly drained flats are dependent for their existence and continuance on the youthful glacial topography. Many a black soil area in the farm land of the Till Plain gives evidence of the former greater abundance of bog communities, some of which were replaced in the normal progress of succession by deciduous forest or, in some instances, by wet prairie, some of which were artificially drained and transformed into fertile farm land.

The most southerly of the well-preserved bog remnants is Cedar Swamp in Champaign County, Ohio (now under state control). Here arbor vitae or northern white cedar (*Thuja occidentalis*) dominates in the forest border and yew (*Taxus canadensis*) is frequent in the undergrowth. The "Great Black Swamp" of northwestern Ohio, now mostly drained and cleared, interrupted the beech–maple forest in a broad band (Shanks, 1939). Numerous bogs in the Lake District in northern Indiana and southern Michigan harbor northern vegetation. In many, Larix and Thuja are present; a white pine stage is discernible in some. In all instances, encroachment of deciduous forest vegetation is evident.[6]

---

[6] For further information on the bogs of this region, see Cowles, 1901; Transeau, 1905; Markle, 1915; Waterman, 1926; Bingham, 1945; also, many of the papers on bog pollens.

Northward in Michigan, the admixture of hemlock in the forest, the increasing number of northern herbaceous and shrub species, and the occurrence of white pine are indicative of the transition to the Hemlock–White Pine–Northern Hardwoods region (p. 305). This same transition is apparent on the Allegheny Plateau of northeastern Ohio where beech–maple soon gives way to the more northern forest type of the high plateau of northwestern Pennsylvania. Again, in southern Ontario, the same transition is seen.

Table 65. Composition of canopy of late developmental stages of *1,* flood-plain succession and *2,* wet glacial clay plain succession, both of which indicate trend toward the forest of moist but drained upland, *3,* which represents an approach to climax in the transitional area between beech–maple, oak–hickory, and maple–basswood (Harms Woods, west of Evanston, Illinois).

	1	*2*	*3*
Number of canopy trees	70	85	130
Acer saccharum	5.7	. .	48.5
Quercus borealis maxima	7.1	21.2	19.2
Tilia americana	31.4	18.8	16.1
Quercus alba	2.8	12.9	7.7
Fraxinus americana	7.1	4.7	3.8
Ulmus fulva	. .	. .	2.3
Ulmus americana	35.7	24.7	.8
Acer rubrum	. .	9.4	.8
Prunus serotina	. .	. .	.8
Juglans nigra	4.3	. .	. .
Quercus macrocarpa	4.3	1.2	. .
Populus deltoides	1.4	. .	. .
Carya ovata	. .	5.9	. .
Quercus ellipsoidalis	. .	1.2	. .

Beech–maple forest, in more or less modified form, extends around the southern end of Lake Michigan, in a narrow band close to the lake. A gap in the distribution of beech in northeastern Illinois[7] eliminates this species from the more mesophytic forests of that area. A transitional type suggests the maple–basswood and maple–basswood–red oak forests of Wisconsin and Minnesota (Chapter 11). Successional development on the several topographic divisions of this band (well represented by the Chicago region) has been considered by Cowles (1899, 1901) who pointed out that all successions lead toward a mesophytic forest climax in which maple or maple and beech are abundant species. In such densely populated and urban sections, no undisturbed primary forest tracts remain, yet the Cook County Forest Preserves supply fair examples of forest types and developmental tendencies in the area. In advanced forest stages where sugar maple, basswood, and red oak are dominant (Table 65), the humus is completely incorporated, thus contrasting with the usual condition of the typical

[7] For map of distribution of beech by counties, see Transeau, 1935.

beech–maple, where the leaf litter is matted and podzolic soils more pronounced. This may be due in part to the dominance of maple and basswood, whose leaves decay more quickly than do beech leaves, and have a neutralizing effect (Alway, Kittredge and Methley, 1933; Alway and McMiller, 1933).

A little farther north (Milwaukee County, Wisconsin), a small area of beech–maple forest is the last outpost of this forest type. It lies entirely to the east of the Kettle Moraine, within the influence of Lake Michigan. Northward, beech–maple soon grades into a more northern forest type, with hemlock, white pine, and yellow birch (Chamberlin, 1877; Brunchen, 1900).

Along the western boundary of the region, the transition is to drier types of forest. Somewhere in this transition, the boundary between the Beech–Maple and Oak–Hickory forests is placed. Beech becomes less and less abundant as the western limit of its range is approached; the forest of mesic sites then begins to resemble the maple–basswood–red oak forest of the Oak–Hickory region (p. 188).

Outside the limits of the lobe of the Oak–Hickory Forest region which extends across northern Indiana into southern Michigan, numerous inclusions of oak forest lie within the Beech–Maple region (p. 309). A variety of xeric oak and oak–hickory communities are normally associated with sandy or gravelly deposits where pervious subsoils result in rapid drying. The area is in a tension zone, where soils and microclimatic factors are determinative. White oak and red oak, together with several species of hickory, comprise the more mesophytic oak communities; while black oak predominates in the drier sites (F. B. H. Brown, 1905; Livingston, 1903; Transeau, 1905). Evidence remains of former oak openings (Moseley, 1928; Wing, 1937), and of prairie communities (Veatch, 1927). Based upon the original land survey, Kenoyer (1930) determined the original extent of the major vegetation types of Kalamazoo County, Michigan, where beech–maple and oak–hickory climaxes, swamp forests, and prairie communities with bur oak are represented. He found no evidence that beech–maple is invading oak–hickory; the boundaries still can be found where they were a century ago. The oak lands were originally rather open, with trees farther apart than in the beech–maple woods.[8]

The more easterly inclusions of oak forest in Ohio are generally represented by open pastured woodlots and scattered trees. One of the largest of these inclusions, centering in Fayette, Madison, and Pickaway counties, and extending to the southeastern limits of the region, contrasts strongly with the beech–maple area which nearly encircles it, and with the diverse vegetation of the Allegheny Plateau border to the east. The absence of beech

[8] Oaks used as bearing trees were on the average farther from section corners than beech trees.

and scarcity of sugar maple (except on ravine slopes), and the general dominance of white oak and locally of bur oak characterize the area. Unlike the white oak and white oak–beech forests of the Illinoian till plain to the south and of certain morainal areas farther north (p. 313), evidence of further development, of progress of the clisere toward beech–maple, is lacking. Prairie inclusions or prairie openings generally occur scattered in the oak forest areas. These are now largely reduced to roadside remnants.

The oak forest and prairie inclusions are stages of a clisere initiated by a dry postglacial period (see Chapters 14, 17). Like the inclusions of northern conifers on bog borders, they are remnants of previous climaxes now completely surrounded by a vegetation more in accord with recent climatic influences.

These transitions and inclusions add to the vegetational diversity of the region without, however, detracting from the essential uniformity of the region as a whole.

Characterization and Contrasts

The Beech–Maple Forest region, as we see it today, is mostly farm country. Those forest communities which occupied the better soils have suffered most. Drainage projects have made available large areas of former swamp forest. The picture of the original vegetation cover, reconstructed from fragmentary areas such as have been described, must be general.[9]

We visualize a large area of flat to rolling country, diversified by occasional fairly deep and youthful valleys in its marginal parts, where transitional forests occur; by sluggish streams controlled by glacial topography and bordered by swampy tracts, often extensive, and occupied by phases of swamp forest, of which the elm–ash–maple is most general; by more or less circular or irregular boggy depressions without outlet, occupied by bog communities and usually remnants of northern coniferous forest vegetation; and by sandy and gravelly morainal and beach ridges and flats with their dry oak forest and prairie plants.

Much of the flat to rolling country, except where too wet, was beech–maple forest, predominantly beech in some places, maple in others (cf. Table 59). In this, in spring, was a vernal flora somewhat comparable in luxuriance, but not in number of species, to that of the mixed mesophytic forest. Greater expanses of a single species, such as squirrel corn (*Dicentra canadensis*) might be seen. Occasional species, as *Maianthemum canadense* and *Aralia nudicaulis,* are prevailingly northern in distribution and are abundant in the beech–maple and mixed communities of the Hemlock–White Pine–Northern

[9] More detailed reconstructions of the original vegetation pattern have been undertaken in a number of Ohio counties by Transeau, his associates, and graduate students, as part of a state-wide reconstruction. Only a few of these are available in the literature (Sears, 1925; Hicks, 1934; Shanks, 1939, 1942). For other states, see Veatch (1928, 1932); Kenoyer (1930).

Hardwoods region. But most are common to this region and to the regions to the south. Among the ferns, the spinulose shield fern (*Dryopteris spinulosa* var. *intermedia*) is conspicuous. Southward of this region, it is confined to deep talus of gorges and steep north slopes, while here and northward, it may appear on the forest floor. Such shrubs as *Sambucus pubens, Lonicera canadensis* and *Diervilla Lonicera* never occur southward of this region except in high altitude communities, and are more plentiful in the region to the north. Thus, we see that no species may be thought of as characteristic of the Beech–Maple Forest region. All are common to this region and the areas southward, to this region and the areas northward, or to all three zones.[10] This is to be expected when we remember that all the vegetation of this younger glaciated area migrated in since the recession of the last ice sheet.

It is the degree of dominance of certain species, and the number of species, which are characteristic. The two species which give name to the region, beech and sugar maple, occur through much of the deciduous forest, yet it is only here that they alone are the dominant species of the climax.[11] Northward they may be codominant in local hardwood communities or share dominance with hemlock. South of the region, they are seldom codominant species, except locally on valley flats or lower north slopes; in both situations beech far exceeds maple. In six out of seven beech–maple communities of the Beech–Maple region for which data are given, beech and sugar maple are the two most abundant species of the canopy; in one, tuliptree is second in abundance (and this is close to the southern margin of the region, cf. Table 59, column 1).[12] The average dominance of beech and maple in the canopy is approximately 80 per cent. These data are in contrast with those pertaining to the same two species in communities of the Western Mesophytic and Mixed Mesophytic Forest regions. The average dominance of beech and sugar maple in the former is approximately 50 per cent, or somewhat higher (60 per cent) in the few communities in which these two species occupy first and second places in abundance. In the Mixed Mesophytic Forest region, beech and sugar maple are abundant species in but a small proportion of the forest tracts. The average dominance of these two species in more or less typical mixed mesophytic communities of this region

---

[10] This is in striking contrast with the regions southward, which have many species not present in the younger glaciated area.

[11] Quick (1924), in a study of the climax association of southern Michigan, called attention to the fact that the predominance of beech is noticeable "only in the more northern localities and in the Alleghenies" and that "there is an increase in the importance of Fagus as regards the climax association as one leaves the southern part of the province [deciduous forest] and approaches the contact with the northeastern coniferous province."

[12] The forest at Turkey Run, Indiana (Table 58) is omitted from this comparison, as that forest is really an outlier of the adjacent Western Mesophytic Forest region.

is approximately 35 per cent. Thus, the degree of dominance of beech and sugar maple in the Beech–Maple Forest region, approximately 80 per cent of the canopy, contrasts strongly with the dominance of these two species in the two regions adjacent to the south.

The number of canopy species in beech–maple (climax) communities of the Beech–Maple region ranges from 3 to 14 (average 9.5) in examples studied.[13] The number of canopy species in the Western Mesophytic Forest region in communities in which beech and sugar maple occur, ranges from 10 to 20 (average 14.5). The number of species in more or less typical mixed mesophytic communities of the Mixed Mesophytic Forest region ranges from 12 to 23 (average 15.7). The average number of canopy species in the Beech–Maple Forest is less than the minimum for the two regions adjacent to the south, and only 60 per cent of the average number in the canopy of typical mixed mesophytic communities. These differences in dominance and in number of species are largely responsible for the noticeable difference in aspect of mesophytic forests of the Beech–Maple and of the Mixed Mesophytic Forest regions.

[13] Usually higher in transitional communities and communities of late stages in the swamp forest succession.

CHAPTER 11

The Maple–Basswood Forest Region

In the northwestern part of the deciduous forest, the climax dominants are sugar maple, *Acer saccharum,* and basswood, *Tilia americana* (*T. glabra*). The region characterized by the Maple–Basswood association is small in comparison with all other eastern forest regions. As interpreted here, it includes the Driftless area (exclusive of its northeastern part which is covered by proglacial lake deposits), the area of old drift adjacent to the west (a hilly strip lying along the west side of the Mississippi River), and an area to the northwest in south-central Minnesota.

The boundaries are almost everywhere indefinite, hence must be placed somewhat arbitrarily. For the southeastern part of the region, they have been placed along the boundaries of the Driftless section of physiographers. Indefinite boundaries and pronounced transitions are to be expected in the North, where little time has elapsed since the last glacial recession. Climatic warming, and alternating moist and dry conditions of post-Wisconsin time have induced general northward movement and alternating migrations of mesic and xeric species.

Local conditions of topography and, in places, direction of streams (because of protection afforded against fire) have influenced extent and location of areas definitely recognized as occupied by maple–basswood communities. Fires set by Indians, and later, logging and farming operations have so decimated the forest vegetation of the upper Mississippi Valley that only remnants remain from which to determine the nature and status of the several vegetation types of the area. Old records, and the few remaining remnants of relatively undisturbed vegetation have been used in determining the extent of the Maple–Basswood Forest region.

A map of "Native Vegetation of Wisconsin" (Chamberlin, 1882) distinguishes three main types of vegetation in southern Wisconsin: (1) the "oak group," including "oak openings," "oak orchards," and the denser oak forests; (2) prairie; and (3) the "maple group." Smaller areas of swamp, of pine forest, and of "dwarf oak and pine" (on sandy outwash plains) are also shown. The "maple group" is defined as mixed forest in which sugar maple on the average is more numerous than other species, oaks generally present, and basswood very numerous. This accords with the present concept of maple–basswood forest. A great lobe of oak and prairie extends northward to about the center of the state, separating the mesic forests near Lake Michigan from two fairly large areas of the "maple group" in the western half of southern Wisconsin. The larger of these two tracts is roughly triangular in shape and lies between the Wisconsin and Kickapoo rivers. The other extends from below the mouth of the Chippewa River northwestward toward the St. Croix River. The first of these lies entirely within the Driftless area; the second is partly within the Driftless area but extends onto adjacent old drifts.

The largest continuous area of maple–basswood forest (in the original forest pattern) is the tract known as the "Big Woods" in central southern Minnesota. This occupies the fine-textured soils on the rolling morainal upland of Younger Gray Drift (Mankato substage of the Wisconsin). The features of the "Big Woods," its boundaries, and the causes of these boundaries have been fully discussed by Daubenmire (1936).

Other small (but not mapped) areas are found scattered over the Driftless area, particularly on sheltered ravine slopes, in the hilly belt bordering the Mississippi River, and on ravine slopes between this and the "Big Woods." Basswood and sugar maple are constituents of many of the mixed oak woods in hilly northwestern Illinois. Throughout, successional relations indicate that maple–basswood is the potential climax.

The maple–basswood forest type is not confined to this region, but is represented elsewhere from central Illinois northward to Lake Superior, eastward to northwestern Ohio, and westward along valley slopes in the Prairie region. In the North, maple–basswood is a well-developed hardwood type of the Hemlock–White Pine–Northern Hardwoods region. Eastward, it is but a variant of the beech–maple forest.

Because the maple–basswood type alternates with other generally more extensive, although perhaps not climax forest types, or is represented by transitional and intergrading communities, it is not possible to exactly delimit the region. Because of its small size and indefinite limits, question may arise as to the desirability of giving to the Maple–Basswood region coordinate rank with the other forest regions. This western expression of mesophytic deciduous forest has at times been considered as a western faciation of the Beech–Maple association. However, it seems best to regard it as distinct because of its isolation from that region, because of its distinctive climax, and also because of its Pleistocene and post-Pleistocene history.

Migrations of oak forests from the south, of prairie from the west, and of mesophytic forest from the east have met and interlocked in the region west of Lake Michigan, resulting in prairie communities, forest communities definitely referable to the oak–hickory type, and more mesophytic forests of sugar maple and basswood, or of sugar maple, basswood, and red oak, together with a number of other species. The Driftless area is a mosaic of unlike types. In those areas of the younger drift suitable for occupancy by mesophytic forest, those mesophytic species which in the course of migration have reached to the northwestern part of the deciduous forest, and have persisted here through xeric periods of post-Wisconsin time, are associated in its most mesophytic forests. The sugar maple–basswood forest of this area has been derived in part by postglacial migrations from older areas to the south, in part from the Driftless area refugium. It is the western counterpart of the more eastern beech–maple forest. These relationships are considered later (Chapters 17, 18).

Although the region is small, it is divided into two sections, the **Driftless** section and the **Big Woods** section.

DRIFTLESS SECTION

An area of some 15,000 square miles in southwestern Wisconsin, southeastern Minnesota, northeastern Iowa, and northwestern Illinois was never covered by Pleistocene ice. Much of this area is submaturely dissected low plateau; a part of it is covered by glacial outwash and lake and stream deposits. To the north and west of this there is an area which, although glaciated in early Pleistocene time, is topographically similar to the unglaciated area which it adjoins. It is included in the "Driftless section" of physiographers. No vegetational features of importance distinguish this old drift border from the Driftless area. The Driftless section as here defined therefore corresponds closely with the physiographic section by the same name, except that an area of lake deposits in the northeast is excluded.

Vegetationally, this is a transition area. Oak or oak–hickory forest is widely distributed, thus relating this area to the Oak–Hickory region. Prairies, some small, some large, were a prominent feature of the vegetation; they indicate former greater expansion of the Prairie Formation in this area. Communities of jack pine on the sand plains, Norway pine on rocky slopes of monadnocks, white pine and hemlock on moist slopes, and larch in boggy spots, are transitional to the Hemlock–White Pine–Northern Hardwoods region which adjoins the Driftless section on the north. Maple, or maple–basswood communities are well represented and apparently climax in the area.

The Driftless area of northwestern Illinois, southwestern Wisconsin (south of the Wisconsin River), and adjacent Iowa is a maturely dissected area with deep ravines and rolling or occasionally flat-topped uplands. Prairie and open oak forest seem to have prevailed over much of the uplands, apparently because of fires set by Indians. Tree invasion has converted some of these former prairies into oak woodland. The ravine slopes appear always to have been forested, the principal trees being oak, accompanied by a variety of other species. Basswood and sugar maple occur more or less frequently, the latter sometimes forming almost pure stands. These are suggestive of the better-developed maple–basswood communities to the north. The profusion of mesic herbaceous species in the richer woods adds to the contrast between these local communities and the more widespread xeric oak woodlands and prairies. On north slopes of deeper valleys, relic outliers of northern plants occur (Pammel, 1905; Pepoon, 1910; Telford, 1926; and Vestal, 1931).

The vegetation of this entire section was greatly modified by yearly fires during the last century of Indian occupancy. Only a few protected slopes and coves escaped the devastating influence of fire (Marks, 1942). Oak scrub, maintained as scrub by fires, dominated much of the upland. Unlike the conditions prevailing through most of the deciduous forest, the vegetation found by the first white settlers was not the natural response to environment. With the cessation of fires soon after settlement, a new impetus was given to the fire-inhibited successions. However, agricultural utilization of the land limited the areas where development could progress undisturbed.

Few of the prairie areas of the upland remain. They have been destroyed by cultivation or encroached upon by forest. Their former location is usually indicated by relic prairie species (Thomson, 1940). Along the bluffs of the Mississippi River, in Wisconsin, Minnesota, Iowa, and Illinois, prairie communities occupy the steep south- and southwest-facing slopes. These "goat prairies" are more or less protected from cultivation and grazing because of their steepness; they are a conspicuous feature of the landscape, as the treeless patches contrast strongly with the generally forested slopes (Shimek, 1924; Thomson, 1940; Marks, 1942; Pammel, 1905).

Hills on the Wisconsin side of the Mississippi River showing forested ravines and the conspicuous "goat prairies" of the more exposed parts of the bluffs. (Courtesy, U. S. Forest Service.)

The oak scrub of the period of Indian occupancy was dominated by red oak, kept in shrubby form by frequent fires. White oak, white birch, aspen (*Populus tremuloides* and *P. grandidentata*), hazel (*Corylus americana*), and prickly ash (*Zanthoxylum americanum*) were probably also present (Marks, 1942). This oak scrub, with the cessation of burning, grew up into red oak woods with an admixture of white oak, bitternut and shellbark hickory, wild cherry, and in the understory, butternut and white elm also (Table 66, area 1). White oak is the dominant species in northwestern

Table 66. Forest communities of the Driftless section: *1,* Red oak woods which has superseded the oak scrub of the Indian period, Coon Valley, Wisconsin; *2,* White oak–Hill's oak community or "upland mixed hardwoods" of Jo Daviess County, northwestern Illinois; *3,* Red oak–basswood community, and *4,* Adjacent basswood–red oak–sugar maple community on north-facing bluffs of the Mississippi River near Winona, Minnesota; *5,* Maple–basswood community of northeast slope, Coon Valley, Wisconsin. (Data for *1* and *5* from Marks, 1942; for *2,* from Telford, 1926.)

	1	*2*	*3*	*4*	*5*
Number of trees	35	[1]	15	29	31
Quercus borealis maxima	85.7	. .	53.0	21.0	. .
Quercus alba	2.8	42.3	. .	. .	. .
Quercus ellipsoidalis?	. .	34.1[2]	. .	. .	. .
Quercus macrocarpa	. .	. .	7.0	. .	. .
Carya cordiformis	5.7	5.2[3]	. .	7.0	6.4
Carya ovata	2.8	. .	7.0	. .	. .
Prunus serotina	2.8	1.1	. .	. .	. .
Ulmus americana	. .	3.3	7.0	. .	9.7
Juglans cinerea	. .	. .	. .	3.0	. .
Juglans nigra	. .	3.2	. .	. .	. .
Fraxinus americana	. .	2.6	7.0	. .	. .
Populus grandidentata	. .	.2	. .	7.0	. .
Gymnocladus dioicus	. .	.07	. .	. .	. .
Tilia americana	. .	6.8	20.0	45.0	25.8
Acer saccharum	. .	1.0	. .	17.0	58.0

[1] "An average acre," based on all trees six inches d.b.h. and up, on 17.4 acres (1414 in all).

[2] Trees called "black oak" by Telford.

[3] Listed as "hickory" by Telford.

Illinois, and Hill's oak (*Quercus ellipsoidalis*)[1] a frequent associate (Pepoon, 1910; Telford, 1926; Vestal, 1931). In this oak community, the two dominants constitute about 75 per cent of the trees (Table 66, area 2). This type may replace the red oak type in development, with basswood and sugar maple as minor constituents. Their presence, however, suggests the ultimate replacement by a maple–basswood community. Red oak and basswood may

[1] *Quercus ellipsoidalis* is not well understood and is often confused with *Q. coccinea.* Thus Pepoon (1910) referred to a "scarlet and white oak association" in northwestern Illinois. It is called "black oak" in Illinois, and is thus referred to by Telford (1926).

be codominant in north slope forests; such communities grade into others in which basswood, red oak, and sugar maple are the predominant species (Table 66, areas 3 and 4).

The most mesophytic forest type of the Driftless section is one in which sugar maple and basswood are the dominant species (Table 66, area 5). This is the climax type which is attained in the most favorable situations in the Driftless section—north and northeast slopes and coves. The most highly developed maple–basswood communities, those with the oldest trees, are so situated that they escaped serious fire damage during the era of Indian burnings. However, many of the younger communities now recovering from fire influence clearly show the trend toward maple–basswood.

The gray-brown forest soil, and good mull humus layer which is seen in many of the red oak–basswood and maple–basswood slope forests which have not been seriously affected by grazing, is favorable to the development of a rich herbaceous layer resembling that of mixed mesophytic forests of the Western Mesophytic Forest region. It contains such widespread species as *Adiantum pedatum, Athyrium thelypteroides, A. angustum, Osmunda Claytoni, Botrychium virginianum, Arisaema atrorubens, Trillium grandi-*

A stand of maple–basswood forest near Richland Center, Wisconsin. (Courtesy, U. S. Forest Service.)

florum, Uvularia grandiflora, Asarum canadense, Anemonella thalictroides, Hepatica acutiloba, Sanguinaria canadensis, Caulophyllum thalictroides, Viola canadensis, Aralia racemosa, Phlox divaricata, Phryma leptostachya, and *Eupatorium rugosum.* A group of northern species emphasizes the transitional position of this section; among these are *Cornus rugosa, Diervilla Lonicera, Actaea rubra, Aralia nudicaulis,* and *Hydrophyllum virginianum.*

The large area occupied by oak–hickory and prairie communities in the Driftless section raises a question as to the proper classification of the Driftless section. Is it to be considered as a part of the Maple–Basswood region, as a part of the Oak–Hickory region, or as a transition region sharing not only characteristics of these two forest regions, but of the Hemlock–White Pine–Northern Hardwoods region and the Prairie region as well? Although its transitional character must be recognized, it is not desirable to distinguish so small an area as a distinct region of the deciduous forest. The area is now in process of development, development induced in part by cessation of burning (development of the past century) and in part by migratory shifts which have followed the climatic changes of post-Pleistocene time—changes in temperature following retreat of the ice, and of humidity in the xerothermic period (or periods) which profoundly affected this area. The climatic succession in progress leads from prairie to oak forest to maple–basswood forest. The climax nature of the maple–basswood community of the Driftless section, and its widespread, although discontinuous distribution in that area, justify its inclusion in the Maple–Basswood region. Under present climatic conditions it seems unlikely that development to the Maple–Basswood climax may be completed except in the most favorable situations.

The history of the Driftless section, lying as it does north of the glacial boundary, yet for the most part unaffected by glacial ice, has had a profound effect upon the vegetation of this area. This history is in part disclosed by the vegetation, for floristically, the Driftless section displays a number of diverse elements of unlike age. The long period of vegetational occupancy in the Driftless area and bordering old drift areas (together comprising the Driftless section) is emphasized by the endemic species which characterize this area. It is reflected in the mature character of the soils which contrast with those of the neighboring young drift areas. It is indicated also by the pollen records of bogs of surrounding areas, which show early entrance of deciduous species, apparently from a dispersal center in Wisconsin (Sears, 1942a). It is shown by bogs in the Driftless area, bogs formed at the time of retreat of the third Wisconsin ice sheet (Cary), in which pollens of deciduous species along with spruce, fir, and pine are present from the beginning of the record (Hansen, 1939).

Relic colonies of northern species are evidence of southward migration of northern vegetation in front of advancing ice. These occupy bogs and certain of the steep north-facing stream bluffs cut in sandstone. They contain such species of the Hemlock–White Pine–Northern Hardwoods region as *Tsuga canadensis* (very local), *Pinus Strobus, Taxus canadensis, Acer spicatum, Diervilla Lonicera, Corylus cornuta, Maianthemum canadense, Cornus canadensis, Clintonia borealis,* and *Coptis groenlandica,* as well as a few of the Driftless area endemics (Pammel, 1905; Marks, 1942). At the southern margin of the region, in northwestern Illinois, white pine occurs on the limestone cliffs of the Apple River (Pepoon, 1910). (A more southerly outlier of this species, in the Oak–Hickory region, occurs on the Rock River.)

The influence of the post-Wisconsin xerothermic period is seen in the numerous prairie relics of the Driftless area. Encroachment of forest on prairie in the subsequent more humid period, and later, agricultural utilization have eliminated many of the prairie communities, although some of the species still persist in areas formerly occupied by prairie (Thomson, 1940).

Invasion of deciduous forest from the south, in part from the Ozark center, in part from farther east by a route leading around the Prairie Peninsula, has resulted in the mixing of forest types evident in this section. However, not all of this mixing was recent. Many of the protected coves and slopes in the strongly dissected areas probably served as refugia for deciduous forest species, not only during the period of prairie expansion, but also earlier, when glacial lobes (Wisconsin) extended southward on one or the other side of the Driftless area (see Chapters 15, 16, 17, 18).

BIG WOODS SECTION

This section includes the "Big Woods" of Minnesota together with isolated stands of maple–basswood forest in an area extending southeastward to the Driftless section.

Quantitative studies by Daubenmire (1936) in the "Big Woods" of Minnesota furnish data on the composition of two well-preserved remnants of the maple–basswood forest—the Minnetonka woods and the Northfield woods. In the first of these, the two dominants comprise 77.4 per cent of the trees over 10 inches d.b.h., as determined from sample plots (75.7 per cent of the basal area of trees in this size class, and 77.16 per cent of the total basal area). In the other, 65.1 per cent of the trees over 10 inches d.b.h. are sugar maple and basswood (65.3 per cent of the basal area); in this tract, red oak is abundant, making up an additional 19 per cent of the trees of this size class[2] (Table 67). Weaver and Thiel (1917) speak of the Minnetonka woods as a "climax maple forest" which is "rather typical

[2] Daubenmire's tables give the density, frequency, and basal area of tree species per 2500 square meters, in each of five d.b.h. classes. The per cent of dominants in the 10 inch d.b.h. class is given here to facilitate comparison with other forest areas for which canopy composition has been given.

Table 67. Composition of the Minnetonka woods, *M*, and of the Northfield woods, *N*, two areas of the maple–basswood type in the "Big Woods" of Minnesota. (Data from Daubenmire, 1936, arranged to show percentages of trees 10 inches or more d.b.h.)

	M	*N*
Acer saccharum	47.2	23.8
Tilia americana	30.2	41.3
Quercus borealis maxima	5.7	19.0
Ulmus americana	9.4	9.5
Ulmus fulva	7.5	. .
Fraxinus pennsylvanica var. *lanceolata*	. .	3.2
Fraxinus pennsylvanica	. .	1.6
Quercus macrocarpa var. *olivaeformis*	. .	1.6

of hundreds of square miles of former forest conditions of south-central Minnesota." They give the composition of this woods, based on transects including nearly a thousand square meters, as follows:

> The dominant, *Acer saccharum,* makes up 78% of the forest; *Ostrya virginiana* 8%; *Ulmus americana* and *Carya cordiformis* each 4%, while *Quercus rubra* and *Tilia americana* complete the list of trees. On the ridge the less mesophytic and also less tolerant red oak is much more abundant than on the slope.

No statement is made as to size of individuals, but apparently even the smallest were included, hence the very high percentage of sugar maple.

In undergrowth, the maple–basswood forest communities resemble more southern forests. Only a few of the shrubs and herbaceous plants are distinctly northern, and these are ones which appear also in the beech–maple forest. *Sambucus pubens* has a fairly high frequency; other of the northern species, as *Aralia nudicaulis, Maianthemum canadense, Actaea rubra,* and *Trillium cernuum* are infrequent. The more abundant species are more southern in distribution. Almost all of them are plants of mull humus, such as *Uvularia grandiflora, Sanguinaria canadensis, Caulophyllum thalictroides, Panax quinquefolium, Eupatorium rugosum,* etc.

Successions in the Big Woods section of the Maple–Basswood region, both primary and secondary, have been considered by Bergman and Stallard (1916) and by Stallard (1929). In small denuded areas, seedlings and sprouts of the climax dominants reproduce a forest similar to the climax. In larger and more disturbed areas where an aspen–birch community has developed, seedlings of maple, basswood, and oak will finally produce a maple–basswood forest.

Climate, soil (especially as it influences ground water), and fire are important factors limiting the extent of this northwestern part of the Maple–Basswood Forest region. It is beyond the geographic limits of beech, and close to the limits reached by sugar maple. Beyond the limits of sugar maple, a basswood–red oak community occupies protected slopes and ravines in the prairie-forest transition in the valleys of the Mississippi Valley section of the Oak–Hickory region (p. 182). Fire has a profound influence on forest composition and forest extent in much of the upper Mississippi Valley; the Indians of that region were accustomed to setting frequent fires which were particularly destructive to forest and favorable to extension of prairies, and/or bur oak groves. Sugar maple, and, to a lesser extent, red oak are fire-sensitive; basswood continues to sprout at the base even after repeated burnings. Many parts of the boundary of the "Big Woods" follow approximately along lakes and streams which may have served as fire barriers, thus protecting the area where the Maple–Basswood climax is best developed. The influence of fire has a limiting effect on the extent of maple–basswood communities and thus obscures the extent of the region. To the northeast, the Maple–Basswood region dovetails into the Hemlock–White Pine–Northern Hardwoods region, in much the same manner as does the Beech–Maple region.

CHAPTER 12

The Hemlock–White Pine–Northern Hardwoods Region

This forest region extends from northern Minnesota and extreme southeastern Manitoba through the upper Great Lakes region and eastward across southern Canada and New England, including, toward the southeast, much of the Appalachian Plateau in New York and northern Pennsylvania (see Maps, pp. 394, 395). Outliers of the region are seen farther south in the Allegheny Mountains of Pennsylvania, Maryland, and West Virginia. The region is characterized by the pronounced alternation of deciduous, coniferous, and mixed forest communities. In primary deciduous communities, sugar maple, beech, and basswood, sugar maple and beech, or sugar maple and basswood are the usual dominants, and yellow birch, white elm, and red maple more or less frequent associates. In secondary deciduous communities, aspen, balsam poplar, paper birch, and gray birch are abundant species. The coniferous communities which occur at intervals almost throughout the region are of two general types: (1) those of more or less dry sand plains and ridges where white pine, red or Norway pine, and jack pine prevail; and (2) those of poorly drained areas, bogs or muskegs, where black spruce, arbor vitae (northern white cedar), and larch prevail. A third type of coniferous community, in which red spruce is dominant, occupies mesic flats and slopes in the northeastern part of the region. The primary mixed communities, except in the extreme western part of the region, are hemlock and northern hardwoods, or, in parts of the Northeast, red spruce and northern hardwoods, principally sugar maple, beech, basswood, and yellow birch, in which there is (or was) usually an admixture of white pine. These, the most characteristic communities of the region, give to the region its name—Hemlock–White Pine–Northern Hardwoods.

The name "hemlock–hardwoods," so frequently used for the northern forest, is not applicable to the westernmost section of the region, for there hemlock is absent and white pine is a climax dominant. Although the climax status of white pine is questionable in much of this vast region, the regional name—Hemlock–White Pine–Northern Hardwoods—takes cognizance of its status in the west, and of its general occurrence throughout.

The grouping of dominants, the prominence of a number of northern shrubs and herbs, the presence of two more or less endemic tree species (red pine, and, in the east, red spruce), and the distinctive secondary communities serve to further characterize the region. Its most abundant dominants, variously grouped in different parts of the region, are hemlock, sugar maple, basswood (*T. americana*), beech, yellow birch, white pine,[1] red spruce (in the east), and, close to the northern limits of the region, white spruce and balsam fir. Its most abundant and characteristic subclimax dominants are white pine, red pine, and jack pine on drier soil, and black spruce on wet soil. The secondary beech–maple or maple sprout forests of this northern region are very striking as beech[2] and maple in the South do not display this phenomenon.

The boundaries of the region are ill-defined, for this is a great tension zone between encroaching more southern species and retreating more northern species. It is a region of interpenetrating climaxes, but a region distinct in the grouping of its climax dominants and in its dry soil physiographic climaxes. From the Atlantic coast in Massachusetts, the southern boundary passes through southern New England (p. 251) and southeastern New York into Pennsylvania where it extends southward in the mountain and plateau region. Only here is the boundary on unglaciated territory. The generalized southern boundary in the Great Lakes region passes from the upper St. Lawrence westwardly along the shores of Lake Ontario and across southern Ontario and the Lower Peninsula of Michigan in the vicinity of Saginaw Bay,[3] and south of Green Bay, northwestwardly across Wisconsin, almost bisecting that state, and almost across northern Minnesota. From this southern boundary, the region extends northward to about 48° north latitude soon giving way to the Northern Coniferous or Boreal Forest. The region as here defined is essentially that discussed by Nichols (1935). It includes the Lake Forest of Weaver and Clements (1938). In Canada, it includes the areas distinguished by Halliday (1937) as Great Lakes–St. Lawrence and Acadian Forest regions. It is approximately coextensive with the Canadian Biotic Province of Dice (1943), except in Minnesota, where Dice includes the Maple–Basswood of the "Big Woods"; and with the "northern hardwood forest" as defined by Frothingham (1915) except in its southern extension in the mountains.

[1] White pine may be considered a constituent of the climax in some cases; more often it is a subclimax species. Its status will be considered in connection with communities of which it is a part.

[2] It has been pointed out previously that American beech is not everywhere the same, that it is made up of distinct genotypical races, or perhaps distinct species. "Red beech" in more or less pure form is the dominant beech of this region. Northward, in the area where fir becomes common, "gray beech" occurs, and introgressions from this contaminate part of the red beech population (from studies by W. H. Camp).

[3] Somewhere in the "Transition Soil Region" of Veatch (1932), as stated in connection with the northern boundary of the Beech–Maple region (p. 305).

Although the boundaries of this region do not coincide with any soil province boundaries, there is considerable correspondence between this region and the area of podzols (see Maps, pp. 24, 26). Variations in the humus layer and in degree of podzolization are displayed. Some humus incorporation has taken place locally in the best hardwood stands where the activity of earthworms is evident; here the soils are podzolic rather than true podzols. Generally, however, the superficial layer of more or less matted organic material—mor—is evident. Bog soils, ground-water podzols or glei soils prevail in the poorly drained areas.

Within the region, there is a decrease northward in the extent of deciduous communities, until toward the northern limits of the region they occur only as isolated patches (Cooper, 1913; Grant, 1934; Halliday, 1937; and Dansereau, 1944, 1944a). In the more southern part of the region there is evidence of further increase in dominance of the deciduous species (Gleason, 1924). Species of the northern conifer forest—*Picea mariana, P. glauca, Abies balsamea, Thuja occidentalis,* and *Larix laricina*—occur more or less frequently throughout the region, but in its southern part are confined to relic communities. Northward, the communities of which they are a part are often so extensive as to control the landscape. As Nichols (1935) has pointed out, the region contains four groups of species: (1) those whose centers of north-south distribution are far to the south (beech, white ash, black cherry, white elm); (2) those whose centers are partly within and partly immediately to the south (sugar maple, basswood, and northern red oak); (3) those whose centers are within the region and whose ranges are more or less coextensive with it or some part of it (hemlock,[4] white pine,[4] yellow birch, Norway pine, and red spruce); (4) those whose centers of north-south distribution are to the north of the region (white and black spruce, fir, larch, balsam poplar, and paper birch). This region is not merely a broad transition belt between the deciduous forest and the northern conifer forest, as might appear from the behavior of communities definitely a part of one or the other, or from the distribution of its dominant species. It has characteristics which define it as a distinct region.

The Hemlock–White Pine–Northern Hardwoods region is made up of two main divisions—a more western part, the **Great Lakes–St. Lawrence Division,** and a more eastern part, the **Northern Appalachian Highland Division.** The former comprises the areas about the Great Lakes and extends eastward across Canada to the St. Lawrence Valley. It is essentially the Lake Forest of Weaver and Clements (1938), exclusive of its small eastern lobe.

---

[4] While these two species in the Great Lakes region and New England have ranges somewhat comparable to the area of the Hemlock–White Pine–Northern Hardwoods Forest, hemlock, and to some extent white pine, are abundantly represented southward in low elevation forests in no way a part of this northern forest region (pp. 45, 483). The possibility of genotypical races such as demonstrated by J. W. Wright (1944) for white ash and indicated for beech by the studies of W. H. Camp is apparent.

It includes all except two sections of Halliday's "Great Lakes–St. Lawrence Forest Region" (the two which lie entirely to the east of the St. Lawrence Valley). The Northern Appalachian Highland Division, as the name implies, covers the northern part of the Appalachian Highland from northern Pennsylvania northeastward across New York, New England, and the maritime provinces of Canada, and extends northward onto the plains of Lake Ontario, merging with the Great Lakes–St. Lawrence Division in the St. Lawrence River valley. It thus includes the "Acadian Forest Region" of Halliday, and parts of his Great Lakes–St. Lawrence Forest where certain transitional features occur.

The vegetational unity of the Hemlock–White Pine–Northern Hardwoods region is emphasized by the nature of the climax communities which vary almost as much locally as regionally. Differences which appear toward the western and northern margins are due to the dropping out of species as the limits of their geographic ranges are reached. Thus west of Marquette, Michigan, beech is not a constituent of the climax; neither is it at the northeastern margin of the region in Canada. In western Wisconsin, the limit of range of hemlock is reached; the number of characteristic dominants is thus further decreased. Yellow birch, edaphically limited in the westernmost part of its range in Minnesota, is not there a constituent of the climax. Basswood, one of the dominants in deciduous phases of the climax towards its western limits, is more infrequent eastward. Red spruce (*Picea rubens*) is an important constituent of some climax communities of the eastern or Northern Appalachian Highland Division; it does not occur (except very locally) in the Great Lakes–St. Lawrence Division. Gray birch (*Betula populifolia*), a dominant in old-field successions over a large part of the eastern division, drops out westward before the limits of this division are reached. This twofold division of the region not only coincides roughly with the topographic break between physiographic provinces, but is supported by species distribution.

However, the boundary between the two divisions is somewhat arbitrary as it is not feasible to limit the Northern Appalachian Highland Division exactly to the area in which red spruce and gray birch (called wire birch by Halliday) occur.

The Hemlock–White Pine–Northern Hardwoods region is related to the Beech–Maple region with which much of the southern margin of the western division is in contact. Hence the western part of the region will be considered first.

GREAT LAKES–ST. LAWRENCE DIVISION

This division of the Hemlock–White Pine–Northern Hardwoods region is divided into four sections: (1) the **Great Lake** section, approximately the northern half of the Great Lake section of physiographers together with a strip across Ontario; (2) the **Superior Upland,** corresponding somewhat with the physiographic province of that name; (3) the **Minnesota** section, which is the northeastern part of the Western Young Drift section of physiographers, together with some contiguous area to the north and east; and (4) the **Laurentian Upland** section, here defined as a Canadian area extending from eastern Lake Superior eastward to the valley of the St. Lawrence River.

On the whole, this is a region of low relief whose topographic features are almost entirely controlled by glaciation. On parts of the Superior Upland and Laurentian Upland this is not the case; the Huron Mountains and the Porcupine Mountains are notable exceptions. It is a region in which secondary forest communities, aspen, jack pine, and sprout hardwoods, are widely distributed and extensively developed. They give little or no indication of the original forest types upon which the emphasis in this book is placed. They do emphasize, however, the distinctness of this region from those adjacent to the south. The bog communities, which are found scattered through the northern half of the Beech–Maple region, here are of frequent occurrence and often extensive—hence, give to the region a northern aspect. Its principal hardwood communities—beech–maple and maple–basswood—are very similar to the climax communities of the Beech–Maple and Maple–Basswood regions.

GREAT LAKE SECTION

This section as here delimited includes three more or less separated areas contiguous to the Great Lakes—northern Lower Michigan, eastern Upper Michigan, and that part of Ontario lying between Lake Huron and Lake Ontario. Its southern boundary is in contact with the Beech–Maple Forest region, to which this section is related. Its western boundary coincides more or less closely with the western limits of beech (excluding a few outlying stations). Its northern boundary (in Canada) is at the margin of the Laurentian Upland (essentially at the contact between Paleozoic sedimentary rocks and Pre-Cambrian mainly crystalline and resistant rocks). Many of its hardwood communities are typical beech–maple forest. In fact, beech–maple as a forest type or as an ecologic climax community is as well illustrated in the northern part of the Lower Peninsula of Michigan or in the eastern end of the Upper Peninsula as it is in northern Ohio or southern Michigan. However, maple is generally more abundant than beech in the more northern hardwood communities, and usually has a higher frequency. Hence the name "maple–beech" so often used is particularly applicable here (Gleason, 1924).

This is a region of low relief, of dominantly glacial topography. Extensive sandy outwash plains afford suitable environment for the pine forests of the region. Morainal ridges and the swells of rolling moraines, if the soil is fine-grained and loamy, are occupied by deciduous forest communities, or mixtures of deciduous species with hemlock and perhaps white pine. The innumerable poorly drained areas, sometimes extensive, sometimes small, and the borders of many of the lakes and ponds are occupied by various of the bog and bog-forest communities.

On the sand dunes, which are especially well developed along the east shore of Lake Michigan, developmental stages in forest succession are well illustrated, and demonstrate the establishment of the regional climax of beech, maple, and hemlock (Cowles, 1899; Whitford, 1901; Coons, 1911; and Waterman, 1917, 1922). The essential features of the succession are the same throughout the entire north-south extent of the dunes, although there is some difference in the species involved. The embryonic dunes of the beach may build up around clumps of *Ammophila*. In many places, the poplars (*P. deltoides* in the south, and *P. tremuloides* and *P. tacamahaca* northward), paper birch, and shrubby willows are instrumental in initiating development on the beach. Windward and lee slopes of dunes differ in development. The windward slopes are usually characterized by stages in which conifers are abundant. Prostrate species of junipers (*J. communis* var. *depressa,* and in the north, *J. horizontalis*), with bearberry (*Arctostaphylos Uvi-ursi*) and *Hudsonia tomentosa* var. *intermedia* colonize the almost bare sand slopes. The pines form a conifer forest stage, in which white pine is usually the most abundant species, although red pine is sometimes the dominant species. Northward, arbor vitae (northern white cedar) and balsam fir mingle with the pines. Occasional deciduous species—red oak and basswood chiefly—mingle with the conifers and suggest the establishment of a deciduous forest. Some old dunes toward the southern end of Lake Michigan (in the Beech–Maple region) are covered by open black oak forests in which a great variety of herbaceous species, including some prairie plants, occur. The lee slopes of the dunes are prevailingly deciduous, particularly those of older dunes. It is in such situations that climax forests of beech, sugar maple, and hemlock develop. Yellow birch may be present; northward, balsam fir sometimes forms a lower layer in the beech–maple communities of the old dunes. Depressions between dunes, if low enough for a permanent water supply, illustrate various stages of hydrarch succession.

Cutting and fire have profoundly modified the original conditions throughout much of the Great Lake section. Jack pine plains and open aspen woods, or open scrubby growths, extend for miles where other forest types once held sway. In fact, in going from the Beech–Maple region into the Hemlock–White Pine–Northern Hardwoods, the most noticeable vegetational feature today is the prevalence of these two secondary communities. Where fire has not been severe in cut-over hardwoods areas, sprout forests arise, unlike anything in the mesophytic forests southward, thus regenerating a beech–maple forest (Woollett and Sigler, 1928).

Virgin or near virgin forest tracts are very few in the Lower Peninsula of Michigan and generally small and isolated from one another. The interrelations of communities may thus be obscured. In the more extensive forest areas in Upper Michigan, the transitions between communities are better shown.[5] Very different soils, and very different soil moisture conditions, have resulted in the development of unlike communities, at least some of which are relatively stable and long-lived physiographic or edaphic climaxes. As these, rather than the hardwood and mixed communities, contrast most strongly with the regions to the south, they will be considered first.

Pine Communities

Jack pine (*Pinus Banksiana*), red or Norway pine (*P. resinosa*), and white pine (*P. Strobus*), singly or in combination, compose the pine forests of the Great Lake section. In Michigan, each of these reaches the southern limits of its more or less continuous range in the "transition soil region" of Veatch (Map, p. 306); jack pine does not enter the Canadian area of this section (Halliday, 1937). White pine is found farther south than the other two species.[6] Original stands are all but gone. Fires, the aftermath of lumbering, have been particularly severe in areas dominantly pine. Even the humus may be completely destroyed, leaving only the mineral soil. Jack and red pine, and sometimes white pine, occupy the sandy soils on which the humus forms a distinct and superficial layer (a mor humus layer). Only when the humus becomes well incorporated, does vegetational development on sand go beyond a pine stage.

Jack pine forms open or closed communities, sometimes miles in extent, on sandy plains. Lumbering and fire have tended to increase the extent of jack pine at the expense of red and white pines, both commercially valuable species. Open jack pine stands may be interspersed with scrubby oaks, or scrubby oaks may dominate on lands which were originally occupied by red pine or by red and white pines (Kittredge and Chittenden, 1929). These scrubby oaks (*Quercus alba, Q. borealis, Q. coccinea, Q. ellipsoidalis,* and *Q. velutina*) are clumps of sprouts which regenerate time and again after fire. "Scrub oak" communities occur most frequently in northern Lower Michigan and in western Wisconsin. Aspen (*Populus tremuloides*)

[5] For a map of the "principal forest types of the Upper Peninsula of Michigan, original and present areas," see Cunningham and White, 1941.

[6] The most southern stands of jack pine are in the dunes of northwestern Indiana and northeastern Illinois; of red pine, in southwestern Michigan. White pine is at the edge of its continuous range in northern Illinois and northwestern Indiana; it occurs in isolated areas in west-central Indiana (Warren, Fountain, Montgomery counties) and in western Kentucky (Todd County); farther east, in the Appalachian Highlands, it is found from eastern Ohio as far south as northern Georgia.

and largetooth aspen (*P. grandidentata*) may be scattered among the pines. The blueberries (*Vaccinium* spp.), the bracken fern (*Pteridium aquilinum*), several species of grasses, together with a scattering of other herbaceous and woody species, form a more or less continuous ground cover. Closed and even-age stands of jack pine are less frequent than the open stands. In these, many of the sun-loving species of the ground layer have been eliminated. In an area near Grayling, Michigan, such a forest has a continuous shrub layer in which *Kalmia angustifolia* is dominant.

Mixed jack pine and red pine communities, sometimes with a scattering of white pine, occupy some of the sandy plains (for example, near Shingleton, in the Upper Peninsula of Michigan). White oak, red oak, and red maple (persisting from sprouts), and pin cherry may be present in mixed jack pine–red pine stands.

A well-developed red pine stand has a striking appearance due to the brilliancy of coloring of the trunks of this species. Red pine is frequently seen along sandy borders of lakes, especially where these rise abruptly above the water table. A representative stand of this sort on the outskirts of Marquette, Michigan, occupies a sandy terrace but a few hundred feet from the shores of Lake Superior.[7] It is a pure stand of red pine, the trees all about 20 inches d.b.h. A middle-size class of trees, forming a lower layer in the forest, is made up chiefly of red pine, paper birch, and red maple with a few largetooth aspen. In the small tree and shrub layer, red maple, red pine, and white pine prevail, with some red oak, balsam fir, paper birch, service berry, and shrubs (*Sambucus pubens, Corylus cornuta, Rubus parviflorus,* and *Diervilla Lonicera*). In the herbaceous layer, *Aster macrophyllus, Clintonia borealis, Oryzopsis* sp., and *Pteridium* are abundant. The evidence from this forest is that the red pine type is temporary and will be superseded by some other forest type.

Mixed red pine–white pine stands are developmentally more advanced than the red pine stands. Such communities are well represented in the Hartwick Pines State Park, northeast of Grayling, Michigan. They will be considered in connection with that area.

The original white pine forest was, very often, a pure stand of this species. Admixtures of red or Norway pine, or of hardwood species were common. White pine was scattered through the hardwood forests and some of the swamp forests. It was sometimes an abundant tree of late stages of bog succession. The southern limit of the area where white pine was originally present in such amount as to be a commercially important tree (Wheeler, 1898) of this section is close to the southern boundary of the Hemlock–White Pine–Northern Hardwoods region. Now white pine is a rare tree

---

[7] This area is on the line of demarcation between the Great Lake section and the Superior Upland, being just within the latter. The pine types do not differ in the two sections.

in much of the area which once was the great pinery of Michigan. Livingston (1905), in describing the white pine forest type, states that large areas, originally covered by pure stands of white pine, have no trees at all. Instead they are areas of dwarfed white oak, red oak, red maple, and shrubs, in which the oaks and maples are perhaps 10 feet tall and burned down every few years; they exist only because of their ability to sprout repeatedly from the roots.

The white pine type in Upper Michigan was described by C. A. Davis (1907) as being more variable than any other forest type of Michigan, because of "the great variety of conditions in which that noble tree will grow and thrive." He included in this type

all the mixtures in which White Pine was the leading species in point of numbers, so that the range is from the two-storied White Pine and hardwood forest in which the old pines rise above the lower growing species, through the pure stand of White Pine, to the mixture of the three species of pines in which the White Pine was the important species numerically. . . . This type did not cover great connected areas as did the Hardwood type, but was confined to relatively small and disconnected areas.

The Hartwick Pines State Park,[8] established to preserve a remnant of the original pine forest of the northern part of the Lower Peninsula of Michigan, is representative of the pine forests of the Great Lake section. It occupies a sandy ridge and a small part of the adjacent flat sandy plain. In the canopy of the area as a whole, white pine prevails.[9] Norway pine is codominant with it in parts of the area; in other parts, hemlock is codominant with the white pine or is dominant in a community almost lacking pines but having a number of deciduous species. In all instances, the middle-size trees (trees of the second layer) indicate a change in forest type. The developmental series indicated here is white pine–Norway pine→white pine–hemlock (with beech and maple)→hemlock–northern hardwoods (Table 68). In a small part of the area, formerly dominated by white pine, a blow-down destroyed the largest trees; the trees of the second layer now fully occupy the space, converting it into a deciduous community with an admixture of hemlock. This conversion to deciduous forest would have taken place finally as a result of the inability of the pines to reproduce successfully in the more advanced developmental stages and of the gradual dying of old pines. The areas with the largest amounts of hardwood species have the richest ground layer; some hemlock areas have an abundant growth of *Dryopteris spinulosa;* while the pine areas have the poorest ground layer. These are condi-

[8] In Crawford County, Michigan, one of the two counties discussed by Livingston (1905). At that time, almost all of these two counties had been lumbered, except the hardwood forest, of which there was very little. As the soils of the two counties are nearly all sandy, pines prevailed in the original cover.

[9] The largest white pine of the area is 150 feet tall, and 50 inches d.b.h.

tions related to the character of the humus layer. The contrast between primary and secondary vegetation of the sand plains is strikingly illustrated here, where this fine remnant of the original forest is surrounded by open jack pine and aspen communities in which patches of spruce, tamarack, and arbor vitae occupy the wet spots. These jack pine and aspen lands were

White and Norway pines in Hartwick Pines State Park, where a remnant of the original pine forest of the Lower Peninsula of Michigan is preserved. (Courtesy, U. S. Forest Service.)

Table 68. Composition of canopy and second layer of communities in the Hartwick Pines State Park, Crawford County, Michigan, determined from transects through the several communities. Note change from white pine–red pine → white pine–hemlock (with hardwoods) → hemlock–hardwoods. In community *5*, pines are almost absent from the canopy and absent from the second layer, which represents the hemlock–northern hardwoods and indicates the ultimate replacement of pines by this climax community. Compare with Table 79.

Canopy	*1*	2	*3*	*4*	*5*
Number of trees	54	102	97	105	37
Pinus Strobus	61.1	73.5	44.3	54.3	2.7
Pinus resinosa	38.9	26.5	1.0	6.6	. .
Tsuga canadensis	. .	. .	25.8	17.1	59.5
Acer saccharum	. .	. .	11.3	7.6	5.4
Acer rubrum	. .	. .	8.2	11.4	13.5
Fagus grandifolia	. .	. .	. .	1.9	8.1
Betula lutea	. .	. .	9.3	.9	8.1
Quercus borealis	. .	. .	. .	. .	2.7
Second layer					
Number of trees	44	60	68	142	58
Pinus Strobus	40.9	28.3	4.4	11.3	. .
Pinus resinosa	9.1	1.8	. .	4.2	. .
Tsuga canadensis	18.2	33.3	51.5	26.0	51.7
Acer saccharum	11.3	3.3	16.2	19.0	8.6
Acer rubrum	13.6	15.0	11.8	16.2	3.4
Fagus grandifolia	6.8	18.2	13.2	19.0	24.1
Betula lutea	. .	. .	2.9	2.8	10.3
Prunus serotina	. .	. .	. .	1.4	. .
Ostrya virginiana	. .	. .	. .	. .	1.7

formerly clothed with white and Norway pine forests comparable to those in the Hartwick Pines area.

Successional development in the several pine communities will lead, ultimately, to the establishment of the regional climax forest of hemlock and northern hardwoods, or of hardwoods alone (see p. 384 for comparable development in Minnesota). This development is exceedingly slow, taking centuries for its completion and is possible only in the absence of fire. It may take place on any type of soil, but is more rapid on fine-grained soils, and slower on sandy soils.

Aspen Communities

Outstanding in the landscape of the Great Lake section, in fact in all of this northern forest region, are the aspen communities which in many places extend for miles over land formerly occupied by other forest types. Gates (1930) has said, "the aspen association is by far the most important secondary association in northern Michigan, revegetating nearly every type

Aspen woodlands are conspicuous in the Great Lake section and in all other parts of the Hemlock–White Pine–Northern Hardwoods region as well. This stand, 45 years in age, is in the Chippewa National Forest in Minnesota. (Courtesy, U. S. Forest Service.)

of location following the removal of virgin growth." The principal trees are *Populus grandidentata, P. tremuloides, Prunus pensylvanica,* and *Betula papyrifera.* The nature of the soil and the amount of soil water largely determine the dominant species—*P. grandidentata* on the drier sandy soil, *P. tremuloides* on moister sandy soil, and *Prunus pensylvanica* on clayey soils. Paper birch is not often a dominant species, but occurs in all situations.

Except in the densest aspen communities, there are a large number of shrub and herb species, most of which belong more properly to the primary communities which the aspens have replaced. Over large areas, the bracken fern (*Pteridium aquilinum*) is a subdominant species, often forming a continuous layer which obscures the lower species. Sweet fern (*Comptonia peregrina*) is a conspicuous shrub of many aspen communities, especially westward.

In the absence of fire, more shade-tolerant species invade the aspen communities, tending slowly to reestablish the pine, bog, or hardwood forest type appropriate to the habitat.

Bog Communities

A consideration of bog communities in a treatment of deciduous forest may at first appear out of place. In the Great Lake section of this northern forest region, bogs and swamps cover a large area, and in many places they dominate the landscape. While most of these are occupied by communities of conifers (or earlier developmental stages), a few are hardwood swamps.

Bogs occupy undrained or very poorly drained areas where there is an abundance of water. Such situations are numerous and often extensive in the northern end of the Lower Peninsula and in the eastern half of the Upper Peninsula of Michigan, where the moraines were deposited beneath the waters of Glacial Lake Algonquin—a Pleistocene predecessor of the Upper Great Lakes (Leverett, 1928). Maps of "the original swamp areas of the Lower Peninsula of Michigan" and of "the original vegetation of the Upper Peninsula of Michigan" (Davis, 1907) readily show why, due to extent alone, the bog communities may control the landscape.

The general sequence of vegetational stages of a developing bog—aquatic, sedge mat, bog shrub, and bog forest—is similar in all bogs. Differences in the course of development may affect or determine the composition of ensuing stages. In northern Lower Michigan, *Chamaedaphne calyculata* is the most abundant and most generally distributed bog shrub. In parts of Upper Michigan, *Ledum groenlandicum* may dominate areas of bog. The

The interior of a white cedar or arbor vitae bog forest in Michigan. (Courtesy, U. S. Forest Service.)

principal trees of the bog forests are *Larix laricina, Picea mariana,* and *Thuja occidentalis.* Any one of these may occur in almost pure stand. Larix bogs, less densely shaded than the others, usually contain an abundant shrub flora. They are always replaced by black spruce or arbor vitae. In Lower Michigan spruce bogs are not well represented; the Thuja bog is the ultimate in bog forest development. The bogs of northern Lower Michigan have been described by Gates (1942) who distinguishes a score or more of developmental stages whose successional relationships are shown in a diagram. Gates states that "the Thuja association is the climax association in boggy areas in the region." Farther north, in the Upper Peninsula of Michigan, black spruce may dominate, or share dominance with arbor vitae or tamarack.

The "coniferous swamp forest" of the Upper Peninsula of Michigan

> covered all of the undrained or poorly drained flat lands, including the peat bogs. The more mature swamps had what has been called. . . . the Cedar–Tamarack–Spruce Association, in which frequently the Arbor Vitae or Cedar, was the dominant species, with a larger or smaller mixture of Black Spruce and Tamarack. The Tamarack, however, was very often abundant or dominant, and the Black Spruce sometimes covered considerable tracts with very few individuals of the other species present. In numerous cases the Cedar was wanting from the association where the other two species were present in about equal numbers. In this type also occurred at times the White Pine, the Jack Pine, and sometimes a few scattering Norway Pines and upon the borders Balsam and White Spruce; Black Ash, White Elm, Balsam Poplar, Red Maple and a few other species of swamp trees also were present in numbers sometimes sufficient to give a mixed character to the type, but this was rather unusual (Davis, 1907).

The very extensive bogs or swamps of this section, as the Manistique and the Tahquamenon (or Taquamenaw) swamps are frequently referred to as muskegs. The Tahquamenon occupies much of the drainage basin of the Tahquamenon River, which carries the waters of the bog to Whitefish Bay in Lake Superior. It is very irregular in outline, approximately 40 miles long in an east-west direction, and in places 10 to 15 miles wide, although usually less. In such large bog areas, the regular zonation of communities so noticeable around small bog lakes is not seen. Instead, patches or bands of bog forest, or scattered trees of spruce, arbor vitae, and tamarack dot the expanse of muskeg. Mounds of bog shrubs—*Ledum, Chamaedaphne,* and *Kalmia* chiefly—rise above the quaking expanse of Sphagnum and sedge bog dotted with cotton grass (*Eriophorum* sp.) and pitcher plants (*Sarracenia purpurea*). Near the streams which concentrate some of the bog waters, the wet but drained habitat is occupied by river swamp communities which contrast strongly with the adjacent bog communities. Shrubby species of willow, alder, and Spiraea form a shrub border; white elm, black ash, and sometimes silver maple dominate in the swamp forest strip. The royal fern (*Osmunda regalis*) is a frequent subdominant. On better drained stream margins, a few additional hardwood species may enter the elm–ash community, suggesting its replacement under good drainage conditions, by mesophytic hardwood forest. The moraines bordering the Tahquamenon Swamp are clothed with hardwood forest or mixtures of hardwoods and conifers (pp. 355–359).

Hardwood and Hemlock–Hardwood Forests

Hardwood forest occupies most of the morainal swells and ridges and some of the sandy areas in the Great Lake section. The term "hardwoods" is locally applied alike to forests made up entirely of deciduous (hardwood) species and to those containing an admixture of hemlock and sometimes of white pine. There is no sharp line of demarcation between the pine forests and the hardwood forest, nor between the spruce or arbor vitae forests and the hardwood forest, although the transition is sharper in the latter case.

The most abundant deciduous species are sugar maple, beech, basswood, and yellow birch. These, together with hemlock and sometimes white pine, are variously combined in mixed deciduous–coniferous communities or in deciduous communities in which sugar maple and beech are dominant. Additional species which are more or less frequent are red maple, white elm, white ash, and red oak. In the understory, hophornbeam is almost universally present, sometimes in considerable quantity; balsam fir is a subdominant in some of the more northern areas.

In northern Lower Michigan and in Ontario, most of the hardwood forests have been reduced to isolated, usually more or less disturbed stands. More extensive tracts in the Upper Peninsula of Michigan, especially in the Tahquamenon Wilderness, better illustrate the interrelations of various communities and the interdependence of forest types and soils.

NORTHERN LOWER MICHIGAN

The forests of northern Lower Michigan have been considered by a number of writers (Clayberg, 1920; H. T. Darlington, 1940; Gleason, 1924; Livingston, 1905; Potzger, 1941, 1944; Gates, 1912, 1926; and Waterman, 1917, 1922, 1922a). Some consider the distribution of forest types in relation to sand or clay soils; some are concerned with successions; some with community composition. Throughout the area, the hardwood forests are more prevalent on clayey soils. Successional development on the dunes leads to the establishment of hardwood forest (with hemlock). Gleason's studies (1924) indicate that hemlock "will practically disappear as a component of the association" and that "the hardwood forest is still young historically and may have attained its full ecological dominance only one generation (about three centuries) ago." Potzger (1941, 1944) finds that on smaller islands in the Great Lakes in this section, the hardwood forest usually, but not everywhere, occupies a central position and is bordered by a peripheral band of northern conifer forest. He considers that this distribution of communities shows "the stress between the replaced boreal forest and the recently invading northern hardwoods," and that "the more rigorous microclimate along the periphery of the islands limits invasion and replacement of the boreal forest complexes by the broadleaved southern species to the more protected central parts."

All statistical data for the hardwood forests of this part of the section illustrate the overwhelming dominance of sugar maple and beech, not only in the forest canopy, but in lower layers as well; the almost universal occurrence of hemlock, sometimes as a codominant; and the abundance of hop-hornbeam (*Ostrya*) among the smaller trees. Clayberg (1920) states that, in the Petoskey-Walloon Lake region, sugar maple composed 70 to 90 per cent of the forest, beech 5 to 30 per cent, and hemlock sometimes 25 per cent. Hemlock is not a constituent of the hardwood forest of Mackinac

Table 69. Composition of canopy and second layer of forest communities in two hardwood stands in northern Lower Michigan illustrating variations in relation to site.

A. Mud Lake hardwoods, Cheboygan County: *1,* Mucky soil of shallow ravines; *2,* Better drained soil of swells.

B. Carp Lake hardwoods, Emmet County: *3,* Forest of old beach ridges; *4,* Typical beech–maple forest.

	A. Mud Lake		*B. Carp Lake*	
Canopy	*1*	*2*	*3*	*4*
Number of trees	277	302	226	118
Acer saccharum	32.1	50.7	12.4	50.8
Fagus grandifolia	15.2	33.4	42.0	47.5
Tsuga canadensis	14.1	9.9	22.1	1.7
Betula lutea	10.8	2.3	6.6	. .
Tilia americana	22.4	3.0	1.8	. .
Ulmus americana	3.2	.3	.4	. .
Acer rubrum	.7	. .	13.7	. .
Fraxinus americana	1.1	.3	.4	. .
Quercus borealis	. .	. .	.4	. .
Fraxinus nigra	.4	. .	. .	. .
Second layer				
Number of trees	26	76	136	84
Acer saccharum	42.3	28.9	22.8	30.9
Fagus grandifolia	15.4	47.4	50.0	65.5
Tsuga canadensis	11.5	18.4	8.1	1.2
Betula lutea	3.8	1.3	3.7	2.4
Tilia americana	19.2	2.6	. .	. .
Ulmus americana	7.7	1.3	. .	. .
Acer rubrum	. .	. .	15.4	. .

Island. In this situation, the hardwood forests have some of the features of more northern forests; balsam fir is the principal subdominant of lower layers, and white spruce not uncommon.

Examples of the hardwood forests of northern Lower Michigan to illustrate canopy composition are chosen from the Douglas Lake region centering around the University of Michigan Biological Station. These are well-known and readily accessible areas, although somewhat disturbed. Variations related to topography and disturbance are evident. All of these are within the limits of the Port Huron morainic system, and in areas submerged by the Glacial Lake Algonquin or the Nipissing Great Lakes (Leverett, 1928, plates 1 and 2). They, in contrast with areas farther south, have had a shorter time for vegetational and soil development. In the Mud Lake hardwoods (Table 69 A), the communities of mucky and hummocky ravines and of swells contrast in amount of hemlock, yellow birch, and basswood, which are more abundant in the lower situations which also contain a few swamp species. On the poorly drained sites, sugar maple and beech make up only 47 per cent of the canopy (nearly 60 per cent of the second layer) and basswood, hemlock, and yellow birch an equal amount. On the better drained soil of the swells, sugar maple and beech comprise 84 per cent of the canopy and nearly the same proportion of the second layer. In the Carp Lake hardwoods (Table 69 B), the hemlock is more or less localized on the old beach ridges, sometimes forming about half of the canopy, more often about one-fourth. It is much less abundant in the second layer. Local areas are essentially beech–maple forest.

Disturbances many years ago have introduced variations in forest composition as illustrated by the hardwood forest on Colonial Point in Burt Lake (Table 70). In no part of the area is sugar maple a dominant tree of the canopy, although its abundance in all smaller size classes indicates its potential dominance. The abundance of red maple may well be correlated with the sandy soil of the area, its low elevation above the waters of Burt Lake, and past disturbance. White pine is locally abundant (26 per cent) in an old secondary area where most of the trees are about 10 inches d.b.h., although a very few *large* beech and yellow birch are present.

The beech–maple or maple–beech communities of the hardwood forests of northern Lower Michigan are, in canopy and lower tree layers, like those of the Beech–Maple region farther south. However, the shrub and herbaceous layers contain northern species which emphasize the northern affiliations. *Acer pensylvanicum, Acer spicatum, Viburnum Lentago, Rubus idaeus* var. *strigosus, Lonicera canadensis, Corylus cornuta,* and *Sambucus pubens* are more or less frequent. *Aralia nudicaulis, Dryopteris spinulosa, Maianthemum canadense,* and *Clintonia borealis* are abundant northern species of the herbaceous layer, the last two especially where conifers are more abundant. *Trientalis borealis, Polygala paucifiora, Chimaphila umbellata,*

Table 70. Variations in composition of the hardwood forest on Colonial Point, Burt Lake, Cheboygan County, Michigan, illustrating effects of past disturbance.

	Semi-virgin Areas			*Secondary*	
	1	*2*	*3*	*4*	*5*
Number of trees	136	95	229	120	63
Fagus grandifolia	72.1	60.0	45.0	15.0	41.3
Quercus borealis	13.2	8.4	4.8	23.3	. .
Acer rubrum	5.9	18.9	8.3	22.5	20.6
Tsuga canadensis	2.2	1.0	35.4	.8	31.7
Acer saccharum	3.7	5.3	1.7	3.3	1.6
Betula lutea	2.9	3.1	3.9	5.8	4.8
Fraxinus americana	. .	2.1	. .	. .	. .
Tilia americana	. .	. .	.4	. .	. .
Ulmus americana	. .	. .	.4	.8	. .
Betula papyrifera	. .	1.0	. .	. .	. .
Pinus Strobus	. .	. .	. .	25.8	. .
Ostrya virginiana	. .	. .	. .	.8	. .
Prunus serotina	. .	. .	. .	1.7	. .

Streptopus roseus, and *Habenaria orbiculata* further emphasize the northern character of the forest. Many of the species, however, as *Botrychium virginianum, Adiantum pedatum, Uvularia grandiflora, Caulophyllum thalictroides, Viola eriocarpa, Osmorhiza Claytoni,* and *Solidago caesia,* range far southward in mesophytic forests.

EASTERN UPPER MICHIGAN

The eastern half of the Northern Peninsula of Michigan (Upper Michigan) is more northern in aspect than is Lower Michigan. This is in part due to great expanses of bog (muskeg) with their coniferous forests, and in part to the greater frequency of northern shrubs and herbaceous plants in the hardwood forests.

In spite of the small number of tree species involved, a number of more or less distinct communities of the hardwood forest are recognizable. These differ from one another in constituent species, in proportional representation of species, and in the character of the ground layer. The Land Economic Survey recognized four distinct types in Upper Michigan: (1) maple–beech–elm–basswood–yellow birch; (2) maple–beech–yellow birch; (3) maple–elm–basswood–yellow birch; and (4) maple–yellow birch (McIntire, 1932). Data on composition of the maple–beech–yellow birch and maple–elm–basswood–yellow birch types (basal area in square feet) were given for four virgin stands of each type in Alger County. Westveld (1933) pointed out the extremely diversified nature of the vegetation of Upper Michigan and gave data on composition of hardwood forests on 25 soil types. Hardwood forest sometimes extends for miles with only minor variations from place to place; again, it is interrupted by low-lying stretches occupied by spruce or by northern white cedar (arbor vitae). Transition communities usually occur between the bog conifers and the hardwood forests of the uplands. In some places, hardwood forest extends to the shores of Lake Superior or of Lake Michigan; in others, a narrow band of conifers occupies the lake slopes. Although hardwood forests are not confined to any particular soil type, but occur on sandy or loamy soils alike, they are less frequent on sand. The sugar maple–beech type seems to be confined to the better-drained sites; basswood is locally abundant in such situations, its abundance apparently correlated with the occurrence of limestone (either beneath shallow drift on the highest parts of the Niagara limestone cuesta, or as boulders in the drift). Combinations with hemlock, yellow birch, and red maple are frequent in moister situations.

The Tahquamenon Wilderness and its bordering moraine on the south (which follows more or less the Niagara cuesta) in Luce and Chippewa counties is outstanding in the opportunities afforded for study of the northern hardwoods. The Miner's Falls area near Munising in Alger County well illustrates the dissected area near Lake Superior, together with the adjacent upland.

In the great Tahquamenon Wilderness, hardwood ridges rise above the vast expanse of bog. These ridges are morainic complexes including swells and depressions. The outer (south) slopes of the ridges are very gradual; the descent from hardwood forest to swamp would scarcely be noticed if it were not for the change in forest composition. The inner (north) slopes of the moraines are steeper, in places almost abrupt—a relief of 100 feet in less than a mile is not unusual. The ridges are not flat-topped, rather they are undulating morainic areas including low swells or mounds and ridges, all well-drained, between which are depressions more or less swampy. One of the morainic complexes borders the Tahquamenon Swamp on the south; another, diverging from it near Eckerman, extends northwestwardly across the Tahquamenon Wilderness toward Lake Superior.[10] The Tahquamenon River, formed by the convergence of several branches which gather bog water from the Tahquamenon Swamp to the north of the more southerly moraine, cuts across the second moraine and into the underlying sandstone in a deep gorge whose slopes are covered with hardwood forest, with here and there patches of white pine.

The variations in forest composition of the morainic complexes along a two and one-half mile transect through virgin forest eleven miles north of Hulbert, about on the Chippewa–Luce county line, and along the highway between Emerson and Eckerman, where the transition from the hardwood ridges to the spruce bog forest is well illustrated, are representative.

[10] Leverett (1928) describes these moraines and points out the correlation of hardwood forest and swamp with moraines and intervening plains.

Additional hardwood areas on the morainic complex to the south of the Tahquamenon Swamp illustrate the forest of the steep inner slope of the moraine and of some of the highest swells.

The various forest communities are related to one another in sequences of decreasing water requirements from the bog to the highest ridges, as shown in the accompanying successional diagram. Although theoretically

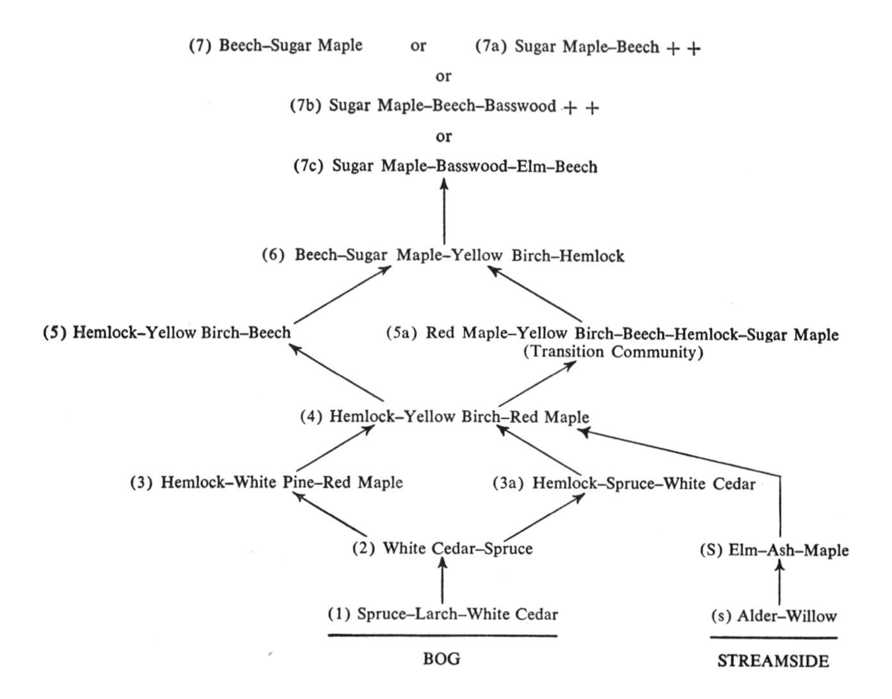

a developmental or successional series, the replacement of bog or transitional communities by hardwood forest cannot take place until the swamps are filled or drained. The conifer physiographic climax of the bogs will maintain itself alongside of the hardwood climax of the moraines. Not all intermediate forest types appear in all the spatial sequences on the moraine. The successional diagram is generalized and inclusive. The canopy composition of these forest communities is given in Table 71.

Communities in situations with excessive water are those of streamsides (s and S) and coastal marshes, and those of bogs (1 and 2). These communities have already been considered (p. 350). Where the outer morainic slope rises very gradually from the bog, a transitional community sometimes appears where hemlock mingles with spruce and white cedar (3a). Occupying equally boggy sites, is a hemlock–white pine–red maple community (3), in which white cedar is more or less localized. The forest floor is mossy and hummocky (moisture-loving mosses, and, with the white cedar, an

Table 71. Canopy composition of forest communities in the Tahquamenon Wilderness, Luce and Chippewa counties, Michigan. Community numbers correspond with those used on the successional diagram (p. 356); the sequence is from river swamp (*S*) and bog (*2*) to beech–maple, maple–beech, or maple–basswood forest of morainal ridges.

	Swamp & Bog		*Intermediate in Position*				*Morainal Ridges*					
Community number	*S*	*2*	*3*	*4*	*5*	*5a*	*6*	*7*	*7a*	*7a*	*7b*	*7c*
Number of trees	53	56	49	114	38	77	123	165	115	133	96	114
Fagus grandifolia	..	..	2.0	.9	23.7	22.1	35.9	53.3	40.0	22.6	19.8	8.8
Acer saccharum	5.7	..	..	.9	5.2	11.6	29.7	44.8	52.2	67.7	61.5	42.1
Tilia americana	..	..	..	..	..	..	..	..	..	..	3.1	14.0
Betula lutea	5.7	10.7	8.2	27.2	31.6	23.4	17.2	1.8	5.2	8.3	7.3	8.8
Tsuga canadensis	..	5.3	49.0	43.9	26.3	15.5	15.6	..	..	..	..	5.3
Acer rubrum	7.5	1.8	10.2	16.6	7.9	27.3	1.6	..	2.6	..	1.0	5.3
Fraxinus americana	..	..	..	..	..	..	..	..	..	..	..	2.6
Abies balsamea	..	3.6	..	..	..	..	..	..	..	..	..	..
Pinus Strobus	..	..	18.4	3.5	2.6	..	..	..	..	..	1.0	x
Picea glauca	..	..	4.1	.9	..	..	..	..	..	..	..	..
Thuja occidentalis	..	55.4	8.2	6.1	2.6	..	..	..	..	..	..	..
Picea mariana	1.9	17.9	..	..	..	..	..	..	..	..	..	..
Fraxinus nigra	7.5	5.3	..	..	..	..	..	..	..	..	..	..
Ulmus americana	69.8	..	..	..	..	..	..	1.5	..	..	6.2	13.1
Betula papyrifera	1.9	..	..	..	..	..	..	..	..	..	..	..
Quercus macrocarpa	x	..	..	..	..	..	..	..	..	..	..	..

abundance of Sphagnum). The herbaceous layer of this community is dominantly northern (*Vaccinium membranaceum, Coptis groenlandica, Cornus canadensis, Chiogenes hispidula, Trientalis borealis, Oxalis montana, Linnaea borealis* var. *americana, Cypripedium acaule, Osmunda cinnamomea,* and sedges).

The hemlock–yellow birch–red maple community (4) is a well marked and common type, occurring frequently as an intermediate community between the bog forests and the hardwood forests. It usually contains a few individuals of relict species of the bog (as *Thuja*), and a few sugar maple and beech, indicative of more advanced successional stages. This forest community is outstanding in the profusion of mosses both on tree trunks and bases, and on hummocks of the forest floor, in which *Oxalis montana* and *Lycopodium annotinum* grow in profusion. Additional herbaceous species, mostly northern in distribution, are *Lycopodium lucidulum, Maianthemum canadense, Cornus canadensis, Coptis groenlandica, Clintonia borealis, Trientalis borealis, Mitchella repens, Medeola virginiana, Aralia nudicaulis, Cypripedium acaule,* and *Dryopteris spinulosa.* The shrubs, too, are mostly northern species—*Sorbus americana, Acer spicatum, Lonicera canadensis,* and *Corylus cornuta.*

With a decrease in the amount of hemlock and an increase in deciduous species, the forest floor loses much of its luxuriance and variety. While many of the same species remain, they are scattered, and the ground cover is more open. The competition of innumerable little sugar maples has crowded out more delicate plants. The forest canopy (5, 5a), varying from place to place, suggests the transitional position of the community; this is upheld by the composition of the second layer which shows a reduction in the proportion of hemlock and yellow birch and an increase in the proportion of beech and sugar maple.

Further increase in the deciduous species results in the establishment of a sugar maple–beech–yellow birch–hemlock community—one of the most widespread types of the northern hardwoods (6). In many instances, the yellow birch and hemlock are much larger than the other trees, thus emphasizing the relation to the preceding stage and suggesting the later, more complete dominance of sugar maple and beech. Either hemlock or yellow birch may persist temporarily, delaying the establishment of the beech–maple forest.

On the higher "ridges" and swells of the moraines, in situations with the best drainage, beech and sugar maple dominate, in canopy, in understory, and in reproduction (7, 7a). Variations in the proportions of these dominants, which together comprise about 90 to 98 per cent of the canopy, may be mere rhythmic variations, or may be developmental. An increase in beech is evident in the second layer and hophornbeam (*Ostrya virginiana*) is a common understory species. The herbaceous layer is sparse (if sugar maple and beech of this layer be excluded), with occasional small patches of *Lycopodium lucidulum,* and scattered *Medeola virginiana, Smilacina racemosa, Trillium grandiflorum, Actaea pachypoda, Botrychium virginianum, Viola eriocarpa, Polygonatum pubescens, Galium triflorum, Athyrium thelypteroides,* and *Dryopteris spinulosa.* The relatively few northern species include *Sambucus pubens, Corylus cornuta, Lonicera canadensis, Trientalis borealis,* and *Habenaria orbiculata.* The accumulation of leaves, and the 2 to 3 inch layer of matted root and leaf mor are unfavorable to the growth-form type which prevailed in early forest stages. The identity of this community with the beech–maple or maple–beech of Lower Michigan is apparent.

A variant of the hardwood ridge forest, in which basswood is more or less abundant (7b, 7c), is encountered where the underlying limestone is not deeply buried, or where the drift contains limestone boulders. It is found on clayey rather than sandy soils. Elm and yellow birch, and sometimes a small proportion of other species (ash, hemlock, white pine) add to the variety of the canopy layer. The herbaceous layer includes a larger number of species than does the typical beech–maple or maple–beech, a feature correlated with the more varied leaf litter and smaller proportion of decay-resisting beech leaves.

Of all these forest communities, the hemlock–yellow birch–red maple forest (4) has the richest undergrowth, the greatest variety of species in the summer herbaceous layer, and the greatest abundance of individuals of these species. Of all, the beech–maple community has the poorest undergrowth, both in species and in number of individuals (if we except the innumerable small sugar maple and beech which in size belong to the herbaceous layer).

In all of these communities except the earlier stages (1 to 3) and the maple–basswood type (7b, 7c), balsam fir (*Abies balsamea*) is more or less abundant in the undergrowth, attaining a maximum size of about 6 inches d.b.h. This lends to the undergrowth of these forests an aspect more northern than is seen in Lower Michigan. Occasional white spruce trees are seen in intermediate stages. These conifers emphasize the approach toward the northern limits of the Hemlock–White Pine–Northern Hardwoods region.

The position of hemlock and of white pine in the climax forest may well be considered. In the series of forest communities just considered, hemlock occupies an intermediate position; it is not a constituent of the beech–maple forest in this area. Nevertheless, it has the qualifications of a climax species (Graham, 1941), and in many places in the Hemlock–White Pine–Northern Hardwoods region, is a constituent of the climax, sharing dominance with beech or sugar maple.

It will be noted upon examination of Table 71, that white pine is an infrequent constituent of several of the communities. In the one community in which it is abundant (3), forming 18 per cent of the canopy, it is not represented in the understory. Its scattered occurrence[11] in most hardwood stands has led to considerable discussion as to its position in the developmental series. It reproduces abundantly only in rather open situations, often as a result of catastrophe, or it replaces red pine in developmental series on sand. It is often present, although not abundant, in the hemlock–hardwood forest and in the beech–maple or maple–beech forest. In these situations, it appears to occupy a climax position. Nichols (1935) has stated that "where represented only by scattered specimens of uneven age, or only very locally, the pine in all probability has originated in very small forest openings brought about by windfall or by some other purely local influence. Openings due to windfall and the like are to be looked upon as a normal incident in the life of the forest." He considers that "white pine is to be regarded as a normal, although minor, constituent of the climatic climax." Graham (1941) does not consider that white pine has the qualifications necessary for a climax species. Whichever way it be considered, it is a constituent, even if accidental, in many of the climax hardwood stands.

In the series of communities just considered, there is an apparent correlation between forest composition and the position of the water table. Wilde (1933), working in the Lake States on the relation of soils and forest vegetation, divided forest sites into (a) those "under the influence of underground water continually;" (b) those "under the partial influence of the water table;" and (c) those "entirely above the influence of the water table." Communities s, S, 1, and 2 of the successional diagram belong in the first category, 3, 4, and perhaps 5 in the second, and 6 and 7 in the third of Wilde's divisions.

Where rock bluffs border the shore of Lake Superior and are indented by young and gorge-like ravines, the slopes afford forest habitats quite unlike those in the Tahquamenon Wilderness. As an example, the Miner's Falls area in the Pictured Rocks district east of Munising, Alger County, Michigan, is used (Table 72). Sugar maple–basswood forest approaches close to the cliff margin, which is fringed with scattered trees of paper birch, mountain ash, basswood, and red pine. White pine, hemlock, and yellow birch occupy concavities in the cliffs. In dissected areas a short distance back from the cliffs, mesophytic hardwood forest prevails, in which sugar maple is dominant and white elm abundant (Table 72 A). There is here no mat of humus, as in the hardwood forests of the Tahquamenon; instead, the humus is incorporated and the soil, therefore, dark for several inches down and gradually changing to brown. It resembles the gray-brown forest soils rather than the podzols. This condition is correlated with the good

---

[11] White pine has been cut from almost all accessible areas, even where no other logging has been done. In consequence, it is now more sparsely scattered in hardwood stands than in the original cover.

Table 72. Composition of canopy and undergrowth of forests near Miner's Falls, Alger County, Michigan.

A. Maple type, with elm and birch, on ravine slopes of dissected area near Lake Superior; similarity in composition of the three layers indicates continuance of this type.

B and *C.* General upland farther from ravines.

B. Maple–birch–beech forest whose second layer indicates further development toward a maple–beech climax.

C. Maple–beech climax, in which permanence is indicated by similarity in composition of the three layers.

	A			*B*		*C*		
Layer	*1*	*2*	*3*	*1*	*2*	*1*	*2*	*3*
Number of trees	168	121	90	63	46	111	85	70
Acer saccharum	79.8	88.4	76.6	55.5	43.5	61.3	51.5	37.1
Fagus grandifolia	1.8	1.7	4.4	11.1	28.3	25.2	31.8	48.6
Betula lutea	6.5	5.0	5.6	23.8	8.7	11.7	5.9	1.4
Ulmus americana	11.3	1.6	5.6	6.3	6.5	1.8	4.7	. .
Tilia americana	.6	1.7	1.1	3.2	2.2	. .	. .	4.3
Ostrya virginiana	. .	.8	4.4	. .	8.7	. .	1.2	2.9
Abies balsamea	. .	. .	. .	. .	2.2	. .	. .	5.7
Betula papyrifera	. .	.8	. .	. .	. .	. .	. .	. .
Acer spicatum	. .	. .	1.1	. .	. .	. .	. .	. .
Sorbus americana	. .	. .	1.1	. .	. .	. .	. .	. .

drainage of the habitat, and the dominance of maple and elm whose leaf litter decays readily. Resulting from this humus condition, the forest aspect is more luxuriant than the usual northern maple–beech forest; the shrub and herb layers contain a large number of species. Northern species are present (*Taxus canadensis, Acer spicatum, Sambucus pubens, Corylus cornuta, Lonicera canadensis, Aralia nudicaulis, Streptopus roseus, Trillium cernuum, Lycopodium lucidulum, Clintonia borealis,* and *Mitella nuda*), but species of wider distribution are represented among the shrubs and prevail among the herbaceous plants (*Cornus alternifolia, Ribes Cynosbati, Adiantum pedatum, Botrychium virginianum, Arisaema atrorubens, Allium tricoccum, Smilacina racemosa, Actaea pachypoda, Ranunculus abortivus, Clematis virginiana, Caulophyllum thalictroides, Dicentra canadensis, D. Cucullaria, Sanguinaria canadensis, Impatiens biflora, Viola canadensis, Viola eriocarpa, Osmorhiza Claytoni, O. longistylis,* and *Eupatorium rugosum*).

On the general upland farther from the ravine slopes, the forest more nearly resembles the forests of the Tahquamenon moraines. Sugar maple and yellow birch are dominant in the canopy in some places, but sugar maple and beech dominate in lower layers (Table 72 B), indicating the replacement of this forest type by one dominated by sugar maple and beech in all layers (Table 72 C). Yew (*Taxus canadensis*) is dominant in the

Airplane photograph of Pictured Rocks on the shores of Lake Superior near Munising, Michigan, showing the hardwood forest which reaches almost to the edge of the lake bluff. Along the forest margin at the brink of the bluffs are scattered white birch, mountain ash, and red pine; occasional white cedars and groups of white pines find foothold in concavities of the cliff face. The two slightly darker and somewhat circular patches cut by the right margin of the photo are bog forest. (Photo Copyrighted by The Detroit News.)

shrub layer; the forest lacks the low sugar maple layer so conspicuous in many areas. Small beech and sugar maple are scattered in about equal numbers, usually occurring in breaks in the Taxus layer. The role of Taxus in relation to reproduction of canopy trees, and its possible inhibition of the pronounced maple dominance in the 1 to 3 foot height class (as seen in most northern forests) may well be considered. Wherever Taxus is abundant, there is no continuous layer of small saplings. There is a larger number of herbaceous plants, which find favorable situations between the Taxus clumps where they do not encounter severe competition with crowded young maples. Taxus was formerly plentiful in many areas where it is now poorly represented. It is severely browsed by deer, which are over-plentiful in many of the northern forests. Maples sprout freely and repeatedly after browsing, while Taxus is soon killed.

The Miner's Falls area is in the northwestern part of the Great Lake section, whose western limit is reached near Marquette at the range limit of beech, which coincides with the geologic and physiographic break at the edge of the Superior Upland. Beech continues as one of the dominants of the forest to within a few miles of the limits of its range.

In the more or less triangular area of Upper Michigan which extends southward between Green Bay and the Menominee River, there are large areas of flat to slightly rolling plain over which many drumlins—long narrow stony clay hills—are scattered, while in the lower land between the drumlins are gravel or sand ridges—eskers.

> The drumlins were covered originally by heavy forests of mixed broad leaved species, of which the Hard Maple, Beech, Basswood, White and Rock Elms and Ironwood were the most common, with usually a considerable per cent of Hemlock and scattering individuals of Balsam and White Spruce and in the northern part of the district, more or less Yellow Birch.

The forest of the gravelly or sandy ridges, sandy stream terraces and sand plains

> was formerly almost pure pine, usually White Pine, but upon the lighter sands, the Norway and Jack Pines often were mixed in considerable abundance. . . . In some areas along the Menominee river, and in other localities as well, where there was an admixture of clay and sand, or where there was a thin layer of sand over a pebbly clay loam, there were "two-storied" forests, the upper story of very tall old White Pines, and the lower one of mixed broad leaved species, with Hemlock, Balsam, White Cedar, and White Spruce, as in the case of the forests of the drumlins. In these mixtures the White Pines tower high above the other species and practically all their branches are above the top of the lower story. . . . In this case, and in the others as well, there was good White Pine reproduction, showing that the type tends to perpetuate itself.

The white pine forests are all but gone and "the areas formerly covered by them are now growing up to Aspen, the large-toothed Poplar, Scarlet Oak, White Birch, and Red or Pin Cherry and to Sweet Fern and Bracken Fern." The poorly drained habitats of this "drumlin region" supported swamp and bog forests. In areal extent, this type prevailed. (Quotations from Davis, 1907.)

From Upper Michigan, the Great Lake section extends southward across the Menominee River through extreme eastern Wisconsin, coming in contact with the northwestern outpost of the Beech–Maple region a short distance south of Green Bay, and east of the Kettle Moraine.

Eastward, on the Canadian side of lower Lake Superior, hardwood communities are so poorly represented and depleted in species that this area is referred to a distinct section—the Laurentian Upland.

ONTARIO

An area lying between Lake Huron and Lake Ontario in the northern peninsula of Ontario is referable to the Great Lake section. This is the

Huron–Ontario section as defined by Halliday (1937). Its forests are more complex than those of the Laurentian Upland to the north, and contain a number of species entirely absent there, as sycamore, butternut, and bitternut hickory. However, hardwood communities are poorer in species than those to the south, and lack the many distinctively southern species which reach their northern limits in the southern tip of Ontario, in the Beech–Maple region (p. 319).

Sugar maple and beech are the dominant species, comprising about three-fourths of the forest.

> With them are basswood, white elm, yellow birch, white ash, and some red maple, and (northern) red, white and bur oak. Small groups of hemlock and balsam fir and an occasional white pine occur within the association, as well as a scattered distribution of large-toothed aspen, bitternut hickory, butternut, ironwood, and black cherry; and blue beech, silver maple, slippery and rock elm, and black ash are found locally on specialized sites. In the southern parts, there is some intrusion of black walnut, sycamore, and black oak. White and red pine stands are found on lighter soils, and were formerly more common (Halliday, 1937).

As a result of the dominance of hardwood species and the somewhat mixed character of many stands, gray-brown and brown forest soils have developed over much of the area, very little of which now retains any remnant of its primary forest.

SUPERIOR UPLAND

This section of the Hemlock–White Pine–Northern Hardwoods region is not quite coextensive at the west and southwest with the physiographic section known as the Superior Upland. It is an area underlain by pre-Cambrian rocks, for the most part heavily drift covered. The resistant character of the underlying rock is, however, responsible for the greater elevation of the section. It extends from Lake Superior southward to the Driftless area (as defined by Fenneman, 1938), and overlaps slightly onto the area of proglacial lake deposits. At Lake Winnebago, its southern boundary leaves the western boundary of the Great Lake section; from thence it extends in a general westerly, thence northwesterly direction toward the head of Lake Superior. This part of its boundary is determined by the limits of range of hemlock.

The area close to Lake Superior is cut by many streams. A number of rugged areas, such as the Keweenaw Peninsula, the Huron Mountains, and Porcupine Mountains, have local relief amounting to 650 to 1400 feet. The morainal area and adjacent outwash plains of the northern Wisconsin upland are dotted with innumerable lakes and swamps. Sandy outwash plains and proglacial lake beds in northwestern Wisconsin offer still another type of topography affecting the nature of occupying vegetation.

Vegetationally, this, as other sections of the Hemlock–White Pine–Northern Hardwoods region, is a mosaic of pine, bog, and hardwood, or hardwood and hemlock communities. It differs from all the more eastern sections in the absence of beech and abundance of basswood. Basswood is codominant with sugar maple in purely deciduous communities, and with hemlock, yellow birch and sugar maple in mixed communities. Red oak may be abundant in the maple–basswood communities of its western part.

The pine communities of the Superior Upland are not unlike those of the Great Lake section. Their general features need not be again considered; some local features, related to the very different topography of this section, will be mentioned in the discussion of specific areas. The bog communities, too, are essentially like those to the east, although black spruce is relatively more abundant. The hardwood communities, an integral part of the Deciduous Forest Formation, may best be characterized by the use of concrete examples located in separate and unlike parts of the section. As elsewhere, the location of remnants of virgin forest must affect the selection of examples. These are located in areas differing topographically and vegetationally from one another: (1) the Huron Mountains and surrounding area; (2) the Keweenaw Peninsula and southwestern extension of the Keweenaw Range; (3) the Porcupine Mountains and adjacent tableland; (4) the "lake region" of northern Wisconsin where marginal or interlobate moraines and outwash plains produce a pitted and undulating topography; (5) the Flambeau River forest.

Huron Mountains and Surrounding Area

Northwest from Marquette, and extending almost to L'Anse at the head of Keweenaw Bay, is an elevated area in most of which pre-Cambrian rock is near the surface. Deep morainal deposits are limited in extent, the strongest moraine extending southwestward from the Huron Mountains. Narrow belts of sandy terrace (beds of glacial lakes) border the area along Lake Superior. Altitudes range from 602 feet, the level of Lake Superior, to 1950 feet in the Huron Mountains. Most of the plateau is between 1500 and 1800 feet. (This plateau continues southwestward across Upper Michigan and includes much of northern Wisconsin.)[12]

Vegetationally, this area as a whole is one in which hardwood forest, with or without hemlock, prevails. In the undissected high southern part (exclusive of moraines), conifers—spruce, white pine, arbor vitae—are more abundant, giving to this part of the area something of the aspect of the

[12] Data from Leverett's map of the surficial deposits of the Lake Superior region (1928).

Table 73. Composition of canopy of selected communities in the Huron Mountains, Marquette County, Michigan.

1→2→3 and *1→2→4*, successional series from stream flats of Salmon Trout River to adjacent slopes and terraces.

5, 6, and *7,* variations in hardwood climax: *5,* On terrace of Lake Superior near Conway Bay; *6,* On mountain slopes along "Mountain Stream;" *7,* On flats near Rush Lake.

8, 9, and *10,* variations in hemlock–hardwoods climax: *8,* On relatively level area near upper Fisher Creek; *9,* On slopes above Pine Lake (near area *6*); *10,* On terrace of Lake Superior, contiguous to area *5.*

11 and *12,* edaphic variations: *11,* Swampy area on "Mountain Stream;" *12,* Pine forest of rock slope.

	Developmental Series of Stream Flats				*Hardwood Climax*			*Hemlock–hardwood Climax*			*Edaphic*	
	1	2	*3*	*4*	5	*6*	*7*	*8*	*9*	*10*	*11*	*12*
Number of trees	66	128	148	79	62	110	61	157	105	102	47	82
Acer saccharum	3.0	25.0	64.2	86.1	69.3	70.0	41.0	59.8	35.2	3.9	14.9	1.2
Tilia americana	3.0	18.7	2.0	..	..	17.3	39.3	..	10.5	..	14.9	.
Betula lutea	4.5	13.3	20.3	5.0	24.2	7.3	8.2	16.6	11.4	24.5	23.4	1.2
Tsuga canadensis	..	.8	9.5	8.9	1.6	5.4	6.6	19.7	40.9	66.6	10.6	15.8
Acer rubrum	1.5	5.5	4.0	..	1.6	..	..	3.2	1.9	3.9	4.2	4.9
Ulmus americana	60.6	21.9	..	..	..	..	4.9	..	..	..	..	..
Pinus Strobus	..	..	..	..	3.2	..	..	.6	..	..	..	58.5
Thuja occidentalis	3.0	4.7	..	..	..	..	..	..	..	1.0	25.5	..
Picea glauca	3.0	.8	..	..	..	..	..	..	..	..	6.2	..
Fraxinus nigra	19.7	9.4	..	..	..	..	..	..	..	..	..	..
Abies balsamea	1.5	..	..	..	..	..	..	..	..	..	..	..
Populus grandidentata	..	..	..	..	..	..	..	..	..	..	..	2.4
Pinus resinosa	..	..	..	..	..	..	..	..	..	..	..	15.8

northern coniferous region. Here bogs occupy the poorly drained spots, while hilly sections (mostly occupied by secondary aspen and white birch communities) are in part coniferous, in part hardwoods (maple, basswood, and red oak) intermingled with spruce and fir. Pine forests, of jack pine, red pine (here usually called Norway pine), and white pine, occupy some of the sandy beds of glacial lakes; red oak is a common species of the jack pine forests. Moist but well-drained areas—moraines, ravine slopes, some of the lake terraces, and much of the mountainous area—are covered with hardwood forest, interrupted on steep and rocky slopes by groves of Norway and white pines, or in swampy ravines by mixed forest in which arbor vitae is abundant.

In the Huron Mountains, and under the protection of the Huron Mountain Club, are extensive areas of virgin forest, and of forest but slightly modified by selective logging, as well as some secondary forests. The highly diversified topography of the area, which includes lake terraces, river and ravine flats and terraces, mountain slopes of differing directions and steepness, as well as cliffs and swamps, results in a number of unlike forest communities together with gradational intermediate communities (Table 73). Because of the small number of species which here make up the northern hardwood forest, communities in unlike habitats often differ from one another only in the different proportions of constituent species.

The open jack pine, jack pine and oak, or jack pine–Norway pine communities of low sandy stretches contrast strongly with the surrounding dense hardwood or hardwood and hemlock forests. Their open shrub and herb layers contain abundant heaths (*Vaccinium, Gaylussacia, Arctostaphylos, Gaultheria, Epigaea*), patches of *Pteridium* and *Lycopodium,* cushions of *Dicranum* and *Cladonia,* and scattered herbaceous plants—all plants not occurring in the climax forests. In equally strong contrast to the prevailing hardwood forests, are the Norway pine and white pine groves occupying very rocky steep slopes.

The successional series on stream flats (well illustrated along Salmon Trout River) includes but few stages: elm–black ash→maple–elm–basswood–yellow birch→maple–yellow birch–hemlock, or→maple. The rank herbaceous layer of the elm–ash community includes an abundance of ostrich fern (*Pteretis nodulosa*). The second stage (clearly indicated by the second layer of the elm–ash community) is a transitional community containing species of the elm–ash forest and of the climax forest; its herbaceous layer is composed of pronounced mesophytes, such as *Athyrium thelypteroides, Athyrium angustum, Impatiens biflora, Circaea quadrisulcata* var. *canadensis,* and *Viola eriocarpa.* Good drainage and the character of the leaf litter result in mull humus on a sandy soil with dark A-horizon.

The lake terraces (on red pre-Cambrian sandstones) are generally occupied by hemlock or hardwood forest. A narrow broken band of coniferous forest, of white spruce, hemlock, white pine, and fir, with scattered white birch and mountain ash, fringes some of the bluffs. The mossy forest floor (*Leucobryum, Dicranum, Hylocomium,* etc.) on which grow *Chiogenes, Linnaea,* and *Coptis,* and the mats of *Lycopodium* (*L. annotinum, L. clavatum, L. obscurum*) emphasize the northern aspect of this localized community. About 100 feet back from the lake bluff, this gives way to the sugar maple–yellow birch forest or to mixed hemlock–hardwoods (Table 73, areas 5 and 10). Where the coniferous band is lacking, and hemlock and hardwoods come to the bluff margin, balsam fir is usually the dominant in the small tree layer at the forest margin.

Hardwood forest, mixed hardwoods and hemlock, and almost pure stands of hemlock occupy most of the Huron Mountains. White pine is scattered through all three forest types, and localized on some of the rocky slopes, and occasionally on flat areas.

In the forest as a whole, sugar maple is the most abundant species—in all layers. In the hardwood forests, it may be the one dominant, or share dominance with yellow birch or with basswood. In the hemlock–hardwoods forests, it may be the dominant species, share dominance with hemlock, or be a subordinate species. Variations from one to another of these forest types may take place irrespective of change in environment from place to place with unlike communities occurring on similar sites, and like communities occurring on dissimilar sites. The status of the several abundant species of these communities (sugar maple, basswood, yellow birch, and hemlock) should be considered. Is there one climax type toward which all forest communities are developing, or are these communities local variations of the climax, destined to retain their identity in succeeding forest generations? Canopy and second layer of these communities are remarkably similar in composition. Where sugar maple is alone dominant in the canopy, it is equally dominant in the understory. Yellow birch in some areas is more abundant, in others, less abundant in the second layer than in canopy; it is present in both layers of every hardwood and mixed community. Graham (1941) does not consider yellow birch sufficiently tolerant to "meet the requirements for the climax although it can establish itself on decaying stumps, logs, and any other accumulation of organic material." Nevertheless it is a constituent of the climax, and appears able to maintain itself indefinitely, although in varying amounts from generation to generation. It certainly should be thought of as one of the dominants of the Hemlock–White Pine–Northern Hardwoods region, because in peripheral parts of this region it alone may be dominant (p. 392). Basswood varies in abundance in different stands; it may be codominant with sugar maple in the canopy, or entirely absent. Canopy and understory abundance seem not at all related; it is infrequent in still lower layers. This may be due in part to overbrowsing by deer, which show considerable preference for this species.

It should be considered as one of the climax dominants, although not constant in this part of the region. Farther west, it assumes greater importance. Hemlock is represented in every community, in some instances more, in others less abundant in second layer than in canopy. In this respect it is like yellow birch. It seems to be periodic in its occurrence here, as it is in the high plateau of northwestern Pennsylvania (p. 398). However, in stands dominantly sugar maple (as area 4), there is little indication that it can ever assume a position of codominance with sugar maple, as competition with the unbroken maple reproduction layer is too severe. This dense layer of maple seems to have resulted from the destructive over-browsing of yew, which was (according to Dodge, 1918) a common plant in the Huron Mountains. Today, it is all but extinct. Its possible role in lessening dominance of small sugar maple has already been considered (p. 362). The status of white pine in the climax forest has also been considered (p. 359). It was the large pine in area 8 (Table 73) to which Nichols (1935) referred in his discussion of the ecologic status of white pine in the climatic climax.

Succession following fire in the Huron Mountains is illustrated by two old secondary communities, one occupying an area burned over in 1894, the other much older. In the younger secondary area, white birch is dominant, with bigtooth aspen (*Populus grandidentata*) an accessory species. Under the shelter of these trees is a more shade-tolerant group of species—sugar maple, hemlock, fir, and striped maple (*Acer pensylvanicum*). In the older area, a few large bigtooth aspen (nearly 2 feet d.b.h.) and large dead white birch are all that is left of the burn community. Sugar maple and red maple (about 10 inches d.b.h.), with abundant hemlock of various sizes, and a few basswood, yellow birch, and fir trees are reproducing a forest that resembles the surrounding hemlock–sugar maple–basswood community.

The high morainal area of central Baraga County, to the west of the Huron Mountains, is prevailingly hardwood forest similar to that of the Huron Mountains. The loose-textured drift is usually deep, although locally bedrock is exposed along streams. The generally sloping terrain combined with loose texture of the drift has resulted in good surface drainage in much of this area, and has permitted development of climax hardwood forests. These are locally interrupted by spruce and white cedar communities in boggy depressions, or by spruce–white pine–yellow birch–white cedar communities on poorly drained flats. Comparison of canopy and second layer suggests considerable stability of the several communities, although there is evidence of an increase in the proportion of sugar maple in the hemlock–birch and slope communities (Table 74).

A comparison of the forests of this morainic region with those of the Huron Mountains discloses certain similarities and differences. Sugar maple is the principal dominant in all but the hemlock–yellow birch communities;

Looking out over an extensive area of virgin hardwoods and hemlock forest in the Sturgeon River valley to the west of the Huron Mountains. White pines at the left. (Courtesy, U. S. Forest Service.)

yellow birch is generally more abundant in the hardwood forests of the moraines than in those of the mountains; red maple, poorly represented in the mountains, is abundant here; white pine is well distributed; fir and white spruce are more or less abundant in the undergrowth. Basswood, a dominant in some of the communities in the Huron Mountains, is not present in any of the four communities of Table 74; it is locally a constituent of the forest of steep ravine slopes where an erosional topography prevails.

The herbaceous layer of the forest of the morainic slopes is richer and more varied than seen in comparable communities of the Huron Mountains. This may be due in part to the absence of the continuous maple reproduction layer. The dark humus overlying a podzolic A-horizon in the hardwood areas is favorable to the growth of a variety of woodland mesophytes, such as *Trillium cernuum, Streptopus roseus, Clintonia borealis, Smilacina racemosa, Sanguinaria canadensis, Asarum canadense, Oxalis montana, Viola eriocarpa,* and ferns (*Osmunda Claytoniana, Dryopteris spinulosa, Athyrium angustum,* and *Phegopteris polypodioides*). Where hemlock is dominant, the herbaceous layer includes *Maianthemum canadense, Clintonia borealis, Mitchella repens, Oxalis montana, Trientalis borealis, Aralia nudicaulis, Lycopodium lucidulum,* and *Phegopteris Dryopteris.*

Table 74. Variations in canopy composition of the forest of the moraines of central Baraga County, Michigan:
1, Level but hummocky upland; *2,* Representative maple–birch community, grading into *3,* A hemlock–birch community which is adjacent to white cedar–black ash swamp; *4,* Slope woods.

	1	*2*	*3*	*4*
Number of trees	46	86	43	87
Acer saccharum	43.5	37.2	4.6	60.9
Betula lutea	26.1	39.5	37.2	21.8
Ulmus americana	. .	. .	. .	11.5
Acer rubrum	26.1	11.6	7.0	. .
Pinus Strobus	2.2	7.0	4.6	. .
Tsuga canadensis	. .	4.6	39.5	. .
Thuja occidentalis	2.2	. .	. .	4.6
Prunus serotina	. .	. .	. .	1.1
Picea glauca	. .	. .	4.6	. .

Locally, on steep ravine slopes, a mull humus has developed; the soil is the gray-brown earth of regions farther south. In these few places, the herbaceous flora also suggests more southern regions. Conspicuous are *Adiantum pedatum, Athyrium thelypteroides, Botrychium virginianum, Uvularia grandiflora, Trillium cernuum* and *T. grandiflorum, Smilacina racemosa, Allium tricoccum, Actaea pachypoda, Caulophyllum thalictroides, Dicentra Cucullaria, Sanguinaria canadensis,* and *Viola canadensis.* The soil development favorable to such an assemblage is limited to well-drained slopes (cf. Miner's Falls area, p. 360). Although the forest canopy may be made up of the same species on all soil types, where topographic development is advanced the herbaceous layer represents a more advanced developmental state.

Keweenaw Peninsula and Southwestward Extension of the Keweenaw Range

The Keweenaw Peninsula is of particular interest because on it is found a northern more or less continuous extension of deciduous forest. Northern outposts occur on the southwestern end of Isle Royale and on the southern side of Michipicoten Island in Lake Superior (Cooper, 1913).

The backbone of the Keweenaw Peninsula is a high ridge of very resistant rocks (735 feet above Lake Superior, and 1337 feet in elevation at its highest point on Brockway Mountain near the tip) flanked by broken bands of morainal deposits. The steeper rocky slopes are occupied by coniferous forest in which white spruce and white pine prevail, with locally a scattering of balsam fir, arbor vitae, sugar maple, and red oak. The lower slopes and flats or depressions are also occupied by coniferous forest in which black spruce, arbor vitae, and tamarack are dominant in the wetter parts, with mixtures of these species and white pine and white spruce in intermediate situations. The sloping areas with deeper soil are occupied by deciduous forest—sugar maple, yellow birch, and red oak, often with a strong admixture of white pine and spruce. Because the deep soil areas are limited in extent in the upper part of the peninsula, the aspect in general is that of northern conifer forest. Yet a deciduous forest climax is able to occupy all better soil sites to the very northern tip of the peninsula. On their map of "Natural Vegetation," Shantz and Zon (1924) show the upper one-fourth of the peninsula as occupied by spruce–fir (northern coniferous) forest; Cunningham and White (1941) map the interior as hardwood forest and show spruce–fir forest as having a marginal and patchy distribution.

Southwestward, the backbone ridge of the Keweenaw Peninsula is less pronounced and less rugged, although it maintains its identity into Wisconsin. Its height or its breadth may be augmented by deep morainal deposits which (where not cleared) support deciduous forest. While toward the northern part of the range, the hardwood forest is generally a sugar maple–yellow birch type in which white pine, elm, basswood, red oak, and hemlock may be scattered, southwestward basswood becomes more abundant, assuming a position of codominance with sugar maple (Table 75). This

Table 75. Hardwood forests on the Keweenaw Range.
A. At Trimountain, toward the basal portion of the Keweenaw Peninsula.
B. East of Wakefield, near the southwestern end of the Keweenaw Range, on the tableland south of the Porcupine Mountains.

	A	*B*	
		Canopy	*Second layer*
Number of trees	56	111	105
Acer saccharum	66.1	27.0	50.5
Tilia americana	*	45.0	18.1
Betula lutea	16.1	8.1	13.3
Ulmus americana	. .	12.6	2.9
Quercus borealis	10.7	. .	. .
Tsuga canadensis	7.2	4.5	14.3
Fraxinus americana	. .	2.7	.9

* *Tilia* present in adjacent but partially cut-over area.

maple–basswood dominance is characteristic of the hardwood forests of the western part of the Superior Upland and of the Minnesota section, areas adjacent to the Maple–Basswood Forest region.

Porcupine Mountains and Adjacent Tableland

The Porcupine Mountains rise over 1400 feet above the shores of Lake Superior to an elevation of 2023 feet. Geologically, they are a more or

"Lake of the Clouds," behind the first range of the Porcupine Mountains. From the rocky cliff in the foreground, one may look northward over Lake Superior, or southward across Lake of the Clouds and the Carp River valley over a vast expanse of virgin hardwood forest in which sugar maple is the dominant species. (Courtesy, U. S. Forest Service.)

less crescent-shaped spur from the Keweenaw Range to the south. Three roughly parallel mountain ranges lie close to Lake Superior; the fringing tableland or plain extends to the south and southwest. The first range (one nearest to Lake Superior) is the most rugged; cliffs extend along its south face, overlooking the valley of Carp River and Carp Lake ("Lake of the Clouds") below. A steep talus slope below the cliffs extends almost to the foot of the slope. This whole area of mountains, moraines and glacial lake plains westward to beyond the Black River, is a vast hardwood forest containing the most extensive primeval tracts remaining on the continent.[13]

Almost throughout this area (Table 76), sugar maple is the climax dominant, with basswood and yellow birch more or less abundant. In many places, hemlock mingles with the hardwood species or forms almost pure hemlock forests. The hemlock forest type occupies the lower part of the lake-shore slope of the first range of the Porcupines, parts of the lower north slopes in the Carp River valley, and many glacial lake terraces cut by streams flowing to Lake Superior (well illustrated along the Black River). White cedar swamps with some balsam fir, white spruce, and paper

[13] A bill signed by the Governor of Michigan on February 29, 1944, authorized the immediate acquisition of 43,000 acres of the Porcupine Mountains wilderness, to be administered by the Michigan Department of Conservation.

birch extend in narrow belts behind the beach ridge along Lake Superior at the foot of the first range. White cedar, with tamarack, spruce, and black ash forms a zone on the flood plain of the Carp River. Groves of white pine or white pine and hemlock occur on the bluffs of some of the rivers.

Pioneer communities occupy the crest and cliffs of the first range of the Porcupines. Lichens and crevice plants in the most exposed places, mats of shrubs (*Arctostaphylos, Vaccinium, Ceanothus*), and widely scattered and gnarled white and red pines form the summit vegetation. Scrubby oaks and aspens form a broken zone near the mountain top. The sugar maple forest extends almost to the top of the north slope. The talus slopes below the cliffs on the south face of the first range are covered by youthful forest communities. The ecologic features of the Porcupine Mountains (particularly of the first range and the valley of the Carp River) have been discussed by Ruthven (1906) and by H. T. Darlington (1931). Ruthven describes the habitat and lists the principal plants and animals at a number of "stations" along several survey lines. Darlington relates the hemlock forest type (Table 76, area 1) of the north slope of the first range to the greater humidity, more uniform temperature[14] near shore, and reduced light intensity of this lower more humid location. However, farther west, along the Presque Isle and Black rivers, hemlock forests are not confined to the immediate vicinity of Lake Superior, but occur also several miles inland. Darlington states that "the maple climax occurs on the higher portions of the first range (except the summit) and on the ranges farther south." Near the mouth of the Presque Isle River, maple forest comes to the edge of the bluff cut in the glacial lake terrace along Lake Superior, with no intervening hemlock community (Table 76, area 6).

Variations in composition of the forests of this area are comparable to those seen in the Huron Mountains. The hemlock forests, sometimes almost pure hemlock, contain trees of all size classes; there is no indication of successional change. Hemlock occurs in mixed stands, in greater or lesser amount, in a variety of habitats. On the slopes of the river gorges it mingles with white pine, arbor vitae, and yellow birch. In the mixed hemlock and hardwood stands, sometimes maple, sometimes hemlock seems to be increasing (as indicated by the lower layers of the forest). Periodic variations in proportion of constituent species are to be expected.

White pine in almost pure stands occupies very steep rock slopes (as on the Black River) or is mixed with hemlock on the sheltered slopes of gorges. It was originally scattered in the hemlock forest of the first range (Ruthven, 1906), but now only the stumps remain. On dry rock exposures, red pine mingles with the white pine.

Morainal slopes are prevailingly hardwoods, with maple the dominant species. In these situations the humus is of a mull type and in places so well

[14] Temperatures are less extreme along the coast than a few miles inland.

incorporated that by late summer no leaf litter remains on the gray-brown forest soil. The slopes of hills on residual areas of the Keweenaw series of rocks are also occupied by hardwood forest, generally the maple–basswood type; here also, mull humus over gray-brown earth is seen. Elsewhere podzol prevails.

Table 76. Canopy composition of forests: (*A*) On the north slope of the First Range of the Porcupine Mountains; (*B*) In the Black River area; and (*C*) Near the mouth of Presque Isle River.

1, The "hemlock climax" of Darlington (1931); *2,* A mixed hemlock–hardwoods area on dissected north slope; *3,* Hemlock–yellow birch area south of Chippewa Hill and about nine miles from Lake Superior; *4,* Level area near Gorge Falls (this mixed forest type alternates with stands of almost pure hemlock); *5,* Hardwood forest of gentle moraine slope; *6,* Maple forest on flat about 100 feet above Lake Superior; *7,* Near *6,* but farther from lake; *8,* Poorly drained area farther from lake bluff than area *6.*

	A. First Range		*B. Black River*			*C. Presque Isle River*		
	1	*2*	*3*	*4*	*5*	*6*	*7*	*8*
Number of trees	*	111	131	61	54	42	111	50
Acer saccharum	13.5	30.6	9.9	41.0	59.3	85.7	64.0	26.0
Tilia americana	1.5	44.1	.8	8.2	9.3	..	.9	..
Betula lutea	4.5	5.4	22.9	21.3	16.6	9.5	17.1	34.0
Tsuga canadensis	80.5	18.9	58.8	26.2	..	4.8	12.6	26.0
Acer rubrum	..	..	6.9	..	..	..	..	..
Ulmus americana	..	..	..	1.6	14.8	..	..	2.0
Quercus borealis	..	.9	..	..	..	..	..	..
Fraxinus americana	..	..	.8	..	..	..	3.6	8.0
Thuja occidentalis	..	..	..	1.6	..	..	1.8	4.0

* Data from H. T. Darlington (1931), based on count of belt transect 220 meters long and 10 meters wide; trees over 1.5 dm. in diameter included in count and hence, canopy and second layer not separated.

The herbaceous and shrub layers of the hemlock forest are usually sparse. In the early part of the century, ground hemlock (*Taxus canadensis*) was one of the principal species of the hemlock forest of the first range (Ruthven, 1906), but has since been almost eliminated by deer. In the Black River area it still forms a more or less continuous layer in mixed forests where hemlock is abundant, and is plentiful on the steep rocky slopes of river gorges. The mixed forest and the hardwood forest contain a large number of shrub and small tree species—*Acer spicatum, Amelanchier arborea, Ostrya virginiana, Sorbus americana, Cornus alternifolia, Cornus rugosa, Ribes Cynosbati, Ribes* sp., *Corylus cornuta, Dirca palustris, Diervilla Lonicera, Lonicera canadensis,* and *Rubus parviflorus,* the last named forming large patches especially in somewhat open spots. This species occurs in all forest types if there is sufficient light. Species of *Lycopodium* (*lucidulum, annotinum, obscurum,* and *clavatum*) are abundant in forests with hemlock, and almost lacking in the hardwood areas, as are also a few other species common to the northern conifer forest (*Cornus canadensis, Oxalis montana, Coptis groenlandica, Chimaphila umbellata,* and *Trillium undulatum*). Many of the plants are geophytes of wide distribution.

A secondary area on the north slope of the first range of the Porcupines (in the "maple climax" studied by Darlington) illustrates the hardwood sprout forest which replaces the primeval forest. Maple, basswood, yellow birch, and red oak sprouts of the original trees, and scattered white birch and aspen, form a dense stand in whose shelter occasional seedling hemlock and fir may thrive.

An old secondary area on the Black River, where "wheat was ripe here in 1847" is occupied by an old maple–basswood sprout forest (basswood 10 inches d.b.h. and over; maple a little smaller), in which are occasional yellow birch and white ash. The understory and herbaceous layers are not unlike those of the mature maple–basswood forest.

Interior of a virgin hardwood forest in the southern part of the Porcupine Mountains area; sugar maple is dominant, and basswood and elm are scattered in the forests of gentle slopes. (Courtesy, U. S. Forest Service.)

"Lake Region" of Northern Wisconsin

A land of innumerable lakes occupies the central portion of the Superior Upland. The underlying pre-Cambrian rocks are here deeply covered by drift. The border moraine of the Superior Lobe of the last ice sheet (fourth substage of the Wisconsin stage) is an intricate network of knolls and depressions sometimes differing in elevation as much as 100 feet. This drift (Young Gray Drift of Mankato substage) is generally more clayey and less stony than that immediately to the south (Young Red Drift of third or Cary substage). Sandy outwash plains, some small, some large, are frequent. The area lies in the headwaters region of streams flowing to Lake Superior, to Lake Michigan, and to the Mississippi River. The general elevation is over 1500 feet. Organized drainage is essentially lacking.

Bogs, with their accompanying northern coniferous vegetation, are frequent and give to the area a more northern aspect than that of the surrounding territory. Tamarack dominates many of the secondary bog areas, but black spruce and white cedar occupy all primary and old secondary bog situations. Spruce mingles with the aspens of surrounding secondary slopes. Morainal knolls which were originally covered by hardwoods generally have a mixture of aspen, spruce, and hardwood sprouts. The less pronounced knolls or ridges supported a mixed forest of hardwoods and hemlock, occasionally with some spruce and white pine.

Jack pine or jack pine and aspen communities on sandy stretches, and aspen communities on every conceivable situation except the wettest bogs generally dominate the landscape. Occasional old but small stands of red pine and of white pine, small tracts of mixed hemlock and hardwoods, and pockets of hardwood forest between the bogs are all that is left of the original forest of the area. No extensive tracts of primeval forest remain.

Sociologic studies of pine and hardwood stands, and pollen studies of bogs in this district (Potzger, 1942, 1946) distinguish two variants of the pine forest, differing in relative abundance of *Pinus Strobus* and *P. resinosa;* two principal variants of the hardwood forest as "determined by greater abundance of either *Acer saccharum* or *Tsuga canadensis;*" and show that maple, hemlock, and yellow birch have the highest fidelity, abundance, and frequency of the tree species of the hardwood forest.

The variations in the hemlock–hardwoods communities of this area are comparable to those seen in other parts of the Superior Upland. Transitions from one to another variant are more frequent, and the extent of any one is less than in the areas with less hummocky topography. A few examples will suffice for illustration of the "Lake Region" (Table 77). Three of these are approximately on the Michigan-Wisconsin line (Vilas-Gogebic counties) in the Gillen Nature Reserve in the border moraine of the Young Gray Drift (Mankato drift) area; the fourth, the Caro Woods west of

Table 77. Forest composition in the "Lake Region" of northern Wisconsin: *1* and *2,* Near Roach Lake and *3,* East of Tenderfoot Lake in the Gillen Nature Reserve,[1] Vilas County; *4,* The Caro Woods in eastern Price County west of Woodruff;[1] *5,* Average of eight acres in Florence, Forest, and Marinette counties.[2]

Canopy	*1*	2	*3*	*4*	*5*
Number of trees	139	181	94	172	41[3]
Acer saccharum	7.2	52.5	52.1	12.2	41.5
Tilia americana	.7	11.7	26.6	16.3	14.6
Betula lutea	47.5	25.4	13.8	32.0	21.9
Tsuga canadensis	43.2	9.9	4.2	38.4	14.6
Ulmus americana	. .	.5	2.1	.6	7.3
Fraxinus americana	. .	. .	1.1	.6	. .
Thuja occidentalis	1.4	. .	. .	. .	. .
Pinus Strobus	. .	. .	. .	x	. .
Second layer[4]					
Number of trees	136	115	69	159	44[3]
Acer saccharum	14.0	33.0	53.6	8.8	31.8
Tilia americana	. .	14.8	16.0	1.2	4.5
Betula lutea	14.0	20.0	8.7	6.3	15.9
Tsuga canadensis	55.1	23.5	18.8	78.0	31.8
Ulmus americana	. .	1.8	1.4	. .	11.4
Thuja occidentalis	11.0	2.6	. .	3.1	4.5
Abies balsamea	5.9	.9	. .	. .	. .
Acer rubrum	. .	. .	. .	1.2	. .
Ostrya virginiana	. .	3.5	1.4	1.2	. .

[1] For additional sociologic data concerning areas in the Gillen Nature Reserve and the Caro Woods, see Potzger, 1946.

[2] Data for area *5* from Table II of Zon and Scholz (1929), where trees are listed by diameter classes. Those 15–30 inches d.b.h. are here listed as canopy trees; those 10–14 inches d.b.h., as second-layer trees.

[3] Number of trees per acre, based on average of eight acres.

[4] Second layer, as here used, corresponds approximately to Potzger's 10–15 inch d.b.h. size class, with at least the smaller of his 16–20 inch d.b.h. class included.

Woodruff, is some miles to the southwest in the area of Young Red Drift (Cary); the fifth is to the east, in Florence, Forest, and Marinette counties (data for this from Zon and Scholz, 1929). The hemlock–yellow birch and sugar maple–birch–basswood communities of Table 77, areas 1 and 2, are adjacent and grade into one another, the former occupying concavities, the latter convexities of a gently undulating surface. The sugar maple–basswood–yellow birch community (area 3) occupies higher knolls which rise above hemlock–hardwoods communities. Locally basswood dominates, hemlock is absent, and a mull humus is developing. The Caro Woods west of Woodruff occupies flat to rolling terrain varied by low narrow ridges. Forest composition shows little variation in relation to these mild topo-

A magnificent hardwood stand in the Nicolet National Forest in northeastern Wisconsin (Forest County). The large trees are (left to right) sugar maple, elm, black ash, and elm. Scattered leatherwoods (*Dirca palustris*), and patches of Trilliums, violets, ferns and ricegrass (*Oryzopsis*) are interspersed among the seedling maples. (Courtesy, U. S. Forest Service.)

graphic variations. Although hemlock leads in number of canopy trees, and is dominant in the second layer, basswood and yellow birch exceed it in size. The Florence–Forest–Marinette counties area represents an average of eight acres in virgin hemlock–hardwoods forest.

FLAMBEAU RIVER FOREST

In a region largely devastated by lumbering and fire, the preservation of a block of 3800 acres of virgin timber and of uncut streamside strips, is particularly fortunate.[15] The area, in eastern Sawyer County, Wisconsin, lies close to the western limits of hemlock and yellow birch. It is a splendid area of hemlock–white pine–northern hardwoods almost at the margin of this forest region.

Table 78. Variations in composition of the virgin forest of the Flambeau River, Sawyer County, Wisconsin. Communities *3, 5, 6,* and *8* are developmentally more advanced than *1, 2, 4,* and *7*. *Betula lutea* maintains its importance in lower layers of the forest (reduced in second layer of area *2* to codominance with *Acer saccharum* and *Tsuga canadensis*). *Acer saccharum* is present in the lower layers of all communities, thus foretelling further development of earlier stages (*1, 2,* and *7*). *Pinus Strobus* is present in the understory only in area *4,* and there is reduced to only one-third its canopy abundance.

	Along River			*Slopes*			*Upland*	
	Wet flat	*Higher*	*Dissected*	*Steep*	*Ravine*	*River*	*Swampy*	*Rolling*
	1	*2*	*3*	*4*	*5*	*6*	*7*	*8*
Number of trees	58	23	24	93	94	72	31	149
Acer saccharum	. .	. .	12.5	16.1	9.6	22.2	. .	3.3
Tilia americana	. .	4.0	12.5	1.1	20.2	22.2	. .	16.7
Betula lutea	24.1	87.0	20.8	29.0	33.0	34.7	22.6	25.3
Tsuga canadensis	48.3	9.0	29.2	20.4	21.3	21.1	74.2	44.0
Pinus Strobus	24.1	. .	. .	33.3	. .	. .	3.2	. .
Ulmus americana	. .	. .	12.5	. .	16.0	. .	. .	9.3
Fraxinus americana	. .	. .	. .	. .	. .	. .	. .	.7
Juglans cinerea	. .	. .	. .	. .	. .	x	. .	.7
Thuja occidentalis	3.4	. .	. .	. .	. .	. .	. .	. .
Fraxinus nigra	. .	. .	12.5	. .	. .	. .	. .	. .

The Flambeau River, now sluggish and with low flat borders, now swift with banks rising steeply on one or both sides, drains the area. Tiny boggy creeks or swiftly flowing brooks interrupt the river slopes. In such a topographic setting, a considerable range of forest communities is seen (Table 78).

A streamside community, in which silver maple (*Acer saccharinum*) is

[15] The Flambeau River State Forest includes "one solid block of 3800 acres held by the State Land Commission" of virgin forest, large areas of cutover land, and "an uncut strip along the river" (Aldo Leopold, 1943).

dominant, fringes the low banks or extends far back from the stream in the occasional river swamps. White elm, black ash, and scattered bur oak, with here and there a butternut (*Juglans cinerea*), mingle with the silver maple. Along the forest margin, and often at the very edge of the river, are clumps of alder, Spiraea, and Cornus, and a rank growth of herbs—tall grasses and sedges, ostrich ferns and sensitive ferns, and such swamp plants as *Boehmeria cylindrica, Asclepias incarnata, Polygonum* spp., *Chelone glabra, Rudbeckia laciniata,* and *Echinocystis lobata.* Instead of this typically streamside community, subject to submergence, yellow birch, white pine, and hemlock may come to the river margin on flats but a few feet above the water, or be situated just back of the silver maple zone. In such situations, white pine grows rapidly, trees attaining a diameter of 30 inches in 130 years. The growth increment of hemlock is only about half as great.

Where the banks rise steeply, trees of the regional forest—white pine, hemlock, and yellow birch, perhaps with an admixture of spruce and fir or of sugar maple and basswood—prevail. While the best of the white pine was removed from most of the Flambeau country at an early date, some large, usually defective, trees and many smaller ones remain. Some of the slopes are now occupied by almost pure stands of white pine. Old stands where white pine is a dominant tree may occupy sites where long ago a natural catastrophe opened up the forest permitting abundant pine reproduction, much as has lumbering in more recent years. In the virgin forest, white pine grew with the hemlock and hardwoods. "Here the cream of the white pine grew on the same acres with the cream of the sugar maple, yellow birch and hemlock. This rich intermixture of pine and hardwoods was and is uncommon" (Leopold, 1943).

The aspect of the virgin forest of the Flambeau is unlike that of other forests of the Superior Upland or of the Great Lake section. The appearance of the forested slopes, and of the rolling upland is almost park-like. There is no brushy layer; instead, there is a carpet of grasses and ferns and other herbaceous plants. The park-like appearance and absence of a brushy layer may be correlated with the lesser abundance of sugar maple. The grassy carpet so conspicuous here is simulated in some of the hardwood forests of northern Upper Michigan where selective logging has reduced the density of the canopy. The grass layer competes with tree seedlings, and the continuity of the reproduction layer is broken. Reduced precipitation in this western part of the mesophytic deciduous forest may account for the wider spacing of trees, root competition becoming keen before crown closure takes place.

Variations in composition of the virgin forest of the Flambeau River are illustrated in Table 78. Yellow birch and hemlock are almost universally abundant species of the canopy and of the second layer, thus demonstrating their importance in this marginal part of the Hemlock–White Pine–

Northern Hardwoods region. Sugar maple is not the most abundant species in any community here illustrated, although it is a constituent of the canopy in all of the developmentally more advanced stages, and is present in lower layers of younger stages from which it may be absent in the canopy. On high ridges (divides between the Flambeau and South Fork, and between the Flambeau and Thornapple rivers) a maple–basswood forest is seen, in which are occasional elm, white ash, and red oak (a forest such as is illustrated by Tables 75 and 80). In the communities of gentle but well-drained slopes, hardwoods form nearly 80 per cent of the canopy, and a higher percentage of the second layer. These, and the forest of rolling uplands best represent the regional climax. The hemlock or hemlock–white pine communities of boggy stream headwaters and swampy upland, and of poorly drained river flats but a few feet above the surface of the river, occupy situations unsuitable for hardwood dominance. On slopes, the abundance of white pine is temporary; it is poorly represented in lower layers.

In the hemlock–white pine communities, the number of shrubby and herbaceous species is fewer than in the mixed communities; for the most part they are species of northern range (as *Phegopteris Dryopteris, Clintonia borealis, Maianthemum canadense, Coptis groenlandica, Mitella nuda, Oxalis montana, Cornus canadensis, Moneses uniflora,* and *Lonicera canadensis*). In the hemlock–hardwoods communities, the undergrowth is richer, although grasses and ferns still predominate. The forest of the rolling upland contains by far the largest number of shrubs of any part of the forest, and two or more times as many herbaceous species as any other community except that of moist but well-drained ravine slopes. Representative woody plants, not present in the dominantly coniferous stands, include *Cornus alternifolia, Corylus cornuta, Dirca palustris, Parthenocissus quinquefolia, Ribes Cynosbati,* and *Viburnum acerifolium.* The herbaceous plants are mostly those of mesic hardwood forests, some northern (as *Trillium cernuum, Streptopus roseus, Viola pallens,* and *Aralia nudicaulis*), some widespread (as *Athyrium thelypteroides, Athyrium angustum, Adiantum pedatum, Botrychium virginianum, Uvularia grandiflora, Asarum canadense, Laportea canadensis, Actaea pachypoda, Hepatica americana, H. acutiloba, Caulophyllum thalictroides, Sanguinaria canadensis, Dicentra canadensis,* etc.). A few species of the coniferous forest also are present (*Maianthemum, Trientalis,* and *Phegopteris Dryopteris*).

The highly modified and secondary character of the forests of most of Wisconsin makes impossible a complete picture of original forest cover. Chamberlin's "Map of the native vegetation of Wisconsin" (1882) distinguishes a "mixed hardwood and evergreen" forest type (the hemlock–white pine–northern hardwoods) which occupies most of the northern half of the state. (That part along Green Bay and Lake Michigan is included in the Great Lake section.) Areas of pine, where "the leading tree is *Pinus Strobus,*" of maple, whose "leading member is sugar maple," of swamp conifer forest (tamarack, arbor vitae, spruce), of "dwarf oak and pine (including the so-called 'Barrens')," are enclosed within the general area where the "mixed hardwood and evergreen forest" prevails, and lobes of oak forest indent its southern border.

Concerning the white pine areas, Gevorkiantz and Zon (1930) wrote:

> The almost insignificant area of essentially pure and even-aged second-growth white pine is in vivid contrast with the 18 million acres over which this species formerly occurred . . . Sometimes white pine occurred in pure stands or mixed with Norway pine, as on sandy areas in Douglas, Bayfield, Jackson, Vilas, Oneida, and Marinette counties. On heavier soils of the north central part of the state, white pine occurred in mixture with hardwoods or not at all.

The Flambeau River area is representative of the section where white pine is mixed with the hardwood species.

MINNESOTA SECTION

The Minnesota section comprises approximately the northeastern half of Minnesota, extreme southeastern Manitoba and adjacent Ontario, and the western part of Wisconsin north of the latitude of St. Croix Falls. It includes all that part of northern Minnesota usually mapped as coniferous forest (Minn. Geol. Survey, Bull. 12, fig. 4; Bergman and Stallard, 1916; Stallard, 1929); the Quetico, Rainy River, and Superior West sections of southern Canada west of Lake Superior (Halliday, 1937); the pinery of northwestern Wisconsin, and a more or less transitional lobe of pine and hardwoods lying between the Chippewa and St. Croix rivers.

The contrast between the appearance of this area and adjacent parts of the Superior Upland section (as exemplified by the nearby Flambeau River forest) is well marked; it is as great as the usual contrast between adjacent regions. Hemlock is lacking, for this section lies to the west of the limit of range of hemlock, and yellow birch is infrequent and generally limited to moist soil sites of stream and lake terraces. Oaks are abundant, especially in the southeastern part of the section, where red oak, white oak, and bur oak now dominate, but where white pine once predominated.

Vegetationally, the Minnesota section is characterized by the prevalence of pines (*Pinus Strobus, P. resinosa,* and *P. Banksiana*) over much of the area, the dominance of northern conifers (*Larix laricina, Picea mariana, P. glauca, Thuja occidentalis,* and *Abies balsamea*) in and around boggy depressions, and the more or less frequent occurrence of hardwood communities in the southern two-thirds of the section. Northward, *Abies* becomes more abundant and more generally distributed, and hardwood species, particularly basswood, sugar maple, and yellow birch become more and more scattered.

The topography is almost entirely depositional. Sandy outwash plains and till deposits are irregularly pitted; lakes are legion, and in and around them are various phases of hydrarch successions. This vegetational section lies mostly in the Western Young Drift section of the Central Lowland, lapping over somewhat onto the western (and more indefinite) part of the Superior Upland as defined by physiographers.

The southeastern part of the section is transitional in nature and vegetationally distinct from the pine area to the north.

Pine Area

Although not everywhere dominated by pine, the forests of this northwestern lobe of the Hemlock–White Pine–Northern Hardwoods region have always contained a large proportion of pine. As pine forest is the most advanced forest type developed over a large part of northern Minnesota and adjacent southern Canada, it is generally considered to be the climax of that area (Bergman and Stallard, 1916; Bergman, 1924; Stallard, 1929). In the northern tier of counties and in adjacent parts of Canada, the climax dominants are Norway (red) pine and white pine, either in mixed or pure stands; on drier soils the former, on moister soils the latter may form almost pure stands. Stallard, referring to unburned islands of Lake of the Woods and Rainy Lake, where these pines were chief dominants, says: "In these places are found the largest and oldest trees in the northern part of the state. They vary in diameter from 1 to 3 feet; in height, from 75 to perhaps 125 feet; and in age, from 65 to 200 years or more." This far north, there are no other species which can compete with the pines.

A little farther south, the possibility of further development toward a deciduous climax, and the frequent occurrence of climax deciduous communities indicate the subclimax character here of the pine forest, although it is still areally dominant. Just as on the Coastal Plain of the South, sandy soil, fire, and youthful topography combine to inhibit the development of the true climax.

Hardwood communities of maple and basswood, principally, are more or less frequent in the southern two-thirds of the area. These hardwood communities, with few exceptions, have developed only on the heavier and better soils. Nevertheless, this maple–basswood community can, in the absence of fire, ultimately replace pines even on sandy soil (Kittredge, 1934). It is the climax of a developmental series (Table 79) which may start with jack pine, and, passing through Norway pine, white pine, and transitional stages, culminate in maple–basswood forest with or without an admixture of other species—white elm, red oak, bur oak, and locally (on shores of Lake Winnebigoshish), bitternut hickory (Johnson, 1927).

The dominants of the subclimax pine forest, Norway pine and white pine, occur mixed or in pure stands. Given time enough without fire, white pine succeeds Norway pine. As white pine is more susceptible to fire than Norway pine, it may be eliminated from habitats entirely suitable to its growth (Corson, Allison and Cheyney, 1929). Jack pine, except locally, is a dominant only in secondary communities. In older jack pine communities, young individuals of Norway and white pine point to replacement of jack pine by Norway pine. The series of forest stages, the composition of each, and the length of the period of influence of each was worked out on Star Island in Cass Lake (Kittredge, 1934). The period of influence of the occupying dominants of a mature jack pine stand was 100 years; of a Norway pine stand, at least 300 years and probably much longer; of white pine, not less than 250 years; of maple–basswood, over 300 years. The dominant vegetation influences the lower vegetation, the forest floor, and the soil. Here on Star Island, where all these forest types occur side by side on one soil type, the soil profile is similar under all, except that there is more organic matter in the upper six inches under the maple–basswood forest, and

Star Island in Cass Lake is of particular interest ecologically because here the series of forest stages and length of period of influence of each have been worked out. (Courtesy, U. S. Forest Service.)

Table 79. Composition of series of developmentally related forest types on Star Island, Cass Lake, Minnesota. (Adapted from tables by Kittredge, 1934, who gives number of trees per acre by species in each diameter class at one inch intervals.)

Trees over 8.5 inches d.b.h.	*1*	*2*	*3*	*4*	*5*
Number per acre	224	200	24	60	144
Pinus Banksiana	94.6	..	..	..	..
Pinus resinosa	5.4	100.0	16.6	8.3	..
Pinus Strobus	..	..	83.3	16.7	..
Quercus borealis	..	..	..	16.7	..
Betula papyrifera	..	..	..	8.3	..
Acer saccharum	..	..	..	50.0	27.8
Tilia americana	..	..	..	..	55.5
Ulmus americana	..	..	..	..	16.7

the acidity is less (Alway and McMiller, 1933). The forest floor shows even more characteristic differences, that of the maple–basswood being much less acid, much richer in lime, and somewhat richer in nitrogen than that in the jack and Norway pine areas (Alway, Kittredge and Metley, 1933). The white pine type is intermediate between the Norway pine and maple–basswood communities in surface soil characteristics, as it is in succession. The occurrence of certain of the undergrowth species may be correlated with specific features of the soils.

For the jack and Norway pine types, *Arctostaphylos uva-ursi, Gaultheria procumbens, Vaccinium pennsylvanicum, Chimaphila umbellata, Melampyrum lineare* and *Lycopodium complanatum* are quite typical and are rarely if ever found with the sugar maple–basswood or with white pine at the stage when those species are invading. . . . Quite different in character from the foregoing species but typical of the maple–basswood forest are *Uvularia grandiflora, Actaea rubra* f. *neglecta, Cornus alternifolia, Polygonatum biflorum, Celastrus scandens, Smilax herbacea,* and *Asplenium filix-femina* (Kittredge, 1934).

Hydrarch successions in the Minnesota section, whether primary or secondary, are similar to those farther east in the region. Here ground water is of sufficient amount to offset the reduced precipitation of this western arm of forest. Tamarack, arbor vitae, and black spruce are the principal dominants of bog forests, with which may be associated balsam fir and white spruce, and occasionally balsam poplar, paper birch, black ash, and red maple (Stallard, 1929). White pine and yellow birch sometimes appear near the ancient shores of filled lakes (Lindeman, 1941). Xerarch successions, except on rock exposures of the northeastern corner of the state, are almost entirely secondary, resulting from denudation by lumbering and fire. Secondary successions, both hydrarch and xerarch, have been described by Bergman and Stallard (1916) and by Stallard (1929). These are essentially like those of the Superior Upland and Great Lake section, except

Forest of white and red pine, with undergrowth of hardwoods. The combined period of influence of red or Norway pine and white pine is five or six hundred years. Star Island, Cass Lake, Minnesota. (Courtesy, U. S. Forest Service.)

Mixed forest on Star Island, with white and red pines, northern red oak, paper birch, and maple. (Courtesy, U. S. Forest Service.)

that aspen communities are much less widespread. The principal dominant of the young forests is jack pine. *Populus grandidentata* may be scattered with the pine, but *Populus tremuloides* is entirely absent from all the drier soils (Stallard, 1929). A Populus–Betula community develops in mesophytic situations following early shrub stages; this is replaced by pine.

The Minnesota section of the Hemlock–White Pine–Northern Hardwoods region extends almost to the western limit of forest in this latitude. Only a narrow broken band of deciduous (oak) forest lies between it and the Prairie region (p. 184). The areal dominance of pine forest is largely due to this western location with its low annual rainfall. The northern location (accentuated by the extremes of winter) limits the northward distribution of deciduous species. Fir and spruce, most abundant in late bog forest stages, are not northward confined to them. A transition to the northern conifer forest such as that of Isle Royale (Cooper, 1913) is suggested by the fir–basswood community which Grant (1934) considers to be the climax community in Itasca County in northern Minnesota. The presence of fir in the understory of red and white pine communities in the northernmost part of the area, further illustrates the transition to the Boreal Forest. Even in northern Minnesota, maple, basswood, and bitternut hickory attain fairly large size and form rich deciduous forest communities (Johnson, 1927). Farther north, the representation of hardwood species is scattered. The occurrence of bur oak in grassy groves in the north suggests the forest border to the west.

Southeastern Transition

A transitional area extends southeastward from the pine area to the Driftless section of the Maple–Basswood region. Here, the vegetation combines some of the features of the regions to the north and the south. Oak woods (with *Quercus borealis* var. *maxima, Q. alba, Q. macrocarpa,* and *Q. ellipsoidalis*) are frequent and occupy areas which once supported large quantities of white pine. The dominance of oaks today is broken only by the hydric communities of lowlands, the young pine stands on sand plains, and the farm country on till soils. Even the secondary aspen communities contain a fair admixture of oaks. The aspect is unlike that of any other part of the Hemlock–White Pine–Northern Hardwoods region and would be separated from it if it were not for the original dominance of white pine or of white and red pines in much of the area. Old white pines, three or four times the height of the oaks and towering above them, are remnants of the former forest.

Considerable alternation of hardwood communities occurs: mixed oak stands (usually with white pine); bur oak, often with red maple and elm, on flatter areas; mixed hardwoods with sugar maple, basswood, elm, red, white and bur oaks, white pine, and other species on more rolling land.

Stands of spruce, white pine, or larch on low soil between the hardwood areas further diversify the aspect. The maple–basswood and maple–basswood–red oak communities relate this area to the Maple–Basswood region adjacent to the south and west. The presence of trees of the more southern deciduous forest (white oak, bur oak, red elm, bitternut hickory, walnut, hackberry) and of more southern shrubs and herbs (as *Celastrus scandens, Zanthoxylum americanum, Rubus occidentalis, Anemone virginiana, Amphicarpa bracteata, Geranium maculatum, Aster laevis, Ratibida pinnata, Desmodium* spp., etc.) point to migrations from the Oak–Hickory Forest region. The occurrence of prairie species suggests the nearness of the grassland to the west. The pines, the northern species in the bogs, the occasional northern shrubs and herbs in the hardwood forest relate this area to the region to the north with which it is here treated.

Statistical data concerning composition of hardwood communities on morainal hills in Washburn County, Wisconsin, are given by Eggler (1938). These data show that sugar maple, basswood and red oak are the dominant species (Table 80). Increment borings made by Eggler in the Hunt Hill

Table 80. Composition of two communities in the "Southeastern Transition" of the Minnesota section: Long Lake and Hunt Hill woods, Washburn County, Wisconsin. (Data from Eggler's tables, 1938, of density and frequency of tree species of the several d.b.h. size classes, here given as percentage composition to facilitate comparison with other tables.)

	Long Lake		*Hunt Hill*	
Inches d.b.h.	10 & over	4–9	10 & over	4–9
Number of trees	75	107	77	88
Acer saccharum	41.3	13.1	27.3	18.2
Tilia americana	34.7	40.2	7.8	23.9
Quercus borealis maxima	18.7	14.0	57.1	38.6
Quercus alba	1.3	. .	2.6	10.2
Betula lutea	1.3	4.7	1.3	1.1
Pinus Strobus	. .	. .	2.6	1.1
Fraxinus lanceolata	. .	.9	. .	2.3
Juglans cinerea	1.3	1.9	. .	. .
Ulmus fulva	1.3	. .	. .	. .
Betula papyrifera	. .	. .	1.3	. .
Carya cordiformis	. .	3.7	. .	. .
Acer rubrum	. .	. .	. .	1.1
Ostrya virginiana	. .	21.5	. .	3.4

Additional species represented in smaller size classes: *Carpinus caroliniana, Populus grandidentata,* and *Populus tremuloides.*

area where the red oaks are abundant "showed the red oaks, and the basswoods to some extent, to be even-aged. This indicates that they were sudden arrivals, probably following fire. Relict survivors of the pre-disturbance forest are largely sugar maples, which points to the fact that in the original forest maple and basswood were more important than oak." Red oak, white pine, yellow birch, and white oak were usually present in the maple–basswood communities.

The larger number of tree species in the Washburn County areas as compared with hardwood communities in the adjacent Superior Upland is noticeable. Three of them (*Carya cordiformis, Quercus alba,* and *Ulmus fulva*) are not represented there. *Carya cordiformis,* here close to the northern limits of its range, never reaches large size, hence does not become a part of the canopy. The low percentage of *Betula lutea* is in marked contrast with its abundance but a short distance eastward.

LAURENTIAN UPLAND SECTION

The southern border of the Laurentian Upland or Canadian Shield lying to the north of the Great Lakes and east of Lake Superior is occupied by the northernmost section of the Hemlock–White Pine–Northern Hardwoods Forest. Its very irregular northern boundary is the boundary drawn by Halliday (1937) between the Boreal Forest and the Great Lakes–St. Lawrence forest.[16] This is largely determined by the limit of range of yellow birch (Halliday and Brown, 1943).

All of the usual dominants of the Great Lakes–St. Lawrence Division occur in this section, but in decreasing amounts northward, and often with pronounced limitation to specific habitats. Hemlock, beech, and red oak generally drop out before the northern limits of the section are reached; sugar maple and yellow birch intrude somewhat into the Boreal Forest to the north where they are confined to the heavier-textured ridges or hilltops. White pine and red pine, characteristic conifers of the region as a whole, only very locally extend northward into the Boreal Forest, and then usually on rocky shores.

The central portion of the section, which is in contact at the south with the Ontario district of the Great Lake section, is the most varied vegetationally. In general character, this is a mixed forest where the dominant community is composed chiefly of sugar maple, yellow birch, hemlock, and white pine, with a variety of other species (basswood, beech, red oak, elm, white ash, red maple, ironwood, white birch, large-tooth aspen, white spruce, and balsam fir), most of which are more common southward.

---

[16] The section outlined here includes most of the "forest sections" distinguished by Halliday—Algonquin, Laurentides, Algoma, Timagami, Haileybury, Saguenay, Upper St. Lawrence, and Lower St. Lawrence sections. The relatively small area of deciduous forest communities, and the transitional position of the section, make subdivision undesirable here. Much of the material for this section is taken from Halliday's paper. Dansereau's studies (1943, 1946) on the Laurentian sugar maple forest, present in great detail features of the easternmost end of the section.

In the western part of the section (Algoma district along the east shore of Lake Superior) the number of deciduous species is fewer; however, sugar maple and yellow birch are well distributed (but soon drop out toward the northeast). Hemlock extends but a short distance northward of Sault Sainte Marie. Mixed deciduous–coniferous forest occupies gentle northerly slopes, and almost pure stands of sugar maple and yellow birch occupy the ridge tops (McCarthy, 1921). The large amount of white spruce (especially on river terraces), balsam fir, paper birch, and the conifer communities of swamps and bogs give much of this area an aspect of boreal forest.

In the eastern part of the section, the mixed forest grades northward into the Boreal Forest, and hardwood groves become more and more isolated. The northern outposts of this vegetational section (reached in the Saguenay district), where restricted stands of deciduous forest, "typical Canadian forest," and "isolated stands of taiga" occur in contact with one another ("interpenetrating climaxes"), are described by Dansereau (1944). Here,

> the last true maple grove occurs at Metabetchouan, on the shores of Lake Saint-Jean itself. Gradually the typical elements fall out: *Acer saccharophorum* a few miles north of the lake, *Acer rubrum* 20 miles, and finally *Betula lutea* 38 miles. The last named, with hardly any admixture of fir (*Abies balsamea*), forms pure stands at the afore-mentioned point of contact.

Here, too, white and red pine reach their northern limits.[17] In an area 88 miles north of Mont Laurier in western Quebec, beech, yellow birch, striped maple, and red oak are found with the more abundant sugar maple (Marie-Victorin and Rolland-Germain, 1942; Dansereau, 1944a). Such isolated areas are beyond the limits of the Great Lakes–St. Lawrence forest.

Toward the southeast, the proportion of pure deciduous forest is greater. There, the climax is (according to Dansereau, 1946) a forest in which sugar maple is dominant, forming pure stands, or accompanied by beech (not over 20 per cent), white ash, and basswood. This is a forest with a rich vernal flora. In addition to the true climax in this area, many other types of hardwood forest may be distinguished, some of which closely resemble the climax except for certain features.[18] Forests containing an admixture of more southern species as hickories (*C. cordiformis, C. ovata*), oaks (*Q. borealis* var. *maxima, Q. alba,* and *Q. macrocarpa*) occur in the area transitional to the Northern Appalachian Highland forest, especially near to Lake Ontario and the upper St. Lawrence Valley.

This section as a whole is decidedly transitional in character. Its chief

[17] Halliday (1937) believes these occurrences are the cause of the usually incorrect mapping of the range of these two pines, where a generalized northern boundary results in the inclusion of large areas of Quebec where these pines do not occur.

[18] For descriptions of the many subclimax, quasiclimax, and disclimax communities, as well as of the climax, see Dansereau's papers on "L'Erablière Laurentienne," 1943, 1946.

interest in a consideration of the Deciduous Forest Formation lies in the development of pure or almost pure deciduous communities to the very northern limits of species ranges. The possible historic significance of the disjunct occurrence of hardwood stands, pointing to cooling climates, adds to their interest (Dansereau, 1946). While podzols prevail over most of the area, local areas of gray-brown forest soil have developed under the influence of predominantly deciduous leaf litter.

NORTHERN APPALACHIAN HIGHLAND DIVISION

This more eastern division of the Hemlock–White Pine–Northern Hardwoods region lies to the east and southeast of the Great Lakes, extending from central and northwestern Pennsylvania northeastward across New York and New England into southeastern Canada. A large part of this area is mountainous; it has more relief, both regional and local, than the Great Lakes–St. Lawrence Division (excluding the Huron and Porcupine mountains). All of the region except the southern lobe in Pennsylvania was glaciated. However, its topography, although modified by glaciation, is to a considerable extent controlled by underlying rock.

The diversity of topography and greater altitudinal range result not only in local diversity of communities but also in distinct altitudinal zonation in the more mountainous districts. Elevations exceed 4000 feet in the Catskills and Green Mountains, 5000 feet in the Adirondacks and in Maine (Mt. Katahdin, 5382 feet), and 6000 feet in the White Mountains. In all of these locations (and in additional northern areas of somewhat lower altitudes) the typical and widespread phases of the Hemlock–White Pine–Northern Hardwoods Forest are modified by an admixture of red spruce or are supplanted by higher altitude or more northern forest types in which red spruce, balsam fir and paper birch predominate.

Portions of the Northern Appalachian Highland Division of the Hemlock–White Pine–Northern Hardwoods region (as interpreted here) have been mapped as "Allegheny Hardwoods–Hemlock" forest type in Pennsylvania and southern New York by Hough and Forbes (1943) (Map, p. 394); as zones C and D by Bray on his map of the vegetation of New York State (1930) (Map, p. 395); as the "Northern Hardwood," "Spruce," and "White Pine" forest regions of New England by Hawley and Hawes (1912) (Map, p. 249); as the "Hemlock–Red Spruce–Northern Hardwoods" and "Hemlock–White Pine–Northern Hardwoods" by Gordon (1937). In Canada, this division includes the Acadian Forest region and two sections of the Great Lakes–St. Lawrence forest lying east of the St. Lawrence River and contiguous to the international boundary (as mapped by Halliday, 1937).

Outliers of the Hemlock–White Pine–Northern Hardwoods occur south of the regional boundary, in the Allegheny Mountains of southern Pennsylvania, Maryland, and West Virginia. These have already been considered in their altitudinal relations in the Mixed Mesophytic Forest region. Within the region, outliers of the more northern Spruce–Fir Forest region of Canada (in which white spruce is a dominant) occur in certain coastal areas and at high elevations in the mountains.

Vegetationally, this division of the Hemlock–White Pine–Northern Hardwoods Forest contrasts with the Great Lakes–St. Lawrence Division in the presence of red spruce (*Picea rubens*) in certain of its northern or higher

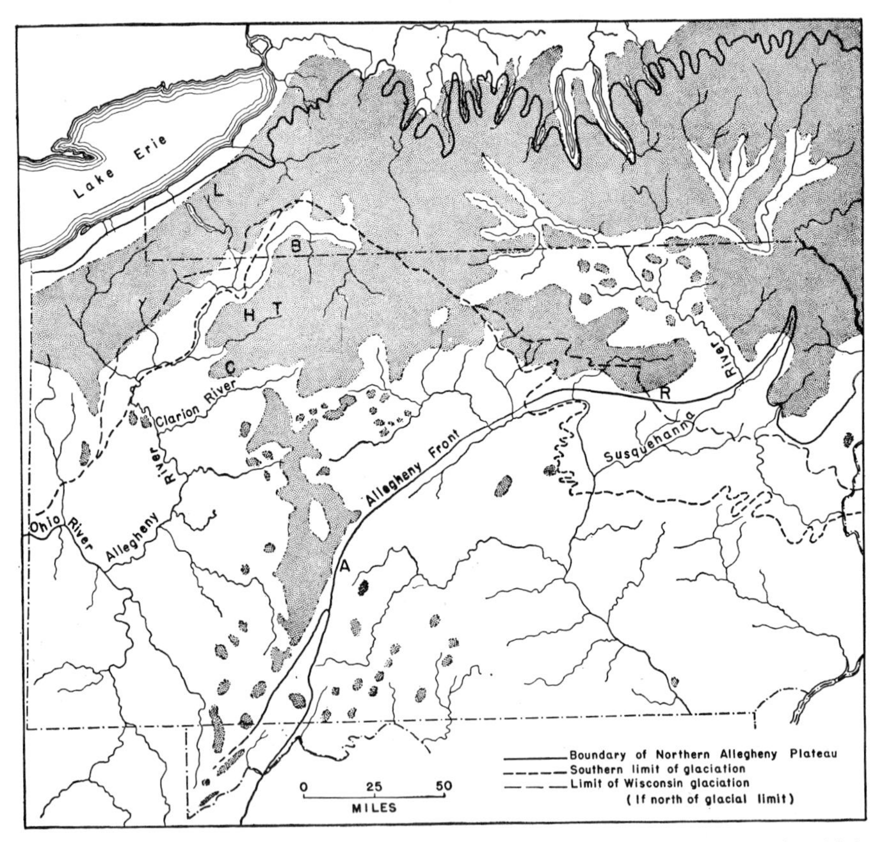

Map of Pennsylvania and southern New York showing (by stippling) areas in which "Allegheny Hardwoods–Hemlock forest types" appear, boundaries of the "Northern Allegheny Plateau," and limits of glaciation (after Hough and Forbes, 1943); limits of Wisconsin glaciation, and Allegheny Front in southern Pennsylvania and Maryland (after Fenneman, 1938). Note occupancy of the plateau almost to its margin by hemlock–hardwoods where the plateau is not deeply cut by streams, and absence of this forest type from the larger river valleys. Initial letters give approximate location of areas mentioned in text: *A,* Altoona; *B,* Big Basin in Allegany State Park; *C,* Cook Forest; *H,* Heart's Content; *L,* Lily Dale; *R,* Ricketts Glen; *T,* East Tionesta Creek Natural Area.

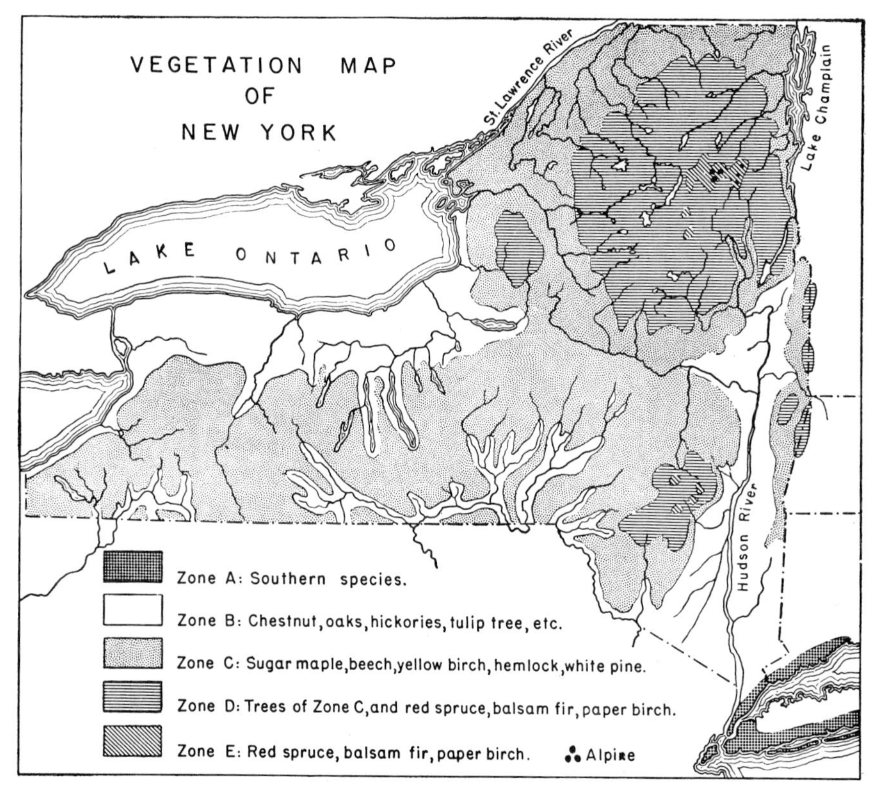

Map of New York State showing distribution of major forest types. (After Bray, 1930, with modifications by House in Hotchkiss, 1932.)

altitude communities; in the abundance of gray birch (*Betula populifolia*) in some of its secondary communities; in the presence of a number of shrubby and herbaceous species absent or poorly represented in the forest communities of the more western division, conspicuous among which are *Viburnum alnifolium, Tiarella cordifolia,* and *Aster acuminatus;* in the absence of jack pine (*Pinus Banksiana*), and relative infrequence of red pine (*P. resinosa*) and arbor vitae (*Thuja occidentalis*) in a large area at the south; in the pronounced admixture of tree species of the central deciduous forest (as oaks, chestnut, cucumber magnolia, tuliptree) in a transition zone along the southern margin of the region and along major streams which flow into other regions to the south (Allegheny River flowing into the Mixed Mesophytic region, and Susquehanna, Delaware, and Hudson into the Oak–Chestnut region). This division might well be subdivided, vegetationally, into a more southern (and generally lower elevation) part, the hemlock–hardwoods or Allegheny hardwoods–hemlock forest, which has certain transitional features that ally it with the central deciduous forest, and which lacks the more northern conifers; and a more northern part, the spruce–hardwoods forest in which red spruce is a dominant in climax communities, and in which other northern conifers (balsam fir, red and jack pines, and arbor vitae) are more or less conspicuous. The northern subdivision is more or less transitional to the Spruce–Fir Forest region. Such a vegetational subdivision would make difficult the delimitation of sections which are geographic or physiographic units.

Three sections are recognized: the **Allegheny** section, including all of the region south of the Mohawk River and south of the lowland adjacent to Lake Ontario; the **Adirondack** section north of the Mohawk River, including the Adirondack Mountains and surrounding area; and the **New England** section which includes most of New England and adjacent areas in Canada belonging to this vegetational region, principally, the Acadian Forest region of Halliday. Of these, the Allegheny section (except the Catskill Mountains) and the southern part of the New England section belong to the hemlock–hardwoods forest; most of the Adirondack section and most of northern New England belong to the spruce–hardwoods forest.

ALLEGHENY SECTION

This section occupies most of the northern half of Pennsylvania and the southern half of New York. Its triangular southerly lobe in Pennsylvania lies between the northern attenuated extension of the Mixed Mesophytic region on the southwest and the northern end of the Ridge and Valley section of the Oak–Chestnut region on the southeast. Through northeastern Pennsylvania and southeastern New York also it is in contact with the Oak–Chestnut region. Along the northwest, it is in contact with the strip of Beech–Maple region bordering Lakes Erie and Ontario.

The Allegheny section includes the Catskill Mountains[19] and most of the Allegheny Mountains of Pennsylvania; the northern end of the Unglaciated Allegheny Plateau in northern Pennsylvania and adjacent New York; all of the Glaciated Allegheny Plateau except its western end (in Ohio and adjacent Pennsylvania).

It has been stated that this forest region in general occupies younger glaciated land. Only on the northern end of the Unglaciated Allegheny Plateau does it occur on unglaciated land. No vegetational boundary is discernible here between the vegetation of the unglaciated and glaciated land. The Unglaciated Allegheny Plateau and Allegheny Mountains doubtless served as refugia during the Pleistocene, and it is quite probable that

[19] The Catskill Mountains are a part of the Appalachian Plateaus Province of physiographers, which province also includes the Allegheny Mountains and Allegheny Plateaus. They are included here in the Allegheny section as they are almost surrounded by the Glaciated Allegheny Plateau.

at least a part of this area was continuously occupied during Wisconsin time by forest vegetation of a sort referable to the Hemlock–White Pine–Northern Hardwoods Forest. From this center, this forest type moved northward and spread laterally over a vast area of glaciated land. The absence of any vegetational boundary[20] coinciding with the Wisconsin glacial boundary on the high plateau of northwestern Pennsylvania is evidence that an equilibrium has been reached with climate.[21]

Along its southern boundary in Pennsylvania where this region is in contact with more southern forest types, northward migrations, especially along the major valleys and their tributaries, have resulted in a modification of the northern forest by the admixture of more southern tree species. This becomes evident when we compare the composition of a typical stand of hemlock–hardwoods forest such as is seen in the East Tionesta Creek Natural Area with the composition of stands near the border of the area, such as Cook Forest on the Clarion River (tributary to the Allegheny River) and Ricketts Glen in a tributary of the Susquehanna River (Tables 81, 84, 86).

On the whole, the forest vegetation of the Allegheny Plateau and Allegheny Mountains portion of the section is quite similar throughout its broad expanse. In the Catskill Mountains, the presence of red spruce as one of the dominant species introduces a distinct forest aspect. Forest types, essentially uniform throughout, and local modifications related to soil or moisture, rather than areal differences, should be emphasized in the Allegheny Plateau and Allegheny Mountains.

Allegheny Plateau and Allegheny Mountains

The forest vegetation of the upland has been so profoundly modified by lumbering and fire, that for the most part it bears little resemblance to the original cover. Today, one may travel for many miles without seeing white pine or hemlock, trees once abundant through much of the area. No vestige of the original forest remains in the most disturbed areas. Thickets and open stands of aspens and pin cherry and, in old fields, of ash and hawthorn, form pioneer communities initiating the slow return to forest. Sprout stands of sugar maple and beech, with red maple in the lower spots, occupy some of the logged but not destructively burned areas. In these, the proportion of

---

[20] Where the glacial boundary approaches the Allegheny River in southwestern New York and western Pennsylvania there is some correspondence between that boundary and a vegetational boundary separating the attenuated mixed mesophytic of valley slopes from the northern forest. Where it crosses the plateau, no vegetational boundary is distinguishable.

[21] That there is a relation to the length of the frostless season may be seen by comparing maps showing average number of days without killing frosts (W. G. Reed, in Atlas of American Agriculture, Part II, Sec. I, U. S. Dept. Agr., 1920) and extent of hemlock–hardwoods forest.

sugar maple is greater than in primary stands, due in part to the more prolific sprouting of sugar maple and in part to the rapid growth of released understory trees. Grazing also influences the composition of stands (Lutz, 1930a). Toward the south, groves of secondary tuliptrees are more or less frequent. Secondary oak stands (especially of white and red oaks, and southward, of chestnut oak), in part at least of sprout origin, are frequent toward the margins of the plateau, especially adjacent to the oak lobe of the Mixed Mesophytic region, and adjacent to the Oak–Chestnut region. The most frequent of all the secondary forests developing after the cutting of beech–maple stands in this section is made up of a group of temporary hardwoods, most characteristic of which are red oak, white ash, basswood, red maple, and cherry. Not all of these will everywhere be abundant; other species, as butternut, bitternut hickory, and elms, may be present. As all of these trees are long-lived they form a fairly permanent type which may occasionally suggest undisturbed forest. The hay-scented fern (*Dennstaedtia punctilobula*) forms large patches or sometimes a continuous ground cover in open aspen and sprout woods, and in pastures. Certain common weeds, especially buttercups (*Ranunculus acris*) and hawkweed (*Hieracium aurantiacum*), because of their abundance, give a distinctive aspect to roadsides and meadows. These secondary communities are characteristically a part of the northern forest, although their aspect little suggests that forest. Cultural use of the cleared land also is distinctive. Timothy meadows, fields of oats and of buckwheat, expanses of red clover (*Trifolium pratense*), and occasional fields of barley lend an aspect not seen in any region to the south.[22]

On abrupt slopes in the plateau, the secondary vegetation is frequently more diversified, and white pines and hemlocks more or less frequent, thus suggesting the original cover. In such situations, there may be interesting local communities dependent upon extremes or peculiarities of the local environment. Interpretation of the many secondary communities, young and old, becomes possible only after primary stands have been examined.

Remnants of the original forest cover remain in only a few places. Fortunately, some of these have been preserved in State Parks and National Forests, while in these and in State Forests, various secondary forest communities and local edaphic communities are safeguarded.

The largest single tract of virgin timber is the East Tionesta Creek Natural Area in the Allegheny National Forest. This is located in one of the higher but dissected parts of the Unglaciated Allegheny Plateau.[23] In general,

[22] Some suggestion of this is seen in the Allegheny Mountains of West Virginia where, in the higher altitude belts, there are outliers of the northern vegetation.

[23] Its vegetation and environmental features have been discussed by Hough (1936) and by Hough and Forbes (1943). These authors give statistical data on forest composition (abundance and frequency of trees, shrubs and herbs), phytographs, periodicity of tree species, tree size and age.

Beech and hemlock in the East Tionesta Creek Natural Area, Allegheny National Forest, Pennsylvania. Hobble-bush (*Viburnum alnifolium*) is seen immediately to the left of the center beech. This shrub is a characteristic species of the Northern Appalachian Highlands Division of the Hemlock–White Pine–Northern Hardwoods Forest and of its outliers to the south. (Courtesy, U. S. Forest Service.)

the Tionesta Creek forest is a hemlock–hardwoods stand in which hemlock, beech and sugar maple (in the order named) are most abundant in the canopy, while beech far outnumbers other species in the subdominant layer and is well represented in all size classes (Table 81). Other more or less

Table 81. Composition of canopy, *a,* and second layer, *b,* of communities in East Tionesta Creek Natural Area, Allegheny National Forest, Pennsylvania: *1,* Forest as a whole; *2,* Hemlock–beech; *3,* Beech–hemlock–sugar maple; *4,* Beech–maple.

	1		*2*		*3*		*4*	
	a	*b*	*a*	*b*	*a*	*b*	*a*	*b*
Number of trees	546	405	283	243	103	69	142	82
Tsuga canadensis	42.3	18.3	63.3	24.7	29.1	10.1	4.9	6.1
Fagus grandifolia	36.6	60.5	27.9	62.1	38.8	55.1	56.3	58.5
Acer saccharum	15.9	12.8	4.9	5.3	19.4	18.8	36.6	31.7
Prunus serotina	2.0	4.2	.3	2.5	7.8	13.0	.7	2.4
Betula lenta	1.5	2.7	1.8	3.3	1.9	2.9	.7	. .
Acer rubrum	.7	1.0	1.4	1.6	. .	. .	. .	. .
Fraxinus americana	.7	.2	. .	. .	2.9	. .	.7	1.2
Betula lutea	.2	.2	.3	.4	. .	. .	. .	. .

frequent species are wild black cherry, sweet birch, yellow birch, red maple and white ash; tuliptree, cucumber magnolia, and basswood also occur. Within the forest local segregation and variations in relative abundance of the dominants result in the development of somewhat distinct communities which may be termed hemlock–beech, beech–hemlock–sugar maple, and beech–maple. The hemlock–beech community is most widespread, occurring on upland, on slopes, and in valleys. Locally, on slopes, sugar maple becomes abundant, resulting in the beech–hemlock–sugar maple type. The beech–maple community is very local, sometimes forming a strip between upland and slope forest. Nowhere does sugar maple assume the dominant role which it has in the Great Lake section, and nowhere does sugar maple reproduction form the unbroken layer so frequent in the Great Lake section and Superior Upland. *Viburnum alnifolium* is the most abundant shrub of this forest area, giving to the undergrowth an aspect not seen in the Great Lakes–St. Lawrence Division. The herbaceous plants are, with few exceptions (as *Tiarella cordifolia*), species widespread in the Hemlock–White Pine–Northern Hardwoods region. Among the most abundant are *Dryopteris spinulosa, Lycopodium lucidulum, Oxalis montana, Mitchella repens,* and *Maianthemum canadense.*

Forest essentially similar to the Tionesta Creek stand originally covered most of the rolling and moderately dissected areas of the plateau except the dry southerly slopes. Remnants of primary (but not virgin) forest are readily interpreted when comparison is made with natural areas. Evidence as to the nature of the original forest cover may also be found in old land surveys, in which surveyor's notes and witness trees give a fair idea of forest composition, and in this way the widespread distribution of the

Wild black cherry (*Prunus serotina*) is frequent in the forests of the Allegheny section, where it reaches large size; it occurs throughout the area of the Northern Appalachian Highlands. Allegheny National Forest, northwestern Pennsylvania. (Courtesy, U. S. Forest Service.)

beech–hemlock forest may be demonstrated (see Lutz, 1930b; Gordon, 1940).

That no difference is discernible between the prevailing forest of the unglaciated and glaciated sections of the plateau is apparent if the Tionesta Creek area (which is about 25 miles outside the glacial boundary) be compared with forests at and within the glacial boundary. The Big Basin area in Allegany State Park, New York, is a few miles within the glacial boundary. Within it a number of forest types are represented (Gordon, 1937; Emerson, 1937), of which the hemlock–beech forest is interpreted as representing "most nearly the natural climax forest of the region." This type occupies the lower slopes and grades into a beech–maple variant above (Table 82). These are mapped together as the prevailing forest of Cattaraugus County, two-thirds of which was glaciated (Gordon, 1940).

Table 82. Canopy composition of forest in Big Basin, Allegany State Park, New York: *1,* Lower slopes with regional climax; *2,* Beech–maple forest on northeast slope.

	1	*2*
Number of trees	101	130
Fagus grandifolia	42.6	63.1
Tsuga canadensis	28.7	.8
Acer saccharum	7.9	33.8
Acer rubrum	7.9	. .
Betula lutea	5.9	. .
Fraxinus americana	3.0	.8
Pinus Strobus, stumps	2.0	. .
Tilia americana	1.0	.8
Prunus serotina	1.0	.8

White pine is not now a constituent of the forest in either of the areas thus far discussed.[24] It was present on the upper slopes of East Tionesta Creek[25] and in small numbers in the Big Basin area, as attested by stumps. In Heart's Content (in Warren County, Pennsylvania, in the Allegheny National Forest) and in Cook Forest it is an abundant constituent of the forest (Tables 83, 84).

Intensive analysis of age distributions in East Tionesta, Heart's Content, Cook Forest, and many second-growth stands, reveals white pine to have originated

[24] White pine was formerly much more abundant in Pennsylvania than it is today. According to Spalding and Fernow (1899), "the best development of the White Pine is usually found along the water courses. Thus. . . . in Luzerne County. . . . along Bear Creek and its tributaries; in Clinton County. . . . on both branches of Hyner Run and along Youngwoman's Creek; in Clearfield County there were 20,000 acres along Sandy Creek and its tributaries heavily timbered with White Pine . . ."

[25] This is shown by Hough and Forbes (1943) in a diagram illustrating the distribution of forest communities. These white pine areas are beyond the limits of the Natural Area and have been logged.

The forest of Heart's Content, Allegheny National Forest, Pennsylvania. The large tree in the foreground is white pine. (Courtesy, U. S. Forest Service.)

as even-aged stands—or in a few cases as successive even-aged groups—as the result of catastrophes which practically wiped out or opened up the preceding stands. Droughts followed by forest fires are known to have been the catastrophic agents in some cases, windthrow in others (Hough and Forbes, 1943).[26]

[26] See also Lutz and McComb (1935) on origin of white pine in virgin stands in northwestern Pennsylvania.

The forests in which white pine is an abundant constitutent are very different in appearance from the more prevalent hemlock–beech type. The dominantly coniferous canopy results in paucity of herbaceous and shrubby growth.

The forest of Heart's Content is a beautiful and easily accessible example of the Allegheny Plateau forest in which white pine mingles with the usual dominants (Table 83). It is but five miles from the deep valley of the

Table 83. Composition of Heart's Content Forest, Warren County, Pennsylvania, in Allegheny National Forest: Columns *1* and *2*, abundance per cent of trees 10 inches d.b.h. and over in *1*, hemlock–beech association, and *2*, hemlock consociation (as given by Lutz, 1930); columns *3–7*, canopy composition of hemlock–beech, hemlock, hemlock–white pine, and intermediate communities (author's data).

	1	*2*	*3*	*4*	*5*	*6*	*7*
Number of trees, or*	163*	126*	76	48	161	49	114
Tsuga canadensis	36.1	81.6	22.4	95.8	50.9	26.5	11.4
Fagus grandifolia	24.0	3.4	30.3	. .	18.0	16.3	36.8
Pinus Strobus	11.1	5.7	9.2	2.1	23.0	53.1	43.9
Acer rubrum	10.6	1.5	21.0	2.1	3.7	2.0	7.0
Castanea dentata	8.8	. .	. .	. .	.6	. .	. .
Betula lenta	2.7	2.7	. .	. .	. .	. .	. .
Magnolia acuminata	2.1	. .	5.3	. .	1.2	. .	. .
Quercus borealis maxima	1.8	. .	2.6	. .	.6	2.0	. .
Quercus alba	1.6	. .	1.3	. .	1.2	. .	.9
Prunus serotina	.8	. .	5.3	. .	. .	. .	. .
Acer saccharum	.3	. .	1.3	. .	.6	. .	. .
Betula lutea	.1	4.2	. .	. .	. .	. .	. .
Fraxinus americana	. .	.8	1.3	. .	. .	. .	. .

* Number of sample plots of .1 acre each.

Allegheny River; however, within the confines of the virgin forest, slopes are gentle and the relief slight, on the whole quite comparable to the upland in the Tionesta area. Lutz (1930) distinguishes a hemlock–beech association and a hemlock association in the forest. In the hemlock–beech community, twelve species are represented in his largest size class (10 inches d.b.h. and over); of these, hemlock and beech comprise 60 per cent of the stand, and white pine, red maple and chestnut an additional 30 per cent. In the hemlock community only seven species are represented in the largest size class, and hemlock is clearly dominant. There is considerable localization of white pine in the hemlock–beech area; this results in a white pine–hemlock community and intermediate communities between this and the hemlock–beech. The climax nature of the hemlock–beech and hemlock types is indicated by similarity of canopy and second layer. The admixture

The hemlock consociation in the Heart's Content Forest. The large-leaved shrub in the foreground is *Viburnum alnifolium*. (Courtesy, U. S. Forest Service.)

of oaks and chestnut[27] suggests the transition from more southern types, a transition which is more noticeable in Cook Forest.

Cook Forest Park on the Clarion River in Pennsylvania is situated at the margin of the Hemlock–White Pine–Northern Hardwoods region. It is a strongly dissected area of steep ravine slopes and river bluffs, and rolling to moderately sloping upland with a variety of forest types including outstanding examples of hemlock–beech and hemlock–white pine forest and also of forest in which oaks (or oaks and chestnut) comprise approximately one-fourth of the canopy trees (Table 84). From a distance, its groves of white pine or of white pine and hemlock on steep slopes look very much like the equivalent community on the Black River in the Porcupine Mountains district of the Superior Upland (p. 374). From within the forest, certain pronounced differences are apparent: beech is abundant and sugar maple almost absent; *Rhododendron maximum* (an Appalachian species) is locally abundant; *Viburnum alnifolium* and *Ilex montana* (characteristic

[27] Lutz's studies (1930) indicate a greater abundance of chestnut than is apparent now.

Table 84. Cook Forest State Park, Clarion and Forest counties, Pennsylvania. Variations in canopy composition: *1*, Hemlock–beech; *2*, Hemlock–beech–red maple; *3*, Hemlock–white oak; *4*, Hemlock–white oak–white pine; *5*, White pine–hemlock; *6*, Hemlock–white pine–red maple.

	1	*2*	*3*	*4*	*5*	*6*
Number of trees	164	182	160	210	152	121
Tsuga canadensis	50.6	30.2	43.8	51.4	35.5	46.3
Fagus grandifolia	18.9	15.9	10.0	. .	9.2	5.0
Pinus Strobus	6.1	8.2	5.6	11.4	47.4	15.7
Acer rubrum	8.0	17.0	10.6	3.3	2.0	11.6
Quercus alba	5.5	8.8	18.1	20.0	3.9	3.3
Quercus borealis maxima	1.8	4.4	. .	2.9	.7	5.0
Quercus montana	. .	. .	. .	1.4	. .	. .
Castanea dentata	5.5	8.8	11.2	4.3	. .	9.1
Prunus serotina	2.4	1.1	. .	. .	. .	.8
Betula lenta	.6	4.4	. .	5.2	1.3	3.3
Magnolia acuminata	.6	. .	.6	. .	. .	. .
Nyssa sylvatica	. .	.6	. .	. .	. .	. .
Acer saccharum	. .	.6	. .	. .	. .	. .

shrub members of the eastern division) are seen. The herbaceous layer, however, is made up almost entirely of species ranging throughout the region. The hemlock–beech and its variant, hemlock–beech–red maple, occupy the most mesic situations, while a more mixed type seems to be transitional between this and the hemlock–white oak–white pine type of warmer (southeast and southwest) slopes. The white pine–hemlock and hemlock–white pine–red maple communities are developmental steps in which the large amount of white pine (which reproduces best in the open) is related to some destructive event, probably drought and fire. In both of these, the high proportion of hemlock and increase in beech in the second layer point to the ultimate replacement of white pine by the regional climax. Oaks are particularly abundant on or adjacent to the river slopes. Some secondary communities are dominantly oak (either white oak or chestnut oak). White pine is often associated with white oak in the young stands. The differences between Cook Forest and the areas previously mentioned are due in part to the large amount of white pine, in part to strong admixture of oaks, and in part to the almost complete absence of sugar maple. As this forest occupies a marginal position it has an admixture of trees common in the nearby regions to the south—the Mixed Mesophytic and Oak–Chestnut.

Everywhere along the southern margin of the region and along valleys indenting the margin some mingling of species from the adjacent regions is seen. In many instances this gives rise to a community resembling the mixed mesophytic. Along the slopes near the Allegheny River these appear to be extensions of, and more or less continuously connected with the Mixed

Mesophytic region. Such attenuated mixed mesophytic communities, although they lack the luxuriance and variety of more typical examples, appear rich in contrast to the regional and dry slope types. At the northern limit of their occurrence along the Allegheny River (in Allegany State Park and adjacent areas in Cattaraugus County, New York) the mixed communities occupy an intermediate position on slopes between the beech–maple below and the oak–chestnut above. The dominants include red oak, beech, chestnut, red maple, black birch, white ash, and black cherry, while cucumber magnolia, white oak, tuliptree, white pine, basswood, bitternut hickory, and sugar maple are less frequent constituents (Gordon, 1937; 1940).

Some suggestion of this mixture is occasionally seen in forests which, at first observation, appear to be typical hemlock–white pine–northern hardwoods. Admixture of the less tolerant species (as cherry, white ash, and white pine) may be due in part to local windfalls opening up the forest canopy, in part to death of old individuals, causing breaks in the canopy. Wherever the canopy contains a variety of species (thus yielding a diversified leaf litter), the ground layer contains more species and is more continuous than in the forest of fewer dominants. Forest understory may indicate trends toward the greater dominance of fewer species. Such features are well represented by a small but beautiful tract of forest at Lily Dale, New York (Table 85) where four species (hemlock, beech, cherry, and red maple) make up about 60 per cent of the canopy in which a dozen species are represented. The composition of the second layer points to the greater abundance of sugar maple and continued abundance of hemlock and beech in the future forest, with continuance of varied canopy. Logging in an

Table 85. Forest at Lily Dale, Chautauqua County, New York, illustrating a diversified forest canopy made up of very large and tall trees, of which hemlock surpasses other species in girth and white pine surpasses others in height. Trend toward greater dominance of sugar maple indicated by second layer.

	Canopy	*Second layer*
Number of trees	177	85
Tsuga canadensis	20.9	21.2
Fagus grandifolia	16.9	11.8
Prunus serotina	12.4	8.2
Acer rubrum	11.3	4.7
Acer saccharum	8.5	36.5
Pinus Strobus	8.5	. .
Magnolia acuminata	5.7	2.3
Quercus borealis maxima	5.7	1.2
Fraxinus americana	4.5	5.9
Betula lutea	2.8	4.7
Tilia americana	1.7	1.2
Carya ovata	1.1	2.3

adjacent tract has resulted in the establishment of a beech–maple community.

Farther east, in the Susquehanna drainage, northward extensions of oak and oak–chestnut forests reach into New York State and are represented on gravelly soils in the valley-head moraine belt south of the Finger Lakes. They are seen frequently in the Finger Lakes district where they may have entered from the south along Pleistocene drainage lines, or, by somewhat more roundabout course, from the plains of Lake Erie. Throughout this area chestnut was a common tree on warmer slopes and on gravelly soil, usually in company with red and white oaks, some hickory, sugar maple, hemlock, tulip, beech, and occasional other species. The disappearance of chestnut has converted some of these stands into "oak–hickory" forest. However, this "oak–hickory" forest is quite different in appearance (if accessory species and undergrowth be considered) from the true oak–hickory of the Mississippi Valley, or even that of the old valley floors of the Ridge and Valley section of the Oak–Chestnut Forest region. It resembles the oak forest of the Appalachian Plateau, but with fewer, although characteristic undergrowth species (*Vaccinium, Azalea, Gaylussacia, Gaultheria,* and *Hieracium venosum*), and occasional species more typical of the northern forest. Abrupt slopes and cliff margins may be occupied by pines, while along the limestone escarpment near the northern base of the Allegheny Plateau, arbor vitae is frequent. Diversity in topography and soils, and the admixture of southern species, gives to the Finger Lakes district more variety in forest aspect than is usual in the Hemlock–White Pine–Northern Hardwoods region.

Northward extensions of oak and oak–chestnut forest are seen along the Delaware River where they reach the southern margin of the Catskills. In the Hudson River Valley the southern influence is seen as far north as Lake George (see Map, p. 395). Southern outliers of the northern forest occur in these same valleys where extreme habitats are unfavorable to the regional vegetation.[28]

These upstream extensions of more southern forest types result in extreme indentation of the southern boundary of northern forest. The boundary of the region—drawn near the Allegheny Front—is thus not everywhere the vegetational boundary (see Map, p. 394). Away from the major drainage lines, or where the plateau is little dissected near its margin, the northern forest occupies the plateau to its southern margin (for example, on the plateau near Eagles Mere) or descends the Front in deep gorges. The forest of Ricketts Glen[29] in such a gorge near Red Rock, Luzerne County,

[28] Harshberger (1909) describes such an area on the Delaware River in Bucks County, Pennsylvania.

[29] Ricketts Glen is the gorge of Kitchen Creek which tumbles down the Allegheny Front in a series of waterfalls.

Pennsylvania, is an excellent example of the hemlock–northern hardwoods forest (Table 86). Here hemlock, beech, basswood, sweet birch, and sugar maple are the most abundant species, comprising nearly 80 per cent of the canopy. The admixture of oaks (red and white), tuliptree, chestnut, and hickory is related to the marginal position of the area. A hemlock consociation occupies areas on the stream flat; a hemlock–yellow birch community occupies steep slopes in narrow gorges. The herbaceous vegetation is typically that of the northern forest, with *Lycopodium lucidulum, Dryopteris spinulosa, Oxalis montana, Maianthemum canadense, Viola rotundifolia,* and *Aster acuminatus* prominent among the many northern species.

Table 86. Composition of forest in Ricketts Glen State Park, Luzerne County, Pennsylvania, on the Allegheny Front. Canopy and second layer of *1*, hemlock–hardwoods community; *2*, hemlock consociation of flats.

	1		*2*	
Number of trees	114	46	95	35
Tsuga canadensis	29.8	34.8	70.5	45.7
Fagus grandifolia	15.8	15.2	8.4	8.6
Tilia americana	13.2	10.9	8.4	14.3
Betula lenta	11.4	10.9	2.1	11.4
Betula lutea	5.3	6.5	3.2	20.0
Acer saccharum	8.8	4.3	5.3	..
Quercus borealis maxima	4.4	4.3	..	..
Liriodendron tulipifera	3.5	4.3	..	..
Fraxinus americana	2.6	..	2.1	..
Castanea dentata, stumps	1.7	..	..	..
Acer rubrum	.9	2.2	..	..
Magnolia acuminata	.9	..	..	..
Carya ovata	.9	..	..	..
Quercus alba	.9	..	..	..
Ostrya virginiana	..	4.3	..	..
Carya sp.	..	2.2	..	..

A general picture of the vegetation of the Allegheny Plateau and Allegheny Mountains must include the rock vegetation and the bogs. Exposures of the resistant Pottsville and Pocono sandstones occur along the Allegheny Front and in gorges cut in the plateau. Xeric cliff margin communities are seen, with scattered and gnarly oaks, with red pine or pitch pine, and perhaps white birch, white pine, and sweet birch, with crevice herbs and shrubs—mountain laurel, sweet fern, blueberries—and lichen-covered sandstone and conglomerate blocks. Shaded and moist exposures in gorges afford suitable habitats for a variety of rock plants. Communities on exposures of calcareous rocks, as in the Finger Lakes district, differ in specific content.

Bogs occur frequently on the Glaciated Allegheny Plateau but are by no means confined to the glaciated area. Some are located in the area of older drift south of the Wisconsin glacial boundary, some farther south in the Allegheny Mountains in outliers of the northern forest in southwestern Pennsylvania and Maryland.[30] Black spruce, tamarack, and balsam fir are constituents of the bog forest in the north; southward these species may drop out, and white pine, hemlock, red maple, and sour gum occupy the bog borders. *Chamaedaphne, Ledum,* and *Nemopanthus* are common bog shrubs. Species of *Vaccinium, Viburnum, Spiraea,* and southward, *Rhododendron maximum* and *Kalmia latifolia* are generally present in the bog border. The bogs are less extensive and generally a less conspicuous feature of the landscape than in the Great Lakes–St. Lawrence Division where a glacial topography dominates. Large cattail and Iris marshes, and elm–silver maple forest in the glacially filled valleys of the northern margin of the plateau are extensions of the swamp communities of the adjacent lake plain.

Catskill Mountains

In the Catskill Mountains, the greater relief and greater range of elevation result in even greater diversity of vegetation than is seen in dissected parts of the Allegheny Plateau (except the Finger Lakes district). Deciduous forest, or deciduous forest with an admixture of conifers (hemlock, and locally, white pine at lower elevations, red spruce and balsam fir at higher) prevails over most of the Catskill Mountains to an elevation of about 3500 or 3700 feet. Oak or oak–hickory forest (formerly oak–chestnut) occupies the lowest elevations.[31]

The forests of the foothills and lowest slopes have much in common with those of the Glaciated section of the Oak–Chestnut region (p. 248) which is contiguous to the southeast. The regional boundary lies somewhere in the foothills belt. Tongues of oak–chestnut forest reach upstream and upslope (to beyond 1500 feet in Kaaterskill Clove) penetrating hemlock–hardwoods communities which usually prevail at moderate elevations. White oak is generally the dominant tree of the low elevation oak forest. With it, in varying abundance related to topography, are red oak, chestnut oak, black oak, scarlet oak, hickories, chestnut (now represented only by old snags and occasional sprouts), sugar maple, white ash, basswood, butternut, and white pine. On low ridges or shale slopes, the chestnut and scarlet oaks become abundant, and white oak may be absent; pitch pine is local. As in the Hudson Highlands (p. 256), some of the communities of the lower mesic slopes are essentially mixed mesophytic, containing beech, chestnut, red

[30] A southern outlier, Cranberry Glades in West Virginia is located in the Mixed Mesophytic Forest region (p. 480).

[31] Literature on the vegetation of the Catskill Mountains is scant. Dwight (1820), Harshberger (1905), and Bray (1930) give some general information which is used here to supplement the writer's observations.

oak, sugar and red maples, basswood, hickory, walnut, butternut, and hemlock. The abundant mountain laurel and the heath layer (of *Kalmia, Azalea, Vaccinium, Gaultheria,* and *Epigaea*) in secondary woods give a pronounced Appalachian aspect to the vegetation. The forest cover may be interrupted by natural openings occupying shallow soil over rock, "the ledges," where herbaceous and shrubby communities border scrubby sprout forests of chestnut oak, scarlet oak, and scrub oak. While the distinctive flora of such openings indicates their primary nature, they have doubtless enlarged as a result of forest cutting.

Above about 1200 feet (except in south-facing valleys) the dominants are sugar maple, beech, yellow birch, hemlock, and white pine. Here, as in the Allegheny Mountains and Allegheny Plateau, a hemlock consociation occurs along the mountain streams. White pine is usually found in second-growth stands on southerly slopes, and is commonly intermixed with the hardwoods on the plateau in the upper Schoharie Valley. Secondary communities of gray birch, aspen, pin cherry, red maple, and ash cover many of the slopes. In general, the hardwood forest of moderate elevations (now represented by old secondary and partially cut-over forests, and remnants of primary stands) is one in which sugar maple appears to be most conspicuous, and with which are intermixed beech, yellow birch, basswood, red oak, butternut, white ash, cherry, and hemlock. Red oak alone of the oaks is present here, but not as abundant as at lower elevations. In the best sites, this hardwood forest has a very luxuriant and varied herbaceous layer dependent on the rich mull humus. Due to the mixed character of these slope forests and their luxuriant undergrowth, the aspect is that of more southern forests, rather than that of the north woods. In character, as in geographic location, they are intermediate.

Sometimes as low as 1800 feet, more often at about 2000 feet, red spruce and balsam fir enter, giving rise to a transition forest. This forest type in which maple, beech, yellow birch, hemlock, white pine, red spruce, and balsam fir are the dominants, is a distinctive feature of the Northern Appalachian Highland Division of the Hemlock–White Pine–Northern Hardwoods region. Southern outliers occur south of the region, for example, at higher elevations in the Allegheny Mountains of West Virginia, in the Mixed Mesophytic region (p. 81). Hobble-bush (*Viburnum alnifolium*) and yew are abundant shrubs of this forest belt. The shrubby and herbaceous plants are for the most part species common to the more northern forests of the region—the Adirondacks and New England, the Superior Upland and Upper Michigan district of the Lake section. Many of them are plants of the spruce–fir forest. At higher elevations, the proportion of spruce and fir increases, and the deciduous species, except yellow birch, gradually drop out.

The highest peaks of the Catskills rise above the mixed or transition forest. Pure stands of spruce and fir, dwarfed and gnarly because of wind exposure, cover the highest slopes. This altitudinal belt (the "Canadian Zone") is poorly represented in the Catskills; in the Adirondacks, it is much more extensive and better developed.

ADIRONDACK SECTION

Much of the area north of the Mohawk Valley is characterized by the presence of red spruce as one of the dominant trees of the forest. This area, together with the surrounding lowlands (where red spruce is absent) is included in the Adirondack section. Physiographically, then, this section is more extensive than the physiographic province of the same name; it includes in addition the Tug Hill Plateau[32] to the west of and separated from the Adirondacks by the Black River Valley; the St. Lawrence Valley and Ontario Lake plains to the north and west; and to the east, that part of the Hudson-Champlain Valley north of the limits of the Oak–Chestnut region. In general it may be thought of as made up of two upland areas (Adirondack and Tug Hill) surrounded by lowland. In the St. Lawrence Valley, this section merges with the Laurentian section of the Great Lakes–St. Lawrence Division.

The entire area is profoundly modified by glaciation. The eastern part of the Adirondacks is truly mountainous with two summits (Mt. Marcy and Mt. McIntyre) above 5000 feet and fourteen others above 4000 feet. The western part of the Adirondacks and the Tug Hill Plateau are dissected plateaus of about 2000 feet elevation. The margins of the Tug Hill Plateau are cut by many deep and narrow gorges. The valleys of the western Adirondack Plateau are often wide and deeply filled with glacial material; much of the intervening surface is rolling, and slopes only moderately steep. The Adirondacks are remarkable for the large number of lakes and ponds which are most abundant in and near the mountains. Extensive proglacial lake deposits are correlated with the retreat of the last ice sheet which disappeared from most of the upland while it still encircled it in the surrounding valleys. These, and the deltas of that date are covered with gravel, sand, and silt. The elevated "benches" so generally occupied by deciduous or mixed deciduous–coniferous forests and the "spruce flats" bordering on lakes and streams owe their origin to the proglacial lake deposits.

The *lowlands* of the section are occupied by extensions of the oak–chestnut forest and of the hemlock–hardwoods forest which prevails over the northern Allegheny Plateau. Their vegetational features ally them with

[32] Tug Hill Plateau is a part of the Mohawk section of the Appalachian Plateaus Province of physiographers. It is geologically distinct from the Adirondacks in that its underlying rocks are Paleozoic sedimentaries.

more southern areas; in the absence of red spruce they are distinct from the uplands whose vegetation may be considered as characteristic of the section.

On sandy soils of delta and lake deposits and on old dunes, gray birch and aspen, jack pine[33] or pitch pine and heath shrub communities suggestive of the Great Lake section are seen. These communities in part at least occupy land once covered by white pine forests. Young secondary white pine stands are not infrequent where commercially important white pine areas formerly occurred.[34]

In wet areas along lakes or slow streams, extensive cattail marshes and Iris–sedge communities are conspicuous with here and there a marginal forest community of silver maple, elm, and cottonwood. Remnants of elm, ash, and red maple woods are indicative of lowland swamp forests (which originally contained arbor vitae, white pine, and hemlock as well). Tall white pines are frequently seen projecting above the secondary hardwood swamp forests. In the Champlain Valley, the admixture of oak in the swamp forest is conspicuous.

The mesic sites of the Ontario–St. Lawrence lowland (most of which have been utilized agriculturally) were originally occupied by hemlock–white pine–northern hardwoods forest. Occasional oaks and hickories represent the northernmost extension of the oak forest element prominent in the Finger Lakes district and on the plains south of Lake Ontario. In the Champlain Valley, oak–hickory communities are frequent on the rolling plains. White oak and shellbark hickory are dominant; with these are red oak, black oak, bur oak, elm, ash, basswood, and occasional sugar maple. Here is an attenuated extension from the northern arm of the Oak–Chestnut region in the Hudson Valley. The regional type, with beech, sugar maple, and basswood dominant, is represented on slopes rising above the rolling plains. Farther north, in Quebec, the oaks and hickories occur with sugar maple and beech in communities which Dansereau (1946) defines as "quasiclimax."

More characteristic of the Adirondack section are the *uplands* (including the valleys of streams dissecting the upland) which support what may be termed a mixed or spruce–hemlock–hardwoods, or spruce–hardwoods forest. This varies in composition in relation to topographic situation (soil and moisture) and in relation to altitude. The principal species are red spruce, balsam fir, yellow birch, sugar maple, and beech, with an admixture of hemlock and white pine.

---

[33] Jack pine is found in abundance in the eastern division of the northern forest only on the Ontario–St. Lawrence plain adjacent to the western Adirondacks, an area transitional to the Laurentian section.

[34] These are shown on the Ontario–St. Lawrence plain and along Lake Champlain on maps by Spalding and Fernow (1899), and Hawley and Hawes (1912).

Tug Hill

The vegetation of Tug Hill is more or less intermediate in character between that of the Allegheny Plateau to the south and the Adirondacks to the east (Hotchkiss, 1932). In the numerous gorges along streams indenting its margin, rock outcrops are frequent, and communities similar to those of the northern escarpment of the Allegheny Plateau are seen. Swamps and bogs are common on the poorly drained summit of the plateau. On all the better drained slopes of the plateau, spruce–hardwood forest prevails, now modified by logging. The virgin climax forest was described by Hotchkiss as

> a beautiful stand of yellow birch, beech, sugar maple, red spruce and hemlock . . . with a dense ground cover of wood sorrel, wild sarsaparilla (*Aralia nudicaulis*), shining club moss (*Lycopodium lucidulum*) and other species. The forest was in general quite open and, except where the crown was more open than usual, hobble-bush and other shrubs were less common than in cut-over forests . . . Many of the trees—except red spruce—had a diameter of three feet or more; the red spruce probably averaged two feet and was equally tall and straight with the others.

The occurrence of red spruce in the mesic forests allies this area with the Adirondacks, although physiographically, it is a part of the Appalachian Plateaus Province.

The Adirondacks

The principal topographic situations of the Adirondacks (exclusive of subalpine and alpine summits) include: (1) swamps and bogs—wet lands bordering slow streams and lakes or occupying the partially filled depressions and low spots in the uneven glacial deposits, (2) moist lower slopes and level to rolling flats in the valleys and around lakes and swamps, (3) gentle slopes and terraces and benches above the flats (the better drained and deep soil areas), and (4) steep slopes with thin, stony soil and the higher mountain slopes.[35] Correlated with these four situations four major forest types have been recognized: (1) swamp forest, (2) spruce flats or mixed wood, (3) hardwood forest, and (4) spruce slope or upper slope forest (Table 87). Red spruce is an abundant constituent of all four. Variations within each type result in a number of distinct communities. The foresters' figures based upon strip counts in each of the several habitats do not emphasize the differences between the four major forest types. Small knolls in the swamp are not occupied by a swamp community but usually by groups of trees more properly referable to the spruce flats. Low spots in

[35] Material for this paragraph is taken largely from Graves (1899) and from McCarthy (Mimeo., 1945).

Table 87. Generalized forest types of the western Adirondacks: *1*, Swamp type; *2*, Spruce flat; *3*, Hardwood; *4*, Spruce slope; *5*, The forest as a whole. (Data from Graves, 1899, based on strip counts in Nehasane Park in Hamilton and Herkimer counties.)

	1	*2*	*3*	*4*	*5*
Number of acres on which counts were based	225	106	442	274	1046
Number of trees per acre	70.93	64.45	78.71	70.17	73.44
Picea rubens	47.94	45.00	36.84	48.45	42.77
Betula lutea	18.33	19.71	19.06	19.52	19.06
Fagus grandifolia	3.38	8.69	20.84	10.26	13.62
Acer saccharum	3.38	3.88	12.83	5.70	8.30
Tsuga canadensis	7.19	8.69	5.08	7.42	6.26
Abies balsamea	13.25	9.00	2.16	4.56	5.72
Acer rubrum	4.23	4.65	2.92	3.56	3.54
Pinus Strobus	.51	.15	.06	.42	.24
Fraxinus nigra	.90	.15	.04	.02	.22
Thuja occidentalis	.85	. .	. .	. .	.16
Prunus serotina	.04	.08	.17	.09	.11

the spruce flats contain swamp trees; on the more pronounced slopes of the rolling surface, the proportion of hardwoods increases. Rugged spots, areas of shallow soil, or low shoulders in the hardwood land have a higher proportion of spruce and birch, and narrow valleys a higher proportion of hemlock than the more representative areas; the best situations (deep and rich soils) have the highest proportion of sugar maple and beech. On the steeper slopes with thin soils and at higher elevations, beech and maple cannot successfully compete with spruce, and spruce becomes dominant, with yellow birch a more or less frequent species.

These four forest types are briefly characterized by McCarthy (1945) in a report to the New York section of the Society of American Foresters as follows:

The swamp type (1) consists of two distinct sub-types—the black spruce and the spruce–balsam mixture. There are occasional spots in which white cedar may be found in practically pure stand, probably due to the occurrence of lime. The mixed-wood type (2) consists primarily of variations in percentage of the following species—red spruce, balsam fir, hemlock, yellow birch, and red maple. These are the five outstanding species. White pine may occur as individuals or small groups, chiefly along the swamp edge.[36] Cedar [Thuja] and black spruce

[36] Spalding and Fernow (1899), in discussing the distribution of white pine state: "In the Adirondacks the pine, now almost entirely removed, fringes with the Spruce and Balsam Fir the many lakes and water courses and keeps to the lower altitudes; mixed in with the Maples, Birches, Beech, and Spruce, it towers 50 to 60 feet above the general level of the woods, with diameters of 30 to 40 inches. Its reproduction under the shade of its competitors, however is prevented, young pine being rarely seen except on old abandoned openings in the forest."

may both appear in this type. The percentage of softwoods is greatest adjoining the swamp and gives way to the hardwoods as the drainage improves up the slope. . . . In the western Adirondacks, I have been accustomed to fix the line of demarcation between this type and the hardwood type (3) where beech appeared and balsam stopped. If they overlap, it is usually in ravines where the balsam extends further up the slope and on ridge spurs where the hardwood extends downward. There is also a distinct separation of these two types in the condition of the upper soil layers, since there is a deep humus layer in type 2 and a leaf litter in type 3. The occurrence of mor, or duff as it was formerly known, is limited quite sharply to the mixed-wood type. . . . The line of demarcation between the hardwood type (3) and the upper slope type (4) will be sufficiently clear since beech and sugar maple with some mixture of yellow birch will become stunted in form and softwoods will take their place.

In the Adirondack section Heimburger (1934) developed a "forest-type classification" based upon ground vegetation, following the plan devised by Cajander (1909). "Ground vegetation as expressing biologically equivalent sites is used as a basis for subdivision of the forests into types." Such types are not equivalent to "cover types" which refer to forest canopy. Heimburger divides the Adirondack vegetation into twenty-two forest types arranged in three forest type series: the subalpine series (including three types), extending from 3200 to 4900 feet; the western series (including nine types) "which occupies the major part of the Adirondacks"; and the eastern series (including ten types) which "occupies the valleys of the extreme eastern part of the Adirondacks." Examples only can be mentioned here: the *Hylocomium–Cornus type* in the subalpine series; the *Vaccinium–Gaultheria type* (a softwood or pine forest type) and *Viburnum type* (a hardwood type) in the western series; the *Vaccinium–Myrica type* (a softwood type limited to outwash plains) and the *Dicentra type* occupying the edaphically most favored localities at the lower elevations in the eastern Adirondacks—sites where rich hardwood forest is found.

TOPOGRAPHIC AND ALTITUDINAL VARIATIONS

No one of the four principal topographic situations of the Adirondacks is sufficiently uniform to be occupied throughout by one of the four major forest types. Minor variations in topography such as small knolls and depressions, variations in steepness or exposure, and in depth of soil accumulation affect composition. Migration and infiltration from contiguous areas have resulted in some differences between the east and west, the north and south parts of the Adirondacks. The influence of altitude is strikingly displayed in the higher rugged eastern half of the Adirondacks.

"Swamp forest," while possessing a certain unity throughout, nevertheless varies greatly from place to place. This is a dominantly coniferous type, and where swamps are extensive, the great areas of spruce (or spruce and balsam fir or arbor vitae or larch) emphasize the boreal aspect and boreal relations of this region. Many areas of spruce swamp forest (really bog forest) are extremely dense; again, the trees may be rather widely placed or clustered in dense groups, the intervening spaces being occupied by bog shrub, bog sedge, or Sphagnum communities. Wherever there is sufficient light in a boggy forest, a dense shrub layer is the rule. Among the common shrubs are *Ledum groenlandicum, Chamaedaphne calyculata, Kalmia polifolia, Aronia melanocarpa, Nemopanthus mucronata,* and species of *Vaccinium* and *Viburnum*. Cranberry and snowberry creep over the Sphagnum mats; cotton-grass (*Eriophorum*) dots the open areas; pitcher plants are frequent. In the drier spruce swamps, hummocks of *Cladonia* and *Hypnum* replace the Sphagnum, and forest herbs as *Cornus canadensis* and *Maianthemum canadense* appear. While black spruce is the most frequent species of the swamp forest, often occurring in pure stands, balsam fir may be associated with it. Occasional yellow birch, red maple, and hemlock enter into the spruce forest, especially along its margins. Arbor vitae or cedar (*Thuja occidentalis*) is far more local in its occurrence than in the Lake section; so also is larch or tamarack (*Larix laricina*). Both of these species are sometimes abundant in the secondary forest of slopes adjacent to the swamp. The occurrence of the former has been correlated with the presence of lime in the soil. Near water courses in the swamp, sedges and grasses prevail, and shrubby willows are common.

Spruce swamps occur over a wide range of altitude, from the lowest basins of the western Adirondack Plateau, to the wet spots in the dwarf subalpine forest. In at least some of the low sites, there may be a temperature correlation as some of these basins are frost pockets.[37]

In general, the less boggy borders of swamps, and the low but not boggy flats along lakes and streams, the "spruce flats," are occupied by forest in which red spruce and yellow birch, or red spruce, yellow birch, and hemlock, are the principal species (about 75 per cent of the forest as a whole). Nevertheless, within an area of "spruce flats" one may see small boggy pockets where spruce is dominant, knolls occupied by hardwood forest in which sugar maple and beech are dominant, and several intermediate types. Variations in a spruce flat forest near Wanakena are shown in Table 88. Local hardwood communities on the spruce flat have a higher percentage of beech and sugar maple (65 to 85 per cent) than is shown in the "hardwood type" (about 34 per cent) where local communities are not distinguished in strip counts (as in Table 87). The ecologist will see each major type as made up of a mosaic of distinct communities with intervening transitions.

Just as canopy composition in the spruce flat varies from place to place, so also does undergrowth. The spruce–yellow birch forest (which may be

---

[37] The susceptibility of young fern fronds to frost and the resultant conspicuous blackening may clearly demonstrate the location of frost pockets if areas are visited soon after a summer frost.

thought of as typifying the spruce flat) has a luxuriant undergrowth in which *Viburnum alnifolium* among the shrubs, and *Dryopteris spinulosa, Oxalis montana,* and *Maianthemum canadense* among the herbs, are most frequent. The ground cover is essentially continuous, and the shrub layer patchy. This is the most beautiful forest type of the Adirondacks—a forest of coniferous and deciduous species, the ground well carpeted, the forest floor generally hummocky with moss-covered logs and boulders bedecked

Forest of red spruce, hemlock, and yellow birch with some white ash. Beneath, balsam fir, striped maple, and hobble-bush, with spinulose shield-fern and a ground cover of Oxalis. (Courtesy, U. S. Forest Service.)

with *Oxalis* and *Maianthemum,* and low spots crowded with ferns. With increase in beech and sugar maple, the fibrous mor of the mixed forest gives way to a mull-like humus and abundant leaf litter. The ground cover is more broken, and the smaller plants less frequent. Instead, *Trillium, Erythronium, Dicentra,* and other of the hardwood forest species appear.

Table 88. Variations in forest canopy composition within an area of the "spruce flat" type near Wanakena, New York: *1,* Typical mixed forest; *2,* Transition to hardwood type; *3,* Maple–beech community. Compare with *2* of Table 87. (Data based on strip counts within the local communities.)

	1	*2*	*3*
Number of trees	113	109	110
Picea rubens	38.9	13.8	8.2
Betula lutea	41.6	13.8	4.5
Fagus grandifolia	2.6	20.2	25.4
Acer saccharum	. .	44.0	60.9
Tsuga canadensis	4.4	5.5	.9
Acer rubrum	12.4	2.7	. .

The "hardwood forest" of the Adirondacks is essentially a sugar maple–beech forest in which red spruce, yellow birch, hemlock, cherry, basswood, white ash, and white elm may occur. In the northern and western parts of the section few of these accessory species, other than spruce and yellow birch, occur in the hardwood forest; maple and beech there attain a high degree of dominance. Sometimes beech, sometimes maple prevails, or the two share dominance. In the eastern and southern parts of the section, the admixture of additional species is pronounced. Some of the richest and most varied hardwood forests occur on slopes of valleys on the eastern slope of the Adirondacks. There, red oak also may appear in the hardwood forest.

Local variations in the hardwood forest type are as well marked as in the spruce flat type. While hardwood communities prevail areally in the type, mixed forest communities (spruce–yellow birch, or spruce–yellow birch–hemlock) frequently interrupt the hardwood stand. The composition of the prevailing hardwood forest of much of the Adirondacks is well represented by the local hardwood areas on the spruce flat. The hardwood forest frequently occupies the low benches bordering lakes. In the mountainous eastern half of the Adirondack section, these benches are less extensive than farther west. Where the benches are indented by ravines, the hardwood forest is interrupted by mixed forest communities, by spruce or spruce–hemlock groves, and occasionally by bands of arbor vitae bordering mountain brooks. The frequent alternation of types on hardwood benches is well illustrated near Heart Lake south of Lake Placid.

Except for the admixture of red spruce (which may be absent entirely) and the generally abundant *Viburnum alnifolium,* the hardwood type of the Adirondack section is essentially the same as the beech–maple community of the Great Lakes–St. Lawrence Division of the region.

Slopes rising above the hardwood benches, or ridges or steep slopes at or below the level of the benches—thin soil or rocky areas—are occupied by the "spruce slope" type. Red spruce in this habitat does not attain the large size of individuals growing on the spruce flats or in the hardwood forest. From a distance such areas appear as dark patches interrupting the hardwood forest. In the more mountainous section, the areal dominance of the spruce slope type is seen to increase at higher elevations, finally entirely replacing hardwood forest. At these higher elevations (about 3200 to 4200 feet), spruce alone is dominant; with it occur yellow birch, balsam fir, and occasional paper birch. Fir, mountain maple (*Acer spicatum*), and mountain ash (*Sorbus americana*) are abundant in the understory; *Cornus canadensis, Clintonia borealis,* and *Maianthemum canadense* are plentiful.

Higher, fir becomes more abundant, and spruce finally drops out. On Mt. Marcy, the fir forest extends from 4250 to 4900 feet (Adams, et. al., 1920).

Heart Lake and Mount MacIntyre. Hardwood forest interrupted by groups of conifers occupies the gentle slopes around the lake. Higher, on the thin soil of steep mountain slopes, red spruce is dominant (the "spruce slope" forest type). The summit of Mt. MacIntyre is treeless and occupied by alpine vegetation. (Courtesy, U. S. Forest Service.)

A forest of red spruce with birch and balsam fir covers the steep slopes above Loch Bonnie. (Courtesy, U. S. Forest Service.)

The spruce and fir zones (3200 to 4900 feet) are classed as subalpine by Heimburger (1934).

The hummocky forest floor at the higher elevations is carpeted with mosses, liverworts, and lichens (*Sphagnum, Hylocomium, Hypnum, Polytrichum, Dicranum, Bazzania,* and *Cladonia*). The herbaceous plants are chiefly those common at lower elevations, as *Cornus, Maianthemum, Clintonia, Oxalis,* etc. *Coptis groenlandica* and *Streptopus amplexifolius* become more abundant. *Dryopteris spinulosa* var. *dilatata* replaces the less wide-leaved varieties common in the lower forests.

Timber line is reached at about 4900 feet. At timber line, the trees are dwarfed, forming a typical Krummholz. Mats of shrubs—*Ledum, Kalmia,*

Vaccinium—occupy spaces between the dwarf firs, while beneath or between the shrubs are dwarfed individuals of the herbaceous species of the subalpine forest.

In the alpine belt, above 4900 feet, dwarf shrubs and herbs prevail. Patches of grass–sedge alpine meadow and of dense dwarf shrub mats occupy all areas with sufficient soil. Lichens (*Cetraria islandica* and *Cladonia rangiferina*) are abundant. Rock crevices and interstices between the granite blocks support an alpine vegetation including some dwarfed individuals of lower elevation species, and some more characteristic species. "Characteristic alpine plants of Mt. Marcy are *Empetrum nigrum, Rhododendron lapponicum, Diapensia lapponica, Solidago Cutleri* and *Vaccinium caespitosum*" (Adams, et al., 1920).

Some departure from the general features of the several forest types as presented here will be noted in the southern and eastern parts of the Adirondacks. For example, the infiltration of more southern species is pronounced in the vicinity of Sacandaga Reservoir and Lake George.[38] The elevations here are low (about 400 to 700 feet). Species from the oak–chestnut forest to the south mingle with the northern hardwoods, or locally replace them. Oak groves (white oak and red oak), or oaks and white pine, with some sugar maple and hickory are seen on gravelly benches. Lower slope forests of white oak, red oak, chestnut oak, basswood, ash, butternut, red maple, hickory, white pine, sugar maple, beech, and occasional remnants of chestnut, are quite different from the less varied hardwood forest of most of the Adirondack area. The prominence of oaks, ash, cherry, and red maple in young secondary stands and the frequent white pine groves further add to the contrasts between the southern margin of the Adirondacks and the interior. The oaks (except red oak) and the chestnut do not extend far into the Adirondacks.

NEW ENGLAND SECTION

This section occupies all of New England north of the Glaciated section of the Oak–Chestnut region.[39] It extends northward into southern Quebec and the Maritime Provinces of Canada, there including the "Acadian Forest region" of Halliday (1937). On the Gaspé Peninsula, it comes in contact with the Spruce–Fir or Boreal Forest. In fact, some areas within this section are outliers of the Spruce–Fir Forest. It is, except at the south, approximately coextensive with the New England physiographic province and extends beyond it northward and eastward. A number of more or less distinct physio-

[38] Bray (1930) has referred to this as "a 'thinned out' extension of Zone B," which is shown in his map of New York as extending northward almost to Lake George (see Map, p. 395).

[39] The boundary between these two sections has already been discussed (p. 251).

graphic areas are included here: the greater part of the New England Upland (except in Connecticut); the Seaboard Lowland from northern Massachusetts northward; the Green Mountains and adjacent higher eastern parts of the Taconic Mountains; the White Mountains and other mountain groups included in the White Mountain physiographic section (the White Mountains proper, Franconia Mountains, Caledonia Mountains, Boundary Mountains, and Katahdin group) together with the intervening and surrounding plateau.

On a basis of prevailing forest types, two principal subdivisions of the New England section should be recognized—the hemlock–hardwoods and the spruce–hardwoods (cf. p. 395). The hemlock–hardwoods forest (with its white pine subclimax representative) covers (or did cover) much of the New England Upland (except in Maine) and the part of the Seaboard Lowland in northern Massachusetts, New Hampshire, and southwestern Maine. The spruce–hardwoods and spruce–fir forest types clothe the mountain areas (except their lower slopes) and much of Maine (eastern end of Seaboard Lowland and most of the New England Upland of that state). Red spruce is an occasional constituent of communities toward the northern part of, or near mountains in the hemlock–hardwoods area; white oak occasionally, and red oak more often extend northward into the spruce–hardwoods area, but there they are confined to low elevations.

The hemlock–hardwoods forest is best exemplified in central New England, the spruce–hardwoods forest in northern New England and eastern Canada. The hemlock–hardwoods area is somewhat comparable in extent to the "white pine region" of Lunt (1938). The soils in the hemlock–hardwoods area are more or less transitional between the gray-brown soils of the deciduous forest to the south and the podzols of the coniferous forest to the north. The soils in the spruce–hardwoods area are strongly podzolized except locally where hardwoods dominate (Lunt, 1938).

The Hemlock–Hardwoods Area

This is a hilly area, locally quite rugged. Low sandy plains border some of the streams. Small depressions occupied by ponds, or by bogs and swamps are frequent in glacially obstructed valleys and in the uneven drift of the upland. The New England Upland part of the area is a more or less maturely dissected upland (upraised peneplain) ranging in elevation from about 1000 to 2000 feet (1500 feet at the border of the White Mountain physiographic section). Stream valleys indent the upland, and monadnocks[40] rise above it. The Seaboard Lowland is, as its name suggests, an area of generally lower elevation adjacent to the coast. The southern part of this

[40] The highest of these monadnocks is Mt. Monadnock in southern New Hampshire, which rises to an elevation of 3166 feet from a plain of approximately 1100 to 1200 feet. Red spruce occupies its slopes from 2300 feet upward, thus producing a small outlier of the prevailing mountain vegetation to the north.

lowland in Rhode Island and southern Massachusetts, and the Boston Basin are included in the Oak–Chestnut region. Near the New Hampshire border, the alternation of gray birch–red cedar and gray birch–white pine old-field communities marks the transition to the hemlock–hardwoods forest which occupies the lowland northward. Here also oak woods are frequent, recalling the oak stands of southern New England.

The hemlock–hardwoods forest, although most widespread in central New England, nevertheless sends tongues northward along the Connecticut Valley and around the Green Mountains, and northeastward into Maine as far as the Bangor Lowland.

When the Pilgrims came to this continent, New England was covered by forest interrupted only where lakes or bogs and river swamps made tree growth impossible; where sand deposits near the coast were unsuitable for closed stands; where fire or windfall had temporarily destroyed the forest; where Indians had burned the forest (especially near the coast); and where rock outcrops occurred in the more rugged sections. Clearing for settlement began at once. The first sawmill in New England was built in 1623 in southwestern Maine (near York), and by 1650 sawmills followed settlement throughout New England. Cutting and burning went on apace for 200 years in an effort at agricultural utilization and expansion, especially through southern and central New England. Then farm abandonment commenced. The lumber industry, meanwhile, began to center farther north. This reached its peak about 1850, then rapidly declined. The greatest period of forest devastation followed the introduction of portable sawmills, which were taken into areas where the earlier cuttings had been more or less selective, and into areas where only small stands remained. Little escaped. The pulp and paper industry farther north was even more destructive. In three centuries, the virgin forest of New England had been reduced from 95 per cent to 5 per cent of the total area (Hawley and Hawes, 1912; Hawes, 1923).

The omnipresent second-growth stands occupy the abandoned fields and pastures, the cut-over and cleared forest lands. Early in this century only a few tracts of virgin forest remained of sufficient size to be worthy of the name forest. Fortunately records are available concerning the composition of a few of these. Smaller areas, mere fragments of forest, still remain to demonstrate the former widespread occurrence of the sort of forest illustrated by the few virgin areas to be considered.

VIRGIN FORESTS

In 1913, Nichols mentioned four areas of virgin forest in northwestern Connecticut, an area of uneven topography with relatively high elevations in or adjacent to the southern end of the Berkshires and Taconic Mountains. These are near the southern margin of the forest section under consideration.

The largest of these virgin forests, at Colebrook, was described in general terms by Hawes (1909) as "a mixture of immense hemlock, beech, yellow birch, sugar maple, fine black cherry, ash, chestnut and oak, with a few giant white pines, and represents the most perfect mixture of the northern and southern New England forest types that the writer has ever seen." Lumbering began in this tract in 1912, and it was doubtless with this in mind that Nichols wrote:

> The ecologist . . . sees in such a group of trees the glorious consummation of long centuries of slow upbuilding on the part of Mother Nature. . . . They represent the survivors of that keen competition and relentless struggle for existence to which their less fit comrades of earlier years have long since succumbed. To precipitate their downfall with the axe seems little short of desecration.

The Colebrook tract represents the type of climax forest which prevailed over at least the greater part of northwestern Connecticut. Hemlock and beech, of about equal abundance, comprised 55 per cent of the entire stand; the proportions of these two species varied locally, hemlock forming as much as 75 per cent in low situations. Sugar maple (12 per cent) and yellow birch (10 per cent) were next in abundance; eight additional species made up the remaining 23 per cent (Table 89). The immense size[41] of these trees would dwarf any of the larger trees of the oldest second-growth stands of New England and emphasizes the pitiful lack of forest illustrating original conditions. "To one accustomed to . . . the ordinary Connecticut woodlands the luxuriance of the underbrush here is a revelation." *Viburnum alnifolium, Taxus canadensis,* and the omnipresent *Kalmia latifolia* were the most abundant shrubs; *Hamamelis virginiana, Viburnum acerifolium, Cornus alternifolia,* and *Lonicera canadensis,* not infrequent; and *Sambucus pubens* (*racemosus*) local. The two small species of maples, *Acer pensylvanicum* and *Acer spicatum,* occasionally formed a distinct layer. Some 54 species of herbaceous plants occurred, making a sparse cover, the two most representative of these being *Oxalis montana* and *Tiarella cordifolia.* It is apparent, from composition of canopy and lower layers of the Colebrook forest, that it was not only representative of the forest of northwestern Connecticut (as stated by Nichols) but also of the hemlock–hardwoods forest of the whole eastern division of the Hemlock–White Pine–Northern Hardwoods Forest.

---

[41] The size and age of most of these species were recorded by Nichols: *hemlock*—average diameter of mature specimens, 90 cm., larger trees, 115, 127, 150 cm. (i.e., 5 feet d.b.h.), average age 275 years, maximum 350 years; *beech*—largest, 85 cm. diameter, 200–225 years; *sugar maple*—diameters over 75 cm. frequent, largest 105 cm., age 250–300 years; *red oak*—fully 1 meter in diameter and 300 years in age; *yellow birch*—commonly 75 cm.; *chestnut*—largest diam. 132 cm., 75–100 cm. frequent, growth rapid, largest individuals not over 150 years; *ash, basswood,* and *cherry* also large.

Three other virgin forest areas of northwestern Connecticut were mentioned by Nichols but only briefly characterized. Two of these were similar to the Colebrook forest. The third (at Cornwall) was an area in which white pine was dominant and of immense size, many of the trees 150 feet tall and 3 to 4 feet in diameter—"a few acres of the most magnificent white pine that can be found in the East" (Hawes, 1909). While most of the mature

Table 89. Virgin forests of the hemlock–hardwoods area of New England.
A. The Colebrook forest in northwestern Connecticut.
B. The Pisgah Mountain area of southwestern New Hampshire: *1,* The forest as a whole (including samples in areas affected by natural disturbance); *2,* The physiographic climax, on high slopes and ridges; *3,* The climatic climax, on low and mid-slopes. (Data for *A,* from Nichols, 1913; for *B,* from Cline and Spurr, 1942.)

	A.	*B. Pisgah Mountain*				
	Cole-brook	*1*	2[1]		3[1]	
			a	*b*	*a*	*b*
Tsuga canadensis	55	37.9	68	49	33	19
Fagus grandifolia		7.4	7	46	37	57
Acer saccharum	12	1.4	. .	. .	12	17
Betula lutea	10	x	. .	. .	. .	x
Quercus borealis maxima	6	11.3	2	. .	. .	. .
Castanea dentata	6	. .	. .	. .	. .	. .
Fraxinus americana	7	.5	. .	. .	5	2
Tilia americana		x	. .	. .	x	x
Prunus serotina	4	. .	. .	. .	. .	. .
Betula lenta		6.4	6	4	9	1
Acer rubrum		3.9	5	1	. .	. .
Pinus Strobus		25.1	12	. .	. .	. .
Betula papyrifera	. .	4.3	. .	. .	. .	. .
Picea rubens	. .	.9	. .	. .	x	. .
Quercus alba	. .	x	. .	. .	. .	. .
"*Others*"[2]	. .	.8	. .	. .	4	4

[1] *a* and *b* of *2* and *3,* "overstory" or canopy, and "middle-story" or second layer, respectively.

[2] Minor constituents grouped; x indicates species included under "others."

trees were white pine, the younger generation was made up mostly of hemlock, with a sprinkling of sugar maple, yellow birch, beech, and other species. This demonstrated the subclimax character of the old-growth white pine and its ultimate replacement by a forest type comparable to that at Colebrook. A somewhat similar white pine forest (located at Ashuelot, Massachusetts) is representative of the white pine type which formerly prevailed in the Connecticut and Champlain valleys (Hawes, 1923).

In the less rugged parts of the New England Upland, the hemlock–hardwoods forest type is more or less confined to northerly slopes. A small

The "climatic climax association" in the Pisgah Mountain area of southwestern New Hampshire. "Hemlock, beech, black birch, and sugar maple are predominant in all crown levels." (Courtesy, Harvard Forest.)

area of old-growth (but not virgin) forest of this sort is located just south of Petersham, Massachusetts.

Cline and Spurr (1942), quoting from Child (1885) say:

> The original forest of Cheshire County [in the southwestern corner of New Hampshire] is described in an early gazeteer as including "dense bodies of the finest white pine timber" along the valleys of the Connecticut and Ashuelot,

"an abundant growth of hemlock" on higher lands, "belts of heavy spruce timber on the highest portions" and maples, beech, birches and red oak on the best-drained tracts.

Remnants of that original forest remaining in the Pisgah Mountain area had been the subject of a series of studies up to the time of the great hurricane of September 21, 1938, which "almost completely destroyed all that remained of the old growth stands. Elevations in the Pisgah area range from about 700 feet above sea level to a little more than 1300 feet. . . . The slopes are generally steep, and often broken by rocky outcrops." Data on the average composition of the overstory of the forest of this area and of the overstory and lower layers of the climax communities (physiographic climax of high slopes and ridges, and climatic climax of low and middle slopes) are given in Table 89, taken from Cline and Spurr (1942). Hemlock, beech, and sugar maple are the most abundant trees in all layers in the climatic climax forest. Hemlock and white pine dominate in the overstory in the physiographic climax of high slopes and ridges, while hemlock and beech prevail in lower layers (hemlock, black birch, and red maple in the very small tree layer, 1 foot or less). The Pisgah Mountain forest differs from the Colebrook forest principally in the higher percentage of hemlock and beech, and in the low percentage of yellow birch. The slight admixture of red spruce in the climax forest of moderate elevations here in central New England[42] suggests the transition to the more northern spruce–hardwoods forest. Communities which have followed disturbance (old partial, old severe, and recent, i.e., within the last 50 years) illustrate the gradual successional development toward the climax communities. They increase the number of cover types located in the Pisgah area—and in any area of forest.

Of central New England, it may be said:

> The primeval forests, then, did not consist of stagnant stands of immense trees stretching with little change in composition over vast areas. Large trees were common, it is true, and limited areas did support climax stands, but the majority of the stands undoubtedly were in a state of flux resulting from the dynamic action of wind, fire and other forces of nature. The various successional stages thus brought about, coupled with the effects of elevation, aspect and other factors of site, made the virgin forest highly variable in composition, density and form.

SECOND-GROWTH FORESTS

The original and present day forest aspects bear little resemblance to one another. Three centuries of human occupancy have left their mark. Every-

[42] Red spruce occurs at higher elevations where these rise above the New England Upland (p. 423), and is frequent in bogs on the upland, when these occupy frost pockets, as does a bog in the Harvard Forest in Petersham.

White pine in old fields. A stand 60 years old, into which young hardwoods are entering. (Courtesy, U. S. Forest Service.)

where are seen the secondary communities which followed repeated cutting and burning of hardwood and mixed forest, and abandonment of farm and pasture land. Since all of the hardwoods sprout if not too old, and the conifers do not, cutting does not dispose of the hardwood species. As they may sprout repeatedly, the same tree keeps coming back, so that many "young" forests are the direct continuation of former old-growth stands. The rapid growth of sprouts soon crowds out any young white pine or paper birch which may have entered, and results in a mixed hardwood stand in which red oak is always prominent. These "young" sprout forests are very different in appearance from young stands resulting from seed, due to the bunching of sprouts around the old stumps.

Except on sand flats, white pine enters in quantity only in old fields, i.e., on land which had been utilized for farm and later abandoned, land from which the hardwoods had been laboriously grubbed out. It usually accompanies gray birch, later crowding it out. Or, if fires have intervened, sprout birch communities may rule. Miles upon miles of these bushy (several stems in a clump) birch communities attest to the use and abuse of large areas. Where fires have not interfered, the white pine stands are gradually invaded by hardwoods and ultimately replaced by them.[43] This old-field succession—shrubs → gray birch and white pine → white pine → mixed hardwoods—is very similar to the old-field succession of southern New England in which gray birch and red cedar compose the initial forest stage (p. 256). That part of New England where white pine tends to dominate early stages of the old-field succession is often referred to as the "white pine forest region" (Maps, p. 250). It does not correspond with the original area of best development of white pine in commercial quantities—"the pinery proper on sandy soils"—which on the whole was farther north (map in Spalding and Fernow, 1899).

Secondary pine forests on sand flats (well illustrated east of Conway, New Hampshire) may be dominantly white pine, or mixed white pine and red pine with some red oak. Younger secondary communities on the sand flats, communities following recent extreme disturbance, may be entirely deciduous, made up of aspen and birch, with pin cherry, red maple, and scrub oak. Remnants of hardwood stands with beech, hemlock, sugar maple, basswood, red oak, and yellow birch testify as to the original occupancy of sand flats (as well as of rolling upland) by a hemlock–hardwoods forest of which hardly a vestige remains today.

Northerly slopes and moist sites afford some advantage to reforestation by coniferous species. Hemlock may be abundant in such situations, pointing to the ultimate establishment of a hemlock–white pine–beech–yellow birch–sugar maple community approximating the regional climax. This contrasts with the hardwood community of red oak, white oak (less frequent), ash, red and sugar maples, and sweet birch, which finally develops in drier situations.[44]

The Berkshire Plateau, because of its rugged topography, affords a variety of sites suitable to a wide range of communities. The usual old-field communities—especially white pine, gray birch, tamarack, and aspen communities—prevail in much of the area. At the higher elevations, a red spruce pasture type comparable to that of the spruce–hardwoods area is seen; white pine and gray birch are uncommon. At the moderate elevations of the plateau, a hemlock–hardwoods community is well represented. At higher elevations, hemlock is scarce, and a spruce–hardwoods type develops (Egler, 1940).

The Spruce–Hardwoods Area

The spruce–hardwoods area of New England and the spruce–hardwoods of the Adirondacks, although separated from one another by the Hudson-

[43] These steps are illustrated by the Harvard Forest Models, halftone reproductions of which were published by Harvard Forest under that title. Cambridge, 1936.

[44] For additional information on second-growth communities and secondary successions, see Spring (1905), Lutz (1928), Cline and Spurr (1942).

Champlain Valley, nevertheless, might well be considered as one vegetational unit.[45] Usage demands their separation, which is justified because of their geographic and physiographic separation. This forest area is often considered to be a southward lobe of the Spruce–Fir Forest region. It is so mapped by Shantz and Zon (1924). However, red spruce, one of the dominant and characteristic species of the area, is not a tree of the Boreal Forest; it reaches the northern limit of its general range south of that forest, in fact, south of the St. Lawrence River, except for "an island of colonization on the lower slopes of the Laurentian Hills . . . ; and outliers as far west as Algonquin Park and Haliburton County, Ontario" (Halliday and Brown, 1943).

Most of northern New England and southeastern Canada, from the coast of Nova Scotia and Maine to Lake Champlain, and all of the higher elevations are occupied by red spruce, or by spruce in combination with the northern hardwoods or with other conifers, chiefly balsam fir. White spruce is abundant near the coast in the northeast and in the northern part of the area where the true spruce–fir forest of Canada extends southward meeting the spruce–hardwoods of the New England section of the Hemlock–White Pine–Northern Hardwoods region. The increasing dominance of conifers northwards is well illustrated by Chittenden's map (1905) of land classification in northern New Hampshire. Much of this area is mountainous, almost all of it is decidedly hilly, for the New England Upland in the north is less of a plain than in central New England. Even the coastal area (Seaboard Lowland) of the eastern half of Maine is often rugged; Mount Desert Island reaches an elevation of 1532 feet only two miles from the coast. The hilly and mountainous topography affords diversity of environment, which is further increased in Maine by the large number of lakes, and by the rugged seacoast strip. Both xerarch and hydrarch successions are abundantly represented.

As in the hemlock–hardwoods area, very few remnants of the original forest remain. The land, although almost completely forested over large areas, is clothed with second-growth stands, some of which are early stages in reforestation, others approach in species, if not in relative abundance of species, the original forest types. Many of these stands can be referred to the comparable original type.

FOREST TYPES

The names locally applied to the forest types are, as is to be expected, similar to those used in the Adirondacks, and are in part descriptive and based upon the occurrence of spruce, the commercially most valuable

[45] The approximate areas of spruce–hardwoods in New York and New England are shown as spruce on the Hawley and Hawes map of "Forest Regions of New England," p. 249.

species of the area (Hosmer, 1902; Murphy, 1917; Dana, 1930; Westveld, 1930). The classification prevalent in the Adirondacks (p. 414) is sometimes used (Chittenden, 1905).

> The names and descriptions given to the various topographically distinct forest types in the literature include: (1) *spruce swamp*—the forest of the lower, more poorly drained areas, composed principally of red and black spruce, balsam fir, tamarack, northern white cedar, soft maple, and black ash; (2) *spruce flats*—the forest of better drained, moist soils along water courses where, in addition to spruce and balsam fir, an admixture of the maples, hemlock, white pine, and the birches may occur; (3) *mixed spruce–hardwood*[46]—the forest of the benches and lower slopes of the mountains, with the deepest, richest forest soils of the region, where red spruce, balsam fir, sugar maple, beech, white and yellow birches predominate, and where normally the best tree growth of the region occurs; (4) *spruce slope*—the forest of the upper slopes, where the soil is thinner, stony, and with variable moisture content, composed of nearly pure stands of red spruce and balsam fir, with locally yellow and white birch trees scattered throughout; and (5) *old-field spruce* (called in the region pasture-spruce)—communities which have established themselves on abandoned pasture or tillage lands and are usually composed of pure stands of spruce, red or white. (Oosting and Reed, 1944.)

Where swamps are few or not extensive, as in the mountainous areas, the spruce swamp type may not be distinguished from the spruce flat. Where lakes and swamps are numerous, as in the northern part of the area, spruce swamp covers vast areas. On gentle bordering slopes, where the "spruce flats" type prevails, tall white pines are often seen rising above the other species of the forest. The "mixed spruce–hardwood" forest or forest of the "hardwood land," is the most distinctive type of the area, now generally represented by fragmentary old-growth but not virgin stands. These may be immediately adjacent to old-field spruce stands with which they contrast strongly because of their large number of deciduous trees. Occasional old sugar maples, yellow birch or hemlock in the old-field spruce stands give evidence of the character of the former forest cover. White birch, or white birch and aspen, almost omnipresent species, frequently dominate considerable areas; the general observer would recognize this as an additional forest type. The composition of the several types as developed in the southern part of the White Mountains and in extreme northern New Hampshire is given in Table 90, taken from Chittenden's studies (1905).

LOCAL VARIATIONS

Additional forest types, less extensive and less generally distributed, or commercially unimportant, occur in distinct habitats or as a result of extreme disturbance. These, rather than the principal types previously mentioned,

---

[46] This is the "hardwood land" of Chittenden (1905).

Table 90. Forest types in the spruce–hardwoods area of New England: *A,* In the southern part of the White Mountains (Waterville tsp.); *B,* In northernmost New Hampshire (Pittsburg tsp.). (Data from Chittenden,* 1905.)

	A			*B*		
	Hardwood Land	*Spruce Slope*	*Upper Spruce Slope*	*Spruce Flat*	*Hardwood Land*	*Spruce Slope*
No. of acres averaged	50	65	5	10	12	10
Trees 10 in. and over	88.28	115.22	42.4	115.7	81.29	120.0
Spruce	27.0	76.0	74.1	76.9	12.2	71.9
Balsam	.2	1.5	16.5	16.8	5.2	12.6
Yellow birch	25.2	14.4	1.4	2.6	27.6	9.1
Paper birch	.5	3.1	8.0	3.5	.2	6.4
Hemlock	3.1	1.6	..	..	..	..
Sugar maple	8.3	1.0	..	.1	51.3	..
Beech	35.0	1.8	..	..	1.5	..
Other species	.6	.6	..	..	2.0	..

* Only the "spruce flat" of *B* was definitely stated by Chittenden to be virgin forest. All, however, are given in connection with the "original forest types."

may control the landscape. This is particularly true of white birch, so characteristic of the northern New England landscape.

White pine, or white pine in mixture with red pine, occupies stretches of low sandy land, for example, along the Androscoggin and Saco rivers. "Originally a heavy stand of white pine covered the whole of the Androscoggin Valley [in New Hampshire], but of this nothing now remains but a few scattered trees" (Chittenden, 1905). Today the type is represented by young stands of white pine, and occasional old trees of white and red pines; birch and aspen, indicative of burns, are here also. In the Saco River basin, the pine type, of white pine, or white and red pines, is conspicuous, and essentially continuous with the pinelands southward. Where the sand is very coarse, pitch pine is also present.

In the coastal area, the vegetation is influenced by proximity to the ocean. Bluffs and exposed headlands seem always to have been occupied by spruce (principally white and black spruce) with zones of dwarf vegetation (as *Juniperus horizontalis, Empetrum nigrum,* and various rock crevice herbs). In many places the coastal strip (both rocky and more favorable situations) is dominantly spruce. However, the amount of spruce has increased as a result of disturbance; there is evidence that the original forest of the lowest elevations near the coast, except on exposed headlands or in a narrow coastal fog belt,[47] was predominantly deciduous at least on the

[47] Along the coast of Nova Scotia, near Yarmouth, the immediate coast is subject to frequent fogs. The forest is composed almost exclusively of black spruce, while three or four miles inland, the forest is mixed and hardwoods are abundant (Transeau, 1909).

White birch, characteristic of the northern New England landscape. (Courtesy, U. S. Forest Service.)

better soil areas (Ganong, 1906; Nichols, 1918; Hill, 1923; Moore and Taylor, 1927). The vegetation of maritime Canada (Nova Scotia and New Brunswick) is highly diversified. The higher elevations in these northern situations are occupied principally by coniferous communities in which red spruce and balsam fir prevail. Barrens and black spruce swamps are extensive in the interior. Slopes are generally occupied by red spruce and hardwoods, chiefly beech, sugar maple, and yellow birch; hemlock may be a constituent of these forests. The presence of a few more southern species (as butternut, red oak, bur oak, basswood, sweet birch, and pitch pine) is a feature of the St. John River Valley (Fernow, Howe, and White, 1912; Halliday, 1937). Farther north, white spruce appears. Of northern Cape Breton Island (the most northeasterly extremity of the region) Nichols (1918) said:

> To one visiting northern Cape Breton at the present day the prevailing aspect of the lowland forest appears to be coniferous; white spruce and balsam fir predominate on every side . . . It has become unmistakably evident that in former times a very large portion of this area was clothed with forests in which the predominant trees were deciduous. It is certain . . . that forests of this sort were developed in practically all edaphically favorable situations.

Beech[48] was the most abundant species (sometimes 65 per cent of the mature trees), and with it were sugar maple, yellow birch, red maple, paper birch, occasional red oak and white ash, and the coniferous trees, balsam fir, hemlock, white pine, and a sprinkling of white and black spruce. Above an elevation of about 700 feet on this northeastern outpost, the hardwoods drop out and a coniferous climax forest, mainly balsam fir, prevails.

DEVELOPMENTAL COMMUNITIES

In any section where lakes and bogs or cliffs and gorges are frequent, developmental communities occupy a large proportion of the area. These, together with the many secondary communities, control the landscape.[49] The most conspicuous of these secondary communities is composed largely of white birch, or birch and aspen.

ALTITUDINAL VARIATIONS

Because of the great range of altitude in some districts of this section, distinct altitudinal sequences occur. Peaks in the three principal mountain areas (the Green Mountains, the White Mountains, and Mount Katahdin) rise above timber line. The spruce slope type gives way to a higher slope type predominantly of spruce and fir, and these to a dwarf forest of balsam fir at timber line. A number of arctic-alpine species are found on the summits of the highest mountains. Among these are a few very dwarf evergreen shrubs and suffruticose plants, some of which are circumboreal in distribution—*Empetrum nigrum, Vaccinium Vitis-Idaea,*[50] *Rhododendron*

[48] The beech of these northern situations is "gray beech" (W. H. Camp, in litt. 1948).

[49] For a discussion of developmental communities see Transeau (1909), Nichols (1918), and Hill (1923).

[50] Represented by the variety *minus* in America.

Forested slopes of the Green Mountains of Vermont. (Courtesy, U. S. Forest Service.)

lapponicum, Arctostaphylos alpina, Phyllodoce coerulea, Loiseleuria procumbens, and *Diapensia lapponica.*

The Green Mountains, stretching from Massachusetts to the Canadian border, rise more or less abruptly above the Taconic Mountains and plains of Lake Champlain on the west, and the New England Upland on the east. Their forest-clad slopes, prevailingly hardwood, contrast with the more open lower lands on either side. Red spruce, mingling with the hardwoods below, becomes dominant on rocky shoulders and higher slopes. Belts of dwarf timber, dwarf shrubs, and alpine vegetation are seen on the highest parts of the mountain range. The alpine slopes of Mt. Mansfield (4393 feet) are extensive.

The forest of the lower elevations exhibits many variations related to topography and soil. On sand deposits, and in areas of outcrop of light (acidic) rock, white pine is abundant. On the richer soils derived from dark (basic) rock, the hardwood forest reaches its best development. In such locations the herbaceous and shrubby vegetation becomes luxuriant, suggesting in quantity and variety, if not entirely in species, the ground layers of mixed mesophytic forest. A wealth of ferns (as *Dryopteris Goldiana, D. noveboracensis, Athyrium thelypteroides, A. angustum, Adiantum pedatum, Phegopteris polypodioides, Onoclea sensibilis, Pteretis nodulosa, Botrychium virginianum*) and of flowering herbs, most abundant and conspicuous of which is *Hydrophyllum virginianum,* form an almost continuous and luxuriant ground cover. In such places a mull humus has developed. The distinctive species of the northern forest (as *Oxalis montana, Clintonia borealis,* and *Cornus canadensis*) are absent or infrequent, or localized under conifers; plants of mull humus prevail.

Sugar maple is locally dominant in the hardwood forest of the lower slopes of the Green Mountains. (Courtesy, U. S. Forest Service.)

The composition of a representative hardwood stand in Gifford Woods State Forest Park, on lower east slopes of the Green Mountains is given in Table 91. Here beech, sugar maple, and yellow birch together comprise somewhat less than 90 per cent of the canopy and over 90 per cent of the second layer. Locally, sugar maple dominates, almost to the exclusion of all other species.

Table 91. Composition of a representative stand of hardwoods in the Green Mountains; Gifford Woods State Park, Vermont.

	Canopy	*Second layer*
Number of trees	126	108
Fagus grandifolia	34.9	29.6
Acer saccharum	34.1	44.4
Betula lutea	18.2	17.6
Tsuga canadensis	7.1	3.7
Fraxinus americana	2.4	1.8
Ulmus americana	1.6	. .
Prunus serotina	.8	. .
Tilia americana	.8	.9
Picea rubens	. .	1.8

Very little of the original forest cover now remains in the White Mountains. Altitudinal sequences are obscured by the excessive amount of lumbering these slopes have suffered.

> Lumbering has brought about a great change in the species. Hemlock and white pine, once common at low elevations and along the valleys, are now of but little importance in the forest. Of the hardwoods, yellow birch, sugar maple, and beech are the commonest, and have greatly increased in numbers on the cut-over land. (Chittenden, 1905).

In the "notches" today some fair stands of hardwoods are seen. Generally, secondary white birch occupies what was hardwood land (the spruce–hardwoods areas) and the spruce slopes as well. Only the higher slopes where the trees are dwarfed have escaped devastation.

Hardwood forest occupied only the lowest slopes, generally below 1800 feet elevation. Spruce forest (the spruce slope type) in which yellow birch, paper birch, and balsam fir are accessory species occupied most of the slopes, generally ranging from about 1800 to 3500 feet in elevation. This spruce forest (commercially valuable) is all but gone; logging and fire have been followed by stands of paper birch. A forest of spruce and balsam fir, in which the trees were small, occupied higher slopes from about 3500 to 4200 feet, gradually giving way to the dwarfed and scrubby growth at timber line. The elevation of timber line varies with exposure, being reached

Small stands of hardwoods remain in some of the "notches" in the White Mountains. (Courtesy, U. S. Forest Service.)

somewhere between 4500 and 5200 feet. The highest slopes support alpine vegetation—crevice and rock plants on granitic exposures, meadows where some soil has formed. The Presidential Range, culminating in Mount Washington (6290 feet) presents the greatest expanses of dwarf timber and alpine slopes.[51]

The altitudinal sequence of forest types in the White Mountains—hardwoods with spruce, spruce with yellow birch, spruce–fir with paper birch—is illustrated in Table 90.

Mount Katahdin, the highest mountain of Maine (5212 feet), rises abruptly on the south from the surrounding lowland whose elevation is approximately 1000 feet. A tableland of about 4500 to 4900 feet extends several miles northward from the highest peak. The upper slopes for 1000

[51] Harshberger (1911) gives a brief description of the alpine vegetation of Mount Washington, distinguishing between alpine canyon, alpine bog, and alpine meadow communities. Best known of the last is "Alpine Gardens" over 1000 feet below the summit. Griggs (1946) discusses timberline problems on Mount Washington.

feet below the tableland are very steep and precipitous, the lower more gradual. Several distinct vegetational belts occupy the mountain. The lowland forest is mesophytic—of spruce, balsam fir, maple, and birches (*Betula lutea* and *B. papyrifera*) with occasional white pine and mountain ash. The typical spruce–fir forest of the lower mountain slopes is made up principally of black spruce and balsam fir; the forest floor is a continuous carpet of mosses—*Hypnum Schreberi, H. cristo-castrensis,* and *Hylocomium splendens*—with *Ptilidium ciliare* and *Bazzania trilobata.* Yew is abundant, and a number of other shrubs represented. The usual herbaceous plants of the spruce–fir forest—*Oxalis, Coptis, Maianthemum, Cornus canadensis, Trientalis, Streptopus roseus* and *S. amplexifolius, Clintonia borealis, Lycopodium lucidulum* and others—are scattered through the moss carpet. Higher, the trees are reduced in size; a typical Krummholz develops of dense and matted spruce and fir.[52] The elevation at which dwarfing becomes pronounced varies greatly on different parts of the mountain. On the south face it is around 3700 feet; in the North Basin, much lower. The upper limit of the Krummholz is very irregular and related to exposure. The steepest slopes and rock slides are essentially devoid of vegetation. An alpine tundra made up of grasses and sedges, cushion mosses, forbs, and dwarf shrubs occupies the tableland and high slopes (Harshberger, 1902; Harvey, 1903).

---

[52] Thoreau, in "Maine Woods," describes this vegetation.

Part III

THE EVOLUTION OF THE PRESENT PATTERN OF FOREST DISTRIBUTION

CHAPTER 13

Problems of Forest Distribution

From the survey of vegetation, particularly of the climax communities, of the several forest regions already presented, it is possible to determine certain distributional facts which are significant in the study of forest distribution and its controlling factors. The nature of these is best suggested by a series of questions:

Where does a particular climax attain its best development?
Where is it most widespread, occupying the greatest variety of sites?
Where is it most circumscribed, most restricted to specific or peculiar sites?
What are its actual limits (in space) and what are its local requirements in marginal habitats?
What are the differences, if any, in composition in central and marginal situations?
What is indicated as regards migratory movements?
What are the underlying causes of these distributional facts?

In seeking the answers to these questions, each climax association will be considered in turn.

The **Mixed Mesophytic association** reaches its best development in the Cumberland Mountains. There it is most luxuriant in aspect and contains a larger number of species than elsewhere. There it displays the greatest variety of types, of association-segregates formed by local shifts in dominance of constituent species. While all of its dominants do not there reach maximum size, all do attain sizes in excess of the average.

The area in which mixed mesophytic communities occupy a variety of sites is more extensive than the area of best development. They are widespread throughout the Cumberland Mountains, at the lower elevations of the Allegheny Mountains of West Virginia, over much of the Cumberland Plateau and the southern part of the Unglaciated Allegheny Plateau. Throughout this area, they occupy a variety of soil and topographic situations, responding to unlike conditions by variations in composition.

The area which has been designated as the Mixed Mesophytic Forest region includes the area of best development, of greatest diversity of habitat, of widespread occurrence of mixed mesophytic communities. When we look beyond the boundaries of this region in any direction, we find mixed mesophytic communities much more circumscribed, much more restricted to particular local environments, and generally containing fewer species or even lacking characteristic species. They do occur more or less frequently in each of the contiguous regions—the Western Mesophytic Forest region, the Oak–Chestnut Forest region, the Beech–Maple Forest region, and at the margin of the Hemlock–White Pine–Northern Hardwoods region (Allegheny section). They do occur rarely in other regions not contiguous to the Mixed Mesophytic—in the Oak–Hickory, the Oak–Pine, and the Southeastern Evergreen Forest regions. Some suggestion of them is seen in the Driftless area, which might vegetationally be referred either to the Maple–Basswood or to the Oak–Hickory region (p. 332). These occurrences of mixed mesophytic communities outside of the Mixed Mesophytic Forest region are always correlated with particular local environments. The actual limits (in space) of mixed mesophytic communities are not far from the limits of deciduous forest (if the Hemlock–White Pine–Northern Hardwoods be excluded from that formation).

The local environments in these extra-regional occurrences of mixed mesophytic forest vary somewhat. However, without exception, they are moist (mesic) well-drained slopes. They have been mentioned in connection with the mixed mesophytic communities of the several regions.

In the Western Mesophytic Forest region, mixed mesophytic communities occupy slopes in hilly areas, areas of mature topography; they are particularly well developed in the Hill section, and are frequent in the hilly Highland Rim adjacent to the Cumberland Plateau (pp. 113, 141, 147, 152). They form an element in the mosaic of unlike communities, of unlike climaxes (climatic, edaphic, and relic) which characterize this region.

In the Oak–Hickory Forest region, mixed mesophytic communities occur but rarely. They are confined to a few slopes in the Boston and Ouachita Mountains (pp. 170, 174).

Mixed mesophytic communities are beautifully represented in the Southern Appalachian section of the Oak–Chestnut Forest region, particularly in the Great Smoky Mountains. There, where they are known as cove forests, they attain a state of development comparable to that in the Cumberlands, although they are limited to particular sites—the coves and ravine slopes of the mountains (pp. 199, 214, 219). They are frequent on ravine slopes of mountain ranges adjacent to the Mixed Mesophytic region, where a distinct transition zone occurs (pp. 228, 231). They have developed in the Glaciated section (p. 255). Elsewhere in the Oak–Chestnut region, mixed mesophytic communities are infrequent and poorly developed; they are confined to valley slopes of the present erosion cycle—valley slopes cut below the Harrisburg peneplain (pp. 240, 245, 247).

In the Oak–Pine and Southeastern Evergreen forest regions, communities which are the Piedmont and Coastal Plain representatives of mixed mesophytic are confined to erosional slopes of the present physiographic cycle (pp. 266, 275, 300). The transitional Beech–Magnolia climax (transitional between the Deciduous and Broad-leaved Evergreen formations) develops in a greater variety of situations, suggesting approach to another vegetation center.

The mixed mesophytic communities of the Beech–Maple region occupy slopes in maturely dissected areas near its southern margin. They are generally poor in species, and might be called attenuated representatives of the Mixed Mesophytic association (pp. 310, 315). Those of the Allegheny section of the Hemlock–White Pine–Northern Hardwoods region are distinctly transitional in character, or are attenuated tongues of mixed mesophytic extending along valleys of the region (p. 406).

The **Oak–Hickory association** reaches its best development in the Ozark Upland. There, some phase of oak-hickory forest covers most of the slopes and ridges. Outside of this area, it is somewhat more circumscribed as to sites, although throughout the area designated as the Oak–Hickory Forest region, it is of widespread occurrence and not limited to particular local environments.

The most extensive extra-regional occurrence of oak–hickory communities is in the Oak–Pine region. In fact, this is a secondary center of oak–hickory and might be termed the Eastern Oak–Hickory region. Oak–hickory is generally regarded as the regional climax on the Piedmont (pp. 259, 265).

In the Western Mesophytic Forest region, oak–hickory communities are well represented, particularly on the Mississippian Plateau and on certain of the ridge crests and southerly slopes of the more hilly sections (pp. 123, 141, 149). In the latter situations, oak–hickory is a physiographic climax, generally alternating with mixed mesophytic communities. In the former, it occupies the low limestone hills which rise above the general level of the plateau, and is invading prairies in the "Barrens."

Many inclusions of oak–hickory forest occur in the Beech–Maple region where they are indicators of the former extent of a prairie lobe. Here they generally occupy areas of immature topography (p. 323).

Oak–hickory communities occur frequently in the Driftless section of the Maple–Basswood region (p. 329).

Oak–hickory communities in the Mixed Mesophytic Forest region are for the most part confined to ridge crests, although southward—in the part of the region transitional to the Oak–Pine—they occupy southerly slopes as well as ridge crests, and some plateau areas. In the Low Hills Belt of the Allegheny Plateau, they are more widespread than in most of the region; this occurrence may be related in part to the extent of proglacial lakes and in part to peneplanation (pp. 93, 97). Ridge-crest oak–hickory communities in the midst of the Mixed Mesophytic Forest are always related to a calcareous substratum. Ridges on noncalcareous rock are occupied by oak–chestnut communities (p. 119).

The oak–hickory communities of the Oak–Chestnut region are generally confined to low hills in the Ridge and Valley section, and to the foothills of the Southern Appalachians. In the Piedmont section, they seem to be somewhat more widely distributed, although some of the present oak–hickory communities of that section have replaced former oak–chestnut communities. The white oak forests of the valley floors and benches of the Ridge and Valley section are closely restricted as to habitat, and always correlated with the Harrisburg peneplain (or Harrisburg and Somerville) which often, but not always, was developed on limestone (p. 238). In the Glaciated section, a white oak or oak–hickory type occurs at intervals throughout the section, in xeric situations. A type of hydro-mesophytic oak–hickory forest occurs on the Piedmont Lowland of New Jersey and adjacent Pennsylvania.

Southeastward, oak–hickory communities extend to the limit of deciduous forest. Successionally, they are more advanced than the subclimax pine which they are capable of replacing (p. 297).

Oak–hickory communities occur almost throughout the deciduous forest. Their distribution indicates two centers, one in the Ozarks, and a secondary center on the Piedmont. Extensions onto younger land surfaces—the Coastal Plain and the glaciated territory—are pronounced. While many of the local habitats are xeric, it is not safe to assume that a correlation always exists between oak–hickory communities and low soil moisture. The species which occur in typical oak–hickory communities are frequently constituents of mesic or even hydro-mesic communities (for example, the white oak stage in the hydrarch succession of the undissected till plains of southwestern Ohio).

The **Oak–Chestnut association** reaches its best development in the Southern Appalachians, where it occupies a variety of sites and extends through a considerable range of elevation (p. 197). It is[1] the most widespread type throughout all of the Oak–Chestnut region except the Appalachian Valley. In contiguous regions, oak–chestnut communities are much more circumscribed. In the Oak–Pine region they occur only in a transitional belt or on the slopes of mountain outliers (p. 267) where their environment is essentially that of the Oak–Chestnut region. Oak–chestnut communities extend locally into the New England section of the Hemlock–White Pine–Northern Hardwoods where they are confined to warmer slopes. They dovetail into the borders of the Allegheny section of the Hemlock–White Pine–Northern

[1] The present tense is used, although all oak–chestnut communities have been changed by the death of chestnut, in order to present the picture of original conditions.

Hardwoods, and in the Susquehanna and Finger Lakes drainage basins, extend across that region to the plains of Lake Ontario in the Beech–Maple region, where the sandy glacial deposits favor the persistence of a more xeric forest than the regional climax. The oak–chestnut communities of the Mixed Mesophytic region occupy ridge crests and dry southerly slopes on noncalcareous substrata.

In the Western Mesophytic Forest region, oak–chestnut communities occupy slopes in the rugged border of the Shawnee physiographic area of the Hill section where the soil, derived from the Pottsville and Cypress sandstones, is sandy and noncalcareous. Chestnut is (or was) an important constituent of the forest in other hilly areas of this region, where it may be associated with beech and other mesophytes (see Table 23). In this region, oak–chestnut communities occupy less xeric sites than do oak–hickory communities.

Oak–chestnut communities range over a smaller proportion of the total area of the deciduous forest than do either of the associations previously considered. Widespread through most of the Oak–Chestnut region, and of frequent occurrence in the Mixed Mesophytic region, they are extremely local beyond the limits of these two regions.

The **Beech–Maple climax** is an association of the younger glaciated area. There, although occupying a relatively small proportion of the total area, it is the culmination of both hydrarch and xerarch succession. Succession tends to increase its area which is limited only because of the youth of the land surface; developmental communities occupy a large area.

While the Beech–Maple region is the center of area of the Beech–Maple association, beech–maple communities are abundantly developed in the Hemlock–White Pine–Northern Hardwoods region from the eastern end of the Upper Peninsula of Michigan eastward, where they occupy moist but well drained sites.[2] West of the Great Lakes section, *i.e.,* beyond the limits of the range of beech, beech–maple communities are replaced by maple–basswood communities. South of the Wisconsin glacial boundary, except in the northern Allegheny Mountains and Allegheny Plateau, beech–maple communities are very local. In the Mixed Mesophytic region, beech–maple is an association-segregate restricted to lower north slopes, and the lower slopes of ravines. Beech–maple communities occur at the higher elevations in the southern mountains (see footnote, p. 201). Although mention of beech–maple communities has been made in the literature in widely scattered localities almost throughout the deciduous forest—as far southeast as the Coastal Plain of North Carolina, as far southwest as the Ozarks of

[2] The beech of the Beech–Maple region is a mixture of white beech and red beech. The beech of the beech–maple communities of the Hemlock–White Pine–Northern Hardwoods region is the red beech; gray beech occurs sparingly in the northern and northeastern parts of the region (fide W. H. Camp).

Arkansas—investigation has shown that these are not representatives of the Beech–Maple association.

The dominants of the Beech–Maple association have extensive ranges; they are constituents of other communities, even of other climaxes. The Beech–Maple association is a climax of relatively recent origin, derived by postglacial migration from the more complex ancestral Mixed Mesophytic association (see Chapters 17 and 18). Its representatives in the unglaciated area are local or altitudinal segregates of that association.

The **Maple–Basswood association** has often been included in the Beech–Maple association, as a western faciation. Only by consideration of its probable derivation, its phylogenetic relationship, can its distinctness be determined (see Chapters 17, 18). It is the climax community of the northwestern part of the deciduous forest, most widespread and attaining best development in the Maple–Basswood region. Farther south it is represented by the maple–basswood and red oak–basswood communities of protected ravine slopes in the prairie–forest border belt and in the Prairie Peninsula section of the Oak–Hickory region. It occurs locally in the Beech–Maple region as far east as northwestern Ohio.

The **Hemlock–White Pine–Northern Hardwoods** forest is usually regarded as a formation distinct from the Deciduous Forest Formation, but here included in the consideration of deciduous forest because of its numerous deciduous communities. Unlike the deciduous forest, it is not made up of a number of distinct associations. It appears to be a formation with only one association which displays a series of faciations of decreasing complexity from east to west.

Considered in its entirety, the Hemlock–White Pine–Northern Hardwoods association is the prevailing forest of all but hydric and xeric situations (which are occupied by developmental communities) of the region by the same name. It occurs frequently in the northern part of the Beech–Maple region, and in the northeastern (Glaciated) section of the Oak–Chestnut region. Its representation by only a few species in extreme habitats (north-facing bluffs and gorges) in the northern part of the Oak–Hickory region and southern part of the Beech–Maple region is significant in the reconstruction of forest migrations. Frequent outliers occur in the higher Allegheny Mountains, where the altitudinal sequence may be from mixed mesophytic through beech–maple or beech–birch–maple to a spruce–hardwoods forest (p. 81). The high altitude forests of the Southern Appalachians, often considered as outliers of the northern association, are, however, distinguished by the presence of the endemic Fraser fir (*Abies Fraseri*), and a number of undergrowth species not occurring in the north. The many hemlock, hemlock–white pine, or hemlock–hardwoods communities of coves and gorges, and of exposed rocky ridges at low elevations in the mountains and plateaus of the Mixed Mesophytic and Oak–Chestnut regions are not here considered as outliers of the Hemlock–White Pine–Northern Hardwoods association. These always contain southern hardwoods and have a lower layer of Rhododendron.[3]

The extent of the area of dominance of any forest association is controlled chiefly by climatic influences. Extremes in annual variations in temperature or precipitation, while they may adversely affect the vegetative vigor or the seed production of individuals of particular species, will not affect the climax as a whole. Even if continued for a number of years, they can have little effect except by causing slight oscillations of margins through their limiting effects on species distribution. However, long-protracted trends or climatic shifts, such as have taken place in the past, have had profound effects upon the extent and distribution of climax associations, and are in part responsible for differences in composition in different geographic areas. Such climatic shifts have brought about mass migrations, or expansions or contractions of vegetation types.

The existing climaxes, or their progenitors, have varied in the locations and extent of area occupied. In part at least, the present distribution of forest climaxes—the mass distribution and the outliers—is indicative of past events; in part it is indicative of present trends, depending on whether outliers are relics or advance stations.

While climate was without doubt determinative in the past extent of climaxes, climate was (and is) affected by the major physiographic features of the continent. These have changed in the past; the land surface has been profoundly affected by processes of gradation and diastrophism involving periodic elevations and depressions of the land surface and shifting of continental margins. Series of erosion cycles, more or less complete, have constantly modified the land surface. They have left evidence of their extent and completeness in the level skylines, the benches, the broad valley floors which are interpreted as remnants of peneplains. A continent with mountains intercepting moisture-bearing winds and a low-lying continent, even though approximately the same in area and outline, cannot be the same in climate. What the climates of the past have been, of the past during which our deciduous forest associations were developing, we have no way of determining except in so far as they may have been correlated with physiographic history or be indicated by the nature of vegetation preserved in the fossil record. We can correlate, or find coincidences between, the extent of forest regions and climaxes, and the physiographic history and erosion cycles of the deciduous forest area.

We may be justified in assuming that the climatic requirements of a given

[3] If the several races of the beech population at different elevations be taken into consideration, the distinction between these low elevation hemlock–hardwoods communities and the northern forest is further emphasized.

association have remained reasonably constant through the ages, that protracted dry periods must have resulted in contraction of mesic associations and corresponding expansion of xeric associations. At the same time, we must certainly admit the possibility of slow progressive changes in requirements or rather adjustments of vegetation. Migration and evolution, not alone of species but of physiologic adjustments of species and hence of vegetation, must have taken place. Although climate must have been determinative, the attempt to reconstruct the past must be based chiefly upon the paleobotanic record, the disjunct distribution of communities and species, and physiographic history (see Chapters 14 to 17).

The problems of forest distribution involve a broad survey of the nature and extent of development, and of the distributional characteristics of the major forest climaxes (Chapters 4 to 12), together with a search for evidence of past distribution and of past influences (Chapters 14 to 16). Their consideration should lead to an orderly, if not complete, story of forest distribution and relationships (Chapters 17, 18).

CHAPTER 14

The Paleobotanic Record

LATE MESOZOIC AND TERTIARY RECORDS

The history of deciduous forest in North America dates back many millions of years to the latter part of the Mesozoic era. The fossil record of this history in the East—within the area now occupied by deciduous forest—is fragmentary. It is confined to local deposits which were laid down along the continental margin, along the slowly emerging coastal plain and Mississippi embayment. There are no records in the eastern interior; the history of the interior or upland forests can only be deduced from the records of the coastal strip, the western interior (the western Great Plains and Rocky Mountain regions), and the Arctic.

THE RECORD IN EASTERN NORTH AMERICA

Late in the Lower Cretaceous (Comanchean) period,[1] angiosperms appeared, at first in small numbers, later abundantly, constituting perhaps one-half of the flora of the Potomac series of the Atlantic States. Even from the beginning of the record in America, a few forms referable to living genera and others with considerable similarity to modern genera occurred. A little later, the representation of modern genera was such as to definitely establish the presence of broad-leaved forest communities.

In the Upper Cretaceous (or Cretaceous) period, the abundance of dicot fossils (in eastern deposits) indicates that by that time broad-leaved forests had become dominant in eastern North America. Those of the Atlantic slope contained representatives of such living genera as *Cinnamomum, Diospyros, Ficus, Juglans, Liriodendron, Magnolia, Myrica, Planera, Quercus, Salix,* and other forms, perhaps ancestral to living genera, as *Celastrophyllum, Cornophyllum, Gleditsiophyllum,* and *Laurophyllum*. Not all of these have continued on in eastern America—some are now confined to tropical and subtropical regions. Gymnosperms, then as now, constituted a part of the forest or perhaps dominated in local areas. Among these were some of the more ancient and now extinct forms; some which are no longer

[1] For geologic time scale, see Chapter 16, p. 490.

a part of the forests of the Northern Hemisphere (as *Araucaria* and *Podocarpus*); and others, as *Pinus, Picea, Thuja, Sequoia,* and *Tumion* (*Torreya*) which still maintain themselves in North America. *Sequoia,* which was commonly preserved in the Upper Cretaceous of the Atlantic coast, is now confined to the far West. *Torreya*[2] has continued on, and is growing now (on the Apalachicola River in northern Florida) not far from a location where its Cretaceous ancestor lived. These Cretaceous fossils were preserved in local lignitic beds or lenses laid down in coastal swamps or estuaries. The nature of the surrounding contemporaneous deposits indicates shallow seas near shore (which was essentially where the Fall Line is today) and an adjacent area (the Piedmont) of considerable relief and swift streams. The climate was mild, but not tropical, and the rainfall plentiful. The flora "is essentially a unit from Alabama to New Jersey, and preserves even in its far northern occurrences on the west coast of Greenland —though differences are perceptible—a facies remarkably similar" (Berry, 1914).

In the Mississippi embayment area of western Tennessee, Upper Cretaceous floras are well represented (Tuscaloosa, Eutaw and Ripley formations). There, as on the Atlantic slope, many genera now living were represented. To those of the deciduous forest already mentioned might be added *Acer, Celtis, Fagus, Kalmia, Populus, Rhamnus,* and *Sassafras. Pachystima* (doubtfully identified in Georgia) is listed for Tennessee (Ripley flora). Additional genera which are now restricted to the tropics or to the southern hemisphere might be added to the examples already given.

Tertiary floras as a whole are more modern in generic composition and aspect than are the Cretaceous. Resemblances to modern species become pronounced, especially in the Miocene and Pliocene, as is indicated by specific names formed by prefixing pre-, mio-, or plio- to the names of living forms. In the Mississippi embayment, where the most continuous plant records are found, a comparison of the uppermost Cretaceous (Ripley) flora with the lower Eocene (Wilcox) flora reveals that the two floras are almost totally unlike, that they have no species in common, although

> the Ripley flora does contain a number of types that became differentiated subsequently and are characteristic elements af the lower Eocene flora in southeastern North America. . . . The Cretaceous cycadophytes and conifers have disappeared in the interval between the two; the flowering plants are distinctly better differentiated and more modern (Berry, 1925).

Early Tertiary Floras

Eocene deposits, some of them bearing plant fossils, extend from the Coastal Plain of North Carolina (local areas of marine sediment farther

[2] In addition to *T. taxifolia* of the Southeast, other species occur in California, Japan, central and western China.

north) southward through the Gulf States and northward in the Mississippi embayment to the mouth of the Ohio River. There was a time interval between the Upper Cretaceous and Eocene in this area, and with this may be correlated the great contrast between the floras of the two periods. (Such a contrast is not seen in the Plains States, where no such interval intervened). Most of the material represented in the Eocene deposits belongs to genera now tropical or subtropical in their distribution; only a few temperate genera are represented and these are chiefly ones whose modern forms do not range far north, and which have relatives in low latitudes. In the Paleocene (represented by the Midway formation),[3] "faunas indicate subtropical bottom temperatures as far north as Paducah, Ky." in the waters of the Mississippi embayment (Berry, 1916). The records of the flora of that epoch are scanty. The Wilcox flora (lower Eocene) is one of the largest "known from a single geologic horizon in a single area." Oscillations in position of the shore line in Wilcox time and generally shallow water in much of the Mississippi embayment favored the development of coastal swamps where plant remains accumulated. In fact, "the known Wilcox flora is almost entirely a coastal flora." However, it contains a vast array of species (over 500 have been listed)[4] and many of its genera have representatives today in inland or upland forests of the warmer parts of the Temperate Zone and in the Tropics. Among its genera with modern representatives in the American deciduous forest are *Aralia, Aristolochia, Asimina, Asplenium, Bumelia, Cassia, Celastrus, Carya, Cercis, Cladrastis, Cornus, Diospyros, Evonymus, Fraxinus, Ilex, Juglans, Liquidambar, Lygodium, Magnolia, Myrica, Nyssa, Rhamnus,* and *Taxodium.* It will be noted at once that these are for the most part genera of southern affinities or genera best represented in lower latitudes. Other genera more southern in distribution (as *Persea, Sapindus,* etc.) are also still represented locally in our deciduous forest. This list contrasts strongly with those of Upper Cretaceous floras in the absence of such common Holarctic genera as *Acer, Fagus, Populus, Quercus,* and *Salix.* The contrast in floras would be further emphasized if the large number of subtropical and tropical genera were mentioned. Conspicuous among these are *Ficus* (with many species), *Cinnamomum, Artocarpus* (breadfruit), *Annona, Engelhardtia, Sterculia, Nipadites,* and *Sabalites* (probably related to the existing *Sabal Palmetto*). Additional examples could be given of genera now exclusively tropical and subtropical; of genera no longer represented in the western hemisphere; of genera whose living relatives are now widely separated geographically.[5] The great contrast between the lower Eocene and Cretaceous floras

[3] Formerly included in the Eocene.

[4] Although its 500 listed species can scarcely be considered to represent nearly that many biologic entities—only about 50 per cent (R. W. Chaney, in correspondence).

[5] Berry (1916, 1930) discusses the composition of the Wilcox flora, and lists its species.

—in the greater proportion of modern genera, in the prevailingly tropical affinities and in the paucity of Holarctic genera in the Eocene as compared with the Cretaceous—suggests that

> after the initial dispersal of the flowering plants in the Cretaceous . . . there was an extensive evolution of forms in low latitudes, and that these forms subsequently during the period of genial climate that culminated in the Oligocene radiated northward . . . (Berry, 1930).

Where there was no considerable time interval between the Cretaceous and the Eocene deposits (as in the western interior), the change in floras was gradual, and the Holarctic element persisted.

The middle Eocene or Claiborne flora is represented by less than 100 described species. Like the Wilcox flora, it is composed almost entirely of genera which are now common in equatorial regions. Among its commonest species were a climbing fern (*Lygodium*), a gymnosperm, a grass (*Arundo*), a strand palm (*Thrinax*), and a number of dicotyledonous forms now represented only in low latitudes. "The Claiborne flora is so distinctly a flora of the coastal region of the warmer part of North America that it has little in common with the Eocene floras . . . from other parts of North America" (Berry, 1924).

Throughout the Eocene, there seems to have been a steady progression toward more tropical conditions. This culminated in later Eocene (Jackson) and early Oligocene (Vicksburg) time, when the floras of the Southeast were still more tropical in character than in the middle Eocene.[6] Only a few genera not now confined to tropical and subtropical latitudes were represented. There was a species of *Castanea* (similar to the modern *C. dentata*) from Georgia;[7] one, or perhaps two, species of hickory (the leaves of one of these resemble the modern pecan); a *Tilia* from Brazos County, Texas (the first *Tilia* in the Tertiary of the Coastal Plain, and one "almost identical with the existing basswood, *Tilia americana*"); and a species of *Liquidambar* (at first referred to *L. Styraciflua,* later given a distinct name).[8]

The floras of the Catahoula sandstone (Oligocene) and Alum Bluff formation (upper Oligocene or basal Miocene) contained abundant palms

[6] Berry (1924), revising earlier statements concerning climate of the Tertiary, states that "none of the fossil floras of the Temperate Zone . . . are in the strict sense of the word 'tropical,'" although they "contain representatives of numerous genera that are now confined to the Equatorial Zone." He further states, concerning the Jackson and Vicksburg floras, that they are "the most nearly tropical known floras of southeastern North America."

[7] Referred to the Claiborne group by Berry in 1914 but later (1924) referred to the lower Jackson.

[8] The *Liquidambar* is from a deposit in the Mississippi River bluffs near Columbus and Hickman, Ky., which was referred to the Pleistocene by Lesquereaux and by Berry (1915) and later (1924) reclassified as not older than lower Jackson, and perhaps younger.

and other species of subtropical shores (Berry, 1916a, 1916d). They were much more tropical than the existing floras of these southern locations.

The flora of the lignite deposit of Brandon, Vermont (in which the fossils are mostly fruits and wood), contains representatives of genera of moderately southern distribution—*Juglans, Carya, Aristolochia, Nyssa, Sapindus, Cinnamomum*—suggesting that the warm temperate early Tertiary floras extended some distance northward.[9] Farther north, in Greenland, a rich flora with abundant dicotyledonous trees prevailed during Eocene time (see p. 457).

Middle Tertiary Floras

With the gradual cooling of climates which became evident in middle Tertiary (Miocene) time, a southward shift of the vegetation zones, which had occupied their most northerly positions in late Eocene and Oligocene time, began. This migration of floras, of vegetation types, has been emphasized in the Pacific Province by Chaney (1938, 1940). It is apparent, but much less well demonstrated on the Atlantic slope. Fossil land plants have been collected only from the Calvert formation (middle Miocene) at Richmond, Virginia, and in the District of Columbia. The flora, as indicated by the few remains,

> was a coastal flora of strikingly warm-temperate affinities, comparable with the existing coastal floras of South Carolina and Georgia . . . The climate . . . , cooler undoubtedly than that of the Apalachicola or preceding epochs, was neither cold nor cool temperate (Berry, 1916b).

Leguminous forms and oaks predominated, although at Richmond, bald cypress was the most common species. Among the genera represented are *Pinus, Taxodium, Salix, Quercus, Carpinus, Ulmus, Planera, Ficus, Platanus, Rhus, Celastrus, Ilex, Nyssa, Vaccinium, Pieris,* and *Fraxinus.* In most of these, the species resemble existing forms.

Late Tertiary Floras

The flora of the Citronelle formation of the Gulf coast (late Pliocene)

> is of a very recent aspect and yet it contains a considerable number of extinct types, some of which represent genera unknown in the Pleistocene or Recent floras of North America . . . Three species—the bald cypress, water oak, and water elm—are abundantly represented in both Pleistocene and Recent floras (Berry, 1916c).

On the whole, the flora closely resembles that now found along the Gulf coast.

---

[9] The Brandon flora was at first referred to Miocene (Perkins, 1905; 1905–06); later Berry (1919a) stated that it was too warm-climate a flora for the Miocene, and referred it to the Eocene (or perhaps Oligocene).

THE RECORD IN THE WESTERN INTERIOR AND THE ARCTIC

The Cretaceous and Tertiary record of vegetation in the Western Interior, where there were more or less extensive inland basins of sedimentation, is comparable to, but differs in certain respects from, the record on the Atlantic and Gulf coasts. The angiospermous flora was a little later in reaching the interior, suggesting its migration from more eastern areas. The record begins near the close of Lower Cretaceous (Comanchean).

Older Floras

In the earliest Upper Cretaceous (Dakota sandstone) an abundant flora, familiar to many because of its beautiful impressions of tuliptree and sassafras leaves,

> includes not only the simpler, but many of the more advanced families of flowering plants. . . . Floras of late Cretaceous age, of which the Laramie of the northern Great Plains is best known, are extensively developed in western America as far north as Alaska. Their families in all cases and their genera in many cases are still surviving, but the latter are largely distributed in warmer parts of the earth at the present time. . . . These Cretaceous floras are essentially like those of the succeeding Eocene (Clements and Chaney, 1936).

In the Rocky Mountain region, the line of demarcation between the latest Cretaceous floras and the earliest Tertiary (Paleocene) floras is placed where a difference in floras is discernible, only about 10 per cent of the species persisting from one to the other (Dorf, 1942). The Mesozoic forms gradually disappear; the modern genera increase in abundance and differentiation.

The older floras of the Tertiary (the Raton of northern New Mexico, the Denver, and the Fort Union of southern Montana, all of which are Paleocene and, therefore, older than the Wilcox flora of the Mississippi embayment) contain a considerable number of species identical with or closely related to Wilcox forms (Berry, 1916). These belong to genera now largely subtropical in distribution. In addition, these floras of the Western Interior contain a considerable percentage of typically temperate genera.

> A comparison of the Fort Union, Denver and Raton floras . . . brings out the increasing percentage of subtropical genera from north to south. Only 15 per cent of the Fort Union genera are subtropical, while 26 per cent of the Denver genera and 41 per cent of the Raton genera are now characteristic of low latitudes. There is a corresponding reduction in percentages of temperate genera from Montana south to New Mexico, over a range of 13½ degrees of latitude (Clements and Chaney, 1936).

Such a latitudinal variation is not displayed by the contemporaneous eastern floras, none of which occur as far north as the Denver and Fort Union deposits, and all of which are coastal instead of interior in location.

Latitudinal, and no doubt altitudinal, variations in forest composition of the early Tertiary eastern forests existed, just as is demonstrated in the Western Interior. The modern forests of the Atlantic and Gulf Coastal Plain and Mississippi embayment contain representatives of subtropical genera and are decidedly more southern in composition than are the forests of the interior. Even if the differences between the Wilcox and Claiborne floras of the coastal strip and the contemporaneous Eocene floras of the Appalachian Upland were no greater than exist today between the Coastal Plain and the Appalachian Upland forests, it could confidently be asserted that in Eocene time a hardwood forest containing a large proportion of the existing genera of our deciduous forest clothed the upland.

While subtropical floras dominated the vegetation of southeastern North America during the later Eocene and earlier Oligocene, temperate genera were abundantly represented in the Western Interior and in the Arctic. The Green River (middle Eocene) flora of the Western Interior contained "an almost equal element of temperate and subtropical types" (Clements and Chaney, 1936). Among its distinctly temperate genera were *Juglans, Quercus, Alnus, Salix, Planera, Celtis, Ulmus, Ilex, Acer,* and others. At about the same time, temperate floras dominated the vegetation of the Arctic, reaching latitudes 20° to 25° north of their present limits. The Greenland flora is credited with nearly 300 species including

> 19 ferns, 28 conifers—*Taxites, Tumion, Ginkgo, Juniperus, Libocedrus, Thuja, Widdringtonia, Taxodium, Glyptostrobus, Sequoia*[10] (6 species), and *Pinus* (6 species)—21 monocotyledons, including two palms, and a vast abundance of dicotyledonous leaves of *Populus, Salix, Myrica, Alnus, Corylus, Fagus, Castanea, Quercus* (15 species), *Ulmus, Platanus, Juglans* (9 species), *Lauraceae* (7 species), *Andromeda* (5 species), *Fraxinus, Viburnum, Cornus, Nyssa, Vitis, Magnolia* (6 species), *Acer* (5 species), *Ilex, Celastrus, Rhamnus, Rhus, Crataegus,* and other forms (Berry, 1924).

Comparable floras occurred in Alaska, in Spitzbergen, and across Siberia. What we now call cool-temperate forest had a circumpolar distribution in early Tertiary time.[11]

Later Tertiary History

Later Tertiary history in the Western Interior, as in the East, is very incomplete; only fragmentary records of Oligocene and Miocene plants are

[10] Both *Sequoia* and *Metasequoia* are represented among the Greenland fossils (Chaney, 1948; 1949, in correspondence).

[11] The exact age of the Greenland flora has been the subject of some discussion. Berry (1924) states that in recent years "it has been generally considered to be of Eocene or Oligocene age."

known. The history farther west—in the Pacific Province—is beautifully recorded in the abundant fossils of that area (Chaney, 1925, 1938, 1940, 1947; Clements and Chaney, 1936; Cain, 1944). This, however, is indicative of trends only, and cannot be used in reconstructing a picture of eastern conditions. In the Pacific area, the general trend in the Miocene, and continuing into the Pliocene, was an elimination of humid types from the interior (coincident with the growth of the Cascade range) and a southward movement of subtropical forms.

> To the east of the Rocky Mountains, increasing aridity reduced the forest vegetation and its opportunities for preservation to such an extent that the fossil record there is limited. A wide extent of grasslands is indicated by the abundance and diversity of grazing mammals . . . The larger valleys were occupied by *Celtis, Fraxinus, Platanus, Populus* and *Ulmus,* all of which genera have survived to the present in more humid situations to the east . . . and the upland areas were covered by grasses resembling *Stipa,* whose seeds have been found in many localities (Clements and Chaney, 1936).

The meager late Tertiary floras of the High Plains (from western Nebraska, Kansas, and Oklahoma) indicate a climate slightly less dry than that of the present. The fossil flora (of the western Oklahoma horizon) resembles that now growing 180 miles or more to the east. The elimination of genera of high moisture requirements[12] from the High Plains is evident (Chaney and Elias, 1936). Evidence of the occupancy of the High Plains in late Tertiary time by a prairie flora is indicated by recent collections of a fossil flora containing 18 species of grasses (Paniceae and Agrostideae), five species of Boraginaceae, one *Celtis,* and two trunks provisionally classed as *Yucca* (Elias, 1935). A continental climate and a pattern of distribution of vegetation types similar to that of the present were initiated. The separation of eastern and western forests had been accomplished. The eastern deciduous forest was forced to retreat eastward from the drying interior, just as it was being forced southward away from high latitudes by the general cooling of climates. The beginning of the differentiation of the eastern deciduous forest doubtless dates back to the climatic shifts of the Tertiary.

THE PLEISTOCENE RECORD

Fossil records of vegetational history during all but the waning stages of the Pleistocene epoch are meager, and, as was true of Tertiary records, are poorly distributed geographically. Pleistocene plant fossils may be considered as belonging to two categories: (1) macrofossils—plant fragments,

[12] Fossil records show that the common genera of mesophytic forests (as *Fagus, Acer, Tilia, Aesculus, Fraxinus, Castanea,* and *Asimina*) were present in the Western Interior in early Tertiary (Fort Union formation).

such as leaves, fruit or seeds, and wood, and (2) microfossils—pollens, spores, and diatoms.[13] The first group of fossils are generally found in interglacial deposits within the glaciated area, in lake deposits adjacent to the ice margin, or in carbonaceous lenses in Pleistocene terrace deposits of the Atlantic and Gulf coasts and Mississippi embayment. Pollens are best preserved in bogs, where the kinds occurring at the different depths (ages) are indicative of the series of vegetational types which have succeeded one another in the general area where the bog is located.

Enforced migrations of vegetation took place not once, but several times, contemporaneously with the growth and advance of the several distinct ice caps. Fossil records do not give conclusive proof, and opinions differ, as to the magnitude of the north-south swing of vegetation belts.

INTERGLACIAL RECORDS

Vegetational recovery in the north, during the long interglacial stages, is shown by buried soils (developed on one drift sheet and covered by a later drift deposit), buried peat deposits, and by occasional plant remains. Outstanding among the latter are those from the Don beds of the Toronto formation (Coleman, 1906, 1927; Flint, 1947).[14] Here was found a varied angiospermous flora, including about 30 tree species, some of which are now northern, some southern in distribution. A few are considered to be specifically distinct from modern species, most are referred to existing species. The presence of such trees as osage orange, locust, hickory, papaw, sycamore, etc., suggests a climate comparable to that of the central part of the deciduous forest; the presence of "northern" species, as *Acer spicatum, Taxus canadensis, Thuja occidentalis, Picea, Larix,* etc., seems on a basis of existing segregation of forest communities to be contradictory. Perhaps climatic segregation of zonal types, which first became apparent in late Tertiary, had not been completed. Perhaps there was an assemblage here of relic or of invading species together with a regional type. Comparable mixtures have been found elsewhere. In central Illinois, buried peat of late Sangamon age contains pollen of *Abies, Picea* and *Pinus,* leaves of *Picea,* and wood of *Larix* (Voss, 1933). Thus, following the retreat of Illinoian ice, or perhaps at the time of Iowan ice advance, a conifer forest

[13] Diatoms are not included here as they are not a part of the forest flora. They are considered by Hanna (1933), by Cocke, Lewis and Patrick (1934), by Patrick (1946), and mentioned by others. Davis (1946) believes that diatoms have little climatic significance.

[14] "The Toronto formation [at Toronto, Canada] has been correlated with the Sangamon in the Mississippi region; but it seems to me to belong farther back in the series, probably in the Yarmouth stage" (Coleman, 1927). Flint (1947) believes that erosion would have removed any such early interglacial deposits, and that they are probably Sangamon in age.

stage developed. How long it persisted, or what it was replaced by, is not known. The peat deposit is overlain by Peorian loess. The cones of white spruce and the wood of black spruce have been found below the Kansan drift, demonstrating the presence of these trees in the Aftonian interglacial stage. Plant remains and pollens recovered from pre-Kansan peat deposits in southeastern Minnesota and in Iowa indicate the presence of mixed coniferous–deciduous forest (or of conifer and hardwood communities) containing *Abies, Picea, Larix, Pinus, Acer, Quercus, Tilia, Betula,* and *Juglans* or *Carya* (Nielson, 1935; Wilson and Kosanke, 1940).

RECORDS FROM PROGLACIAL LAKES

The advance of ice into the valleys of northward-flowing streams ponded such streams, forming proglacial lakes. Plant remains have been collected from terrace deposits of the Monongahela River near Morgantown, West Virginia, which were laid down when an ice sheet blocked the then northward flowing master stream near where Beaver, Pennsylvania, is now located (White, 1896; Leverett, 1902; Fenneman, 1938, map, p. 317). This was an event of early Pleistocene, probably Jerseyan, age. Among the plant remains (identified by Knowlton, 1896) are river birch (*Betula nigra*), beech, oak (*Quercus falcata,* stated by Berry, 1907, probably to be *Q. predigitata*), chinquapin (*Castanea pumila*), elm (*Ulmus racemosa*), sycamore, and sweet gum. This is an assemblage which might be found today in the same general area; it certainly emphasizes the similarity of the earliest Pleistocene and modern forest of the Appalachian Plateau. No northern species were noted from this deposit, which indicates that the climatic influence of the ice front had not at this time extended far to the south of the ice.[15]

PLANT FRAGMENTS OF ESTUARINE AND FLUVIATILE DEPOSITS

South of the glacial boundary, Pleistocene vegetation is recorded in part by plant fragments, and in part by pollens in bogs. The former are confined almost entirely to coastal and river terraces; the latter are more widely distributed, some on the younger terraces, some in older areas farther removed from the coast.

In a number of places along the Atlantic and Gulf Coastal Plain, plant

[15] Leverett (1902) states that "diversion of the old Monongahela system to the Middle Ohio system took place as a result of the first glaciation, though it may have been brought about some time before the ice sheet had reached its farthest limits." The distance between the location of the plant remains and the edge of the ice may thus have been somewhat more than is indicated by maps showing the position of the southern margin of the ice sheet.

remains have been collected, generally in localized spots in estuarine or fluvial terrace deposits.[16] The ages of these deposits are uncertain. Some have been definitely correlated with certain of the marine terraces—Pamlico, Chowan, or Wicomico; others are located only geographically. At least one (on the Neuse River in Wayne County, North Carolina) is stated to be postglacial. Pleistocene and recent leaf beds (stranded and buried rafts) in flood plains cannot always be distinguished from one another (Berry, 1906a). There has been an attempt to correlate the several Pleistocene terraces of the Coastal Plain with interglacial stages. If these terrace deposits are of interglacial age, their plant remains tell nothing as to conditions during the glacial stages. Flint (1940) considers that from the James River northward, the sediments were deposited chiefly by streams; and that from the James River to Florida, at least up to a present elevation of 100 feet, sediments were deposited chiefly under marine conditions, and are interglacial. However, Berry (1925) states that it is not likely that all the Pleistocene terraces are interglacial, particularly as they contain what appear to be ice-borne boulders. He further states that there is no element in the Pleistocene of North Carolina which can be considered northern, that none of the evidence in the Pleistocene floras of these low altitudes shows an appreciable lowering of temperature. This is in contrast with statements by Buell based on pollen records (see p. 471). The plant fragments—leaves, fruit, and wood—of these coastal plain deposits represent a large number of tree species, many of wide distribution, others generally confined to estuarine swamps. *Taxodium* is widely represented (from New Jersey to Florida and Mississippi), sometimes by stumps in situ. Logs, roots and stumps in situ of *Chamaecyparis thyoides* are preserved in bog deposits on the lowest coastal terrace[17] in North Carolina (Dachnowski-Stokes and Wells, 1929). *Fagus* is known from a variety of horizons in the Pleistocene from Massachusetts to Texas. Oaks are abundantly represented, generally with a goodly proportion of swamp species. Pines, hickories, walnut, magnolias, tuliptree, sweet gum, redbud, maples, gums, and many others are recorded and point conclusively to the general persistence throughout the Pleistocene, at least in coastal regions, of forest communities comparable to those of today. Even the estuarine deposits of Chesapeake Bay, which must have been affected by melt-

---

[16] Papers by Berry (1906, 1907, 1907a, 1909, 1909a, 1910, 1910a, 1912, 1917, 1924a, 1925a, 1933, 1943) give localities and nature of collections made. Knowlton (1919) lists Pleistocene flora.

[17] The stratigraphy indicates that, following withdrawal of the sea (probably with advance of the Late Wisconsin ice sheet), marsh vegetation developed, and was followed by a pure stand of *Chamaecyparis*. An advance of the sea over this area (correlated with glacial melting) killed and buried the trees. Recent emergence (to present elevation of six to eight feet) permitted development of the present evergreen shrub bog.

water from the ice front in the upper Susquehanna Valley, give no indication of glacial influence.[18]

At some time in the Pleistocene, a flora similar to that of the Atlantic Coastal Plain appears to have occupied the Great Valley. Near the west base of the Blue Ridge, in Rockbridge County, Virginia, is the only inland station for *Taxodium; Quercus alba* and *Q. predigitata,* together with a few other species, were also found in the same locality (Berry, 1912).

Records from peninsular Florida (at Vero, about midway on the east coast) include a number of more southern species than those generally found as fossils on the Atlantic Coastal Plain. Among these may be mentioned *Pinus caribaea, Serenoa serrulata, Sabal Palmetto, Quercus virginiana,* and *Annona glabra.* "The live oak is one of the most abundant trees in the Pleistocene deposits of our southern states, and its ancestors are already well defined in the Pliocene of this region" (Berry, 1917a).[19] Latitudinal influence, somewhat comparable to that of today, is suggested.

Conditions in the lower Mississippi Valley appear to have been slightly different from those of other parts of the Coastal Plain, if the fossils preserved in the different areas are equally representative of surrounding forest. "The remains of typically northern species such as white spruce, larch, and white cedar have been found with such southern plants as oaks, bamboo, hard pine, mulberry, hickories, persimmon, tulip poplar, and elms" (Brown, 1938). In one of three exposures of Pleistocene fossil-bearing deposits north of Baton Rouge, Louisiana, *Taxodium* is present and represented by stumps. Among other deciduous forest and southern species listed are southern white cedar, walnut, beech, cucumber tree, box elder, Styrax, ash, and elderberry. The remains of northern and southern species are intermingled, and there is no evidence of stratification or reworking of sediments so as to mix the remnants of two climatic stages in the deposits. Rather, the indications are that northern and southern species both grew in the lower Mississippi Valley, doubtless, however, occupying unlike sites.

Submarine deposits in the Gulf of Mexico (Trask, Phleger, and Stetson, 1947) exhibit two subarctic faunal layers between which is a cool-water faunal layer. No information is available as to the lateral extent of these

---

[18] Berry (1933) states that the terminal moraine of the Illinoian glacier and the valley train on the north and west branches of the Susquehanna River are practically continuous with the Wicomico terrace on Chesapeake Bay. Plant remains from a fossil swamp deposit in Washington, D.C., of Wicomico age (Wentworth, 1924; Berry, 1924a) prove the existence then of a cypress swamp similar to those of the present time. If any temperature effects were felt below the mouth of the Susquehanna, they were not extreme enough to be felt to the limits of tidewater on the Potomac arm of the Bay.

[19] The Vero deposits are considered by Berry (1917a) to be of "late Pleistocene age." Hay, who studied the vertebrate remains of the deposits, considers that the upper bed is of middle Pleistocene age, while the lower beds "belong to the early part of the epoch."

beds, or as to their possible correlation with the effects of water discharged from the Mississippi River.

It must be remembered that the Mississippi River carried a tremendous volume of melt-water from the ice front, that it was an overloaded stream depositing here, eroding there, constantly modifying its broad valley somewhat as does a braided stream. Thus new and bare areas and river swamps were being formed into which migration was more rapid than onto occupied land. Under the influence of the cooler melt-water, and in the absence of competition of closed communities, it is readily conceivable that a few species migrated from the ice front almost to the Gulf. Their presence does not imply great cooling of climate; in fact, the presence of the southern species in larger numbers precludes that possibility. Neither does their presence here imply migration southward into heavily forested areas of the southeastern interior (see Chapter 17).

THE POLLEN RECORD

In the north, the history of vegetational recovery and migration during and following the retreat of the last ice sheet is recorded by the pollens preserved in the numerous bogs within the glaciated territory. South of the glacial margin, relatively few pollen records, widely separated in space and occasionally of doubtful age, afford paleontologic evidence of the composition of forests, of migrations of forest species on the unglaciated lands.

Bogs in which pollens are preserved may have formed in depressions on the drift mantle, in scoured rock basins blocked by drift, in coastal areas as a result of subsidence or emergence, around lakes which occupied depressions in emerging marine terraces, or in otherwise obstructed drainage lines, or even as a result of beaver dams. The character of the peat differs under different modes of formation and different plant communities.[20] Interruptions in the record may have occurred as a result of drowning or other catastrophe, as fire. The uppermost layers, particularly, have suffered in the latter manner.

The pollens found in bogs are generally taken to indicate that the species of trees from which they came grew in the surrounding forests, at no great distance from the site of preservation. Sears (1935) states that "pollen from remote sources seems within most of the regions studied to be a theoretical, rather than a practical, consideration." Carroll (1943), studying recent pollen deposition in bryophytic polsters and mats in the spruce–fir forest of Mount Collins in the Great Smoky Mountains, found that "pollen from several species foreign to the spruce–fir forest was present as a contamination which averaged 23 per cent," and that "in some cases the source of this pollen was several miles distant and several thousand feet

[20] See Waksman (1943) for a classification of types of peat.

lower in elevation." Erdtman (1943) gives data on distances to which pollen grains are carried, based in part upon their collection on an ocean voyage. He also notes instances of pollen, especially of pine and spruce, carried 500 or 600 miles from its source, and of well-defined pollen rains during strong winds. He concludes that "it is extremely probable that pollen grains of *Betula* and *Pinus, Quercus* and *Salix,* as well as of certain sedges and grasses, are carried in great quantities by the wind for more than 1000 kilometers into the middle of the ocean."

Exceedingly large numbers of pollen grains may be found in a single gram of dry peat, the numbers sometimes ranging from 50,000 to 1,000,000, of which the most important are those of trees (Waksman, 1942). The percentages of the various kinds of pollens present in the different levels from bottom to top of the bog are usually calculated from counts of 150 to 250 pollen grains (Cain, 1939). A few workers use 500 or even 1000 grains. In many (perhaps most) instances, it has not been possible to distinguish species within a genus.[21] Thus, for example, all oaks are lumped under *Quercus,* although this genus today contains southern and northern, eastern and western species, swamp species, xeric species, and mesophytes. The same is true of *Pinus* (with certain exceptions). Pollen profiles from a single locality might illustrate forest sequences resulting from successional development as well as those induced by climatic change.[22] When a considerable number of profiles from different localities display comparable trends, these are interpreted as demonstrating climatic influences. Especially do the appearance and disappearance of a climax species (as beech) suggest climatic change.

Records in the Glaciated Area

From a study of a large number of pollen profiles of bogs in the glaciated area, it has been possible to determine certain climatic sequences of post-glacial time. These periods, as outlined by Sears (1942) are:

V. The present—probably cooler and with more available moisture than in IV.

IV. A warm dry period—maximum of oaks and hickories, minimum of beech.

---

[21] Species within a genus are sometimes distinguishable because of size differences of the pollen grains. However, there is considerable range in size, and size-frequency curves overlap. For example, Cain (1948) states: "It is not claimed that individual pollen grains of spruce can be identified and that percentage composition of the species can be determined for use in pollen spectra, but it does seem that by a combination of size and form characteristics, when numerous grains are available from a particular sediment, the presence of certain species can be detected."

[22] The Dismal Swamp profile (Lewis and Cocke, 1929; Cocke, Lewis, and Patrick, 1934) is one of the former; all profiles in the area of Wisconsin glaciation illustrate climatic change, probably complicated by successional development.

III. A more humid, also warm, period—maximum of beech, and, in places, of hemlock.
II. A dry, probably warmer, period—maximum of pine, often with oak.
I. A moist cool period—maximum of fir and spruce.

Recently, a more detailed correlation with Late Wisconsin substages has been attempted (Sears, 1948). Twelve phases, including six distinct advances and subsequent retreats of ice, are distinguished. Bogs near the margin of Wisconsin drift record vegetational changes accompanying eight of these phases.

At the bottom of most bogs of the glaciated area, *i.e.,* in period I, pollens of *Abies* and *Picea* are fairly abundant. In many instances, pollens of other tree species are present, and often in larger amounts. Pollen profiles from bogs in northern Ohio and Indiana generally illustrate the pronounced spruce–fir maximum of period I. In the East (New Jersey and southern New England in particular) pine is conspicuous in the early conifer period.

Gradually in some areas, more abruptly in others, dominance is shifted to *Pinus*. This change marks period II. The pines of this period are generally *P. Banksiana* and *P. resinosa;* oaks almost always accompany the pines. *Pinus*[23] continues in large amounts, even to the topmost levels, in profiles of bogs in northern Wisconsin, western Upper Michigan, and Minnesota (Potzger, 1946); in variable, but generally large amounts throughout the profiles of bogs in New Jersey and Connecticut (Potzger and Otto, 1943; Potzger, 1946; Deevey, 1939).

An increase in pollen representation of genera of higher moisture requirements—as *Fagus, Tsuga*—and a decline of *Pinus,* characterizes period III. This is shown by profiles of Ohio bogs.

The subsequent warm-dry period, IV, was marked by an oak maximum, often accompanied by *Carya*. The more mesic genera decline. This period, also, is illustrated by Ohio bogs. It is very prominent in certain of the Indiana bogs, where *Quercus* retains its dominance into the upper levels. This warm-dry period indicated by pollen frequencies corresponds with the so-called "xerothermic" period postulated by Gleason (1923) on a basis of relic colonies. It coincides with the time of maximum extension of the Prairie Peninsula. In mid-western locations, for example, in Iowa, a marked increase in grass and other herb pollens is evident at this time.

Indications of climatic shift to a somewhat moister and cooler climate mark period V. This is shown in many bogs by a decrease of *Quercus* and *Carya,* an increase in mesic genera, both in number and in pollen representation. It is shown best in the North by a recent slight increase in *Picea;* in the East, this period may be marked by an increase of *Tsuga* and *Castanea.*

[23] It must be remembered that several species of *Pinus* may be involved. Sears (1942) states that *P. Strobus* is more abundant upward. In New Jersey and Connecticut, *P. rigida* is doubtless present.

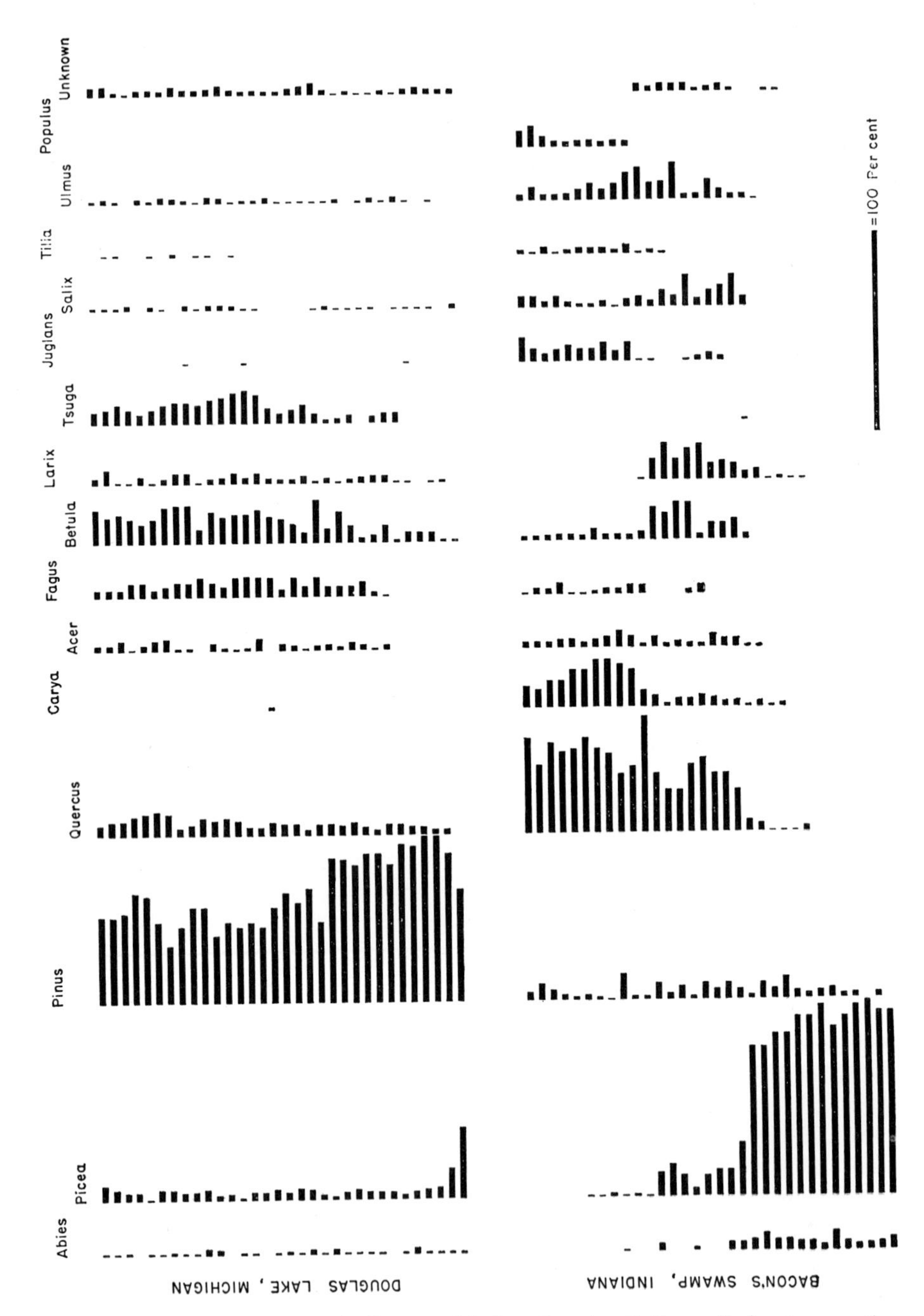

Pollen profiles for Bacon's Swamp, Marion County, Indiana (below) near the southern limits of Early Wisconsin glaciation, and for Douglas Lake (above) in northern Lower Michigan. Percentages of various pollens at one foot intervals (to depth of 32 feet in Bacon's Swamp, from depth of 70 to 101 feet beneath Douglas Lake), indicated by length of bars. In the Indiana bog, note long period of *Picea* dominance, strong *Quercus* and *Carya* maxima, weak representation of *Fagus,* and

 (*Continued on facing page.*)

For specific features of profiles of different bogs, the reader must turn to the abundant literature;[24] especially helpful are those papers which present information on bogs of several regions, thus facilitating comparisons (see Sears, 1935a, 1942a, 1948; Potzger, 1946).

No brief generalization can do justice to the array of evidence of forest composition and of the migration of genera which the large number of individual pollen profiles affords. Neither will it be correct for all parts of the glaciated area. Bogs of the north are much younger than those near the southern borders of the Wisconsin ice sheet; these bogs started to develop later, so that their basal sediments do not correspond in time to the basal sediments of more southern bogs. Bogs in or near the Prairie Peninsula emphasize the warm-dry period IV. Those in the northeast give the poorest evidence of a drier period. Differences in vegetational response in relation to climatic periods as interpreted by Deevey (1939) are shown in the accompanying chart.

Throughout the series of vegetational changes of Late Wisconsin and postglacial time, successional development, soil development, and topographic development were in progress. Certainly the raw soil material and the complete lack of a drainage system (resulting in water-logged soils) must have influenced the nature of the dominant vegetation, particularly of the earlier postglacial stages. The vegetational changes may, perhaps, be correlated with the widespread edaphic factors as well as with climatic factors. The strong contrast between bog records within and beyond the glacial boundary in New Jersey, with lack of evidence of pronounced climatic changes outside the glacial boundary, emphasizes the probable importance of the edaphic factors within the glacial boundary.

Records From South of the Glacial Boundary

Pollen records from bogs south of the glacial boundary[25] afford information as to forest composition during some portion of glacial or postglacial

[24] No attempt is made here to cite a large number of papers, as literature lists in those which are cited will furnish adequate material.

[25] Such records are available from New Jersey (Potzger, 1945; Waksman, *et al.*, 1943), Pennsylvania (Sears, 1935; Schrock, 1945), Tennessee (Sears, 1935), Virginia (Lewis and Cocke, 1929; Cocke, Lewis, and Patrick, 1934), North Carolina (Buell, 1945, 1946), South Carolina (Cain, 1944a), Florida (Davis, 1946), Arkansas (Sears and Couch, 1932), Oklahoma (Sears, 1935), and Texas (Potzger and Tharp, 1943, 1947).

almost complete absence of *Tsuga*. These features (except of *Picea*) are related to location in the Prairie Peninsula. In the Michigan profile, note strong representation of *Pinus* throughout the section, comparatively weak *Quercus* representation, absence of *Carya* (except at one level), good *Fagus* representation almost throughout the record, and good representation of *Tsuga* which is in strong contrast with its absence in Indiana. (Data for profiles from Potzger, 1946; records previously published by Otto, 1938, and Wilson and Potzger, 1943.)

Proposed Correlation of Climatic and Vegetational Changes in Eastern North America

Climate	*Eastern Canada* (Auer)	*Connecticut* (Deevey)		*New York* (McCulloch)	*Ohio* (Sears)	*Wisconsin* (Voss)
		Coastal	*Inland*			
Moister and/or Cooler	Spruce	Oak–Chestnut	Spruce–Hemlock	Oak–Hemlock	Oak–Beech	Spruce
Warm, Dry	Oak–maximum	Oak–Hickory		Oak–maximum	Oak–Hickory	Oak–Mixed Deciduous
Warm, Moist	Hemlock–Oak	Oak–Hemlock		Oak–Hemlock	Oak–Beech	
Warmer, Dry	Pine	Pine		?	Pine	Pine
		Spruce–maximum (last glacial advance?)				
Cool	Spruce–Fir			Spruce–Fir	Spruce–Fir	Spruce–Fir
		Spruce–Fir				
Deglaciation		←		Tundra Flora? →		

time. Records from buried soils and buried peats may be earlier (p. 471). The length of the record differs in different bogs; the time of initiation of bog sedimentation is often uncertain. Some of the records are very short. The trends in forest composition which are well marked in the glaciated area, and are correlated with climatic fluctuations or sequences of climatic periods are faintly displayed or absent in bog records from south of the glacial border.

In New Jersey, where records within and south of the glacial border are available within a short distance of one another, there is a striking difference in the history of vegetation. Whereas north of the glacial boundary, a series of changes, of forest migrations, is indicated, south of the glacial boundary the record shows "weakly expressed succession, and a heterogeneous composition of the forest, which was uniformly the same for the time during which the peat accumulated" (Potzger, 1945). A slight infiltration of northern species is apparent, particularly in those bogs nearest the glacial boundary; fir and spruce (*Picea glauca* and *P. mariana*) are represented at the same time that more southern trees, as *Castanea* and *Liquidambar,* were present. *Pinus* and *Quercus,* as today, were the dominant genera almost throughout the history of these bogs, and "*Castanea, Carya,* and *Tsuga* were present when deposition began . . . while in northern New Jersey [in the glaciated area] considerable succession had taken place in the forest composition before they entered." This would seem to indicate that the mixed forest was not displaced south of the glacial boundary. Potzger suggests that "the Pine Barrens constituted a refugium during Pleistocene times."

Two records from Pennsylvania are situated at altitudes of 2400 feet and 1800 feet, the former in the Allegheny Mountains section of the Appalachian Plateaus, the latter in a high valley in Seven Mountains in the Ridge and Valley Province. The vegetational (and climatic) sequence evident in bogs of the glaciated area is suggested by the Allegheny Mountain bog, which is located in the transitional belt between the Mixed Mesophytic Forest region and the southern lobe of the Hemlock–White Pine–Northern Hardwoods. The warm-humid period (III) is marked by a hemlock maximum. The short record from Seven Mountains (the bog formed in a beaver-made lake), with fir prominent at the lower levels (spruce and fir are still present), shows a marked increase in *Quercus* in the upper half of the record. Oaks are now dominant over many of the mountain slopes. Records from Cranberry Glades, West Virginia, indicate that there has been very little change in vegetation of this area since the beginning of peat deposition. The lowest layer (13½ foot level) shows spruce and fir as most abundant, but with hemlock, birch, and oak also present; subsequent levels indicate that the surrounding vegetation was similar to the present vegetation (H. C. Darlington, 1943). One Tennessee record, a short one in a beaver-made

Pollen profiles for Bulltown bog (below) in the unglaciated New Jersey Pine Barrens, and for White Lake (above) in glaciated northern New Jersey. Percentages of various pollens at one foot intervals (one-half foot intervals in lower two feet of White Lake) indicated by length of bars. Note contrast in amount of *Abies* and *Picea* pollen in the two bogs, absence of *Castanea* in the lower half of White Lake bog, and its presence throughout the period of sedimentation in the Bulltown bog. *Carya* (accidentally omitted from Bulltown profile) is present in small amounts at 2, 3, 4, and 5 foot levels. (Data for profiles from Potzger, 1946; records previously published by Potzger, 1945, and Potzger and Otto, 1945.)

lake, is similar to the Seven Mountains profile in Pennsylvania, differing chiefly in the larger amount of *Picea* in the lower part, in the prominence of *Acer* above, and in the lesser abundance of *Quercus.*

The Dismal Swamp records in Virginia demonstrate successional development from open marsh to closed forest. No climatic changes are indicated.

In a bog on the Coastal Plain of North Carolina (Bladen County), which is surrounded today by sand-hill vegetation of turkey oak and longleaf pine, *Abies* is present in the bottom one foot. The size range of its pollen is such that either *A. balsamea* or *A. Fraseri* is indicated. Six tree genera are important in the record: *Quercus, Pinus, Abies, Carya, Betula,* and *Acer.* The pine size frequency indicates *P. Banksiana.* Buell (1945) believes that the ancient lake was surrounded by jack pine on the coarse, sandy ridge, while on the sand plain to the west was mesic forest with oak and fir predominating. It should be noted that this bog developed on a marine terrace newly exposed as a bare land area at the time of the last glacial advance; competition of established vegetation was lacking.

Pollen analyses of certain highly organic buried soils in the Piedmont of South Carolina show pollen of *Abies* and *Picea,* and of a pine which, on the basis of size frequency of the grains, may be *Pinus Banksiana;* "the *Abies* is not one of the modern species of Eastern North America." The time of deposition is unknown, although it was probably during the Pleistocene. Cain (1944a) considers that the general vegetational pattern is as follows:

> The rolling uplands were covered by a climax of oak–hickory–chestnut; ravines and protected slopes contained stands of mixed mesophytes; several places where small streams were impounded postclimax spruce–fir grew on and around bog-like basins; over the upland the prevailing climax was interrupted by stands of pine and pine–hardwood mixtures representing various stages of secondary succession.

Florida records (Davis, 1946) show the occurrence of spruce and of fir in peat deposits buried beneath marine deposits. The age of these deposits is uncertain. However, they are not post-Wisconsin, and their records may later be found to correlate with those of buried soils in South Carolina.

The records from the 22 foot deep Patschke Bog in Lee County, Texas, are of particular interest (1) because of the presence in small percentages of *Abies, Picea glauca,* and *P. mariana* (the three together totalling 0.5 to 3.5 per cent, except in the lowest foot-level, where they comprise 6 per cent of a total of only 50 pollen grains) in the lower part of the profile; and (2) because of the high percentage of *Castanea,* especially in the upper half. This genus disappeared just before the close of the record; its range now terminates about 100 miles to the east. Throughout much of the record, *Quercus* is a dominant; grass pollens are very abundant except for a short time near the middle of the record. An early pine maximum is evident. An infiltration of northern trees is suggested, here into a region where the early regional type appears to have been a pine–oak–grass or oak–grass savannah. A more mesic period is suggested later by the greater abundance of *Castanea*.

The very short Arkansas and Oklahoma records give no evidence of development. They show composition in keeping with their geographic location—*Pinus* and *Quercus* prominent, and in Arkansas, *Nyssa* also.

About the only generalizations which can be made from pollen profiles of bogs south of the glacial boundary are: (1) in certain instances, a slight infiltration of northern genera early in the record; (2) general agreement of the records with existing surrounding vegetation; (3) a period more humid than the present is indicated in the west, which perhaps may be correlated with Sears' period III.

Some information on probable location of glacial refugia, and courses of migration may be gained by observation of the horizon of first appearance of a genus in each available bog profile both north and south of the glacial boundary. Such evidence, worked out by Sears (1942a), will be used in tracing the final stages of development of the existing forest pattern of northern eastern United States (see Chapter 17).

CHAPTER 15

Disjunct Occurrences of Species and Communities and Their Significance in Forest Migration

---

Disjunct occurrences of species and communities afford information as to migrations or previous extent of vegetation types to which the species or communities belong. Disjunct distribution may result from the isolation of relic stations due to shifting vegetational boundaries. However, not all disjuncts are relics; some are "advance guards" in migrations still in progress. While it may not always be possible to ascertain whether disjunction is due to retreating or advancing migrations, evidence in most instances is satisfactory. Diagnostic features indicating the nature of the migration are: (1) the place of the disjunct species or community in successional development of vegetation; (2) the nature of the habitat occupied (physiographically youthful or old); (3) the climatic affiliations displayed; and (4), in the case of disjunct species, the location of related species and varieties, hybrids and genotypes. The element of chance may occasionally enter in; unusual instances of dispersal are known.

There must be a fundamental cause when disjunct communities are composed of a considerable number of species, when a number of similar disjunct communities occur and under similar environmental conditions, and when several unrelated species have the same distribution pattern. Deductions based on relic occurrences are sometimes supported by evidence from the fossil record.[1] Such support lends weight to the principles involved in the cause and effect relationship.

In all instances of disjunction the present habitat must be suitable for the continuance of the disjuncts. Usually the habitat is marked by peculiar edaphic conditions which tend to favor continuance of the disjuncts, or to exclude the surrounding regional vegetation. But the location and extent of the area with controlling edaphic factors may have been determined by a

---

[1] For example, the post-Wisconsin xerothermic period was postulated by Gleason (1923) on evidence afforded by relic colonies. As a result of pollen studies of bogs, the existence of such a dry period was again demonstrated (see p. 465). For discussion of the xerothermic theory, see Sears, 1942b.

complex series of past events. When these are deciphered, then the distribution pattern becomes significant as a source of information concerning past vegetational migrations. Such disjuncts are relics. If location and extent of the area of controlling edaphic factors is determined by recent events, events still in progress, disjuncts are indicative of advance migration, of possible further expansion of range.

It is not within the province of this book to discuss the principles involved in the interpretation of discontinuous distributions.[2] Concrete instances of disjunctions will be presented, together with the attendant implications.

The age, that is, the possible period of continuous vegetational occupancy of any land surface where disjunct communities occur, must be considered in interpreting such communities.[3] So also must the possibility of and time of former land connections or continental extensions be kept in mind in explaining migrations and disjunctions.

Relic communities within the area of Wisconsin glaciation are postglacial in origin. They give evidence of vegetational migrations during and after the retreat of the last ice sheet. On areas of older drift not reached by later glaciation, disjunct occurrences could, but not necessarily do, antedate the later glaciations. The age of vegetation of unglaciated spots north of the glacial boundary (the Driftless area is one of these) is not necessarily fixed by the age of nearby glaciated land; it may be much older.

Areas south of the glacial boundary may be young areas—areas as youthful as the latest till sheet; old land areas—areas antedating the origin of the angiospermous flora; or land areas of intermediate age—areas which emerged from the sea at some time during the Tertiary, during the period of rapid deployment of modern floras. On all of these there are disjunct communities. Age of the land surface will not alone suffice to fix the age of these communities. In the older areas particularly, physiographic history has played an important role. The modification of land surfaces by erosion, the degree of maturity of topography, the completeness or extent of peneplains, all affect the nature of vegetation and hence the extent of different types of communities. Expansions of some, contractions of other vegetation types go hand in hand with the progress of the erosion cycle. In the past, as now, drainage systems have served as migratory pathways. Changes in drainage systems, reversals of direction of flow, realignment of tributaries have taken place, some as a result of advance of early ice sheets. Relics dating back to Tertiary peneplains, and to Tertiary drainage systems (pre-

[2] For discussion, see Cain (1944) and the many references there cited.

[3] It is not to be understood that relic communities have maintained themselves with unchanging floristic composition, or continuously in the same spot. Rather, the community has retained its characteristic physiognomy and at least an assortment of forms whose lineage carries back to the time of the establishment of the vegetation type to which the present relic belongs. Its area and exact location have shifted as the extent and location of the controlling edaphic factors have shifted.

Pleistocene) still exist. Disjunct communities of the other sort (advance guards), apparently keeping pace with the progress of the present erosion cycles, are occupying areas of erosion topography in the youngest areas.

Continental extent and elevation, as well as topography, affected vegetational boundaries. A former coastal strip, now the submerged Continental Shelf, must be reckoned with if disjunctions in the North Atlantic States and maritime provinces of Canada are to be explained.

Climatic changes of the Tertiary, which resulted in successions of floras (see Chapter 14) and in shifting boundaries of vegetational types, have left their imprint on the present pattern of forest distribution.

From among the hundreds of instances of peculiar distribution of species and communities which bear the imprint of past events, selected ones are mentioned here which are of value in tracing the evolution of the present pattern of forest distribution and the relationships of climaxes. Many others could be cited. There are also numerous, as yet inexplicable, cases which must await further information on species distribution, local physiographic history, and genetic variability. The examples used are not arranged chronologically, but in relation to the physiographic and age areas in which they occur.

DISJUNCTS IN THE AREA NORTH OF THE GLACIAL BOUNDARY

Area of Wisconsin Glaciation

Many bogs in the southern part of the area of Wisconsin glaciation still harbor boreal species. In the more southern, few of the northern conifers remain; northward the number of species increases, and disjunction decreases, until finally bog communities are not climatically out of place in the surrounding vegetation. These bog communities indicate clearly, as do the pollen profiles in their sediments, the northward trend of postglacial forest migrations.

A few boreal relics remain on cliffs, in areas unsuitable for species belonging to the surrounding regional vegetation. *Thuja* and *Taxus* occur more or less frequently in such situations. Hemlock, and hemlock and white pine colonies in Indiana have been interpreted as boreal relics. Based upon the lateness of appearance of *Tsuga* pollen in pollen profiles of bogs in the general area of the Prairie Peninsula, these hemlock colonies must have become established late in postglacial time—after the close of the xerothermic period, which was unsuitable to *Tsuga* (Sears, 1942a).

Prairie relics are numerous on the Wisconsin drift; small and more or less isolated at the eastern limits of their occurrence in Ohio, they increase in number and extent westward, until finally they merge with the prairie. Together they indicate the extent of the Prairie Peninsula. In the same general area, and frequently surrounding or supplanting relic prairie colonies, oak or oak–hickory forest interrupts the Beech–Maple Forest. Oak communities continue prominent eastward of the prairie colonies. This eastward lobe of prairie and oak–hickory forest dates from the warm-dry postglacial period, the oak maximum, clearly indicated in the pollen profiles of bogs (p. 465).

A number of coastal plain species occur in local edaphic situations in central New York, in the Great Lakes region, and in the sand plains of northwestern Wisconsin. These are indicative of former more or less continuous suitable pathways for migration from the northern Atlantic Coastal Plain westward.

Outliers of southern species in the glaciated area are indicators of advance migrations. Some of these may have reached their present northward limits late in postglacial time. *Castanea,* which has outposts as far north as southwestern New Hampshire, reached its maximum abundance (according to Connecticut bog records) in late postglacial time and has recently declined slightly (Deevey, 1939, 1943). More decidedly southern species, as *Chamaecyparis thyoides,* in bogs of glaciated northern New Jersey and southern New England, *Pinus rigida* on the sand plains of southern and southeastern New England, in the Hudson-Mohawk barrens, and in the St. John Valley, *Ilex opaca* on Sandy Hook and at the east end of Long Island, and *Diospyros virginiana* at its northern outpost at Lighthouse Point, New Haven, Connecticut, point to the efficacy of the Continental Shelf—exposed during late Pleistocene or early postglacial time—as a migratory pathway from areas south of the glacial margin. Antevs (1928) believes that "part of the now submerged coastal plain may still have been land when the temperature had risen higher than that now prevailing." Some coastal plain or pine-barren species which occur in Nova Scotia or Newfoundland are believed to have reached these places by northward migration along the now submerged Continental Shelf (Fernald, 1929, 1931), still further demonstrating the availability and use of this migratory route. The frequent occurrence of such outliers in New England, together with other types of evidence, indicates environmental changes in relatively recent time—within the past 3000 years. Raup (1937) believes that the southern element was once more widely distributed, that the present shift is southward, and that the scattered distribution of the southern or Appalachian element is due to recent restriction of range.

Older Glaciated Areas

Areas of older drift from Indiana eastward are small; none is far enough removed from the Wisconsin glacial boundary to have been entirely beyond the climatic influence of the last glacial advance. Areas of older drift from Illinois westward, i.e., in the forest-prairie transition, are extensive.

Boreal communities are few and not pronounced, except in northeastern Iowa, where a few colonies of *Abies,* always accompanied by *Taxus,* still persist (King, 1902; Conard, 1932, 1938). A few northern species (for example, *Spiraea alba, Viburnum Lentago,* and *Sambucus pubens*) are scattered over glaciated northern Missouri (area of Kansan and Nebraskan drift). Dry grassland relics of the Illinoian drift area in Illinois reflect the intensity of the xerothermic period in this longitude. This is also suggested by the absence from the area of such mesophytes as beech, tuliptree, cucumber tree, and dogwood (Gleason, 1923). Northward migration of southern species from the unglaciated area has progressed somewhat onto the Illinoian drift.

Driftless Area

In or near the Driftless area of southwestern Wisconsin, a few endemic species occur which are generally interpreted as members of a relic flora (Fassett, 1931). The forest vegetation also suggests the antiquity of the area, and the possibility of a Pleistocene refugium in the area. This is supported by order of postglacial appearance of forest genera in the neighboring areas as shown by pollen profiles (see Sears, 1942a). It must be remembered that the Driftless area was at no time surrounded by ice, that the glacial deposits to the east and to the west are of different age (or different substage), and that it was never bordered, simultaneously, by ice on both sides (see map, p. 498). Beyond a doubt its vegetation suffered by proximity to the ice, and by the shifts enforced by ice alternately on one or the other side. The area of earliest glaciation bordering it on the west has had time to assume (topographically and vegetationally) many of the features of the Driftless area, and is included in the Driftless section of physiographers and in the Driftless section of the present forest regions classification. At the time of the later Wisconsin ice advances, the refuge area centered in the rugged topography bordering the Mississippi River on either side, the area where now the richest vegetation occurs. The eastern part was temporarily covered by a proglacial lake.

DISJUNCTS IN THE AREA SOUTH OF THE GLACIAL BOUNDARY

South of the glacial boundary, the more varied physiographic history of the older areas is reflected in the nature and distribution of disjunct communities among which are some of the oldest relics on the continent. In the younger areas, there is evidence of relatively recent migrations brought about by the rising and falling of sea level. All four of the great physiographic areas of unglaciated eastern America are at some place in contact with the glacial boundary. The northern part of the Appalachian

Highlands was glaciated, the boundary between glaciated and unglaciated cutting across the plateaus and mountains in Ohio, Pennsylvania, and New Jersey. The northern limit of the Interior Highlands (Ozarks and Ouachita Mountains) and of the Interior Low Plateau is the glacial boundary. The northern end of the Atlantic Coastal Plain was glaciated, and the northern end of the Mississippi embayment section of the Coastal Plain is within a few miles of the southern limit of Illinoian glaciation in Illinois. Three of the areas (Appalachian Highlands, Interior Highlands, and Interior Low Plateau) were land areas before the rise of angiosperms, and have been subjected to alternating uplift and peneplanation. As the type of topography affects the distribution of communities, the degree of completeness of peneplanation and the persistence of more or less rugged areas are important.[4] Examples of disjuncts will be located and related so far as possible to initiating cause. Many more, distributed in each category, are known to occur.

Disjuncts in the Appalachian Highlands

Near the glacial boundary, certain disjunct occurrences appear to be related to pre-Pleistocene drainage. A major northward-flowing stream, which has been called the Teays, collected the waters of all of what are now the Kanawha and Big Sandy rivers, and a number of lesser streams including some which have since been incorporated into the Ohio River between Wheeling, West Virginia, and Manchester, Ohio; it flowed northward in Ohio in somewhat the position of the present Scioto River reversed. The advance of early Pleistocene ice ponded this stream, forming an extensive proglacial lake whose shore line for some time approximated the 900-foot contour (Leverett, 1902; Tight, 1903; Rich, 1934; Transeau, 1941; Wolfe, 1942; Ver Steeg, 1946). That this lake persisted for a long time is indicated by the depth of lacustrine deposits. A considerable list of plants has been assembled, which in this area are generally confined to elevations above 900 feet; or if below 900 feet, to areas near higher "islands" which have subsequently assumed some of the edaphic features of the non-flooded areas (Wolfe, 1942). Some of these, as *Lygodium palmatum, Styrax grandifolia, Chionanthus virginicus, Magnolia macrophylla* and *Pachystima Canbyi* are more or less pronounced disjuncts; others, as *Ilex opaca, Magnolia tripetala, Rhamnus caroliniana, Rhododendron maximum,* are here at the northern limits of ranges which terminate near the glacial boundary. It is believed that these species (and the forests of which they are a part) have persisted here throughout the Pleistocene.

There is a decided contrast between the vegetation of the areas covered by the proglacial lake and the vegetation of the higher more rugged areas, and between the former and areas separated from the proglacial lake by

[4] These features are considered in Chapters 16 and 17.

Pleistocene divides (p. 97). The former are characterized by mixed oak forest, the others by mixed mesophytic forest insofar as habitats are suitable. Persistence was favored by the ameliorating local effects of a large body of water on climate. A number of northern plants have been noted in the hills of Hocking County, Ohio, at the extreme northern end of the ponded area, and along its western margin in Adams County—both near the glacial boundary (Griggs, 1914; Braun, 1928, 1928a). A very few occur on southern arms of the lake.

On Tygarts Creek in Kentucky (which was one of the southern arms) *Taxus canadensis* is abundant, and conceivably, it may have been rafted in on icebergs which are believed to have been carriers of glacial erratics not far distant. In places it is mingled with *Pachystima Canbyi,* which is very abundant here. *Acer spicatum,* separated about 200 miles from its southern outposts in Ohio, and *Sullivantia ohionis,* whose range is usually correlated with the glacial boundary, are very local in the area. The extremely local and rare *Thaspium pinnatifidum*[5] occurs in the area. *Viola tripartita* is a pronounced southern disjunct occurring here. Southern and typically Appalachian species, together with the usual widespread species of the Mixed Mesophytic Forest region, prevail in the area.

Disjunct occurrences of northern species, other than those of high elevations or of bogs, are not infrequent in the northern part of the Appalachian Highland. *Thuja* occurs south of the glacial boundary in Ohio (at the western edge of the plateau), in the Alleghenies, and in the Ridge and Valley Province from Pennsylvania southward through Maryland, West Virginia, Virginia, to Tennessee, always at rather low elevations, and on dolomitic limestone. *Pinus resinosa* and *Betula papyrifera* have an outlying station on North Fork Mountain in Pendleton County, West Virginia (J. H. Buell, 1940; Allard, 1945).

The spruce–fir forests of the higher elevations of the more southern mountains are usually considered as isolated occurrences of a northern forest, indicative of southward migration in the Pleistocene, when, presumably, altitudinal belts in the mountains were sufficiently lowered as to afford continuous southward migratory routes. Later, with ameliorating climates, forest belts in the mountains pushed upward, and only the highest summits remained suitable for spruce and fir, thus bringing about disjunction. That there are many species in common between the northern spruce–fir forests and their mountain counterparts is certain. However, there is a considerable number of species in the mountain spruce–fir forest which are not a part of the northern forest. Among these should be mentioned *Abies Fraseri, Rhododendron catawbiense, R. carolinianum, Leiophyllum Lyoni, Vaccinium erythrocarpum, Menziesia pilosa, Diervilla sessilifolia, Saxifraga*

[5] The common *Thaspium barbinode* var. *angustifolium* is often misidentified as *T. pinnatifidum.*

leucanthemifolia, Chelone Lyoni, and *Solidago glomerata.*[6] If the disjunction of the Southern Appalachian spruce–fir forest was an event of the Pleistocene, when did it acquire the distinct species of *Abies?*[7] How did it acquire so large a number of species not common to the northern spruce–fir? Some of course are capable of living in more than one altitudinal belt. Some, however, are endemics, or species of such peculiar and disjunct ranges that an earlier localization is clearly indicated. The endemic Rhododendrons have their closest relatives in Asia, and these, together with some other Southern Appalachian endemics, date back to mid-Tertiary, to the residual mountainous and rugged areas of the Schooley peneplain (p. 506). The facts of distribution suggest the probability of some Pleistocene infiltration of northern species into already established mountain communities. Farther north in the Allegheny Mountains of West Virginia (p. 86) the summit spruce–hardwoods forest is almost an exact replica of the present Adirondack spruce–hardwoods. That the northern forest extended into the Alleghenies of West Virginia but no farther south, is definitely indicated. The highest elevations (over 4000 feet) in the Cumberland Mountains, connected without pronounced break with the Alleghenies, have no remnant of spruce–fir forest and *very few* species which can be considered northern. Whether *Picea rubens* migrated south in the Pleistocene or was in the mountains together with the endemic fir, cannot now be ascertained.

Bog communities in the Allegheny Mountains and Unglaciated Allegheny Plateau of Pennsylvania are similar to those to the north on the Glaciated Allegheny Plateau. They are located in or near the southern border of the Allegheny section of the Hemlock–White Pine–Northern Hardwoods Forest; disjunction is not pronounced. More isolated bog communities in the Allegheny Plateau of West Virginia are southern outposts of northern bogs. Cranberry Glades,[8] in Pocahontas County, is representative; the Sphagnum and Polytrichum–Cladonia mats, the cottongrass (*Eriophorum*), the cranberries (*Vaccinium Oxycoccus, V. macrocarpon*), bog rosemary (*Andromeda glaucophylla*), and *Menyanthes trifoliata* var. *minor* emphasize the northern aspect of the area. A spruce–hardwoods border, where red spruce and the trees of the surrounding deciduous forest are intermingled, contains a number of northern shrubs—*Taxus canadensis, Acer spicatum, Viburnum alnifolium,* and *V. cassinoides*—as well as the southern *Rhododendron maximum.*

[6] Examples taken from lists of species of spruce–fir communities of the Great Smoky Mountains (Cain, 1930, 1935a).

[7] Note that pollen records from a buried soil of doubtful age in South Carolina (p. 471) show the existence there of a species of *Abies* distinct from *A. Fraseri* and from *A. balsamea.* Ancestral firs (*Abietites*) were living in the Southeast in Cretaceous time, and while there are no records between then and the Pleistocene, it should not be assumed that they were absent from the southern mountains during the Tertiary.

[8] See H. C. Darlington, 1943, for discussion of Cranberry Glades.

Bear Meadows in the Seven Mountains in central Pennsylvania, is one of the few boreal bog relics of the Ridge and Valley Province. It is not far from the boundary of the Hemlock–White Pine–Northern Hardwoods Forest on the Plateau to the northwest, and its northern or bog species (as *Picea rubens, Abies balsamea, Nemopanthus mucronata, Chamaedaphne calyculata, Coptis groenlandica, Drosera rotundifolia*) are not pronounced disjuncts.

Boggy areas farther south almost always harbor disjunct coastal plain species, among which are *Lorinseria areolata, Uniola laxa, Andropogon glomeratus, Orontium aquaticum, Habenaria flava,* and *Viola primulifolia.* Colonies of coastal plain plants are often found in the boggy heads of unrejuvenated streams on the Cumberland Plateau (Schooley peneplain), in the Cumberland Mountains, and in the Blue Ridge. Such outlying stations of coastal plain plants are generally interpreted as relics—remnants of a more widespread vegetation type of the Schooley peneplain, which became circumscribed in the old area and expanded onto the emerging coastal plain (Fernald, 1931; Braun, 1937, 1937a).

A number of endemic species occur in the southern Appalachian Highlands. Some of these range over a considerable area, some occur in disjunct areas, and a few are very local. Many have relatives in Asia. These are generally interpreted as Tertiary relics which have found continuously favorable environments in the unreduced areas of the Schooley peneplain. This interpretation is emphasized by a pattern of distribution which includes one or more of the monadnock, hilly or mountainous areas of the Schooley peneplain. Outstanding examples among the more local species are *Buckleya distichophylla, Shortia galacifolia,* and *Diphylleia cymosa,* each with one or two rather local Asiatic relatives. Among the less rare disjuncts of the Appalachian Highlands are *Rhododendron catawbiense, Stewartia ovata* (*S. pentagyna*), and *Clethra acuminata,* which range generally through areas where the Schooley peneplain was never perfected; *Pachystima Canbyi,*[9] a large portion of whose stations are in the old (Tertiary) Teays River drainage, and which genus is known in the fossil record from the Upper Cretaceous (Ripley) flora of the Mississippi embayment (p. 452); *Gaylussacia brachycera,* from separated stations in seven states (not all Appalachian Highland)[10] and locally abundant in West Virginia and Kentucky; *Cladrastis lutea,* in the Appalachian Highlands of Tennessee and Kentucky (Great Smoky Mountains, Cumberland Mountains, Cumberland Plateau), in the Interior Low Plateau (where locally abundant), and in the Ozarks.

Buckleya, a shrub parasitic on the roots of *Tsuga canadensis,* is a member of the Santalaceae, a family now almost entirely tropical. The ancestry of

---

[9] Massey (1940) lists the known localities for *Pachystima Canbyi;* to these should be added Adams County, Ohio (Braun, 1941).

[10] See map by Wherry, 1934.

Buckleya probably dates back to the Eocene, when many of what are now warm temperate or subtropical genera were well represented in the United States (see Wilcox flora, p. 453). This of course points to the Tertiary ancestry of the host plant, *Tsuga canadensis,* in the southern Appalachians.

Conradina verticillata, known from only two localities along the western edge of the Cumberland Plateau (Braun, 1936), belongs to a genus whose other three representatives are on the Coastal Plain (one on the east coast region of Florida, two in the Gulf coast region). The Appalachian Highland–Coastal Plain affinity dating back to middle or early Tertiary time is indicated. Two very local endemics of this area—*Eupatorium Luciae-Brauniae* (*E. deltoides*) and *Solidago albopilosa*—belong to genera without Asiatic species; the former is a member of a genus best developed in warm climates.

Further evidence of antiquity of certain elements of the flora of the Appalachian Highlands is afforded by *Lygodium palmatum,* which belongs to a genus now largely tropical, and which was common in the Middle Eocene (Claiborne flora, p. 454). This plant is abundant near the western margin of the southern Allegheny and Cumberland plateaus, an area which still retains many of its mid-Tertiary features.

The areas most favorable to the preservation of the endemics and local disjuncts of the Appalachian Highlands are the areas which were most favorable to continuous occupancy by the ancestral mixed mesophytic deciduous forest. Its best development is in areas where unreduced remnants of the Schooley peneplain still persist and where no later peneplain was extensively developed. Where a later peneplain (Harrisburg) was extensively developed, mixed mesophytic communities, if they exist at all, are relics (in coves of the mountain slopes), or advance disjuncts (on slopes formed in the present erosion cycle). (See Chapter 17.)

Disjuncts in the Interior Low Plateau

The erosional (and probably climatic) history of the Interior Low Plateau differs markedly from that of the Appalachian Highlands to the east. This is reflected in the pronounced difference in the number and kinds of disjunctions. The area as a whole, which embraces most of the Western Mesophytic Forest region, is characterized by a mosaic of unlike climaxes, among which mixed mesophytic, oak–hickory, oak–chestnut, and prairie are prominent. All, except perhaps the first, in the eastern part of the region, and the second, in the western part of the region, are disjuncts, more or less.

Northern disjuncts are almost absent in this area, even along the glacial boundary. The more or less rugged Indiana "Knobs," and stream gorges of southern Indiana and northern Kentucky would appear to be favorable for their persistence, yet they are not here. *Betula lutea* var. *macrolepis,* "on the sides of two deep, rocky ravines . . . in Crawford County," Indiana (Deam, 1940), and in deep gorges in Edmonson and Todd counties, Kentucky, is about the only species which can unquestionably be called northern. In both Kentucky situations it is a constituent of mesophytic forests, and in the former it is associated with such southern species as *Magnolia macrophylla, M. tripetala,* and *Ilex opaca.* The communities are in no way northern. A few mildly northern species, as *Spiraea tomentosa, Aronia melanocarpa,* and *Vitis Labrusca,* do occur in the Interior Low Plateau but their generally wide distribution and community associates do not indicate that their southern stations have any relation to Pleistocene migrations. *Pinus Strobus,* which occurs in western Kentucky (Todd County) may better be considered as an Appalachian outlier than as a northern disjunct,[11] and like hemlock, to date back to the Tertiary forest.

Outliers of mixed mesophytic forest occur in the more rugged parts of the Interior Low Plateau, especially where the summit elevations are somewhat above the general level, i.e., in areas which were not reduced in the last major erosion cycle and probably not in the Schooley cycle (p. 494). These communities may be considered as relics of the retreating migration of late Tertiary time (p. 509).

The local distribution of oak–chestnut communities in the Knobs and Shawnee section, and with them, of an attenuated Allegheny flora, suggests restriction coincident with the reduction of sandstone areas, with which such communities are commonly associated.

The most conspicuous disjunct communities of the Interior Low Plateau are the prairies and cedar barrens, which, before the white man's profound modification of natural conditions, were very extensive (pp. 151, 155). Comparable in some respects are the cedar glades of Tennessee (p. 131). That these prairies and cedar barrens are not relics of the postglacial xerothermic period is suggested by the presence of well-marked endemic species (especially in the Nashville basin) and by the pronounced admixture (in the Kentucky "barrens") of southeastern species. On the other hand, the rapid changes which have taken place since disturbance of the Kentucky "barrens" point to a short period of dominance of prairie vegetation.

Coastal plain species are few (in contrast to the adjacent Cumberland Plateau), except those which inhabit alluvial flats. The cypress swamps of the Wabash and Green River valleys are disjuncts which, from their location upstream of the area of continuous distribution of bald cypress,

[11] White pine (*P. Strobus*) occurs in the Appalachian upland as far south as Georgia; it does not frequent the higher elevations of the southern mountains, but is common at low elevations along the east side of the Blue Ridge and in the foothills in Tennessee. It is locally common on the Appalachian Plateau in Kentucky and adjacent Tennessee. It is represented in the highlands of Mexico (Sharp, 1946) by a doubtfully distinct variety.

suggest recent advance. However, if the history of alluviation be considered, there is evidence that these swamps are not post-Wisconsin in age.

As sedimentation progressed in the northern part of the Mississippi embayment, additional land surface was added along the western margin of the Interior Low Plateau. This area was henceforth subjected to the erosional and climatic history of the adjacent plateau, and its vegetation belongs with that of the plateau (as a part of the Western Mesophytic Forest region). Like the plateau, it has its prairie communities ("barrens"), and oak–hickory woodlands. One of its most distinctive vegetational features is the mixed mesophytic forest of the bluffs east of the Mississippi River, and, to a lesser extent, of Crowley's Ridge on the west side of the river (pp. 159, 161). This mesophytic forest is interpreted as a Tertiary relic preserved in ravines of the bluffs and ridge where fertile soil favors its persistence. The Mississippi River, until some time in the Pleistocene, is believed to have followed a course to the west of Crowley's Ridge; thus, this ridge was less isolated than now. Tuliptree (*Liriodendron*) in Arkansas is confined to Crowley's Ridge.

Disjuncts in the Interior Highlands (Ozark and Ouachita Mountains) and Westward

The Interior Highlands have had a physiographic history somewhat similar to the Appalachian Highlands and, like the latter, may be expected to harbor relics of an ancient flora. But, unlike the latter, this area has been subjected to the drier climates of late Tertiary and postglacial times and probably of Pleistocene time. Hence in the Interior Highlands, mixed mesophytic forest communities occur only as relics while in the Appalachian Plateau they are widespread and characterize the forest region of that area. These relic mixed mesophytic communities, which occur in the Boston Mountains in the Ozark Province and on Rich Mountain in the Ouachita Province, have been described in connection with the vegetation of those areas (pp. 170, 174). They are particularly significant in indicating former extent of mixed forest (Chapter 17).

This area, like the Appalachian Upland, has representatives of ancient genera, some of which are endemic species or species with peculiarly disjunct ranges. Among the endemics may be mentioned *Castanea ozarkensis, Hamamelis vernalis,* and *Andrachne phyllanthoides,* the first two with eastern relatives (as well as Asiatic), the last with only Asiatic relatives; among those with generally disrupted ranges, are *Cotinus americanus* and *Cladrastis lutea*. It has been suggested that these forms may have originated somewhere in this old land area (Palmer and Steyermark, 1935). *Cladrastis* was represented in the lower Eocene (Wilcox flora) of the Mississippi embayment (p. 453). A number of Appalachian species, or species of rather wide distribution to the east of the Mississippi occur here as local disjuncts, for example, *Fagus grandifolia, Magnolia acuminata, M. tripetala, Chionanthus virginicus, Oxydendrum arboreum,* and *Tilia heterophylla* var. *Michauxii.* These disjuncts are interpreted as relics of a former more widespread mesophytic forest (Chapter 17).

A few species of the lowlands of the Mississippi embayment (as *Crataegus viridis, Ilex decidua,* and *Forestiera acuminata*) occur as disjuncts in the Ozarks, suggesting a former more widespread occurrence on the late Tertiary peneplain (Palmer and Steyermark, 1935). Coastal plain relics, comparable to those of the Appalachian Plateau, are lacking, although a few coastal plain species are found.

While the Ozark Province is in contact at the north and northeast with glaciated country (Kansan and Illinoian drift), it is considerably removed from the area of Wisconsin glaciation. If its vegetation was affected by southward migration, little evidence remains today. A few local communities, made up partly of Appalachian and partly of northern species (as *Zygadenus elegans, Cypripedium reginae, Galium boreale* var. *hyssopifolium,* and *Campanula rotundifolia*), may record the influence of early ice advances (see Palmer and Steyermark, 1935).

Xeric communities of dry ledges, glades, and knobs contain many species of southwestern range, and may be relics of the postglacial xerothermic period.

West and south of the Interior Highlands, in the oak savannah and forest–prairie transition area, such mesic forest communities as occur in canyons (p. 172), are relic communities indicative of former more humid conditions such as are suggested by a Texas pollen record (p. 471). Eastern mesophytes occur well beyond the limits of deciduous forest, in canyons in the Edwards Plateau of Texas, there associated with species of more western or southwestern range, with species closely allied to eastern species, and with more or less local endemics (Palmer, 1920). *Quercus Muhlenbergii* maintains far-western outposts in the Guadalupe Mountains of Texas and the Capitan Mountains of New Mexico (in the Sacramento physiographic section). Also in this section (in Ruidoso Canyon in the Sierra Blanca Mountains) there are deciduous forest communities in which the remarkable resemblance of many of the trees and shrubs to those of the East, and the large representation of eastern herbaceous genera in the woods strongly suggests an ancient connection of this area with the Interior Highlands. Any such connection must have been broken by the increasing aridity of the Great Plains in late Tertiary time, as indicated by the fossils of that area (p. 458).

Disjuncts in the Mountains of Mexico and Guatemala

It is significant that some of the trees and undergrowth species of eastern American forests occur in the highlands of Mexico and Guatemala; some are represented by closely similar species; some by varieties. Sharp

(1946) lists a number of these plants. The woody species as named by Sharp are: *Pinus Strobus, Smilax glauca, Liquidambar Styraciflua, Prunus serotina, Cercis canadensis, Ptelea trifoliata, Rhus radicans, Acer Negundo, Parthenocissus quinquefolia, Ascyrum hypericoides, Cornus florida, Nyssa sylvatica,* and *Gelsemium sempervirens.* Some of these are referred by Standley (1920–26) to other species: *Prunus capuli*—"a closely similar species" (to *P. serotina*); *Acer orizabense*—"This is closely related to *A. Negundo* L. . . . and may not be distinct"; *Cornus urbiniana*—"doubtful whether it is more than a mere form of *Cornus florida.*" Sharp also refers to the many species closely related to plants of eastern United States. These Mexican deciduous forest communities are widely separated from the main area of forest to which they are related. The many paired species and the more or less distinct varieties indicate a long period of separation. Such specific segregation is not (as was pointed out by Fernald, 1931) found in recently separated floras. During early and middle Tertiary time floras were free to migrate northward or southward. The warmth of the Eocene and Oligocene epochs may have resulted in withdrawal of temperate forms to higher elevations, in eastern United States and in Mexico. During the Pliocene, a climatic barrier—the dryness of the Plains—developed. Even if withdrawal to higher elevations was accomplished earlier, a final separation would have been effected by the climatically dry belt, and contemporaneously with the late Tertiary shrinkage of mesophytic floras.

Disjuncts on the Atlantic and Gulf Coastal Plain

Although the Coastal Plain is marked by a high degree of endemism and numerous instances of pronounced disjunction, the majority of species involved are not plants of deciduous forest communities and afford little evidence as to the history of deciduous forest on the Coastal Plain. Some of the bog and pine barren species occur also in the Appalachian Highlands (as *Cleistes divaricata, Orontium aquaticum,* and *Schwalbea australis*); others are related to, rather than identical with, species of the Appalachian Highlands (species of *Sarracenia,*[12] *Leiophyllum,* etc.). Many of the bog and pine barren species (at least some of which are pronounced disjuncts) belong to groups largely tropical or subtropical in distribution (Fernald, 1931).

Instances of disjunction of greater significance in deciduous forest history are to be found in the mesophytic forest communities of low river bluffs and ravines indenting the bluffs, and of certain of the higher and hilly areas (pp. 298, 300). The occasional broad-leaved mesophytic forest communities of low river bluffs occupy soils unlike the prevalent sandy soils of more or

[12] The ranges of species of *Sarracenia* have been interpreted by Wherry (1935) on a basis of migration onto the emerging Coastal Plain; no account is taken, however, of Pleistocene submergence.

less flat interstream areas; often they are but little above flood levels. These communities suggest advance (wherever suitable habitats develop) from upstream and inland regions, and may be referred to as "advance disjuncts." At least some of them are occupying land submerged during one or more of the interglacial stages; all of them are on slopes of the present erosion cycle, hence they are relatively young communities, geologically speaking, although successionally advanced. Most of these communities contain beech. Groves of *Fagus* on slopes in Polk County, Texas (Tharp, 1926, Plate X, A) are far removed from the general range of beech. But until the forms (genotypes?) of beech are distinguished, the presence of beech does not afford evidence as to the probable time of disjunction, or direction of migration (p. 501).

Forest communities of the Apalachicola River bluffs, the Tallahassee Red Hills, and the Marianna Red Lands (pp. 301, 303) are pronounced disjuncts on hilly islands surrounded by gently rolling country with more typically coastal plain vegetation. Kurz (1928, 1930–1932) considered these to be "northern disjuncts." In contrast to the surrounding vegetation, they are northern; they do partake of the character of the deciduous forest of the Appalachian Highland (see Harper, 1914, for lists of species). But with the deciduous forest species are others, as *Magnolia grandiflora,* which relate them to the Warm Temperate and Subtropical Broad-leaved Evergreen Forest. Still others, particularly *Torreya taxifolia, Taxus floridana,* and *Croomia pauciflora* are local endemics of ancient ancestry. *Torreya* (or *Tumion*) was present on the southern Atlantic Coastal Plain in Cretaceous time (p. 452); there are no records in the long interim, but somewhere through these 60 million years, representatives of the genus lived on, giving rise (on the American continent) to *Torreya taxifolia* of Florida and *T. californica* of California (two Asiatic species are known, also). *Taxus* has had a similar history, and is represented in America now by a Floridian species, a species of the Pacific Northwest, and the widespread northern *T. canadensis. Croomia,* with one Asiatic species in addition to the one of the Gulf Coastal Plain,[13] is also of ancient ancestry. A community containing such highly local endemics of ancient ancestry has had a long and complicated history, although some of its constituents may have entered recently. However, there is evidence that its *Fagus* may be genetically distinct (fide W. H. Camp).

The extent of area suitable for these disjuncts, of favorable edaphic environment (the Oligocene limestone soils) was probably more extensive in late Tertiary and early Pleistocene time. The Sunderland terrace, bespeaking the extent of submergence in the Yarmouth interglacial stage, abuts the hills where the relic disjuncts now occur. The analogy of Kurz was

[13] A locality in the Appalachian Valley, 130 miles within the Fall Line, is recorded by Harper, 1942a.

then a reality: "The rugged areas in question may then be considered hilly islands more or less completely surrounded, even to the north, by an ocean of more flatly rolling or flat topography." The rugged areas rose above the Yarmouth sea. They afforded a refugium for a vegetation still represented here.

A comparable, but less old, group of hills in southern Mississippi and adjacent Louisiana—the Tunica Hills—is occupied by mesophytic forest including in the undergrowth a number of species here isolated from their general range (C. A. Brown, 1938). The presence of an endemic species, *Physalis Carpenteri,* an endemic variety, *Magnolia acuminata* var. *ludoviciana* Sarg., and an isolated occurrence of *Pachysandra procumbens,*[14] are of particular interest. The forest in this area, as in the hilly "islands" of northern Florida, contains trees of the central deciduous forest and the evergreen forest to the south. It represents, as do the Florida areas, a Tertiary forest extension, which in modified form, has persisted in this area of mature topography.

[14] Also in the Marianna Red Lands previously mentioned, and in the Appalachian Valley locality in which *Croomia* occurs; and abundant on the Mississippian Plateau in the Western Mesophytic Forest region (p. 124). Not a mountain species; in fact, very few of the species listed by Brown, referring to Cocks (1914), are "mountain species."

CHAPTER 16

A Brief Physiographic History of Eastern United States

Since the topography of the land surface, its altitude, the nature of the substratum, and the development of soils are intimately related to, in fact partly determine the nature of occupying vegetation, a knowledge of the history of development of the land surface is requisite for an understanding of its vegetation cover. Paleozoic history is not significant in the study of modern forests. We may start with the development of the land surface on which the earliest known angiosperms of eastern America grew, i.e., with the Appalachian Revolution at the close of the Paleozoic era.[1]

The Appalachian Revolution was a time of mountain making in the East, of intense folding and faulting of the accumulated sediments of the great Appalachian geosyncline. The structural features of what is now known as the Ridge and Valley Province were then produced. This folding, with subsequent erosion, is responsible for the configuration of the land surface—the longitudinal ranges on resistant rock, and the valleys on weaker rock. Mild folding, or slight tilting of the strata, is the structural basis for the different land surface immediately to the west—now called Appalachian Plateau. The same period of folding was responsible for uplift of the Pennsylvanian series of rocks in what is now the western part of the Interior Low Plateau (Shawnee physiographic section). The Ozark region, too, was finally raised above the sea in the Appalachian Revolution as an asymmetric dome. Intense folding in the adjacent Ouachita province at that time makes the Ouachita Mountains analogous to the folded Appalachians, with which, via Alabama, they may have been continuous. These two provinces together constitute the "Interior Highlands."

In the East, the old land area, Appalachia, with which the Piedmont and possibly the Blue Ridge are identified, became greatly changed. Its eastern and southern parts sank and are now beneath the sedimentary platform of

[1] Facts used in the construction of this physiographic history are taken chiefly from Fenneman (1938) unless otherwise stated. Pre-Pleistocene physiographic history north of the glacial boundary is but briefly mentioned as it is of little significance in the study of forest development. Only those physiographic features of significance in vegetational development are included.

Cenozoic	Quaternary	Recent			A series of coastal terraces (or marine plains and wave-cut scarps) and of fluvial terraces or plains
		Pleistocene	Wisconsin	Mankato	
				Cary	
				Tazewell	
				Peorian	
				Iowan	
			Sangamon		
			Illinoian		
			Yarmouth		
			Kansan		Late partial peneplains (Somerville, Nashville, Coosa, Parker Strath)
			Aftonian		
			Nebraskan (incl. Jerseyan)		
	Tertiary	Pliocene			Uplift of Harrisburg peneplain Harrisburg cycle of erosion (including Allegheny and Lexington)
		Miocene			Uplift of Schooley peneplain
		Oligocene			Schooley cycle of erosion (including Cumberland and Kittatinny)
		Eocene			
		Paleocene			
Mesozoic	Upper Cretaceous				Uplift
	Lower Cretaceous (Comanchean)				Deposition on submerged Fall Zone Peneplain
	Jurassic				Sinking
	Triassic				Fall Zone Peneplain (Pre-Cretaceous)
Paleozoic					Appalachian Revolution

the Atlantic. Along its western side, much new land was added, and the "Appalachian Highland" assumed something of its present area. Farther west, the old land area, "Llanoria" (in the Gulf Coast region), which probably had been continuous with Appalachia, sank, while crustal movements formed the Ouachitas and Ozarks. The continuity of land between the Appalachian Highlands and Interior Highlands (or nearby land), from the closing Paleozoic to the present (first southward by way of Appalachia and Llanoria, later northward, around the Mississippi embayment), is of fundamental importance in vegetational development. In the North, were the extensive Laurentian Highlands.

Erosion began its work of reducing the great mountains and plateaus formed during the Appalachian Revolution. The earliest record of Mesozoic erosional history in the East is that preserved in the Fall Zone peneplain of pre-Cretaceous age. It is preserved near the Fall Line only because it was long-covered by coastal plain sediments. This was a seaward-sloping surface which extended eastward and southward of the present inner margin of the Coastal Plain—how far is not known—and doubtless westward above the height of any remaining peneplain remnants. Sinking of the Fall Zone peneplain took place in time for it to receive Cretaceous sediments. The extent of submergence of the Fall Zone peneplain is not definitely known; the seacoast then was somewhere *within* the present limits of the Appalachian Highland, landward of the present margin of Cretaceous deposits of the Coastal Plain, and perhaps inland of the folded belt. This would probably place it in the neighborhood of the Allegheny-Cumberland Front (the eastern escarpment of the Appalachian Plateau), as "the dominant drainage after the Appalachian uplift was probably toward the Mississippi" (Fenneman, 1938). This extensive submergence of the Fall Zone peneplain is postulated by Douglas Johnson (1931) in an attempt to explain stream patterns on the Atlantic slope. It is referred to as the superposition hypothesis (Fenneman, p. 259). If the superposition hypothesis is accepted, the Appalachian Highland was reduced in size approximately to what is now Appalachian Plateau.

Meanwhile, the epicontinental seas of the Western Interior had increased in size, and the sea had entered from the Gulf, covering a wide belt between the Rocky Mountains and the Ozark-Ouachita uplift. The Mississippi embayment extended inland about to what now is the mouth of the Ohio River. The Cretaceous land area was small and for the most part low-lying, although some areas of considerable relief are indicated by the nature of deposits on the southern Atlantic coast (Berry, 1914).

There may have been a cycle of erosion which ended in a Cretaceous peneplain. The Fall Zone peneplain, on which Cretaceous sediments were deposited is older than Cretaceous; the next peneplain of which there is positive evidence is younger than Cretaceous. The so-called "Cretaceous peneplain" of earlier writers is younger than Cretaceous.

Uplift of the Cretaceous continent initiated the Schooley cycle of erosion. In this uplift, the previously submerged Atlantic slope emerged, and a coastal plain (covered with Cretaceous sediments) came into existence. If the maximum Cretaceous submergence postulated by the superposition hypothesis is granted, then, in early Tertiary time, a coastal plain bordered the Appalachian Plateau (or perhaps only its northern half) on the east. The Early Tertiary continent in the east would then have been in part upland, in part coastal plain. On the west, the margin of the upland was far to the west of the present western escarpment of the Appalachian Plateau. The Mississippi embayment was somewhat smaller than in Cretaceous time. The vast epicontinental sea of the Western Interior was reduced in size, and its continuity and ocean connections were broken. The areas of Carboniferous rocks of eastern New Mexico and central Texas were now probably connected with the Interior Highlands; the highlands of Mexico were no longer isolated by a water barrier.

The Schooley cycle of erosion, initiated by the post-Cretaceous uplift, cul-

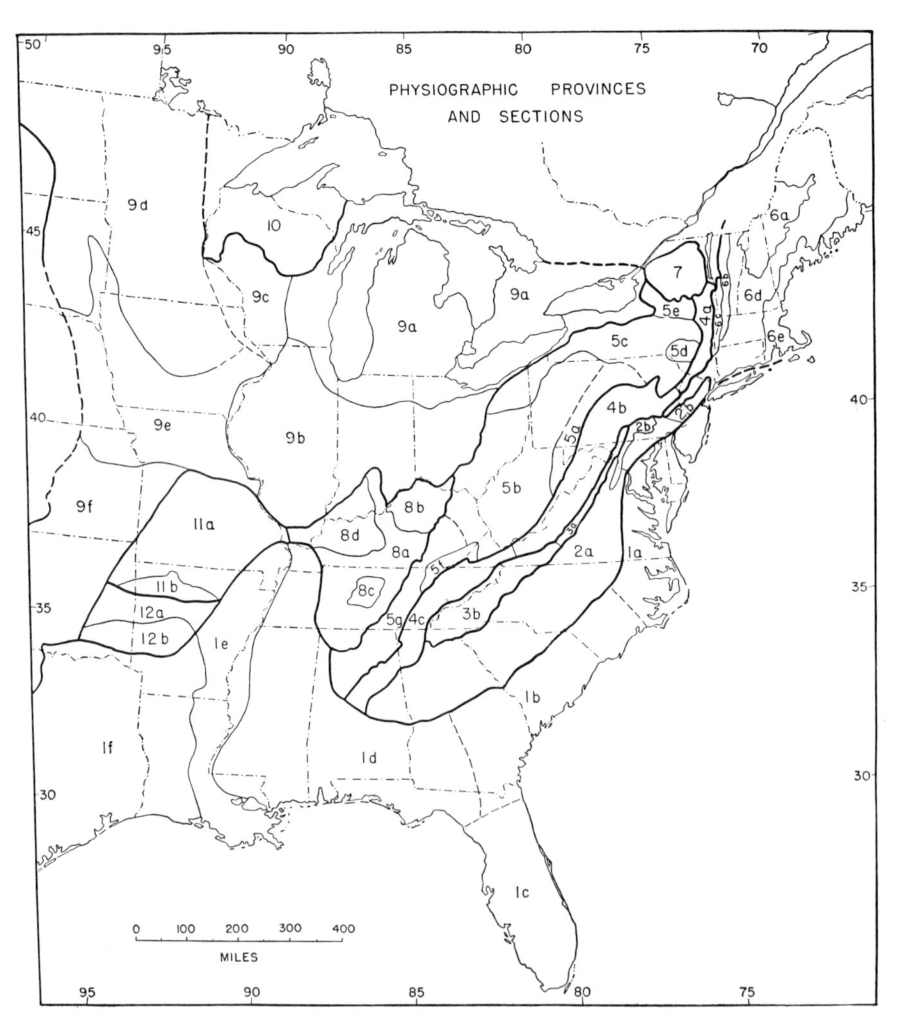

Map of Physiographic Provinces and Sections. (After Fenneman, 1938.)

(Legend continued on facing page.)

(*Continuation of legend for figure on facing page.*)

1. Coastal Plain
   - *a* Embayed section
   - *b* Sea Island section
   - *c* Floridian section
   - *d* East Gulf Coastal Plain
   - *e* Mississippi Alluvial Plain
   - *f* West Gulf Coastal Plain

Appalachian Highlands:

2. Piedmont Province
   - *a* Piedmont Upland
   - *b* Piedmont Lowland
3. Blue Ridge Province
   - *a* Northern section
   - *b* Southern section
4. Ridge and Valley Province
   - *a* Hudson-Champlain section
   - *b* Middle section
   - *c* Southern section
5. Appalachian Plateaus Province
   - *a* Allegheny Mountain section
   - *b* Unglaciated Allegheny Plateau
   - *c* Glaciated Allegheny Plateau
   - *d* Catskill Mountains
   - *e* Mohawk section
   - *f* Cumberland Mountains
   - *g* Cumberland Plateau
6. New England Province
   - *a* White Mountain section
   - *b* Green Mountain section
   - *c* Taconic section
   - *d* New England Upland (including Reading Prong)
   - *e* Seaboard Lowland
7. Adirondack Province

8. Interior Low Plateau Province
   - *a* Highland Rim section
   - *b* Bluegrass section
   - *c* Nashville Basin
   - *d* Shawnee section
9. Central Lowland
   - *a* Great Lake section
   - *b* Till Plains section
   - *c* Driftless section
   - *d* Western Young Drift section
   - *e* Dissected Till Plains
   - *f* Osage section
10. Superior Upland—a part of the Laurentian Upland

Interior Highlands:

11. Ozark Plateaus Province
    - *a* Ozark Plateau (including Salem and Springfield Plateaus)
    - *b* Boston Mountains
12. Ouachita Province
    - *a* Arkansas Valley
    - *b* Ouachita Mountains

minated in Miocene time in the Schooley peneplain, the oldest surviving peneplain and the most complete of which there is record. While the exact time at which the Schooley cycle was interrupted by uplift is open to question, it is (according to Fenneman, 1938) "doubtful that it was raised enough for erosion to begin cutting it away before the end of the Miocene." While a large part of the area had doubtless reached peneplain status earlier, much of the area did not feel the effects of rejuvenation until later. In fact, some areas are still undissected.

The extent of the Schooley peneplain and its correlatives[2] (the Kittatinny and Cumberland) is a matter of concern in the interpretation of forest history. Certainly it affected the entire Appalachian Highland. It must have developed farther west, over what is now the Interior Low Plateau. Because its surface everywhere declined toward the coast and toward major streams still in existence, it sloped westward to the Mississippi embayment. As later peneplains also declined toward the Mississippi, the Schooley and later surfaces finally converged or almost converged. The ridges which have survived a later peneplain (the Lexington or Highland Rim) probably survived also on the Schooley (Cumberland) peneplain. The rugged areas still apparent are Muldraugh's Hill, the hills along the Dripping Springs escarpment, and a height of land roughly connecting the two and lying north of the Green River in Kentucky. In the Ozark and Ouachita provinces, an older and a younger peneplain are distinguished. Only the Boston Mountains in the Ozarks and Rich Mountain in the Ouachitas remain to indicate the old surface, which may be more or less equivalent to the Schooley peneplain of the Appalachian Highland.

Although the Schooley peneplain was the most complete peneplain of which there is record, it was not everywhere a plain. Remnants of the peneplain where it was best developed indicate a relief of several hundred feet in 40 or 50 miles. In the Blue Ridge of Virginia, knobs and rounded hills reached a thousand feet above the peneplain surface. In the Southern Appalachians, subdued[3] mountains several thousand feet in height remained. The Cumberland and Allegheny Mountains, and the western escarpment of the

---

[2] Physiographers are not in complete accord as to the number of peneplains which should be recognized, or as to the exact correlation of surfaces on the eastern and the western slopes of the Appalachian axis, if indeed, exact correlation is possible. Since peneplanation works headward, the time in which a given surface is reduced is a function of the length of the stream. Streams on the Atlantic slope are generally short as compared with those to the west of the axis. The writer follows the conservative views expressed by Fenneman, who said "all evidences of extensive peneplains in the entire Appalachian Highland may best be referred to a very few major cycles" (p. 188). Thus the Cumberland and Kittatinny peneplains may be considered as more or less equivalent to the Schooley, and the Lexington and Highland Rim (and perhaps Allegheny), to the Harrisburg peneplain.

[3] The term "subdued" is used "to designate a stage in the cycle when height and steepness are so far lost that a mantle of decayed rock is general" (Fenneman, 1938, p. 174).

Appalachian Plateau retained considerable relief. In fact, wherever in the Appalachian Highland resistant rocks outcropped, smoothness of the land surface was interrupted, and some topographic diversity maintained. This was true in the Adirondack and New England mountains as well as southward, but as glaciation later destroyed all vegetation, the pre-Pleistocene physiographic history has little significance in a consideration of forest development.

Progress of the Schooley cycle of erosion in early Tertiary time, and its culmination in mid-Tertiary, resulted in a relatively low-lying continental mass whose Atlantic and Gulf shore line was in general not far from the present inner margin of the Coastal Plain. As time went on, the shore line retreated seaward, and northern Florida emerged. The Mississippi embayment became progressively smaller. Epicontinental seas in the Western Interior became smaller and discontinuous, and consequently more land was exposed. Contemporaneously with these physiographic changes, the great deployment of angiosperm forests took place (see Chapter 17).

Uplift and warping of the Schooley surface (the highest axis was near the Allegheny Front) gave renewed cutting power to the streams. A new cycle of erosion, the Harrisburg, was initiated, a cycle which consumed only a fraction of the time of the Schooley cycle. This was a late Tertiary cycle, interrupted about the close of the Pliocene while still far from complete. Because of its incompleteness, accordant levels over wide areas were not produced. Developing peneplains (of the same cycle) were growing in each of many separate drainage basins. Some of these became confluent, some did not. Some are at one level, some at another. Some are definitely Harrisburg peneplain, others at least approximate it in age (see footnote, p. 494).

Lowering of the land surface progressed most rapidly in downstream areas and over areas of weaker rocks. Outcrops of strong rocks were but little lowered while adjacent weak rocks were rapidly worn down. Thus in what is now the Ridge and Valley Province, narrow and steep-sided ridges on resistant rock were left projecting above the neighboring strips of weak rock. In many of these ridges, strata are strongly inclined or almost vertical, and little soil can accumulate. The lowlands on weak rocks progressively increased in width at the expense of the adjacent slopes. The Great Appalachian Valley became a major topographic feature. Its surface best represents the Harrisburg peneplain level. The Piedmont Plateau drained by short streams flowing to the Atlantic (or at the southern end to the Gulf), was lowered far below the Southern Blue Ridge whose streams follow a more circuitous route to the sea; it too, is Harrisburg peneplain. The Northern Blue Ridge, where the belt of resistant rock is narrow, became a relatively narrow mountain range rising abruptly above the Harrisburg levels, the Piedmont on the east and the Great Valley on the west.

Extensive areas underlain by resistant rock, as the Appalachian Plateaus, the Southern Blue Ridge (Southern Appalachians), the Adirondacks and New England Upland, were dissected but not greatly reduced. They were changed less than were areas of weak rock, or areas near the coast or the larger rivers. They came to resemble the areas of greater relief (the unreduced parts) of the Schooley peneplain.

The Schooley peneplain over the resistant rocks of the Appalachian Plateaus was not destroyed, although the extent of rugged land was increased by dissection. Lower levels do exist, particularly in the drainage basin of major streams, giving evidence of a possible younger partial peneplain. This lower level, observable in West Virginia and eastern Ohio, has been called the "Allegheny peneplain," but here (following Fenneman) is considered as more or less equivalent to the Harrisburg of the eastern slope. Whether the Allegheny Plateau surface be referred to "Allegheny" or Harrisburg peneplain makes little difference in the interpretation of forest features. Of importance is the fact that its western margin remained as a rugged and elevated strip representing the Schooley level, somewhat lowered. Most of the Cumberland Plateau, also, retained its Schooley level, somewhat reduced.

Westward, over the Interior Low Plateau, the extensive development of the Lexington or Highland Rim peneplain, "provisionally correlated with the Harrisburg," profoundly modified the topography except in certain areas (p. 494). With reduction of the Interior Low Plateau, the Cumberland Plateau escarpment became a prominent topographic feature. The areas underlain by massive limestones were underdrained, as they are today, although the present surface is locally cut below the Lexington peneplain. In general, the Interior Low Plateau of late Tertiary time was a plain of low relief sloping westward to the Ohio-Mississippi Valley.[4] Sluggish streams wandered over the plain, frequently shifting their positions as they deposited their load, which ultimately came to form a more or less continuous mantle of unconsolidated material over the western part of the plain. These gravels have, by some, been interpreted as indicative of a drier climate; their significance, however, appears to be doubtful.

At about this time, the Ozark and Ouachita provinces were probably reduced, except for such areas as are now seen to rise above the younger of the recognizable levels.

The Coastal Plain of late Tertiary time approached the present Coastal Plain in size. Its earlier emergent parts had been subjected to all or to parts of the erosion cycles of the Tertiary. Some of the highest hilltops, among them Pontotoc Ridge, in northeastern Mississippi, appear to

---

[4] Until late in the Pleistocene epoch, the Mississippi River maintained its course near the western side of the valley, and the Ohio River flowed southward near the east bluffs. Crowley's Ridge was the divide between these two major streams, which united somewhere near where Helena, Arkansas, is now located.

represent the Highland Rim peneplain (Harrisburg correlative) and probably have been continuously hilly since the time of their first dissection. The upland levels in some of the hilly areas appear to represent a slightly younger surface—the Coosa peneplain. As all peneplains declined toward, and converged toward, the coast, and as the coast-line has gradually moved outward (except at times of Pleistocene submergence), any evidence of pre-Coosa hilliness would have been destroyed. The physical conditions of the substratum which now result in hilliness would no doubt have so resulted earlier. This suggests the essential continuity of environmental conditions over a long period of time (see p. 509). Later erosion has in most places obliterated all trace of Tertiary cycles.

In the Western Interior, the continued rising of the land which had earlier drained off the epicontinental seas culminated in the final regional uplift of the Rocky Mountains. The climatic effects of this newly formed mountain range were far-reaching. Aridity resulted in the interior of the continent, perhaps extending far eastward.

Uplift of the Harrisburg peneplain (and its correlatives) in late Pliocene time initiated new cycles of erosion which have resulted in the formation of a number of local peneplains or straths. No one is extensive enough to be far-reaching in its vegetational effects, although some have been determinative in local vegetational features (for example, the Cedar Glades of the Nashville Basin, which is the "Nashville peneplain"). The Coosa peneplain on the Gulf Coastal Plain has almost completely obliterated all trace of earlier peneplains. It extends into the Great Valley from the south (along the Coosa River), and it is here that coastal plain vegetation extends northward. Uplift in some places resulted in deep entrenchment of streams, a feature of only local significance in forest vegetation. Most of the topographic features of the Interior Low Plateau have resulted from erosion of the Lexington (or Highland Rim) peneplain. Broad, often abandoned valleys, are features of the younger cycles. Occasionally these may be of value in suggesting possible time of vegetational migrations.

In the north, the advance of the continental ice sheets of the Pleistocene epoch interrupted the progress of post-Tertiary erosion cycles. Alternating with the ice advances (glacial stages) were interglacial stages with climates comparable to the present. The several ice caps differed from one another in area, and in southward extension (Map, p. 498). Hence the southern boundary of glaciation is a composite line, here following one, there another of the glacial margins. Outwash deposits and proglacial lakes carried the topographic influences of glaciation beyond the limits of the ice caps. During the periods of recession, marginal features were produced in many places. Temporary overflow channels for glacial waters developed, only to be abandoned later leaving wide pathways of sand and gravel. Some of these have influenced the course of subsequent migrations.

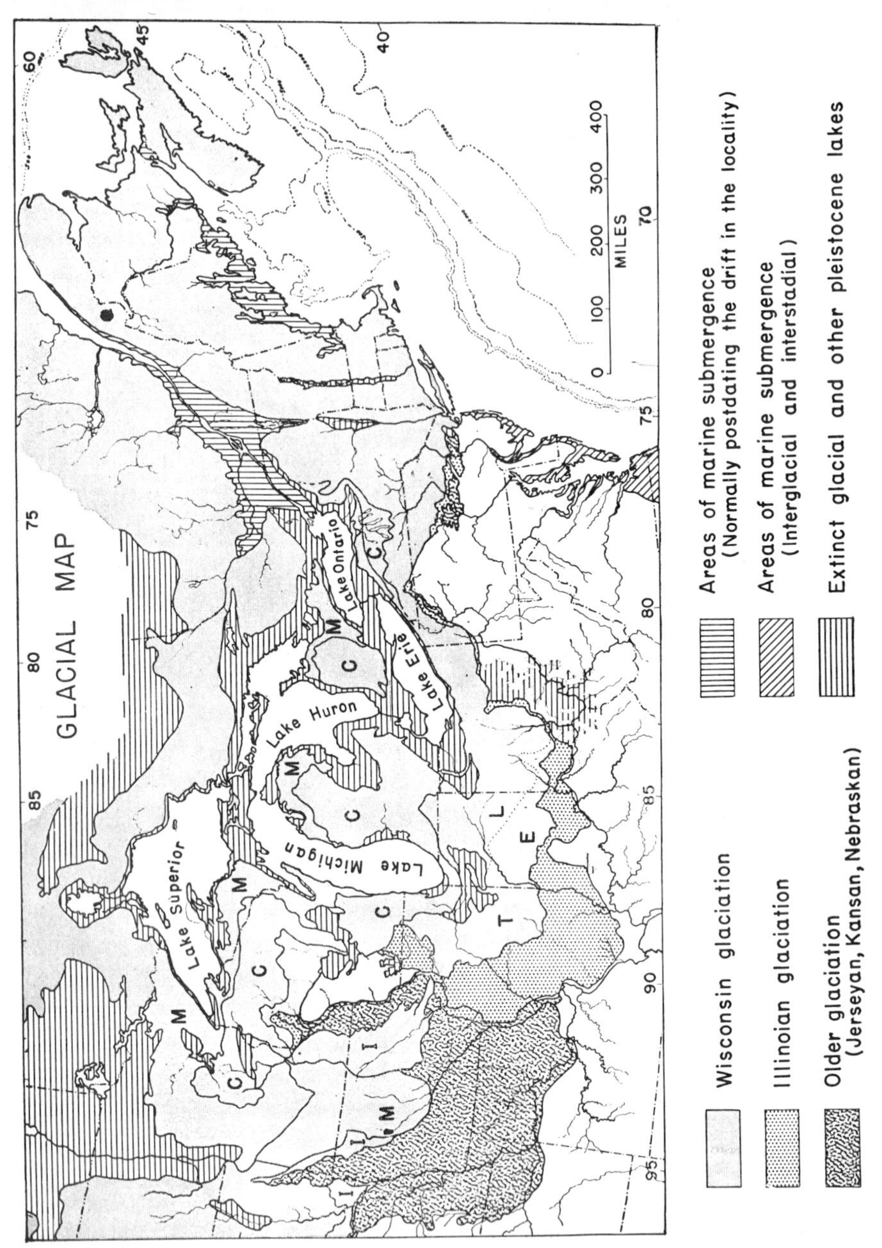

Map showing extent of glaciation, and of submergence by extinct glacial or other Pleistocene lakes and by the ocean. Substages of the Wisconsin stage (where distinguished) are indicated by initial letters: *I*, Iowan; *T*, Tazewell; *C*, Cary; *M*, Mankato. In Indiana and Ohio, where Early and Late Wisconsin are terms commonly used, the letters *E* and *L* are used, with dotted (doubtful) boundary between the two areas. Where no letters and boundaries appear within the area of Wisconsin glaciation, substages have not been distinguished. Kansan and Nebraskan drift west of the Mississippi River not separated; older drift in Pennsylvania and New Jersey is Jerseyan. (Adapted from map by Flint and Committee of the Geological Society of America, 1945; with addition of extinct Lake Tight, in southern Ohio and adjacent Kentucky and West Virginia, from Wolfe, 1942.)

Glaciation modified the topography of all surfaces reached by the ice. Initially rugged areas, as the Adirondack Mountains and much of New England, the Catskills, the Huron Mountains, etc., retained much of the diversity of erosional topography. To this, distinctly glacial features were added, as outwash plains, obstructed valleys, bogs, etc. Areas of lower relief—the Central Lowlands—were covered by deep deposits of drift over wide areas, forming till plains whose undulating surface is still poorly drained except in the oldest areas. These unlike sets of physical conditions are reflected in vegetational features.

While glaciation was determining or modifying topography in the north, fluviatile terraces were being formed along many streams to the south, and coastal terraces, at least in part marine, were developing. Sea-level is known to have varied at least 400 feet (perhaps much more) during the Pleistocene. During each glacial stage, tremendous quantities of water were locked up in glacial ice, and shores receded. During interglacial stages, the ocean level rose. The history of these alternate transgressions and recessions is in part recorded by a series of coastal terraces. No map is available to show the entire extent of such terraces, or their correlation from place to place. Neither is the number of terraces recognized everywhere the same. Marine features are recognized up to an elevation of 100 feet in the Carolinas and 160 feet in Georgia (Flint, 1940), and up to 160 feet in Louisiana (Holland, 1944). During the glacial stages when the water receded below the present shore line, additional coastal land was added. While glacial ice covered New England, the present continental shelf was drained of water. An ice-free coastal strip extended along the North Atlantic coast probably as far as Newfoundland. Each new terrace, each new exposure of continental shelf afforded new land areas onto which or along which vegetation could migrate.

The omnipresent forces of erosion are at work on the new lands and the old lands alike, and, hand in hand with the modification of land surfaces, the development of soils and of vegetation proceeds.

CHAPTER 17

A Chronologic Account of Forest Development Correlated with Physiographic History

The development of deciduous forest in America began many millions of years ago with the earliest appearance of angiospermous trees on the continent. It has continued through Tertiary and Quaternary time to the present, always in response to environmental conditions. A history of this development involves the physiographic history of the land—the changes of land surfaces, peneplanation, uplift, and dissection; it involves climatic changes; it involves the inherent capacity of the species to migrate and the flexibility or rigidity of habitat requirements—physiologic responses—of species; it involves the inherent genetic constitution of species.

The lines of evidence used to build the chronologic story of the development of the present pattern of forest distribution—the pattern described in Part II—are as follows:

1. The paleobotanic record.
2. Disjunct distribution of species and communities.
3. Species relationships, hybrids, genotypes.
4. Physiographic history—available land areas, their age, and contour.
5. Known climatic shifts, or climatic conditions.

Each of these in itself is entirely inadequate; used in conjunction, they may support one another.

The facts derived from the paleobotanic record afford a time scale for the chronologic account. Plant remains are positive evidence of the existence of the plants of which they are a part. Insofar as the record goes, it is definite; but the gaps in time and space are great. The completeness of the record depends to a great extent upon the presence of suitable areas of deposition. The paleobotanic record has been treated in Chapter 14.

Disjunctions in species and community ranges are facts; the interpretations based on these disjunctions are deductions made in accordance with other lines of evidence, especially the third, fourth, and fifth. Disjunctions have been considered in Chapter 15.

The third line of evidence is a promising approach, but one in which little work has been done. The common ancestry of related species is a generally accepted fact. Present separated occurrences have resulted from migrations and restrictions imposed by environment, most of which are incident to changing physiography, and dependent upon gene variability. In the words of Mason (1946):

> The dynamics of the geographic distribution of any species involves the interaction between the environment, the physiological processes of the individuals of a population, and the genetic processes that fix tolerances and maintain or elaborate the population and preadapt individuals to environmental fluctuations.

Hybridization cannot occur unless the ranges of the parent species have met or overlapped. The occurrence of hybrids, then, when parent ranges are remote, is evidence of past migrations. The recognition of genetic races, of diploid and polyploid species, will afford evidence as to the relative age and direction of migration of species.[1] One example will suffice to illustrate the possible significance of this line of approach. Evidence is accumulating to demonstrate that the American beech is far from uniform throughout its range, that it is made up of a number of more or less distinct types. One of these, "gray beech," is

> characteristic of the high elevations from Tennessee and North Carolina northward along the crest of the mountains, coming to near sea-level in Nova Scotia; this same one marks the *northern limits* of our beech population westward to approximately the Great Lakes. Another major type ["white beech"], differing markedly from the high altitude-high latitude beech in leaf shape, margin, texture and pubescence type, as well as in flower shape, mature fruit, bark and twig characteristics and in general stature and conformation, is developed most typically on the southern coastal plain (or the low hills adjacent to it) and has northward extensions along the coastal plain and Piedmont to about the glacial boundary; it is also very common in the interior and apparently was common in the drainage system adjacent to the old Mississippi embayment. A third type ["red beech"] is characteristic of mid-elevations of the Southern Appalachians [and of the Hemlock–White Pine–Northern Hardwoods region] . . . The fourth beech, *Fagus mexicana,* is found in a few localities in Central Mexico in the mountains; admittedly it is a close relative of our high-mountain beech.[2]

The problem is much more complicated than the above statements would indicate; many introgressions occur. For example, over a large territory in the interior there is a genetic blending of white and red beech. When more is known about beech and others of our widespread tree species, more

---

[1] For a discussion of the principles, and citations to the literature, see Cain, 1944.

[2] Information concerning beech was furnished by W. H. Camp in conversation and in correspondence. Work on the problem was interrupted by the war, but has been resumed.

evidence will be available concerning the development of deciduous forest.

The use of physiographic history as a line of evidence in reconstructing the past history of forest development is based upon the interdependence of habitat area and species or community area. It is built upon observable topographic and structural features, with gaps sometimes filled by analogies and deductive reasoning. As given in Chapter 16, it is independent of other lines of evidence.

Little is known about climates of the past. Such statements concerning climate as are based upon the kind and character of plant or animal life depend for their correctness on the essential similarity of requirement of similar modern and ancient forms of life. Where a variety of forms resembling modern ones from cool and warm climates are assembled in a heterogeneous mixture, either lack of differentiation of ancestral forms is indicated, or mixing of remains from climatically distinct communities. Then too, a fossil horizon may represent a long period of time so that the mixing may be apparent rather than real. Inferences concerning climate which are based upon the nature of deposits are not influenced by other lines of evidence; they may support the inferences based upon paleobotanic evidence.

Forest expansions and contractions, and changes in forest composition, must go hand in hand with physiographic and climatic changes. It is not to be understood that physiography of itself controls vegetation. Rather, it has a profound effect upon climate, and upon soil development, and it determines the nature, extent, and diversity of local habitats. The shifts in vegetation which must have taken place under the impulse of changing climate are modified by physiography (habitats) of the land surface.

In constructing the chronologic story of forest development, an attempt is made to use all lines of evidence, to reconcile the sometimes seemingly contradictory evidence. With the evidence at hand, it is not possible to construct an infallible story. Some will disagree with one interpretation or use of evidence, some with another. Any interpretation, to be acceptable, must make allowance for all evidence produced; it must not, in the interpretation of one phenomenon, of one fossil or vegetational feature, build up an impossible or unlikely case for some other known features.

MESOZOIC HISTORY

Geologic and physiographic events between the time of the Appalachian Revolution and the completion and later submergence of the Fall Zone peneplain determined the nature and extent of the continent on which the earliest recorded angiosperm forest development was taking place (p. 489). The first records of this development, in the Potomac series of the Lower Cretaceous, are in sediments deposited on the depressed and partially submerged Fall Zone peneplain. These early forests were composed in part of gymnosperms, in part of angiosperms, the latter increasing in proportion to the former in the later part of the period, demonstrating the presence of broad-leaved forest communities (p. 451).

Increase in the number of genera represented, and in the abundance of remains, in Upper Cretaceous time, points to the probable dominance of broad-leaved forests in eastern North America at that time. Based upon the somewhat later occurrence of certain genera in the Western Interior, migration from east to west is indicated. That the Cretaceous eastern American deciduous forest had a modern aspect is suggested by the presence of such genera as *Acer, Fagus, Quercus, Liriodendron, Magnolia, Diospyros, Juglans, Celtis, Populus,* and *Salix*. Gymnosperms, also, were present in the eastern American forests, perhaps segregated into distinct communities, perhaps intermixed with the angiospermous species. Among the gymnosperms were representatives of genera which now have widely differing ranges, as *Pinus, Picea, Thuja, Sequoia, Tumion* (*Torreya*), *Araucaria,* and *Podocarpus*. Genera which we now think of as belonging in unlike climatic belts are mingled in the records. However, no one of these genera is unquestionably cool temperate or warm temperate. Most of them, even today, have representatives in widely different climatic areas; hence, they give no definite climatic implication. The early origin of two of the genera now represented in the East by local endemic species or species with disrupted ranges (*Torreya* and *Pachystima*) is of interest in establishing the antiquity of these lines (p. 452).

Coincident with the developing deciduous forest, soil development, with the incorporation of humus, must have been taking place. When, then, the Cretaceous continent was uplifted (p. 491), old areas and young areas (the emerging coastal plain) were separated by differences of the edaphic environment. How long such differences persist is conjectural, but we know that large areas of the Coastal Plain of today still retain soils in which the nature of parent material is apparent. To some extent, at least, the environmental differences between newly emergent coastal plain and older forest-covered upland must have influenced early Tertiary forest development. If the superposition hypothesis (p. 491) is correct, differences still apparent between the vegetation of the Appalachian Plateaus and land to the east may have been initiated on the early Tertiary continent and maintained by continuing differences between the adjacent areas.

TERTIARY HISTORY

EARLY AND MIDDLE TERTIARY

The records of early Tertiary forest development are preserved chiefly in the Mississippi embayment and Western Interior (pp. 452, 456). Although the aspect of the Eocene forest of the Mississippi embayment was comparable to that of present inland and upland areas of the warmer parts of the Temperate Zone (because of the large number of its genera now confined to lower latitudes), it did contain many genera with modern representatives in eastern American forests, as *Aralia, Aristolochia, Asimina, Bumelia, Celastrus, Cercis, Cladrastis, Cornus, Diospyros, Evonymus, Ilex, Liquidambar, Magnolia, Rhamnus,* and *Taxodium.* A large proportion of these are shrubs or understory trees in our modern forests. Again, some genera now represented by species of peculiarly disrupted ranges (as *Cladrastis* and *Lygodium*) are among the constituents of this early American flora. The common Holarctic genera represented in the Upper Cretaceous floras are not recorded in Eocene floras of southern United States, although they were still present in the northern part of the Western Interior (pp. 453, 457).

If, now, we recall that post-Cretaceous uplift initiated the long Schooley cycle of erosion, and that in the final stages of this cycle, residual mountains several thousand feet in height remained (p. 494), it is evident that large areas in the Appalachian Highlands must have stood at high altitudes in the early part of the cycle. Altitudinal and latitudinal differentiation of the deciduous forest of Eocene time was no doubt somewhat comparable to the altitudinal and latitudinal differentiation of today. The red spruce, white pine, fir, and hemlock of the southern mountains may have progenitors as old as those of *Pachystima, Cladrastis,* and *Lygodium,* with remarkably disrupted ranges. White pine is represented in the mountains of southern Mexico and Guatemala, along with a considerable number of deciduous forest genera (p. 485). The warm Eocene epoch would have brought about retreat of cool temperate forms to the highlands, and consequent disruption of ranges, as well as retreat northward, with consequent latitudinal segregation, as demonstrated by the older Tertiary floras (Raton, Denver, and Fort Union) of the Western Interior (p. 457).

Based upon evidence afforded by fossils, by physiography, and by relationships and ranges, the early Tertiary broad-leaved forest of eastern North America extended over much of the continent from the present location of the Rocky Mountains eastward (and northward also). In its southern and lower elevation areas, it contained a greater number of warm temperate and tropical forms than now occur; broad-leaved evergreen and deciduous genera were represented. Northward, and at higher elevations, the forest was temperate, and essentially modern in aspect. At the same time, temperate forests thrived in Greenland, Alaska, Spitsbergen, and Siberia (p. 457).

The warm temperate or subtropical floras of the southern lowlands of early Tertiary time culminated in Oligocene time. Subsequently, the gradual cooling of climates, indicated by the nature of the fossil record, brought about southward migration of forest zones (p. 455). The Miocene forests of coastal South Carolina and Georgia, although they displayed striking warm temperature affinities, nevertheless contained an array of genera to be found in those forests today (*Pinus, Taxodium, Quercus, Planera, Ilex, Nyssa*) and species similar to existing forms. By this time, most of the common genera of trees and shrubs of the deciduous forest had been represented in the fossil record. The constituents of the modern forest are known to have been present.

The nature and extent of the continent during the time of forest development recorded near the coast by early and middle Tertiary fossils must have greatly affected the development of forest upon it. During this time, the Schooley cycle of erosion was in progress. Progress in the cycle at first increased topographic diversity due to dissection of the upraised peneplain. Later, as old age in the cycle was approached, more and more of the land again became reduced to peneplain status (the Schooley peneplain). Since development of a peneplain proceeds from the coast inland and upstream, the coastal areas—where the record of early and middle Tertiary vegetation was preserved—were reduced to peneplain status while yet the interior was mountainous. The middle and later Eocene lowland floras had large areas over which to expand, while the area suitable for upland floras was lessened. Even within the Appalachian Highland, peneplanation was evident, and the plant communities favored by topographic diversity of maturity in the cycle became restricted, while those which found the poor drainage of the peneplain, or the acid swamps of sandy low plateaus most favorable expanded. Plants and plant communities of hilly and mountainous lands were of necessity restricted to the unreduced areas (p. 494).

Three features of present forest distribution (on unglaciated land) appear to be significant. These are:

1. The most typical examples of mixed mesophytic forest, those displaying the characteristic wealth of species, occur in places where the Schooley peneplain was never perfected, namely, in the Cumberland Mountains, along the western edge of the Cumberland and southern Allegheny plateaus, and in coves in the Southern Appalachians.

2. Mixed mesophytic forest prevails over the great area where the upland levels represent the reduced Schooley peneplain, and where no later peneplain was extensively developed, that is, over the southern half or more of the unglaciated Appalachian Plateaus.

3. Relic disjunct mixed mesophytic forest communities occur almost everywhere south of the glacial boundary where isolated remnants of the Schooley peneplain are found, except to the east of the Allegheny and Cumberland fronts (or more exactly, to the east of the eastern boundary of the Mixed Mesophytic Forest region). Advance disjuncts are the rule in that direction.

There appears to be a cause and effect relationship. This should not be interpreted as meaning that the distribution of mixed mesophytic forest is independent of present environment, but that suitable environmental conditions have for hundreds of thousands of years been in part controlled by a topography persisting from middle Tertiary time. The mixed temperate forest of the Tertiary became restricted to areas of diverse topography, areas where slopes prevailed; its range became discontinuous.

Not only are there correlations between forest distribution and early and middle Tertiary physiography, but the many now highly local disjunct and endemic species of these areas can only be explained on a basis of Tertiary history and development of the Schooley peneplain. Many of the ancient forms became restricted as peneplanation rendered large areas unsuitable for their continuance. Thus, today they exhibit relic distributions correlated with these always unreduced areas of the Schooley peneplain (pp. 481, 494). Most of these, and many of the commoner species, have relatives in Asia, thus, carrying them, in ancestry, back to a period of widespread Arcto-Tertiary floras. Some, which belong to families now largely tropical or subtropical in distribution, probably date back to the early Tertiary period of supremacy of subtropical forms in southern United States.

More or less contemporaneously with the restriction of one group of species, another group expanded, spreading over the flats of the peneplain. For evidence, note the relic distribution of what we now call coastal plain plants, and of colonies of coastal plain plants, on the relatively unmodified remnants of the peneplain, swampy unrejuvenated stream heads, or sandy plateaus dotted with low monadnocks (p. 494).

When later, elevation took place and dissection of the peneplain began, the mixed forest again expanded. However, many of the restricted species were no longer able to extend their ranges.[3] Other species, at that time more widespread, were forced into smaller and smaller areas by changing environment—dissection of the peneplain. Many of the relic disjuncts of today belong to one or another of these groups.

The progenitor of what we now call the Mixed Mesophytic association was present in Miocene time, or even earlier, and has continued to the present. Eastern North America is not alone in this respect. A similar relationship exists between modern and Miocene forests of eastern Asia as has been pointed out by Chaney and Hu (1940). A comparison of the present "mixed deciduous forest" and the Shanwang flora of Shantung Province shows that "the modern forests of Korea, and also of Japan, have much in common with Miocene vegetation of North China."

[3] Limitation of range is due to a complex of limiting environmental factors and reduced biotype variability.

LATER TERTIARY: HARRISBURG CYCLE

Climatic changes, and the more or less contemporaneous physiographic changes resultant upon uplift of the Schooley peneplain and initiation of the Harrisburg cycle of erosion, brought about pronounced changes in forest distribution and forest composition in late Tertiary time. The southward shifts of vegetation zones which first became apparent in the north during Oligocene and Miocene time continued.[4] While temperate forests were widespread in the uplands, and in northern latitudes in general, in Miocene time, it was not until Pliocene time that the flora of the Gulf Coast of southeastern United States came to closely resemble the present flora of that area (p. 455). The continued southward trend in the Pliocene brought about a zonal distribution comparable to that of today. Increasing dryness in the Western Interior, brought about by the gradual rise of the Rocky Mountains, reduced the forest vegetation of that area, and favored the development of grassland. Flood-plain forests comparable to those now found somewhat farther east occupied the valleys in the High Plains, while grassland covered the intervening space, and genera of higher moisture requirements disappeared from the area, or, in small numbers, remained in restricted favorable habitats (pp. 458, 485). A pattern of distribution of grassland and deciduous forest similar to that of the present resulted. The mixed forest of the middle Tertiary was forced to retreat eastward from the drying interior; the more mesophytic forest species were most curtailed, being eliminated entirely from the western part of the deciduous forest, or restricted to the few areas of diversified topography which remained on the rising Schooley peneplain. Thus the Oak–Hickory association arose and the more mesophytic species of the mixed forest found only a few spots in the area west of the Mississippi suitable for their survival. It is in these areas, in the Boston and Ouachita Mountains, that the relic mixed mesophytic forest communities occur today (pp. 176, 484). At this time, too, the final separation was effected (because of dryness) between the forests of eastern United States and the Mexican Highlands, where a number of trees of the eastern deciduous forest (or species closely related to them) still survive (p. 485).

To the east of the Mississippi, changes in forest composition and in extent of area occupied by the mixed forest of the Tertiary accompanied progress in the erosion cycle initiated by uplift of the Schooley peneplain. Differential erosion (controlled largely by structure and lithologic character

[4] This latitudinal shifting of Tertiary forests is well shown by the fossil records in the Pacific Northwest (Chaney, 1940) and by the fossil records in Eastern Asia, whose forests resemble those of eastern America (Chaney and Hu, 1940; Kryshtofovich, 1929). It has been pointed out in Europe by Reid and Chandler (1933).

of substratum) soon delimited distinct physiographic areas—the Interior Low Plateau and the several provinces of the Appalachian Highland (p. 495).

In the areas underlain by resistant rocks, areas which were dissected but not greatly reduced, the extent of land suitable for mixed mesophytic forest was increased. Thus, in the Appalachian Plateaus particularly, mixed forest expanded, spreading out, as habitats became suitable, from the always hilly areas.

Where the new peneplain (Harrisburg) developed on weak rocks, the complex of environmental factors was different. The Ridge and Valley Province, with its broad valley flats (Harrisburg peneplain) and steep-sided mountain ranges, developed just to the east of the Appalachian Plateaus, to the east of the Allegheny and Cumberland fronts. One of the most pronounced regional forest boundaries approximates this line (p. 194). That vegetational differences along this line may have been initiated at the close of the Cretaceous (if the superposition hypothesis is accepted) has already been suggested (p. 503). This line (again?) became a pronounced topographic boundary in the Pliocene, and may have been one throughout the Schooley cycle.[5] Some vegetational differences may always have existed along this boundary. With the development of the topographic features which now characterize the Ridge and Valley Province, vegetational differentiation must have taken place. The infertile soils derived from the resistant rock of the ridges, and the shallow soil over the steeply inclined strata are not (and doubtless were not) favorable to a mixed mesophytic forest; neither were the flats of the peneplain. Oaks and chestnut probably then gained ascendancy, and have kept it, until recently,[6] except where erosion in the present cycle has produced habitats favorable for the invasion of species of the mixed mesophytic forest.

The more massive residual mountains—the Southern Blue Ridge or Southern Appalachians—retained the mixed Tertiary forest in their coves, while their more exposed slopes became occupied by oak–chestnut forest,[7] thus initiating the Oak–Chestnut association.

The oak, oak–hickory, or oak–pine forest of the Piedmont is related to the Harrisburg peneplain, and the development of low elevation, low latitude regional soils. That it is not just a recent development is demon-

[5] It may have been a boundary at the beginning of the Tertiary, at the time of uplift initiating the Schooley cycle; it probably was a boundary during a large part (or all) of the Schooley cycle (because of difference in the rate of erosion of adjacent outcrops); it assumed the prominence that it now has with the development of the Harrisburg peneplain.

[6] The elimination of chestnut, as a dominant in the forest, is an event of the past two or three decades. For discussion of the term, oak–chestnut, see p. 192.

[7] Certain problems in connection with the dominance of oak–chestnut forest on slopes of the Southern Appalachians may be related to the severity of occasional summer droughts.

strated by Pleistocene pollen records (p. 471). Wherever suitable habitats have developed on the older parts of the Coastal Plain (Atlantic and Gulf), this vegetation has occupied the Coastal Plain.

To the west of the Appalachian Plateaus, where the Harrisburg cycle of erosion lowered the great area extending from the plateau escarpment to the Mississippi, a complicated vegetational history has been in progress. The mosaic of vegetational types of the Western Mesophytic Forest region (p. 122) has been in large part a result of late Tertiary history. Wherever, in the formation of the Highland Rim or Lexington peneplains (Harrisburg equivalents), the land was not reduced and hills remained, edaphic conditions were favorable for the persistence of the mixed Tertiary forest, if, as seems probable, these areas had retained some topographic diversity throughout the Schooley cycle (p. 494). Thus there are in these areas, forest communities best referred to the Mixed Mesophytic association. They reflect, in the absence of certain species common to the east, the stress of alternating expansions and contractions of area (suitable edaphic environment) and of late Tertiary decrease in humidity. At the same time that climatic conditions restricted the mixed forest to the slightly higher areas of diverse topography with compensating effects of microclimates and slightly higher rainfall due to condensation in an unstable column of air, the expansion of oak forests over the peneplain was favored. Underdrainage, evident over large areas in the Interior Low Plateau because of a karst topography, probably accentuated the climatic effects. Whether the grassland, which had been developing on the High Plains since late Miocene time, reached this area during the late Tertiary is not known. However, the prevalence of oak forests, or of mixed forests in which oaks are abundant and the most exacting mesophytic species absent may well date from the closing epoch of the Tertiary.

The Coastal Plain, from the time of its first emergence as a result of post-Cretaceous uplift to the end of the Tertiary, was gradually enlarging. New land areas onto which vegetation could expand were continually being formed. In part, at least, these were populated by migration from older peneplain areas, areas which during subsequent erosion were so decreased in size that little remained of vegetation from whence the coastal plain migrations had come. In places where age, elevation, or lithologic character of the substratum favored diversity of topography and development of fertile soils,[8] mixed mesophytic communities occur, always with a pronounced admixture of more southern species, as *Magnolia grandiflora*. Such areas, in part at least, are above the level of Pleistocene submergence; they are areas where the essential continuity of environmental conditions probably dates back into the Tertiary (p. 496). Their forests are indicative of Tertiary ex-

[8] Usually areas of Orangeburg, Memphis, or Grenada soils, all of which are derived from calcareous material (Marbut, 1935).

pansion. The Apalachicola River bluffs and the Tunica Hills are outstanding examples, marked by disjuncts and relic endemics, among them *Torreya taxifolia* (pp. 300, 303, 487).

CONDITIONS AT CLOSE OF TERTIARY

Let us now picture the forest conditions at the close of the Harrisburg cycle of erosion. The forest pattern seems to have been one related to climate and to physiographic cycles. The mixed Tertiary forest (modified perhaps by latitudinal segregation) continued to occupy the areas where the ancient Schooley peneplain, although dissected, was not replaced by a later peneplain (Appalachian Plateaus, Southern Appalachians, and perhaps the New England Uplands and Adirondacks). The rising West affected the climate of the whole interior of the continent, thus eliminating the more mesophytic species from much of this area. The Oak–Hickory association, centering in the Ozarks, was a result of this climatic segregation. At the same time, segregation, induced by physiographic change and contemporaneous with the progress of the Harrisburg cycle of erosion, resulted in the establishment—on the new or greatly enlarged areas—of forests derived from the mixed Tertiary forest, forests poorer in species and with fewer dominants than the forest of dissected areas of the Schooley peneplain. To what extent climate entered in is problematic. The similarity of response to the east and the west of the Appalachian axis suggests that both areas were almost equally dry. At any rate, forests dominated by oaks, or by oaks and chestnut spread over most of the Harrisburg levels, dominating the forest aspect of the Ridge and Valley Province, the Piedmont Plateau, and much of the Interior Low Plateau. Thus, the eastern Oak–Hickory association of the Oak–Pine region, the Oak–Chestnut association, and the white oak forest type became dominant. Due to the peculiar reactions of an occupying oak forest upon the soil, continuance of oak types is favored until such time as new soils are formed as a result of stream rejuvenation. A physiographic pattern, and a pattern of forest distribution similar to that of the present had been attained by the close of the Tertiary in the area which is now unglaciated eastern America. What the forests to the north were at the end of Tertiary time we have no way of knowing.

PLEISTOCENE HISTORY

The next significant and far-reaching event in the land history of eastern America was the initiation of glaciation. Uplift had already interrupted the Harrisburg cycle of erosion, and valleys were being deepened (p. 497). With the growth and advance of ice caps the vegetation of the North was destroyed. Southward migration in advance of the ice (in response to climatic change) no doubt took place, the newcomers invading the established vegetation insofar as suitable habitats were available or weakened competition of more southern forms permitted. However, evidence is entirely inadequate to enable us to determine what were the nature and extent of migratory waves of plant and animal life. We do know that in Europe the effects of glaciation were more pronounced than in America or Asia. That there, many genera of trees (as *Juglans, Carya, Magnolia, Sassafras,* etc.) which had been present in Tertiary time, became extinct as a result of glaciation, for the east-west mountain ranges and bodies of water prevented southward migration.

We have no evidence that there was a band of spruce–fir forest in the North before the advent of the first continental glacier. Neither spruce nor fir are represented in the fossil record of the Tertiary of the East, although *Picea* is found in the Upper Cretaceous, and *Abietites* in the Lower Cretaceous deposits of the Atlantic and Gulf coastal plains, where they are associated with broad-leaved and conifer genera, some of which are now tropical and warm temperate in distribution. Records of *Picea* in the far northwest and Alaska in Eocene and Miocene deposits indicate that it was associated with genera which we now think of as indicative of a mild climate. The segregation of a northern spruce–fir forest came late in forest history. It may have been pre-Pleistocene, altitudinal or latitudinal, or both; it may have resulted from the climatic stress of glacial climate, and been aided by the vast expanse of area of raw soil material unsuited to invasion by broad-leaved species; it may have been perfected only in late Pleistocene time. Evidence from buried soils in the Piedmont (probably Pleistocene in age) shows that a species of *Abies* not referable to *A. balsamea* or *A. Fraseri* was then in existence (p. 471). Did this antedate the final segregation in the East of high altitude, high latitude species of *Abies?* In the West, there are species of both genera in warm and in dry climates. *Abies* and *Picea* are not necessarily indicative of a cold temperate climate, although certain species of these genera are. It is doubtful if positive identification of evolving species can be made from fossil pollen grains; it is certain that genotypes (such as those of *Fagus,* with distinctive ranges) could not be distinguished by fossil pollen grains.

The earliest unquestionable records of forests in which spruce and fir were dominant are in the lower levels of post-Wisconsin bog deposits in the glaciated territory. However, as white spruce has been found in most of the Pleistocene deposits of Minnesota which have been examined, Rosendahl (1948) concludes that white spruce was preponderant in the interglacial forests of that area. Voss (1933) considers that there is evidence of forests of fir, spruce, pine, and larch in late Sangamon or early Peorian time. Remains of spruce and fir in pre-Kansan peat deposits of Iowa and Minnesota are associated with a variety of deciduous species (p. 460).

The first glacial advance (Jerseyan or Nebraskan) reached southward into New Jersey, Pennsylvania, southern Ohio, and the adjacent northernmost tip of Kentucky. This destroyed all forest in its path, and forced (by climatic influence) some southward migration before it. But migration is possible only where habitats are suitable. This early glacial boundary crossed the Appalachian Plateaus (late Tertiary haven of mixed forest) from central Ohio through northern Pennsylvania, swinging slightly southeastward in eastern Pennsylvania, and then crossed the southern lobe of the New England Upland (here known as the Reading Prong and New Jersey Highlands). The glacial boundary was not far north of the southern boundary of the Plateau as it curves eastward around the northern end of the Ridge and Valley Province in Pennsylvania (see map, p. 394). Thus, the mixed forest was largely restricted to the unglaciated parts of the Appalachian Plateaus. Farther east, a small area—the Reading Prong and New Jersey Highlands—remained unglaciated, separated, however, from the main mass of mixed forest by the whole east-west expanse of the curved northern end of the Ridge and Valley Province. The frequency of mixed mesophytic communities in the northern part of the Piedmont section in Pennsylvania, and the southwestern part of the Glaciated section of the Oak–Chestnut region is correlated with the small eastern refugium of mixed Tertiary forest not reached by Pleistocene ice, but isolated from the main area of that forest on the Appalachian Plateaus and deprived of some of its species because of the nearness of glacial ice and the lack of suitable edaphic environments adjacent to the south on which to expand. This much decimation of the mixed Tertiary forest must have resulted from the first glacial advance.

To and fro migration took place, not once, but with each advance of ice. Vegetational occupancy of glacially denuded land in the interglacial stages is positively shown by plant remains in interglacial deposits (p. 459). Exposures of old drift in areas not reached by later glaciations may have been continuously occupied by vegetation since the first establishment of vegetation on the newly exposed drift. Although their vegetation may have been later affected by the approach of another ice margin, the longer period of time (with consequent topographic and soil development, and cumulative effects of occupying vegetation) has had an influence on present vegetation. Migrations following the last ice sheet have determined the nature of the vegetation of all the area covered by that ice sheet.

VEGETATION SOUTH OF ICE MARGIN

One of the biggest problems in connection with Pleistocene vegetation is the extent to which the vegetation south of the ice margin was affected by glacial climate, was modified by invasions of more northern plants. Two lines of evidence are available:

1. Fossil evidence, including postglacial pollen records preserved in bogs south of the glacial boundary, and plant fragments preserved in proglacial lake, fluviatile, and estuarine deposits (pp. 460, 463).
2. Evidence afforded by present distribution of species, disjunct occurrences, and the like (p. 477).

Fossil Evidence

From fossil evidence we may conclude that nowhere far beyond the glacial boundary was climate during the glacial stages sufficiently severe to displace occupying vegetation. Such infiltration of northern species as is demonstrated in certain instances by pollen records or plant fragments was made possible in part by slightly cooler climates, in part by the lack of competition on newly exposed Pleistocene coastal terraces and river deposits. It is true that there is conflicting evidence, and that no interpretation based on present meager evidence will remain unchallenged.

In interpreting the fossil evidence, which is almost entirely from the Coastal Plain, we must keep in mind that the great mass of coastal plain vegetation is still in developmental or subclimax stages; few climax communities occur. Only in the climax do vegetational reactions tend to exclude invaders; developmental stages are constantly changing and the opportunity for extraregional invaders to enter is far greater than in climax communities. Migration southward on the Coastal Plain was facilitated during each glacial advance by the withdrawal of the sea (water locked up in glacial ice) which resulted in a continuous similar habitat from the glacial margin southward to the Gulf along which rapid migration could take place. In addition, the discharge of glacial melt-water and icebergs into the Atlantic as far south as New York and Long Island and of melt-water into Delaware and Chesapeake bays, may well have resulted in offshore cold waters and a fog belt along the coast. Ocean drifts could not have been where they now are, with an emergent continental shelf. In the Mississippi Valley, the overloaded stream of glacial stages built up many bare areas open to invasion from the north. Microclimates may have played a part. Evidence from the Patschke Bog, near Austin, Texas, and from buried peats of Florida (p. 471) is difficult to interpret in the light of present knowledge. Pollen of northern species (*Abies,*[9] *Picea glauca,* and *P. mariana*) is reported in small amount from the lower part of the Texas bog (Potzger and Tharp, 1943; 1947). These trees must have been present in favorable edaphic situations in an area where the regional type (as indicated by the bog pollen) was pine–oak–grass or oak–grass savannah. Although an infiltration of northern species is evident here, unless the few pollen

[9] Species of *Abies* not stated.

grains were carried in by strong northerly winds of glacial anticyclones (pp. 464, 518), the deciduous forest species were not eliminated from the region; pollen of deciduous trees (*Quercus, Carya, Castanea, Betula, Tilia, Acer, Ulmus, Nyssa,* and *Salix*) occurred at the same levels as the spruce and fir pollens (not all at any one level, although all were present during the faint early spruce–fir maximum. Glacial anticyclones and plains "northers" similar to but more extreme than those of the present, and increased humidity are possible factors permitting the southward migration of northern species to the west of the Mississippi.

Almost all of the records, however, are in marginal positions, and give little or no evidence as to glacial influences in the Appalachian Highlands and Interior Low Plateau. For this extensive area, the evidence afforded by species and community distribution is all that is available.

Distributional Evidence

Vegetational features of significance in connection with Pleistocene migrations (or lack of migrations) south of the ice sheet in the eastern interior may be outlined as follows:

1. Position and nature of northern boundaries of species ranges and of communities with reference to the glacial boundary.
2. Nature and composition of different parts of communities occupying continuous environments extending in north-south lines away from the glacial boundary.
3. Contrasts between the vegetation of areas flooded by ponding of streams during early Pleistocene ice advances and adjacent nonflooded areas.
4. Disjunct occurrences of southern species along preglacial drainage lines.
5. The contrast in species composition between the more northern and the more southern of the mountain or plateau bog and swamp communities.
6. Relic occurrences of northern species and communities south of the glacial margin.

Each of these features will be considered along with examples.

1. A number of forest species have ranges whose northern limits appear to bear some relation to the glacial boundary, particularly to the Wisconsin boundary. The two most characteristic species of the canopy of the Mixed Mesophytic forest, *Aesculus octandra* and *Tilia heterophylla* (including var. *Michauxii*) are abundant almost to the glacial boundary and locally within the area of Illinoian glaciation in southwestern Ohio and southeastern Indiana, but have scarcely invaded into the area of Wisconsin ice. In neither of these is there any evidence that a climatic limit has been reached at the north, and in neither is the limit of range at a soil boundary. They are forest trees growing in areas of melanized soils with mull humus. Mistletoe

(*Phoradendron flavescens*) has a comparable range which cannot be related to soil and bears no relation to the ranges of its several host trees. No recognizable climatic boundary corresponds with the limits of range of this species. Among additional woody species whose northern limits appear to be correlated with the Wisconsin glacial boundary are *Aralia spinosa* and *Rhamnus caroliniana.* The first of these has a few outposts well within the younger glaciated area in Ohio (showing that it is not climatically limited) but is abundant only south of the glacial boundary. *Rhamnus* is a calciphile, occurring abundantly along the outcrop of Silurian dolomites at the western edge of the Allegheny Plateau, to within 25 miles of the Wisconsin glacial boundary. There it stops, although its edaphic environment continues to and within the glacial boundary. Numerous examples among herbaceous species could be added. It would seem (as Gleason, in 1923, suggested) that the ranges of these species were abruptly terminated during the Pleistocene, and that they have not as yet greatly re-extended their ranges.

The boundary between the Mixed Mesophytic and Beech–Maple associations is approximately at the Wisconsin glacial boundary from central Indiana to western Pennsylvania. There has been time enough for mixed mesophytic forest communities to develop where habitats are suitable, but not to dominate, in the older, Illinoian, area.

2. Because of the physiographic configuration of the East, and the more or less general north-south trend of rock outcrops in the Appalachian region, strips of continuous similar environments extend in a north-south direction away from the glacial boundary. One of the most distinct of these is the rugged western margin of the Appalachian Plateau extending from the glacial border in south-central Ohio, southward through Kentucky and Tennessee. Almost throughout its extent it harbors excellent mixed mesophytic forest which is as rich in species in all layers near (not at) the glacial boundary as it is 200 miles farther south.[10] At the glacial margin in Hocking County, Ohio, there are a few northern species in the undergrowth, but there is a greater number of southern species here at their northern limits. In an area in Highland County, Ohio, reached by Illinoian but not by Wisconsin ice, the paucity of *Tilia heterophylla* and abundance of *T. americana,* the absence of *Aesculus octandra,* and the presence of local patches of *Taxus canadensis* on the cliffs, and the absence of distinctly southern[11] species (some of which are present farther north in unglaciated Hocking County) distinguish the forest from areas farther south. This rugged western border of the Plateau (specifically the Cliff Section and the Knobs Border area of the Mixed Mesophytic Forest region) is also marked by its many southern disjuncts, highly local species, endemics, and species whose dis-

---

[10] Compare the several areas in Tables 14 and 15.

[11] Unless *Pachystima Canbyi,* on cliffs in the vicinity, be considered as southern.

junct ranges correspond with the unreduced remnants of the Schooley peneplain (pp. 481, 482).

A similar although less well-marked instance of community continuity from the north southward is to be seen along the Allegheny Front where the typical mixed mesophytic forest along this strip in southern Pennsylvania is continued northward, but in a somewhat attenuated form, into the latitude of the southern lobe of the Hemlock–White Pine–Northern Hardwoods Forest on the Plateau.

Still another example, also within the Mixed Mesophytic Forest region, is to be seen along the slopes bordering the Allegheny River, where a tongue of mixed mesophytic forest extends northward to the glacial boundary in southwestern New York (p. 406).

The white oak forest of the Harrisburg peneplain levels of the Ridge and Valley Province is essentially the same from central Pennsylvania southward into southern Virginia and Tennessee. White pine is a more frequent constituent in the north, but is present also in oak woodlands of the section farther south.

3. The marked contrast between the vegetation of areas occupied by proglacial lakes during the early Pleistocene, and of surrounding higher lands suggested the possibility that the vegetation of the higher lands was not completely displaced (p. 478).

4. Drainage lines of the past have served, as do present drainage lines, as pathways of migration along which lowland species might advance in the bottomlands, slope species advance along the valley sides, and ridgetop species advance along divides. The Teays River (p. 478) was the largest of the glacially obstructed streams, and the one whose tributaries came from farthest south (headwaters in the Southern Appalachians of North Carolina). Parts of its old course to the north of the Ohio River are now occupied by southward flowing streams (the Scioto River and other smaller streams); parts of it now contain no streams of appreciable length. Yet in this old valley have been found "remarkable isolated assemblages of species growing in habitats that apparently are not unlike those in many other parts of unglaciated Ohio, but in which comparable assemblages are lacking" (Transeau, 1941). It is believed by this author and his associates that these are remnants of the Tertiary vegetation which had migrated downstream (northward) at least this far before the advance of early Pleistocene ice, and before the reversal of drainage which took place in the early Pleistocene.

5. Bog or swamp communities are found widely scattered and at various elevations in the Appalachian Highlands (p. 480). The more northern of these (such as Cranberry Glades in West Virginia and Bear Meadows in Pennsylvania) are more or less typical northern bog communities. The more southern generally harbor coastal plain relics and very few, if any, mildly

northern forms; in these boggy spots, surviving from the mid-Tertiary peneplain, the pre-Pleistocene flora has not been displaced. Pleistocene migration did not reach as far south as the Cumberland Plateau.

6. A number of northern species[12] occur more or less isolated south of the glacial border, and in only a few instances do we find these grouped into distinct communities. They usually occur individually, as if by accident, or localized in some extreme habitat not suitable for the regional vegetation. This is particularly true of *Thuja occidentalis* on limestone cliffs. Northern communities occurring south of the glacial boundary include the bog communities of the northern part of the unglaciated Appalachian Highlands (p. 480), and certain of the mountain-top spruce–hardwoods, spruce, or spruce–fir forest communities (p. 479). The mesophytic deciduous forest communities of the Gulf Coastal Plain (which by some authors have been referred to as northern disjuncts) have been correlated with late Tertiary migrations (p. 488).

A pronounced mingling of floristic elements is noted in almost all localities where northern species occur, except in the more northern mountain bog and mountain top communities. If the northern relics are interpreted as remnants left from a widespread migration and shifting of northern vegetation southward, then the explanation of the many southern species, and particularly of the southern disjuncts and local disjuncts, becomes extremely difficult if not impossible. Local occupation of suitable habitats by northern species took place while southern species continued to occupy most of the area.

As a whole, the evidence afforded by species and community distribution south of the glacial boundary indicates that southern species and communities persisted not far from the ice front during the glacial stages of the Pleistocene. Some lowering of altitudinal belts is indicated which permitted southward migration of northern vegetation. There is little evidence that this extended south of West Virginia along the Allegheny-Cumberland axis, nor south of Virginia in the Blue Ridge. Away from mountain axes, there is some indication of local infiltration of northern species.

Fossil and vegetational evidence are in accord, both indicating that the deciduous forest was not displaced south of the glacial boundary, except in a band of varying width along the ice front, thus permitting continuance of the pattern of distribution which had been attained by the close of the Tertiary (p. 510), modified of course by climatic influences of the Pleistocene. The width of this band may have been affected by elevation, by degree of diversity of the topography, by direction of drainage (toward or away from the ice front), by the presence of proglacial lakes (with their local

[12] "Northern species" is here interpreted as species usually occurring in northern forest regions, particularly in the Hemlock–White Pine–Northern Hardwoods region.

ameliorating climatic effects), and by the direction of prevailing winds and winds off the ice. The severity of freezing effects such as have been noted in certain periglacial soils may be helpful in determining the extent of climatic shift.

Probable Effects of Climate

Estimates of the severity of periglacial climates vary. Some would raise serious doubt as to the possibility of any but tundra vegetation[13] being able to persist near the ice, and would insist that forest vegetation was much curtailed. Bryan (1928) believes that an arctic climate must have bordered the ice at each advance; however, he states that the plains of the interior of North America allowed a free sweep of winds from the southwest, and that in consequence, the zone of periglacial climate may have been very narrow. The margin of a glacier is held stationary or recedes because wasting equals or exceeds the rate of ice advance. The extensive outwash deposits of gravel and sand of the Wisconsin stage are indicative of periods of rapid melting, as are also the well-developed terminal moraines. Earlier glaciers (Illinoian, Kansan) apparently wasted away, depositing their loads chiefly as ground moraine. During wasting, surfaces become covered with detritus, which when thick enough, may temporarily insulate the frozen mass within. In this condition, the margin of a glacier is not greatly different from a permanently frozen soil mass covered by a layer of soil which thaws annually. Vast areas in the North in which the subsoil is permanently frozen support vegetation, in many places, mature spruce forests. The narrow margin of Greenland adjacent to the Greenland ice cap supports a vascular flora of several hundred species, and there, because of high latitude, summers are short and the sun's rays very oblique. The possibility of more varied vegetation adjacent to ice at lower latitudes is apparent. Within the drift area, lack of true soils and poor surface drainage temporarily might have excluded species for which the temperature conditions were not unsuitable, and thus favored extensive development of spruce or spruce–fir forests.

Due to glacial cold, atmospheric circulation was affected. Permanent high pressure areas developed over ice caps, hence storm tracks (courses followed by the "highs" and "lows" so characteristic of middle latitudes) were crowded to the south of the ice; the thermal gradient was increased, and storminess increased. The extra-tropical belts of high pressure remained essentially where they are but were less dry. The prevailing westerlies con-

[13] No evidence of tundra plants was found in the many interglacial and postglacial deposits examined in Minnesota; rather, most of the plant remains belong to species of the existing flora of Minnesota (Rosendahl, 1948). However, the general absence of evidence of tundra may be due to the fact that bog basins generally retained masses of stagnant ice after glacial recession and, therefore, the bog records did not begin for some time after the surrounding surface was ice-free (Sears, 1948).

tinued to control the circulation of the air in middle latitudes south of the ice front. Except locally and at intervals, the glacial anticyclone could not have overcome planetary circulation of the atmosphere, although it injected into it an extreme influence. Some lowering of temperature throughout the country must have taken place, in part due to the actual influence of the ice, in part to the fact that the land, relative to the ocean, was at a slightly higher elevation. That this lowering of temperature was not sufficient to bring about extinction of the more sensitive of eastern American species is attested by their continued presence on the continent. That it did not cause pronounced shifts in the position of vegetation of the Cumberland Plateau and Southern Appalachian Mountains is indicated by the wealth of ancient relic endemics of Asiatic relationship which could not conceivably have migrated away from, and returned to the few areas retaining their Miocene topography. Many plants of the Southeastern Evergreen Forest region are known to have been present in preglacial and interglacial time. While some could have migrated a little southward on the emergent continental shelf of glacial stages, those with edaphic requirements met only by specific sites could not so move. The deciduous forest as a latitudinal forest belt could not have moved southward, although curtailed at the north. In all the time since the Coastal Plain first began to emerge there are few areas where an edaphic environment has developed which can be occupied by the more mesophytic climax species of the central deciduous forest. They did not, under climatic stress, occupy a sandy coastal plain. The deciduous forest did not, en masse, move far south, and there is nothing in the evidence which has been set forth to indicate that it was ever completely displaced by more northern vegetation types except close to the ice front. How close may have depended on the nature of the topography. In areas of diversified topography, local differences in temperature are pronounced (Wolfe, Wareham, and Scofield, 1943, 1949). Although the ridge-top or plateau may have been subjected to the full intensity of the glacial anticyclone, many parts of sheltered ravines were warmer. Thus the more rugged areas, even near the ice front, may have served as refugia for the more exacting forest species, permitting their persistence through the various cold cycles of the Pleistocene in the general areas in which they are now found, while in less rugged country—rolling land and plains—the glacial anticyclones and accompanying cold may have been more effective in curtailing deciduous forest vegetation and permitting southward penetration of northern species (as seems to be indicated by fossil records from west of the Mississippi River).

If there had been any great infiltration of northern species at times of glacial advance there should be more evidence (relics) remaining in regions near to the margin of more recent glaciation than near the margins of earlier ice sheets. Thus, no ice sheet has advanced close to the Ozark area since early Pleistocene, while in south-central Ohio at least two and perhaps three ice advances have reached almost the same limits. Yet there is almost as much evidence of northern infiltration in Missouri as in Ohio.

Effects of Withdrawals and Transgressions of Ocean on the Coastal Plain

The problem near the coast is somewhat different due to alternating withdrawals and transgressions of the ocean (p. 499). Although the Coastal Plain (except northward) was removed from the immediate effects of glacial cold, there were enforced to and fro migrations, the newly exposed land being populated *during* glacial stages, the area of vegetation being curtailed during interglacial stages. That southward migrations parallel to the shore (along newly exposed strips) took place during glacial advance is indicated by the pollens in bogs on the Coastal Plain (p. 471). But can we be sure that different biotypes were not involved? Or that spruce and fir, in preglacial time, or even in the last great interglacial stage, had the same geographic distribution that they now have? White spruce is a variable species with several geographic races and at least one variety (Halliday and Brown, 1943). Isolation of biotypes, due to migratory shifts, no doubt took place, and their physiologic requirements probably were not identical. Such biotypes or genotypes could not be distinguished in the pollen record. The climatic implications of spruce and fir pollen in South Carolina or Florida deposits may not be as extreme as usually interpreted, or any more certain than are the climatic implications of "beech," one genotype of which is a low-elevation southern coastal plain inhabitant, another, a high-elevation and far northern inhabitant (see pp. 501, 511). That northward migration parallel to the shore took place during glacial retreat, and before submergence of the continental shelf, is indicated by the disjunct occurrences of coastal plain species as far north as Newfoundland, and of southern trees near the coast of New England (p. 476).

The extent to which migration affected established vegetation is another question. In New Jersey, only a short distance beyond the glacial boundary, the fossil record shows that the deciduous forest species persisted, although some northern species entered the area. This was on land not submerged by interglacial invasions of the sea, hence not laid bare for later mass invasions. What may have been taking place on the older parts of the south Atlantic Coastal Plain (parts not reached by Pleistocene transgressions of sea) is entirely conjectural. The present pattern of vegetation indicates the presence of southern pine forests in all areas where this edaphic climax could be maintained by sandy soils; the expansion of oak, or of mixed oak–pine forests whenever developmentally possible; and the continued development of mesophytic forest in some edaphically suitable areas. As new coastal strips became available for forest occupancy, pine forests advanced over them, and as erosion modified the younger terraces, and stream deposition produced strips of better soil across the terraces, hardwood forests advanced downstream, their composition dependent upon the sources of migration.

POSTGLACIAL MIGRATIONS

Postglacial migrations of forests onto the area covered by Wisconsin ice (migrations which have resulted in the present configuration of forests in this youngest glaciated area) are recorded in pollen spectra of bogs, and are in part indicated by disjunct distribution of species and communities (pp. 463, 475). The character of the topography (pp. 309, 499) has been a factor in determining the relative proportions of the land now covered with developmental and climax (or physiographic climax) communities, and the persistence of boreal bog or cliff relics.

Upon retreat of Wisconsin ice, a land surface covered by ground rock and rock flour and entirely devoid of humus was opened up. Onto this, plants began to migrate. If it is possible to form any conclusions from vegetational advance as observed near receding glaciers of the North today, early entrants were scattered, and limited to such species as could exist in the raw soil material. The entrance of nitrogen-fixing species (among which alders are important) paved the way for the advance of spruce. Unlike the climax species of the deciduous forest, spruce is able to play a pioneer role in forest advance, however, not everywhere bringing about a forest of the same aspect (Raup, 1941a). Entrance of deciduous species, other than the occasional pioneer, must have been delayed by lack of soil development, and by the refrigerated and doubtless deeply frozen ground. Spruce forests are able to develop where the subsoil is permanently frozen, and migration may not have lagged far behind the retreating ice margin.

The long-continued dominance of spruce in the southern part of the area of Wisconsin glaciation, and the establishment of a broad band of northern conifer forest in this area is indicated by the pollen spectra of bogs and may be correlated with the slow and halting retreat and several readvances of the ice which are indicated by the several recessional moraines, and with slow soil development. We have no evidence that a spruce–fir forest existed south of the glacial boundary (except in the mountains), although its species (?) are known to have been present in the south (see p. 520).

An early conifer forest stage in which *Pinus* is abundantly represented is indicated by pollen records from New Jersey and southern New England. This may be correlated with the large areas suitable for pine in unglaciated southern New Jersey and on the emergent continental shelf, and from which early migrations could take place, as well as with the initially better drainage of much of this eastern area.

The sequence of forest communities and the approximate direction of migration may be determined from pollen spectra. The order of appearance of pollen of certain of the deciduous forest genera is a basis for postulating time and direction of migration (Sears, 1942a). Bearing in mind the climatic sequences of postglacial time (p. 464), the differences which spectra of different longitudes display, and the order of appearance of genera in different geographic areas, it is possible to make some generalization concerning the origin of the several forest associations of the glaciated area, and the development of forest regions.

Climax Associations in the Glaciated North

The climax associations in the glaciated North owe their origin to postglacial migrations. The boundaries of the forest regions which they characterize have not everywhere reached a position of climatic equilibrium, hence are poorly defined.

The migration of oaks into the glaciated area took place early, soon after the period of spruce dominance, or in some instances while conifer forests were still dominant (this is shown in pollen profiles). At first pines were associated with the oaks, and in many places continued to be associated almost throughout postglacial time. In the northeastern and the northwestern parts of the deciduous forest this association still continues. The oak dominance of the xerothermic period (climatic period IV) is in some places scarcely more pronounced than the earlier oak dominance of period II.[14] Throughout much of the area of the Prairie Peninsula, in Ohio, Indiana, and southwestern Michigan, the continued dominance of oaks characterizes forest stands of the Prairie Peninsula section of the Oak–Hickory Forest region. The mass invasion into this area probably came from the Ozarkian center (established in Pliocene time), although some invasion of oaks may have come from mixed forests south of the glacial border, as indicated by their early entrance in all profiles. Furthermore, a number of species of oak are able to occupy sites unsuitable for climax mesophytes and establish long-enduring developmental stages which retain dominance until topographic and soil development permit successional advance and the ultimate dominance of more shade-tolerant species.

To the east of the mountain axis (as to the west) oaks entered early, being present in the pine stage, and continued as one of the dominant genera through subsequent stages (see chart, p. 468). An ample source for oak migration in the east existed to the south of the glacial boundary in eastern Pennsylvania, New Jersey, and on the then emergent continental shelf. The period of greatest northward extension of oak, oak–hickory, and oak–

[14] The lumping of all oaks as though they were ecologically equivalent has doubtless led to errors in interpretation. The genetic constitution of the "*Quercus borealis—maxima* complex" is under consideration by W. H. Camp, who believes that a northern red oak (not *Q. borealis* or *Q. ellipsoidalis*) has entered into the gene composition of our red oaks. This northern oak may have been responsible for the early oak maximum of period II (Camp, in correspondence, 1948).

chestnut communities into central New England may well have corresponded with the time of maximum eastward extent of prairie and of oak forest in the Prairie Peninsula (as suggested by Raup, 1937). The northward movement in the East was not, however, derived from the same sources as the movement to the west of the Appalachians, but came from the late Tertiary oak forest derivatives on the Harrisburg peneplain. There is no evidence of continuity across the mountains, although there is evidence of northward migration across the Allegheny Plateau by way of the upper Susquehanna River valley to the Finger Lakes of New York State.

Both to the east and the west of the Appalachians, postglacial migrations under the impetus of drier periods (II and IV) resulted in expansion of previously established forest associations without giving rise to pronounced differences between those parts of the association on glaciated and on unglaciated territory. Thus the Oak–Hickory region to the west and the Oak–Chestnut region to the east both have "sections" on the younger land areas (pp. 179, 248).

The Beech–Maple association arose as a result of postglacial migration from the south. It is derived from the Mixed Mesophytic association by segregation of those climax dominants which have been able to advance most rapidly onto the youthful topography of the area of latest glaciation. Its southern boundary is correlated with an age boundary—the limits of Wisconsin glaciation. It is undergoing change in its southern part, and is expanding in its northern part as rapidly as topographic development permits.

The development of the Beech–Maple Forest was not the result of uninterrupted migrations from south of the Wisconsin glacial border. The aspect of mixed forest communities of ravine slopes, and the nature of the soil of the Driftless area and adjacent very old drifts to the west suggest that this hilly area may have been a Pleistocene refugium for an attenuated mixed mesophytic forest which was isolated hereabouts in late Tertiary time, or at latest, in pre-Wisconsin time. From this refugium, early post-Pleistocene migration took place. This is indicated by the early appearance of *Fagus* and *Tilia* in records of nearby bogs. During the mid-postglacial humid, warm period (period III), *Fagus* reached much farther west than its present limits. The subsequent xerothermic period (IV) and expansion of oak forest curtailed its range in the west, restricting it to the immediate vicinity of Lake Michigan. That northward migration of beech (and probably also of maple[15]) across Ohio and Indiana was delayed by the early warm dry period (II) is indicated by its relatively late appearance in most Ohio and Indiana bog records. After expanding in period III, its range was again curtailed to approximately the area of the present Beech–Maple For-

[15] As maples are insect-pollinated, they are always very poorly represented in bog records, and the absence of maple pollen in any record cannot be construed as positive evidence of the absence of maple.

est. Thus two migrations, one from the south and one from the Driftless area, met to the north of the Prairie Peninsula.

The Maple–Basswood Forest is also young and at least in part postglacial in origin. It appears to have been derived by climatic modification of the late Tertiary or interglacial forest of the Driftless area and by post-Pleistocene migrations from this refugium, migrations involving *Tilia* and *Fagus,* and doubtless *Acer,* as well as *Quercus.* The subsequent climatic elimination of *Fagus* has left, in the northwestern part of the deciduous forest, the Maple–Basswood, or Maple–Basswood–Red Oak association.

The Hemlock–White Pine–Northern Hardwoods Forest of the Lake region and the Northeast has resulted from postglacial expansion of a forest persisting throughout the Pleistocene in mountainous portions of the East, particularly on the Allegheny Plateau and Allegheny Mountains of Pennsylvania, where this forest still maintains itself on unglaciated land. From this Pleistocene stronghold it moved rapidly into New York and New England (as shown by the early appearance there of *Tsuga* pollen in the bog records), and westward through the Great Lakes region. Potzger (1946), in tracing the postglacial history of the Lake Forest, points out that "*Tsuga* is very sparingly represented in all pollen profiles from Indiana," and that it "played only a minor role in forests south of a line crossing Farwell and Shelby, Michigan." That is, it did not reach Michigan by northward migration from the south, but by westward migration from Pennsylvania. Further evidence of the source and direction of spread of this forest formation is afforded by the ranges of its species, many of which have few disjunct stations to the south except in the mountains. Furthermore, its beech (except at the northern limits of the region) is the "red beech" of the mountains to the south, not the beech with the mingled red and white genes found at lower elevations in the interior (see p. 501). These facts, together with the evidence afforded by bog records, emphasize the distinctness of the Hemlock–White Pine–Northern Hardwoods Forest region (with its included maple–beech communities) from the Beech–Maple Forest region just to the south of it, in the Great Lakes region.

CHAPTER 18

The Relationships of Climaxes and the Climax Elements of the Forest Regions

The chronologic account of forest development in eastern North America emphasizes the relationships of the several parts of the Deciduous Forest. It explains the derivation of the major associations of the Deciduous Forest Formation—Mixed Mesophytic, Oak–Hickory, Oak–Chestnut, Beech–Maple, and Maple–Basswood associations—and the manner and time of origin of the characteristic topographic configuration of the land surface of the forest regions. Each of these major associations is, in some part of the deciduous forest, a regional climax, there occupying a variety of sites, and thus generally emphasizing climatic control, rather than topographic or edaphic control. Yet each of these associations occurs in favorable situations outside of the area of its dominance, outside of the region it characterizes. Communities representative of different associations, and even of different formations, may occur side by side. This is particularly true of the Western Mesophytic Forest region. This raises a question as to the status of these major associations. Each is recognized as a climax, yet all have been shown to be genetically related. The climax concept calls for the recognition of an ultimate or final stage of development which is the result of climatic control, which has certain unifying characters making recognition possible, and which has some more or less defined spatial relations. But if the time element is considered, each climax is seen to be a result of gradual development—not from seral stages, but from an older established climax. The climax is not static, although its period of endurance may be thousands or hundreds of thousands of years. Slow changes, which may be due in part to gradual evolution of species and changing magnitude of adaptability, in part to changing climate resulting in change in range of species, in part to continuing migration of species, modify each climax. Although each has its unifying characters, no climax is the same throughout its area. Spatial variations emphasize the variable responses of the climax or illustrate the long-time development which is possible. The recognition of five major associations (climaxes) in the Deciduous Forest Formation is an illustration of man's attempt to classify forest vegetation. Each is a more or less artificial unit. In recognizing these units, undue emphasis may be placed on characteristics and stability. That all are variable is abundantly demonstrated by the Tables of forest composition. That intermediate or transitional types of climaxes may occur is seen in the Southeastern Evergreen Forest region, where climax dominants are representative of two forest formations. That all are changing or have resulted from changes in the past, has been shown in the preceding chapter.

In summary, the status of each of the associations of the Deciduous Forest, and of certain neighboring associations to the north and south is given. Each forest region is characterized and related to the chain of events which has already been traced.

THE CLIMAX ASSOCIATIONS

The **Mixed Mesophytic association** is the lineal descendant of the mixed Tertiary forest. In the course of its long and complicated history, that mixed forest which had been in existence since early Tertiary time suffered a number of alternating expansions and contractions, correlated with the progress of erosion cycles and with climatic change. Its mid-Tertiary extent was correlated with the Schooley peneplain. The mixed forest was preserved on the unreduced parts of this peneplain. With uplift and dissection of this peneplain, the mixed forest expanded. Later peneplains (particularly the Harrisburg and its correlatives), and the drier climate of late Tertiary time curtailed the area of mixed forest on the east and on the west. Pleistocene glaciation destroyed any mixed forest which had formerly existed to the north of the glacial boundary. Throughout the Pleistocene, as now, the southern half of the Appalachian Plateau was the stronghold of Mixed Mesophytic Forest.

The **Oak–Hickory association** was derived from the mixed Tertiary forest as a result of increasing aridity in the Western Interior, which eliminated species of higher moisture requirements from the western part of the deciduous forest. At about the same time, comparable segregation was in progress on the peneplaned southern and eastern borders of the Appalachian Highlands, resulting in the Oak–Pine forest, with its oak–hickory climax communities and pine subclimaxes (mostly on the Piedmont and inner Coastal Plain), and the White Oak forest (of Harrisburg peneplain levels of the Ridge and Valley Province). Question may arise as to the desirability of including in the Oak–Hickory association all variants of comparable origin. Such an inclusive interpretation would place the Ozarkian Oak–Hickory, the Piedmont or eastern Oak–Hickory, and the White Oak forest in one great association, each perhaps to be designated as faciations. This question would not arise if the artificiality of the climax concept were recognized; each of these "faciations" would represent a pause, a temporary state of stability in the great complex of deciduous forest. Migrations from the first of these extended onto the glaciated area of the Central Lowland and into the Great Lakes region; migrations from the two latter populated parts of the glaciated area of southern New England and the southern half of the Hudson-Champlain trough, and sent tongues northward over the Allegheny Plateau in the Susquehanna River-Finger Lakes area.

The **Oak–Chestnut association** was derived from the mixed Tertiary forest. Its derivation appears to be correlated with late Tertiary forest segregation coupled with the influence of substratum. This is indicated by its development as an association-segregate in the Mixed Mesophytic Forest region, where it is always related to a noncalcareous substratum. The reasons for its dominance in the Southern Appalachians and the limitation there of mixed mesophytic communities to coves are not clear. They appear to be related to the higher incidence of summer drought. Postglacial migrations extended the Oak–Chestnut association into southern New England. Due to the ravages of the chestnut blight, this association in recent years has been undergoing profound modification. The end is not yet predictable (see p. 192).

The **Beech–Maple association** was derived from the Mixed Mesophytic association by postglacial migration into the area of Wisconsin drift. Such migration need not have come entirely from what is now the Mixed Mesophytic Forest region; migration northward from the Western Mesophytic Forest region doubtless took place. The Beech–Maple association is somewhat comparable in mesophytism to its progenitor, the Mixed Mesophytic association. However, it is an association which has developed on the immature glacial topography of the Central Lowland and southern Lake region. That is, its areal extent within the region it characterizes is determined by existing physiographic characteristics of the youthful landscape. It is questionable whether it can maintain itself when erosional topography replaces glacial topography. The forests of dissected areas show evidence of change, again suggesting the artificiality of the climax concept. The Beech–Maple association is the climax in its region because it is the end stage of local hydrarch and xerarch successions. Yet it is also a stage in a climatic succession which is still in progress, and which has not had time enough, because of topographic control, and slowness of migration, to progress beyond a beech–maple forest.

The **Maple–Basswood association** is the result of progressive deterioration of a remnant of mixed Tertiary forest marooned in and near the Driftless area during the Pleistocene, and of postglacial migrations from the late Pleistocene refugium. Because of its origin, and the early post-Pleistocene elimination of beech from its area, it is a distinct segregate of the mixed forest of the Tertiary or of the pre–Pleistocene or pre-Wisconsin mixed mesophytic forest.

The **Hemlock–White Pine–Northern Hardwoods,** the **Hemlock–Hardwoods,** and the **Spruce–Hardwoods** forests of the Great Lakes area and the Northeast, together with the attenuated western representatives in the Minnesota section, illustrate progressive expansion and migration from a Pleistocene refugium in the northern Allegheny Mountains and Allegheny Plateau. Derivation from the forests of the Tertiary is obscure and may have been due to altitudinal or latitudinal segregation of which no record remains. The direction of migration is suggested by pollen records of bogs. Early westward movement of red spruce is demonstrated by bog records from southeastern Michigan (Cain, 1948). Later climatic trends have eliminated it from all but the humid to superhumid climates of the Northeast. Evidence of former more northward extent of the hardwood communities is suggested by their locations and disjunctions. These hardwood communities, of which beech–maple is prominent, are not made up entirely of migrants from other regions to the south. Again, the genetic constitution of beech enters the picture. The beech of these hardwood communities is dominantly "red beech"; in part, it is "gray beech," a tree "typically developed where the deciduous forest comes into contact with and actually commingles with the spruce–fir association" (W. H. Camp, in litt. 1948). It is distinct from the beeches of the Beech–Maple region. The Hemlock–White Pine–Northern Hardwoods forest is best interpreted as a formation distinct from the Deciduous Forest Formation. It is a formation whose "climax associations" are still in a state of flux, a formation in which the influence of migrations and of climate, past and present, have resulted in differences across its vast east-west extent. It occupies a position intermediate between the Deciduous Forest Formation to the south, and the Northern Coniferous or Spruce–Fir Forest to the north, and contains communities referable to both neighboring formations.

At the southern border of deciduous forest in eastern United States (and within the Southeastern Evergreen Forest region), a transitional climax type—the **Beech–Magnolia association**—occurs. Its derivation is obscure. It suggests early and probably continuous occupancy since early Tertiary time of parts of the Coastal Plain by warm-climate species. It suggests, because of some of its constituents, migration from the Appalachian Upland. Its beech is "white beech" (pp. 298, 501). Early and distinct origin of this southern climax is indicated.

THE FOREST REGIONS

The pattern of forest distribution, the configuration of forest regions, is a result of environmental influences of the present and of the past. The development of the land surface has been slow, and it has been shown that many of its features are reminiscent of the past. The development of soil is in part dependent upon topography and climate, in part on the reactions of occupying vegetation, which in turn is influenced by topography and climate. Changes in climate, in topography, or in soil bring about changes in vegetation. These, however, do not always keep pace with the environmental change. Vegetation is not everywhere in complete equilibrium with its present environment. This is attested by its lack of uniformity even within a single forest region.

The **Mixed Mesophytic Forest region** is characterized by the prevalence of mixed mesophytic communities indicating climatic equilibrium. Developmental communities, of course, are seen. Such inclusions of vegetation belonging to other climaxes as do occur are in equilibrium with their edaphic environment or persist because, in their edaphic situation, competition is not severe. Thus the oak–hickory and oak–chestnut communities of dry slopes and ridges are physiographic or edaphic climaxes in the Mixed Mesophytic region and there illustrate local segregation from the mixed forest, while in regions to the west and east these are climatic climaxes and the prevailing type. In the Mixed Mesophytic region, local inclusions of prairie vegetation, of northern vegetation, of coastal plain plants are remnants related to past climates or past physiography, and are persisting because competition is not too severe in their peculiar habitats, but are gradually being crowded out by encroachment of regional vegetation. The region as a whole has suffered less profound changes during the progress of erosion cycles, and less pronounced climatic changes, than have adjacent regions. Continuity of occupation since early Tertiary time accounts for the antiquity of its vegetation; lack of extreme changes, for the continuance of one climax type; and a consistently humid climate (with rare lapses to subhumid) for the prevalence of this climax.

The **Western Mesophytic Forest region** is characterized by a mosaic of unlike climaxes. The region is, to a considerable extent, a tension zone where the compensating effects of local environments permit unlike climaxes to exist close to one another. It is impossible that all of these are in equilibrium with the regional climate. Their presence is indicative of past events—of climatic shrinkage of the mixed Tertiary forest, of depauperization of mixed mesophytic communities, of expansion of oak–hickory forest, of prairie migrations, of migrations of the southern Mississippi embayment flora, and, least pronounced, of migrations of northern vegetation. The older Tertiary vegetation was greatly circumscribed, being now confined to a few hilly areas, or was so depauperized that present mesophytic forest communities are for the most part quite distinct from true mixed mesophytic communities. The later Tertiary oak–hickory segregate expanded to occupy a large proportion of the area, especially toward the west. Prairie communities have been able to persist, in part because of pronounced underdrainage in areas of karst topography, in part because of the trend of escarpments, which places small areas at their eastern bases in the rain-shadow of hills to the west.

Migrations of the southern Mississippi embayment flora—well exemplified by bald cypress—brought this vegetation far into the Western Mesophytic Forest region where it is seen today in the alluviated lands of the Green River Basin. These migrations may have preceded the uplift of the Lexington peneplain or have been early interglacial. (Peculiarities of distribution argue against their postglacial age.) The time of the great southern prairie expansion, which reached almost to the escarpment of the Appalachian Plateau and in places onto it, and which populated the Nashville Basin, there to give rise to certain of the endemics of the Cedar Glades, is in doubt. There is evidence that this was an event of one of the interglacial stages, probably Yarmouth, when very mild temperatures prevailed in the north and in consequence, storm tracks were spread latitudinally, thus decreasing storminess in the interior, and the probability of frequent summer rains. The post-Pleistocene xerothermic period may also have influenced the distribution of xeric communities of the region. The last century and a half has witnessed the almost complete replacement of prairie by woodland (where cultivation has not prevented natural succession). Almost all evidence of Pleistocene infiltration of northern species is gone, suggesting that it could not have been pronounced. The Appalachian floral element, weakly represented in a number of areas, shows no tendency to advance. Vegetational equilibrium with climate has not been reached.

The **Oak–Hickory Forest region** is characterized by the prevalence of oak–hickory climax communities. Almost throughout the region, a state of equilibrium with climate appears to have been reached. The segregation of the Ozarkian Oak–Hickory association was an event of late Tertiary time, correlated with increasing aridity in the Western Interior. Existing climate continues to control the development of this climax throughout the Southern Division of the Oak–Hickory region. However, inclusions of other climaxes do occur; isolated remnants of mixed mesophytic forest, indicative of the mixed Tertiary forest; stranded southern lowland species in the western Ozarks, explained as dating back to a time previous to the uplift of the Ozark (late Tertiary) peneplain when southern lowland vegetation would have been more widespread than now; northern plants, isolated during some glacial advance; and xeric southwestern and prairie species on "bald knobs," rocky glades and dry slopes, species dating from a time or times of eastward extension of the southwestern and prairie floras. Existing climate controls the development of the Oak–Hickory climax through much of the Northern Division of the region. However, its boundary in Indiana and southern Michigan is indefinite, because this area lies in a tension zone between oak–hickory and beech–maple, or oak–hickory and attenuated mixed mesophytic forest. That this lobe of oak–hickory developed as a result of increasing dryness in postglacial time is established by pollen records. It maintains itself at its eastern extremity in part because of microclimates, where the contact of the more xeric and more mesic climaxes is in hilly areas, and in part because of immaturity of topography, where drainage or soil preclude the entrance of the Beech–Maple climax. In the north it either advanced into the relic more mesophytic forest of the Driftless area, or developed in that area simultaneously with development in the Ozarks. Extensions westward in the river valleys along the western border of deciduous forest are late postglacial, following climatic amelioration after the xerothermic period.

The **Oak–Chestnut Forest region** is characterized by the prevalence of some variant of the Oak–Chestnut association[1] over most of its slopes, and by the dominance of white oak forest on valley floors of Harrisburg peneplain age. Physiographically, the region assumed its present configuration as a result of the Harrisburg cycle of erosion. Climatic changes contemporaneous with the development of the Harrisburg peneplain may have favored the extension of the eastern Oak–Hickory association, here represented by the white oak forest over the valley floors, while the Oak–Chestnut association occupied most of the slopes of the region. Mixed mesophytic forest communities are beautifully developed in the coves of the Southern Appalachians, where the evidence indicates continuance from the mixed Tertiary forest. The elimination of mixed mesophytic forest from most of the region may be a result of conditions produced by the Harrisburg cycle of erosion, or may possibly date back to earlier differentiation between the Appalachian Plateaus and land to the east. The presence of mixed mesophytic communities in the northern part of the region (at the northern end of the Piedmont) and in the Glaciated section of southern New England appears to be a result of migration from eastern areas of mixed mesophytic forest which persisted south of the ice border. These mixed mesophytic communities of the Glaciated section may have been more extensive previous to the post-Wisconsin xerothermic period; they may again extend their areas. Wherever in the region stream rejuvenation is taking place, habitats suitable for mixed mesophytic forest develop. Communities representative of that association are developing in such places, pointing toward the potential ability of the Mixed Mesophytic association to increase its extent. Theoretically, complete dissection of the Harrisburg peneplain might result in displacement of oak and occupancy by mixed mesophytic forest. However, the frequency of summer droughts favors the continuance of the present regional types.

The **Oak–Pine Forest region** is somewhat transitional in character. The climax over most of the rolling land is an oak or oak–hickory forest, earlier

[1] See Chapter 7 for present status of the Oak–Chestnut association, and discussion of future possibilities.

referred to as the eastern Oak–Hickory association, or as a faciation of that association. Like the Ozarkian Oak–Hickory, this is a product of late Tertiary climatic and erosional history. The pronounced admixture of pine, particularly in developmental and subclimax stands, represents a climax element apart from the deciduous forest, but one not recorded in the fossil record. The Oak–Pine region includes isolated stands of what were oak–chestnut communities (usually on hills rising above the general level). As in the Oak–Chestnut region, mixed mesophytic communities are occupying slopes cut in the present erosion cycle. Climatic conditions, however, do not favor any great expansion of this forest type, because of the frequency of summer droughts, and because of the open winters, which result in leaching and the development of the lateritic soils of the South.

The **Southeastern Evergreen Forest region** occupies a transitional position between the Deciduous Forest Formation and the Subtropical Broad-leaved Evergreen Forest, and because of this contains communities related to one or the other formation, as well as its own distinctive communities. The history of the region has in general been progressive, commencing with the first emergence of the Coastal Plain at the close of Cretaceous time. The ancestry of certain relic species of the Apalachicola River bluffs dates from that period. The Tertiary history of the region is largely a history of warm temperate and subtropical coastal floras, becoming, in Pliocene time, similar to that of the present. The presence of mixed mesophytic communities in certain areas of diverse topography which survived the last general coastal plain peneplain (the Coosa) suggests occupancy of these areas by the mixed Tertiary forest. Tertiary migrations from peneplaned areas of the Appalachian Highland account for a considerable part of the Coastal Plain vegetation. The vegetational occupancy of the lower terraces of the Coastal Plain (of marine origin, Pleistocene in age, and progressively younger toward the coast) took place during glacial stages, as a result of direct migration from the inland margin of the terrace and of migrations parallel to the coast, the latter accounting for entrance of the northern element recorded in the pollen record. Mixed forest, evidently derived in part from the Mixed Mesophytic association, is advancing onto ravine slopes and river bluffs cut in the lower terraces. The relatively stable and long-enduring, but distinctly seral, swamp communities are mainly intra-regional as are also the subclimax pine communities. Because of the latitudinal extent of the region, and its transitional position, the climax broad-leaved forest communities vary in the degree to which they contain elements of the two bordering formations. Because of the nature of the soil of much of the region, vegetational development is held indefinitely in subclimax stages.

The **Beech–Maple Forest region,** which lies entirely within the area of the last glacial stage, was populated entirely by postglacial migrations. It is characterized by the development of the Beech–Maple climax association.

But because of the youth of the land surface, a large part of the area is still occupied by developmental stages; because of the changing climate of postglacial time, inclusions of relic boreal forest, prairie, and of oak–hickory communities occur frequently. Advancing migrations from the regions to the south are beginning to modify the southern border of the Beech–Maple region, and especially those places where stream dissection has brought about a mature topography. The Beech–Maple association there assumes the status of a stage in the postglacial clisere.

The **Maple–Basswood Forest region** is small, yet it has two distinct subdivisions, one occupying ancient land of the Driftless area, the other on land covered by Late Wisconsin ice. Oak–hickory forest is prevalent in much of the older area, although successional development points to the climax character of the Maple–Basswood association. Many features of the more mesophytic communities of the Driftless area suggest the persistence here of deteriorated remnants of mixed forest, dating back to pre-Wisconsin time or earlier. Prairie and oak–hickory communities are the result of the post-Wisconsin xerothermic period. Early post-Wisconsin migrations in a northwesterly direction from the Driftless area resulted in expansion of mesophytic communities (at first containing beech, as shown by the pollen record), and the establishment of the Maple–Basswood association in what is now known as the "Big Woods" area of Minnesota.

The **Hemlock–White Pine–Northern Hardwoods Forest region** is, throughout its extent, a mosaic of hardwood, conifer, and mixed communities, whose composition varies with the east-west extent of the region. Limits of the ranges of species in part account for the differing composition of climax communities of different parts of the region. Progressive development of topography and soil may increase the proportion of hardwood communities, and may enable such communities to extend their range northward as the edaphic environment becomes suitable for the entrance of maple, beech, and hemlock. The vegetation of this region is the result of post-Wisconsin migrations, which brought about an expansion from the Unglaciated Allegheny Plateau and northern Allegheny Mountains without pronounced modification of type. The climax elements are almost entirely a result of these expanding migrations.

The forest regions of eastern North America thus give visual expression to the controlling environmental forces which, although continually changing, have been continuously effective in controlling forest development since the first appearance of angiospermous trees on the continent.

Bibliography

Adams, Charles C., George P. Burns, T. L. Hankinson, Barrington Moore, and Norman Taylor: Plants and animals of Mount Marcy, New York. *Ecol.* **1**:71–94; 204–33; 274–88, 1920.

Aikman, J. M.: Distribution and structure of the deciduous forest in eastern Nebraska. *Univ. Nebr. Studies* **26**:1–75, 1927.

Aikman, J. M., and A. W. Smelser: The structure and environment of forest communities in central Iowa. *Ecol.* **19**:141–50, 1938.

Akerman, Alfred: The forests of Surry County, Virginia. *Va. Geol. Commission, Va. Forestry Publ.* No. **37**, 1925.

Allard, H. A.: Length of day in relation to the natural and artificial distribution of plants. *Ecol.* **13**:221–34, 1932.

———: A second record for the paper birch, Betula papyrifera, in West Virginia. *Castanea* **10**:55–57, 1945.

———: Shale barren association on Massanutten Mountain, Virginia. *Castanea* **11**:71–124, 1946.

Allard, H. A., and E. C. Leonard: The vegetation and floristics of Bull Run Mountain, Virginia. *Castanea* **8**:1–64, 1943.

Alway, F. J., J. Kittredge, and W. J. Methley: Composition of the forest floor layers under different forest types on the same soil type. *Soil Sci.* **36**:387–98, 1933.

Alway, F. J., and P. R. McMiller: Interrelations of soil and forest cover on Star Island, Minnesota. *Soil Sci.* **36**:281–94, 1933.

Antevs, E.: The last glaciation with special reference to the ice retreat in northeast North America. *Amer. Geogr. Soc. Research ser.* No. **17**, 1928.

Ashe, W. W.: Forests of North Carolina. *N. C. Geol. Surv. Bull.* **6**:139–224, 1897.

Atlas of American Agriculture. *U. S. Dept. Agr.* (See authors of parts or sections.)

Auten, John T.: Porosity and water absorption of forest soils. *Jour. Agr. Res.* **46**:997–1014, 1933.

Baldwin, J. T., Jr.: Chromosomes of Cruciferae—II. Cytogeography of Leavenworthia. *Bull. Torr. Bot. Club* **72**:367–78, 1945.

Beaven, G. H., and H. J. Oosting: Pocomoke Swamp: a study of a cypress swamp on the eastern shore of Maryland. *Bull. Torr. Bot. Club* **66**:367–89, 1939.

Bergman, H. F.: Flora of North Dakota. *Sixth Biennial Report, N. D. Soil and Geol. Surv.* 1912, pp. 145–372.

———: The composition of climax plant formations in Minnesota. *Papers Mich. Acad. Sci.* **3**:51–60, 1924.

Bergman, H. F., and H. Stallard: The development of climax formations in northern Minnesota. *Minn. Bot. Studies* **4**:333–78, 1916.

Berry, E. W.: Pleistocene plants from Virginia. *Torreya* **6**:88–90, 1906.

———: Leaf rafts and fossil leaves. *Torreya* **6**:246, 1906a.

———: Contributions to the Pleistocene flora of North Carolina. *Jour. Geol.* **15**:338–49, 1907.

———: Pleistocene plants from Alabama. *Amer. Nat.* **41**:689–97, 1907a.

———: Additions to the Pleistocene flora of North Carolina. *Torreya* **9**:71–73, 1909.

———: Juglandaceae from the Pleistocene of Maryland. *Torreya* **9**:96–99, 1909a.

———: Additions to the Pleistocene flora of New Jersey. *Torreya* **10**:261–67, 1910.

———: Additions to the Pleistocene flora of Alabama. *Amer. Jour. Sci.* Ser. 4, **29**:387–98, 1910a.

———: Pleistocene plants from the Blue Ridge in Virginia. *Amer. Jour. Sci.* Ser. 4, **34**:218–23, 1912.

———: The Upper Cretaceous and Eocene floras of South Carolina and Georgia. *U. S. Geol. Surv. Prof. Paper* **84**, 1914.

———: The Mississippi River bluffs at Columbus and Hickman, Kentucky, and their fossil flora. *U. S. Nat. Mus. Proc.* **48**:293–303, 1915.

———: The Lower Eocene floras of southeastern North America. *U. S. Geol. Surv. Prof. Paper* **91**, 1916.

———: The physical conditions and age indicated by the flora of the Alum Bluff formation. *U. S. Geol. Surv. Prof. Paper* **98E**, 1916a.

———: The physical conditions indicated by the flora of the Calvert formation. *U. S. Geol. Surv. Prof. Paper* **98F**, 1916b.

———: The flora of the Citronelle formation. *U. S. Geol. Surv. Prof. Paper* **98L**, 1916c.

———: The flora of the Catahoula sandstone. *U. S. Geol. Surv. Prof. Paper* **98M**, 1916d.

———: Pleistocene plants in the marine clays of Maine. *Torreya* **17**:160–63, 1917.

———: The fossil plants from Vero, Florida. *Fla. Geol. Surv. Ann. Rep.* **9**:19–53, 1917a.

———: Upper Cretaceous floras of the eastern Gulf region in Tennessee, Mississippi, Alabama, and Georgia. *U. S. Geol. Surv. Prof. Paper* **112**, 1919.

———: The age of the Brandon lignite and flora. *Amer. Jour. Sci.* Ser. 4, **47**:211–16, 1919a.

———: The Middle and Upper Eocene floras of southeastern North America. *U. S. Geol. Surv. Prof. Paper* **92**, 1924.

———: Organic remains, other than diatoms, from the excavation [at the Walker Hotel site, Connecticut Avenue and DeSales Street, Washington, D. C.] *Jour. Wash. Acad. Sci.* **14**:12–25, 1924a.

———: The flora of the Ripley formation. *U. S. Geol. Surv. Prof. Paper* **136**, 1925.

———: Pleistocene plants from North Carolina. *U. S. Geol. Surv. Prof. Paper* **140C**, 1925.

———: Revision of the Lower Eocene Wilcox flora of the Southeastern States. *U. S. Geol. Surv. Prof. Paper* **156**, 1930.

———: New occurrences of Pleistocene plants in the District of Columbia. *Jour. Wash. Acad. Sci.* **23**:1–25, 1933.

———: A Talbot cypress swamp at Greenbury Point, Maryland. *Torreya* **34**:85–91, 1943.

Billings, W. D.: The structure and development of old field shortleaf pine stands and certain associated physical properties of the soil. *Ecol. Monog.* **8**:437–99, 1938.

Bingham, Marjorie T.: Flora of Oakland County, Michigan. *Cranbrook Inst. Sci. Bull.* **22**, 1945.

Blizzard, Alpheus W.: Plant sociology and vegetational change on High Hill, Long Island, New York. *Ecol.* **12**:208–31, 1931.

Blodgett, Frederick H.: Ecological plant geography of Maryland, Midland Zone, Upper Midland district. In "Plant Life of Maryland" pp. 221–74. Johns Hopkins Press, Baltimore, 1910.

Bonck, Juanda, and W. T. Penfound: Plant succession on abandoned farm land in the vicinity of New Orleans, Louisiana. *Amer. Midl. Nat.* **33**:520–29, 1945.

Braun, E. Lucy: The physiographic ecology of the Cincinnati region. *Ohio Biol. Surv. Bull.* **7**, 1916.

———: The vegetation of the Mineral Springs region of Adams County, Ohio. *Ohio Biol. Surv. Bull.* **15**, 1928.

———: Glacial and postglacial plant migrations indicated by relic colonies of southern Ohio. *Ecol.* **9**:284–302, 1928a.

———: The undifferentiated deciduous forest climax and the association-segregate. *Ecol.* **16**:514–19, 1935.

———: The vegetation of Pine Mountain, Kentucky. *Amer. Midl. Nat.* **16**:517–65, 1935a.

———: Forests of the Illinoian till plain of southwestern Ohio. *Ecol. Monog.* **6**:89–149, 1936.

———: Some relationships of the flora of the Cumberland Plateau and Cumberland Mountains in Kentucky. *Rhodora* **39**:193–208, 1937.

———: A remarkable colony of coastal plain plants on the Cumberland Plateau in Laurel County, Kentucky. *Amer. Midl. Nat.* **18**:363–66, 1937a.

———: An ecological transect of Black Mountain, Kentucky. *Ecol. Monog.* **10**:193–241, 1940.

———: A new station for Pachystima Canbyi. *Castanea* **6**:52, 1941.

———: Forests of the Cumberland Mountains. *Ecol. Monog.* **12**:413–47, 1942.

———: "An annotated catalog of spermatophytes of Kentucky." Published by the author, Cincinnati, 1943.

Bray, William L.: The development of vegetation in New York State. *N. Y. State Coll. For., Syr. Univ. Tech. Publ.* **29**, 1930.

Brenner, Louis G., Jr.: The environmental variables of the Missouri Botanical Garden wildflower reservation at Gray Summit. *Ann. Mo. Bot. Gard.* **29**:103–35, 1942.

Britton, W. E.: Vegetation of the North Haven sand plains. *Bull. Torr. Bot. Club* **30**:571–620, 1903.

Bromley, S. W.: The original forest types of southern New England. *Ecol. Monog.* **5**:61–89, 1935.

Brown, Clair A.: The flora of the Pleistocene deposits in the Western Florida Parishes: West Feliciana Parish and East Baton Rouge Parish, Louisiana. *La. Dept. Cons. Bull.* **12**:59–94, 1938.

Brown, D. M.: The vegetation of Roan Mountain: A phytosociological and successional study. *Ecol. Monog.* **11**:61–97, 1941.

Brown, Forest B. H.: A botanical survey of the Huron River valley. III. The plant societies of the Bayou at Ypsilanti, Michigan. *Bot. Gaz.* **40**:264–84, 1905.

Bruncken, Ernest: Notes on the distribution of some trees and shrubs in the vicinity of Milwaukee. *Bull. Wis. Nat. Hist. Soc.* n.s. **1**:31–42, 1900.

Bruner, W. E.: The vegetation of Oklahoma. *Ecol. Monog.* **1**:99–188, 1931.

Bryan, Kirk: Glacial climate in non-glaciated regions. *Am. Jour. Sci.* **16**:162–64, 1928.

Buell, Jesse H.: Red pine in West Virginia. *Castanea* **5**:1–6, 1940.

Buell, Murray F.: Late Pleistocene forests of southeastern North Carolina. *Torreya* **45**:117–18, 1945.

———: The age of Jerome Bog, a "Carolina bay." *Science* **103**:14–15, 1946.

Buell, Murray F., and Robert L. Cain: The successional role of southern white cedar, *Chamaecyparis thyoides,* in southeastern North Carolina. *Ecol.* **24**:85–93, 1943.

Butters, Frederic K., and C. O. Rosendahl: The distribution of the white oak in Minnesota. *Minn. Studies in Plant Sci., Biol. Sci.* No. **5**, 1924.

Cain, Stanley A.: An ecological study of the heath balds of the Great Smoky Mountains. *Butler Univ. Bot. Stud.* **1**:177–208, 1930.

———: Ecological studies of the vegetation of the Great Smoky Mountains of North Carolina and Tennessee. I. Soil reaction and plant distribution. *Bot. Gaz.* **91**:22–41, 1931.

———: Studies on virgin hardwood forest. I. Density and frequency of the woody plants of Donaldson's Woods, Lawrence County, Indiana. *Proc. Ind. Acad. Sci.* **41**:105–122, 1932.

———: Bald cypress, Taxodium distichum (L.) Rich., at Hovey Lake, Posey County, Indiana. *Amer. Midl. Nat.* **16**:72–82, 1935.

———: Ecological studies of the vegetation of the Great Smoky Mountains. II. The quadrat method applied to sampling spruce and fir forest types. *Amer. Midl. Nat.* **16**:566–84, 1935a.

———: Studies on virgin hardwood forest. III. Warren's Woods, a beech–maple climax forest in Berrien County, Michigan. *Ecol.* **16**:500–13, 1935b.

———: Synusiae as a basis for plant sociological field work. *Amer. Midl. Nat.* **17**:665–72, 1936.

———: The composition and structure of an oak woods, Cold Spring Harbor, Long Island, with special attention to sampling methods. *Amer. Midl. Nat.* **17**:725–40, 1936a.

———: Pollen analysis as a paleo-ecological research method. *Bot. Rev.* **5**:627–54, 1939.

———: The Tertiary character of the cove hardwood forests of the Great Smoky Mountains National Park. *Bull. Torr. Bot. Club* **70**:213–35, 1943.

———: "Foundations of Plant Geography." Harper & Brothers, New York, 1944.

———: Pollen analysis of some buried soils, Spartanburg County, South Carolina. *Bull. Torr. Bot. Club* **71**:11–22, 1944a.

———: Palynological studies of Sodon Lake: I. Size-frequency study of fossil spruce pollen. *Science* **108**:115–17, 1948.

Cain, Stanley A., Mary Nelson, and Walter McLean: Andropogonetum Hempsteadi: a Long Island grassland vegetation type. *Amer. Midl. Nat.* **18**:334–50, 1937.

Cain, Stanley A., and Wm. T. Penfound: Aceretum rubri: the red maple swamp forest of central Long Island. *Amer. Midl. Nat.* **19**:390–416, 1938.

Cajander, A. K.: Ueber Waldtypen. *Acta-forestalia Fennica* 1. (Also in *Fennica* 28), 1909.

———: The theory of forest types. *Acta-forestalia Fennica* 31, 1926.

———: The theory of forest types. Review in *Ecol.* **8**:135–136, 1927.

Camp, W. H.: The grass balds of the Great Smoky Mountains of Tennessee and North Carolina. *Ohio Jour. Sci.* **31**:157–65, 1931.

Carroll, Gladys: The use of bryophytic polsters and mats in the study of recent pollen deposition. *Amer. Jour. Bot.* **30**:361–66, 1943.

Chamberlin, T. C.: Native vegetation of eastern Wisconsin, in Geology of Wisconsin, Survey of 1873–1877. **2**:1877.

———: General map of native vegetation of Wisconsin, 1882. In *Atlas of the Geol. Surv. of Wis.*

Chaney, Ralph W.: A comparative study of the Bridge Creek flora and the modern redwood forest. *Carn. Inst. Wash. Publ.* **349**:1–22, 1925.

———: The Mascall flora—its distribution and climatic relation. *Carn. Inst. Wash. Publ.* **349**:23–48, 1925a.

———: Paleoecological interpretations of Cenozoic plants in western North America. *Bot. Rev.* **9**:371–96, 1938.

———: Tertiary forests and continental history. *Geol. Soc. Amer. Bull.* **51**:469–88, 1940.

———: Redwoods around the Pacific basin. *Pacific Discovery* **1**:4–14, 1948.

Chaney, R. W., and M. K. Elias: Late Tertiary floras from the High Plains. *Carn. Inst. Wash. Publ.* **476**:1–46, 1936.

Chaney, Ralph W., and Hsen Hsu Hu: A Miocene flora from Shantung Province, China. *Carn. Inst. Wash. Publ.* **507**, 1940.

Chapman, A. G.: An ecological basis for reforestation of submarginal lands in the central hardwood region. *Ecol.* **18**:93–105, 1937.

———: Forests of the Illinoian till plain of southeastern Indiana. *Ecol.* **23**:189–98, 1942.

———: "Original Forests," in "Ohio's Forest Resources." *Ohio Agr. Exp. Sta., Forestry Publ.* No. **76**, 1944.

Chapman, Charles S.: A working plan for forest lands in Berkeley County, South Carolina. *U. S. Dept. Agr. Bur. For. Bull.* **56**, 1905.

Chapman, H. H.: Is the longleaf type a climax? *Ecol.* **13**:328–34, 1932.

Child, Hamilton: Gazetteer of Cheshire County, N. H. Syracuse, N. Y., 1885.

Chittenden, A. K.: Forest conditions of northern New Hampshire. *U. S. Dept. Agr. Bur. For. Bull.* **55**, 1905

———: The red gum. *U. S. Dept. Agr. Bur. For. Bull.* **58,** 1905a.

Chrysler, M. A.: The ecological plant geography of Maryland, Coastal Zone, Western Shore district. In "Plant Life of Maryland" pp. 149–97. Johns Hopkins Press, Baltimore, 1910.

———: The origin and development of the vegetation of Sandy Hook. *Bull. Torr. Bot. Club* **57**:163–76, 1930.

Clark, E. D.: Report on an examination of the timberland of Mrs. Harriet S. Turner, Dyer County, Tennessee. *U. S. Forest Service,* typed report, Oct., 1908.

Clayberg, H. D.: Upland societies of Petoskey-Walloon Lake region. *Bot. Gaz.* **69**:28–53, 1920.
Clements, F. E.: Plant Succession. *Carn. Inst. Wash. Publ.* **242**, 1916.
———: "Plant Succession and Indicators." H. W. Wilson Company, New York, 1928.
———: The relict method in dynamic ecology. *Jour. Ecol.* **22**:39–68, 1934.
———: Nature and structure of the climax. *Jour. Ecol.* **24**:252–84, 1936.
Clements, F. E., and R. W. Chaney: Environment and life in the Great Plains. *Carn. Inst. Wash. Suppl. Publ.* **24**:1–54, 1936.
Clements, F. E., and V. E. Shelford. "Bio-ecology." John Wiley & Sons, New York, 1939.
Cline, A. C., and S. H. Spurr: The virgin upland forest of central New England. *Harvard Forest Bull.* **21**, 1942.
Cobbe, Thomas J.: Variations in the Cabin Run Forest, a climax area in southwestern Ohio. *Amer. Midl. Nat.* **29**:89–105, 1943.
Cocke, E. C., I. F. Lewis, and Ruth Patrick: A further study of Dismal Swamp peat. *Amer. Jour. Bot.* **21**:374–95, 1934.
Cocks, R. S.: Notes on the flora of Louisiana. *Plant World* **17**:186–91, 1914.
Coile, T. S.: Soil reaction and forest types in the Duke Forest. *Ecol.* **14**:323–33, 1933.
———: Soil changes associated with loblolly pine succession on abandoned agricultural land of the Piedmont Plateau. *Duke University School of Forestry Bull.* **5**, 1940.
Coker, W. C.: The plant life of Hartsville, S. C. *Jour. Elisha Mitchell Sci. Soc.* **27**:169–205, 1911.
Coleman, A. P.: Interglacial periods in Canada. *Compte-Rendu, Congrès geologique international,* Mexico, 1906.
———: An estimate of post-glacial and inter-glacial time in North America. *Étude faite à la XIIe session du Congrès geologique international, reproduite de Compte-Rendu,* 1914.
———: Glacial and interglacial periods in eastern Canada. *Jour. Geol.* **35**:385–403, 1927.
Conard, Henry S.: Tree growth in the vicinity of Grinnell, Iowa. *Jour. For.* **16**:100–106, 1918.
———: A boreal moss community. *Ia. Acad. Sci.* **39**:57–61, 1932.
———: The plant associations of central Long Island. *Amer. Midl. Nat.* **16**:433–516, 1935.
———: The fir forests of Iowa. *Ia. Acad. Sci.* **45**:69–72, 1938.
Coons, G. H.: Ecological relations of the flora. Sand dune region on the south shore of Saginaw Bay, Michigan. *Mich. Geol. & Biol. Surv. Publ.* Ser. 2, **4**:35–64, 1911.
Cooper, William S.: The climax forest of Isle Royale, Lake Superior, and its development. *Bot. Gaz.* **55**:1–44; 115–40; 189–235, 1913.
Core, Earl L.: Plant ecology of Spruce Mountain, West Virginia. *Ecol.* **10**:1–13, 1929.
Corson, C. W., J. H. Allison, and E. G. Cheyney: Factors controlling forest types on the Cloquet Forest, Minnesota. *Ecol.* **10**:112–25, 1929.
Coulter, S. M.: An ecological comparison of some typical swamp areas. *Mo. Bot. Gard. Ann. Rep.* **15**:39–71, 1904.
Cowles, Henry C.: The ecological relations of the vegetation on the sand dunes of Lake Michigan. *Bot. Gaz.* **27**:95–117; 167–202; 281–308; 361–91, 1899.

———: The physiographic ecology of Chicago and vicinity. *Bot. Gaz.* **31**: 73–108; 145–82, 1901.

———: The causes of vegetative cycles. *Bot. Gaz.* **51**:161–83, 1911.

Crafton, W. M., and B. W. Wells: The old field prisere: an ecological study. *Jour. Elisha Mitchell Sci. Soc.* **49**:225–46, 1934.

Crandall, A. R.: Report on the timber growth of Greenup, Carter, Boyd, and Lawrence counties in eastern Kentucky. *Ky. Geol. Surv.* **1**:9–26, 1876.

Cunningham, R. W., and H. G. White: Forest resources of the Upper Peninsula of Michigan. *U. S. Dept. Agr. Misc. Publ.* **429**, 1941.

Curran, H. M.: The forests of Garrett County. *Md. Geol. Surv., Garrett Co.,* pp. 303–29, 1902.

Dachnowski-Stokes, A. P., and B. W. Wells: The vegetation, stratigraphy, and age of the "open land" peat area in Carteret County, North Carolina. *Jour. Wash. Acad. Sci.* **19**:1–11, 1929.

Dana, S. T.: Timber growing and logging practice in the Northeast. *U. S. Dept. Agr. Tech. Bull.* **166**:1–112, 1930.

Dansereau, Pierre: L'Érablière Laurentienne I. Valeur d'indice des espèces. *Canadian Jour. Res.* **21**:66–93, 1943.

———: Interpenetrating climaxes in Quebec. *Science* **99**:426–27, 1944.

———: Les érablières de la Gaspésie et les fluctuations du climat. *Contr. Inst. Bot. Univ. Montréal* No. **51**, 1944a.

———: L'Érablière Laurentienne II. Les successions et leur indicateurs. *Canadian Jour. Res.* **24**:235–91, 1946.

Darlington, H. C.: Vegetation and substrate of Cranberry Glades, West Virginia. *Bot. Gaz.* **104**:371–93, 1943.

Darlington, H. T.: Vegetation of the Porcupine Mountains, northern Michigan. *Papers Mich. Acad.* **13**:9–65, 1931.

———: Some vegetational aspects of Beaver Island, Lake Michigan. *Papers Mich. Acad. Sci., Arts, Letters* **25**:31–37, 1940.

Daubenmire, R. F.: The relation of certain ecological factors to the inhibition of forest floor herbs under hemlock. *Butler Univ. Bot. Studies* **1**:61–76, 1930.

———: The "Big Woods" of Minnesota: its structure, and relation to climate, fire, and soils. *Ecol. Monog.* **6**:233–68, 1936.

Davis, Charles A.: Peat: Essays on its origin, uses and distribution in Michigan. *Report Mich. Geol. Surv.* for 1906:93–395, 1907.

Davis, J. H., Jr.: Vegetation of the Black Mountains of North Carolina: an ecological study. *Jour. Elisha Mitchell Sci. Soc.* **45**:291–318, 1930.

———: The natural features of Southern Florida, especially the vegetation, and the Everglades. *Fla. Dept. Cons., Geol. Bull.* **25**, 1943.

———: The peat deposits of Florida. *Fla. Dept. Cons., Geol. Bull.* **30**, 1946.

Deam, Charles C.: "Flora of Indiana." *Dept. Cons., Div. Forestry,* Indianapolis, 1940.

Deevey, Edward S., Jr.: Studies on Connecticut lake sediments. *Amer. Jour. Sci.* **237**:691–724, 1939.

———: Additional pollen analyses from southern New England. *Amer. Jour. Sci.* **241**:717–52, 1943.

Demaree, D.: Submerging experiments with Taxodium. *Ecol.* **13**:258–62, 1932.

Dice, Lee R.: "The Biotic Provinces of North America." Univ. Mich. Press, Ann Arbor, 1943.

Diller, O. D., and Eugene D. Marshall: The relation of stump height to the sprouting of Ostrya virginiana in northern Indiana. *Jour. Forestry* **35**:1116–19, 1937.

Dodge, C. K.: Contributions to the botany of Michigan. II. *Univ. Mich., Mus. Zool., Misc. Publ.* **5**, 1918.

Dorf, Erling: Upper Cretaceous floras of the Rocky Mountain region. *Carn. Inst. Wash. Publ.* **508**, 1942.

Drew, William B.: The revegetation of abandoned crop-land in the Cedar Creek area, Boone and Calloway counties, Missouri. *Univ. Mo. Coll. Agr., Agr. Exp. Sta. Res. Bull.* **344**, 1942.

Dunston, C. E.: Preliminary examination of the forest conditions of Mississippi. *Miss. Geol. Surv. Bull.* **11**:6–76, 1913.

Dwight, Henry E.: Account of the Kaatskill Mountains. *Amer. Jour. Sci. and Arts* **2**:11–29, 1820.

Dyksterhuis, E. J.: The vegetation of the Western Cross Timbers. *Ecol. Monog.* **18**:325–76, 1948.

Eggler, Willis A.: The maple–basswood forest type in Washburn County, Wisconsin. *Ecol.* **19**:243–63, 1938.

Egler, Frank E.: Berkshire Plateau vegetation, Massachusetts. *Ecol. Monog.* **10**:145–92, 1940.

Elias, M. K.: Tertiary grasses and other prairie vegetation from High Plains of North America. *Amer. Jour. Sci.* Ser. 5, **29**:24–33, 1935.

Emerson, F. W.: A botanical survey of Big Basin in the Allegany State Park. *N. Y. State Mus. Handbook* **17**:89–150, 1937.

Erdtman, G.: "An Introduction to Pollen Analysis." Chronica Botanica Co., Waltham, Mass., 1943.

Erickson, Ralph O., Louis G. Brenner, and Joseph Wraight: Dolomitic glades of east-central Missouri. *Ann. Mo. Bot. Gard.* **29**:89–101, 1942.

Esten, M. M.: A statistical study of a beech–maple association at Turkey Run State Park, Parke County, Indiana. *Butler Univ. Bot. Studies* **2**:183–201, 1932.

Ewing, J.: Plant succession of the brush-prairie in northwestern Minnesota. *Jour. Ecol.* **12**:238–66, 1924.

Fassett, Norman C.: Notes from the herbarium of the University of Wisconsin—VII. *Rhodora* **33**:224–28, 1931.

Fenneman, N. M.: "Physiography of Eastern United States." McGraw-Hill Book Co., New York, 1938.

Fernald, M. L.: Some relationships of the flora of the northern hemisphere. *Proc. Intern. Congress Plant Sci.* **2**:1487–1507, 1929.

———: Specific segregations and identities in some floras of eastern North America and the Old World. *Rhodora* **33**:25–63, 1931.

———: Local plants of the inner Coastal Plain of southeastern Virginia. *Rhodora* **39**:321–66, 1937.

Fernow, B. E., C. D. Howe, and J. H. White: Forest conditions of Nova Scotia. *Commission of Conservation,* Ottawa, Canada, 1912.

Flint, R. F.: Pleistocene features of the Atlantic Coastal Plain. *Amer. Jour. Sci.* **238**:757–87, 1940.

———: "Glacial Geology and the Pleistocene Epoch." John Wiley & Sons, New York, 1947.

Flint, R. F., et al: Glacial map of North America. *Geol. Soc. Amer.,* New York, 1945.

Fogg, John M., Jr.: The flora of the Elizabeth Islands, Massachusetts. *Contr. Gray Herb. Harvard Univ.* **91**, 1930 (reprinted from *Rhodora* **32**).

Fosberg, F. R., and E. H. Walker: A preliminary check list of plants in the Shenandoah National Park, Virginia. *Castanea* **6**:89–136, 1941.

Foster, J. H.: Forest conditions in Louisiana. *U. S. Dept. Agr. Forest Service Bull.* **114**, 1912.

Freeman, Chester P.: Ecology of the cedar glade vegetation near Nashville, Tennessee. *Jour. Tenn. Acad. Sci.* **8**:143–228, 1932.

Friesner, Ray C.: Indiana as a critical botanical area. *Proc. Ind. Acad. Sci.* **46**:28–45, 1937.

Friesner, R. C., and Charles M. Ek: Correlation of microclimatic factors with species distribution in Shenk's Woods, Howard County, Indiana. *Butler Univ. Bot. Studies* **6**:87–101, 1944.

Friesner, R. C., and J. E. Potzger: Studies in forest ecology I. Factors concerned in hemlock reproduction in Indiana. *Butler Univ. Bot. Studies* **2**:133–44, 1932.

———, and ———: Studies in forest ecology II. The ecological significance of Tsuga canadensis in Indiana. *Butler Univ. Bot. Studies* **2**:145–49, 1932a.

———, and ———: Soil moisture and the nature of the Tsuga and Tsuga–Pinus forest associations in Indiana. *Butler Univ. Bot. Studies* **3**:207–9, 1936.

———, and ———: Contrasts in certain physical factors in Fagus–Acer and Quercus–Carya communities in Brown and Bartholomew counties, Indiana. *Butler Univ. Bot. Studies* **4**:1–12, 1937.

———, and ———: Survival of hemlock seedlings in a relict colony under forest conditions. *Butler Univ. Bot. Studies* **6**:102–15, 1944.

Frothingham, E. H.: Second-growth hardwoods in Connecticut. *U. S. Dept. Agr. For. Serv. Bull.* **96**, 1912.

———: The northern hardwood forest: its composition, growth and management. *U. S. Dept. Agr. Bull.* **285**, 1915.

Frothingham, E. H., and R. Y. Stuart: Timber growing and logging practice in the southern Appalachian region. *U. S. Dept. Agr. Tech. Bull.* **250**, 1931.

Fuller, Geo. D., and P. D. Strausbaugh: On the forests of La Salle County, Illinois. *Trans. Ill. State Acad. Sci.* **12**:246–72, 1919.

Fulling, E. H.: Abies intermedia, the Blue Ridge fir, a new species. *Jour. So. App. Bot. Club* (*Castanea*) **1**:91–94, 1936.

Galloway, J. J.: Geology and natural resources of Rutherford County, Tennessee. *Div. Geol. Bull.* **22**, 1919.

Gano, Laura: A study in physiographic ecology in northern Florida. *Bot. Gaz.* **63**:337–72, 1917.

Ganong, W. F.: The nascent forest of the Miscou beach plain. Contributions to the ecological plant geography of the Province of New Brunswick, No. 4. *Bot. Gaz.* **42**:81–106, 1906.

Garren, Kenneth H.: Effects of fire on vegetation of the southeastern United States. *Bot. Rev.* **9**:617–54, 1943.

Garver, R. D., J. B. Cuno, C. F. Korstian, and A. L. MacKinney: Selective logging in the loblolly pine–hardwood forests of the Middle Atlantic Coastal Plain with special reference to Virginia. *State Comm. on Conserv.* and *Devel., Va. Forest Service Publ.* **43**, 1931.

Gates, Frank C.: The vegetation of the region in the vicinity of Douglas Lake, Cheboygan County, Michigan. *Ann. Rep. Mich. Acad. Sci.* **14**:46–106, 1912.

Gates, Frank C.: Plant succession about Douglas Lake, Cheboygan County, Michigan. *Bot. Gaz.* **82**:170–82, 1926.

———: Aspen association in northern Lower Michigan. *Bot. Gaz.* **90**:233–59, 1930.

———: "Flora of Kansas." *Agr. Exp. Sta., Kan. State Coll.,* 1940.

———: The bogs of northern Lower Michigan. *Ecol. Monog.* **12**:213–54, 1942.

Genaux, Charles M., and John G. Kuenzel: Defects which reduce quality and yield of oak–hickory stands in southeastern Iowa. *Agr. Exp. Sta., Ia. State Coll. Res. Bull.* **269**, 1939.

Gevorkiantz, S. R., and Raphael Zon: Second-growth white pine in Wisconsin. *Wis. Agr. Exp. Sta. Res. Bull.* **98**, 1930.

Gleason, H. A.: The vegetation of the inland sand deposits of Illinois. *Bull. Ill. State Lab. Nat. Hist.* **9**:21–174, 1910.

———: An isolated prairie grove and its phytogeographical significance. *Bot. Gaz.* **53**:38–49, 1912.

———: The relation of forest distribution and prairie fires in the Middle West. *Torreya* **13**:173–81, 1913.

———: The vegetational history of the Middle West. *Ann. Assoc. Amer. Geogr.* **12**:39–85, 1923. Also, *Contr. N. Y. Bot. Gard.* **242**, 1923.

———: The structure of the maple–beech association in northern Michigan. *Papers Mich. Acad. Sci.* **4**:285–96, 1924. Also, *Contr. N. Y. Bot. Gard.* **266**, 1925.

———: Is the synusia an association? *Ecol.* **17**:444–51, 1936.

Gordon, R. B.: The primary forest types of the east-central states. *Abstr. Doctor's Dissertations,* No. **8**, Ohio State Univ. Press, 1932.

———: A preliminary vegetation map of Indiana. *Amer. Midl. Nat.* **17**:866–77, 1936.

———: The botanical survey of the Allegany State Park. *N. Y. State Mus. Handbook* **17**:23–88, 1937.

———: A botanical survey of the southwestern section of the Allegany State Park. *N. Y. State Mus. Handbook* **17**:199–247, 1937a.

———: The primeval forest types of southwestern New York. *N. Y. State Mus. Bull.* **321**, 1940.

———: The natural vegetation of West Goshen Township, Chester County, Pa. *Pa. Acad. Sci.* **15**:194–99, 1941.

Gould, Frank W.: Plant indicators of original Wisconsin prairie. *Ecol.* **22**:427–29, 1941.

Graham, Samuel A.: Climax forests of the Upper Peninsula of Michigan. *Ecol.* **22**:355–62, 1941.

Grant, Martin L.: The climax forest community in Itasca County, Minnesota, and its bearing upon the successional status of the pine community. *Ecol.* **15**:243–57, 1934.

Graves, Henry S.: Practical forestry in the Adirondacks. *U. S. Dept. Agr. Div. of Forestry, Bull.* **26**, 1899.

Gray, Asa. See Sargent.

Griggs, Robert F.: A botanical survey of the Sugar Grove region. *Ohio Biol. Surv. Bull.* **3**, 1914.

———: The timberlines of northern America and their interpretation. *Ecol.* **27**:275–89, 1946.

Hall, T. F., and Wm. T. Penfound: A phytosociological study of a Nyssa biflora consocies in southeastern Louisiana. *Amer. Midl. Nat.* **22**:369–75, 1939.

———, and ———: A phytosociological analysis of a cypress–gum swamp in southeastern Louisiana. *Amer. Midl. Nat.* **21**:378–95, 1939a.

———, and ———: Cypress–gum communities in the Blue Girth Swamp near Selma, Alabama. *Ecol.* **24**:208–17, 1943.

Halliday, W. E. D.: A forest classification for Canada. *Forest Service, Dept. Int. Canada, Bull.* **89**, 1937.

Halliday, W. E. D., and A. W. Brown: The distribution of some important forest trees in Canada. *Ecol.* **24**:353–73, 1943.

Hanna, G. Dallas: Diatoms of the Florida peat deposits. *Fla. Geol. Surv., 23d–24th Ann. Rep.,* 1933.

Hansen, H. P.: Postglacial vegetation of the driftless area of Wisconsin. *Amer. Midl. Nat.* **21**:752–62, 1939.

Harper, R. M.: A phytogeographical sketch of the Altamaha Grit region of the coastal plain of Georgia. *Ann. N. Y. Acad. Sci.* **17**:1–415, 1906.

———: Economic botany of Alabama. Part I. Geographical report, *Geol. Surv. Ala. Monog.* **8**, 1913.

———: Geography and vegetation of northern Florida. *Fla. Geol. Surv.* Sixth Ann. Report: 163–437, 1914.

———: A botanical bonanza in Tuscaloosa County, Alabama. *Jour. Elisha Mitchell Sci. Soc.* **37**:153–60, 1922.

———: The cedar glades of middle Tennessee. *Ecol.* **7**:48–54, 1926.

———: Economic botany of Alabama. Part II. Catalogue of the trees, shrubs and vines of Alabama. *Geol. Surv. Ala. Monog.* **9**, 1928.

———: A depressed outlier of the Cumberland Plateau in Alabama and its vegetation. *Castanea* **2**:13–18, 1937.

———: Palms in the Southern Appalachian region in Alabama. *Castanea* **3**:19–24, 1938.

———: Natural resources of the Tennessee Valley region in Alabama. *Geol. Surv. Ala. Special Report* **17**, 1942.

———: Croomia a member of the Appalachian flora. *Castanea* **7**:109–13, 1942a.

———: Forests of Alabama. *Geol. Surv. Ala. Monog.* **10**, 1943.

———: Hemlock in the Tennessee Valley of Alabama. *Castanea* **8**:115–23, 1943a.

Harshberger, J. W.: A botanical ascent of Mount Katahdin, Me. *Plant World* **5**:21–28, 1902.

———: An ecological study of the flora of mountainous North Carolina. *Bot. Gaz.* **36**:241–58; 368–83, 1903.

———: A phytogeographic sketch of extreme southeastern Pennsylvania. *Bull. Torr. Bot. Club* **31**:125–29, 1904.

———: The plant formations of the Catskills. *Plant World* **8**:276–81, 1905.

———: The plant formations of the Nockamixon Rocks, Pennsylvania. *Bull. Torr. Bot. Club* **36**:651–73, 1909.

———: The vegetation of the Navesink Highland. *Torreya* **10**:1–10, 1910.

———: "Phytogeographic Survey of North America." G. E. Stechert & Co., New York, 1911.

———: "The Vegetation of the New Jersey Pine Barrens." Christopher Sower Co., Philadelphia, 1916.

———: Slope exposure and the distribution of plants in eastern Pennsylvania. *Bull. Geogr. Soc. Phila.* **17**:53–61, 1919.

Harvard Forest Models. Harvard Forest Tercent. Celebr., Harvard Univ., 1936.

Harvey, L. H.: A study of the physiographic ecology of Mount Ktaadn, Maine. *Univ. Me. Studies,* **5**, 1903.

Hawes, A. F. See Hawes and Hawley, 1909.

———: New England Forest in retrospect. *Jour. Forestry* **21**:209–24, 1923.

Hawes, A. F., and R. C. Hawley: Forest survey of Litchfield and New Haven counties, Connecticut. (Litchfield Co. by Hawes; New Haven Co. by Hawley.) *Conn. Agr. Exp. Sta. Bull.* **162**, 1909.

Hawley, R. C. See Hawes and Hawley, 1909.

Hawley, R. C., and A. F. Hawes: "Forestry in New England." John Wiley & Sons, New York, 1912.

Hazard, James O.: The trees of the Reelfoot Lake region. *Jour. Tenn. Acad. Sci.* **8**:55–60, 1933.

Heiberg, S. O.: Nomenclature of forest humus layers. *Jour. For.* **35**:36–39, 1937.

Heiberg, S. O., and R. F. Chandler, Jr.: A revised nomenclature of forest humus layers for the northeastern United States. *Soil Sci.* **52**:87–99, 1941.

Heimburger, Carl C.: Forest-type studies in the Adirondack region. *Cornell Univ. Agr. Exp. Sta. Mem.* **165**, 1934.

Hicks, L. E.: The original forest vegetation and the vascular flora of northeastern Ohio—Ashtabula County. *Abstr. Doctor's Dissertations,* No. **13**, Ohio State Univ. Press, 1934.

Hilgard, E. W.: Report on the geology and agriculture of the State of Mississippi. *Miss. State Geol. Surv.,* 1860.

Hill, A. F.: The vegetation of the Penobscot Bay region, Maine. *Proc. Portland Soc. Nat. Hist.* **3**:305–438, 1923.

Holland, W. C.: Physiographic divisions of the Quaternary lowlands of Louisiana. *Proc. La. Acad. Sci.* **8**:11–24, 1944.

Hollick, Arthur: The relation between forestry and geology in New Jersey. *Amer. Nat.* **33**:1–14; 109–16, 1899.

Hosmer, R. S.: A study of the Maine spruce. *Ann. Rep. Forest Commission State* of *Me.* **4**:65–108, 1902.

Hotchkiss, Neil: A botanical survey of the Tug Hill Plateau. *N. Y. State Mus. Bull.* **287**, 1932.

Hough, A. F.: A climax forest community on East Tionesta Creek in northwestern Pennsylvania. *Ecol.* **17**:9–28, 1936.

Hough, A. F., and R. D. Forbes: The ecology and silvics of forests in the high plateaus of Pennsylvania. *Ecol. Monog.* **13**:299–320, 1943.

Hursh, C. R., and F. W. Haasis: Effects of 1925 summer drought on Southern Appalachian hardwoods. *Ecol.* **12**:380–86, 1931.

Hutchinson, A. H.: Limiting factors in relation to specific ranges of tolerance of forest trees. *Bot. Gaz.* **66**:465–93, 1918.

Jennings, O. E.: Classification of the plant societies of central and western Pennsylvania. *Proc. Pa. Acad. Sci.* **1**:23–55, 1927

Jillson, W. R.: Topography of Kentucky. *Ky. Geol. Surv.* Ser. VI, **30**, 1927.

———: "Filson's Kentucke—a facsimile reproduction of the original Wilmington Edition of 1784." John P. Morton & Co., Louisville, 1930.

Johnson, A. M.: The bitternut hickory, *Carya cordiformis,* in northern Minnesota. *Amer. Jour. Bot.* **14**:49–51, 1927.

Johnson, Douglas: "Stream sculpture on the Atlantic slope." Columbia Univ. Press, New York, 1931.

Kearney, T. H.: Report on a botanical survey of the Dismal Swamp region. *U. S. Dept. Agr., Contr. U. S. Nat. Herb.* **5**:321–550, 1901.

Keller, Carl O.: An ecological study of the Klein Woods, Jennings County, Indiana. *Butler Univ. Bot. Studies* **8**:64–81, 1946.

Kellogg, Charles E.: Development and significance of the great soil groups of the United States. *U. S. Dept. Agr., Misc. Publ.* **229**, 1936.

Kenoyer, L. A.: Ecological notes on Kalamazoo County, Michigan, based on the original land survey. *Papers Mich. Acad. Sci., Arts, Letters* **11**:211–18, 1930.

King, Charlotte M.: A summer outing in Iowa. *Plant World* **5**:222–24, 1902.

Kittredge, Joseph: Evidence of the rate of forest succession on Star Island, Minnesota. *Ecol.* **15**:24–35, 1934.

Kittredge, J., and A. K. Chittenden: Oak forests of northern Michigan. *Lake For. Exp. Sta. Bull.* **190**, 1929.

Knowlton, F. H. (in I. C. White), 1896.

———: A catalogue of the Mesozoic and Cenozoic plants of North America. *U. S. Geol. Surv. Bull.* **696**, 1919.

Korstian, C. F.: The indicator significance of native vegetation in the determination of forest sites. *Plant World* **20**:267–87, 1917.

———: Natural regeneration of southern white cedar. *Ecol.* **5**:188–91, 1924.

Korstian, C. F., and W. D. Brush: Southern white cedar. *U. S. Dept. Agr. Tech. Bull.* **251**, 1931.

Korstian, C. F., and T. S. Coile: Plant competition in forest stands. *Duke Univ. School of Forestry Bull.* **3**, 1938.

Korstian, C. F., and William Maughan: The Duke Forest: a demonstration and research laboratory. *Duke Univ. Forestry Bull.* **1**, 1935.

Korstian, C. F., and Paul W. Stickel: The natural replacement of blight-killed chestnut in the hardwood forests of the Northeast. *Jour. Agr. Res.* **34**:631–48, 1927.

Kryshtofovich, A. N.: Evolution of the Tertiary flora in Asia. *New Phytol.* **28**:303–12, 1929.

Kuenzel, J. G., and C. E. Sutton: A study of logging damage in upland hardwoods of southern Illinois. *Jour. For.* **35**:1150–55, 1937.

Kurz, Herman: Northern aspect and phenology of Tallahassee Red Hills flora. *Bot. Gaz.* **85**:83–89, 1928.

———: Northern disjuncts in northern Florida. *Fla. Geol. Surv., 23d–24th Ann. Rep.,* pp. 49–53, 1930–1932.

———: Florida dunes and scrub, vegetation and geology. *Fla. Dept. Cons., Geol. Bull.* **23**, 1942.

Kurz, Herman, and Delzie Demaree: Cypress buttresses and knees in relation to water and air. *Ecol.* **15**:36–41, 1934.

Leopold, Aldo: Flambeau—the story of a wild river. *Amer. Forests* **49**:12–14; 47, 1943.

Leverett, Frank: Glacial formations and drainage features of the Erie and Ohio basins. *U. S. Geol. Surv. Monog.* **41**, 1902.

———: Moraines and shore lines of the Lake Superior basin. *U. S. Geol. Surv. Prof. Paper* **154A**, 1928.

Lewis, Ivey F., and E. C. Cocke: Pollen analysis of Dismal Swamp peat. *Jour. Elisha Mitchell Sci. Soc.* **45**:37–58, 1929.

Liming, Franklin G.: The range and distribution of shortleaf pine in Missouri.

U. S. Dept. Agr. Forest Service, Centr. States For. Exp. Sta. Tech. Paper **106**, 1946.

Lindeman, Raymond L.: The developmental history of Cedar Creek bog, Minnesota. *Amer. Midl. Nat.* **25**:101–12, 1941.

Little, E. L., Jr.: The vegetation of the Caddo County canyons, Oklahoma. *Ecol.* **20**:1–10, 1939.

———: "Check list of the native and naturalized trees of the United States including Alaska." *U. S. Forest Service,* 1944.

Little, E. L. and Charles E. Olmsted: Trees and shrubs of the southeastern Oklahoma Protective Unit. *Proc. Okla. Acad. Sci.* **16**:52–61, 1936.

Livingston, B. E.: The distribution of the upland plant societies of Kent County, Michigan. *Bot. Gaz.* **35**:36–55, 1903.

———: The relation of soils to natural vegetation in Roscommon and Crawford counties, Michigan. *Bot. Gaz.* **39**:22–41, 1905.

Lobeck, A. K.: "A Physiographic Diagram of the United States." A. J. Nystrom & Co., Chicago.

Lowe, E. N.: Plants of Mississippi. *Miss. State Geol. Surv. Bull.* **17**, 1921.

Lunt, Herbert A.: Forest soil problems in New England. *Ecol.* **19**:50–56, 1938.

Lutz, Harold J.: Trends and silvicultural significance of upland forest successions in southern New England. *Yale Univ. School of Forestry Bull.* **22**:1–68, 1928.

———: The vegetation of Heart's Content, a virgin forest in northwestern Pennsylvania. *Ecol.* **11**:1–29, 1930.

———: Effect of cattle grazing on vegetation of a virgin forest in northwestern Pennsylvania. *Jour. Agr. Res.* **41**:561–70, 1930a.

———: Original forest composition in northwestern Pennsylvania as indicated by early land survey notes. *Jour. For.* **28**:1098–1103, 1930b.

———: Relation of forest site quality to number of plant individuals per unit area. *Jour. For.* **30**:34–38, 1932.

———: Ecological relations of the pitch pine plains of southern New Jersey. *Yale Univ. School of Forestry Bull.* **38**, 1934.

———: A geological explanation of the origin and present distribution of the New Jersey pine barren vegetation. *Ecol.* **15**:399–406, 1934a.

Lutz, Harold J., and Robert F. Chandler: "Forest Soils." John Wiley & Sons, New York, 1946.

Lutz, Harold J., and A. L. McComb: Origin of white pine in virgin forest stands in northwestern Pennsylvania as indicated by stem and basal branch features. *Ecol.* **16**:252–56, 1935.

Macoun, J.: The forests of Canada and their distribution, with notes on the more interesting species. *Trans. Roy. Soc. Canada* **4**:3–20, 1894.

Malott, C. A.: The physiography of Indiana. pp. 59–256 in "Handbook of Indiana Geology." *Dept. Cons. Div. of Geol. Publ.* No. **21**, 1922.

Marberry, W. M., J. D. Mees, and A. G. Vestal: Sample-plot statistics in University Woods. *Trans. Ill. Acad. Sci.* **29**:69–71, 1936.

Marbut, C. F.: Soils of the United States. Pt. III of Atlas of American Agriculture. *U. S. Dept. Agr., Bureau of Chemistry and Soils,* 1935.

Marie-Victorin, F., and F. Rolland-Germain: Premières observations botaniques sur la nouvelle route de l'Abitibi. *Contr. Inst. Bot. Univ. Montréal* **42**:1–49, 1942.

Markle, M. S.: The phytecology of peat bogs near Richmond, Indiana. *Proc. Ind. Acad. Sci.,* pp. 359–75, 1915.

Marks, John B.: Land use and plant succession in Coon Valley, Wisconsin. *Ecol. Monog.* **12**:113–33, 1942.

Mason, Herbert L.: The edaphic factor in narrow endemism. *Madrona* **8**:209–26; 241–57, 1948.

Massey, A. B.: Discovery and distribution of Pachystima Canbyi Gray. *Castanea* **5**:8–11, 1940.

Mattoon, W. R.: Forest trees and forest regions of the United States. *U. S. Dept. Agr. Misc. Publ.* **217**, 1936.

McCarthy, E. F.: Variations in northern forest and their influence on management. *Jour. For.* **19**:867–71, 1921.

———: Report to committee on cutting practices, New York section, *Soc. Amer. For.* 1945.

McDougall, W. B.: The forests of Vermilion County. *Trans. Ill. State Acad. Sci.* **12**:282–89, 1919.

———: Forests and soils of Vermilion County, Illinois, with special reference to the "striplands." *Ecol.* **6**:372–79, 1925.

McInteer, B. B.: The Barrens of Kentucky. *Trans. Ky. Acad. Sci.* **10**:7–12, 1942.

McIntire, G. S.: Theory and practise of forest typing, with special relation to the hardwood and hemlock associations of northern Michigan. *Papers Mich. Acad. Sci., Arts, Letters* **15**:239–51, 1931.

Miller, R. B., and Geo. D. Fuller: Forest conditions in Alexander County, Illinois. *Trans. Ill. State Acad. Sci.* **14**:1–17, 1921.

Mohr, Charles: The timber pines of the southern United States. *U. S. Dept. Agr. Div. Forestry Bull.* **13**, 1896.

———: Plant life of Alabama. *U. S. Dept. Agr., Contr. U. S. Nat. Herb.* **6**, 1901.

Moore, Barrington, and Norman Taylor: Vegetation of Mount Desert Island, Maine, and its environment. *Brooklyn Bot. Gard. Mem.*, **3**, 1927.

Moseley, Edwin L.: Flora of the oak openings. *Ohio Acad. Sci. Sp. Paper* **20**, 1928.

Murphy, L. S.: The red spruce: its growth and management. *U. S. Dept. Agr. Bull.* **544**, 1917.

Nash, Robley W.: Winter killing of hardwoods. *Jour. For.* **41**:841–42, 1943.

Nichols, G. E.: The vegetation of Connecticut. *Torreya* **13**:89–112; 199–215, 1913; **14**:167–94, 1914. *Bull. Torr. Bot. Club* **42**:169–217, 1915; **43**:235–264, 1916; **47**:89–117; 511–48, 1920.

———: The vegetation of northern Cape Breton Island, Nova Scotia. *Trans. Conn. Acad. Arts and Sci.* **22**:249–467, 1918.

———: A working basis for the ecological classification of plant communities. *Ecol.* **4**:11–23; 154–79, 1923.

———: The hemlock–white pine–northern hardwood region of eastern North America. *Ecol.* **16**:403–22, 1935.

Nielsen, E. L.: A study of a pre-Kansan peat deposit. *Torreya* **35**:53–56, 1935.

Olmsted, C. E.: Vegetation of certain sand plains of Connecticut. *Bot. Gaz.* **99**:209–300, 1937.

———: Growth and development in range grasses. V. Photo-periodic responses of clonal divisions of three latitudinal strains of side-oats grama. *Bot. Gaz.* **106**:382–401, 1945.

Oosting, H. J.: Plants occurring on calcareous rock outcrops in North Carolina. *Torreya* **41**:76–81, 1941.

Oosting, H. J.: An ecological analysis of the plant communities of Piedmont, North Carolina. *Amer. Midl. Nat.* **28**:1–126, 1942.

———: Tolerance to salt spray of plants of coastal dunes. *Ecol.* **26**:85–89, 1945.

Oosting, H. J., and L. E. Anderson: Plant succession on granite rock in eastern North Carolina. *Bot. Gaz.* **100**:750–68, 1939.

Oosting, H. J., and W. D. Billings: Edapho-vegetational relations in Ravenel's Woods, a virgin hemlock forest near Highlands, North Carolina. *Amer. Midl. Nat.* **22**:333–50, 1939.

———, and ———: Factors effecting vegetational zonation on coastal dunes. *Ecol.* **23**:131–42, 1942.

Oosting, H. J., and John F. Reed: Ecological composition of pulpwood forests of northwestern Maine. *Amer. Midl. Nat.* **31**:182–210, 1944.

Otto, James H.: Forest succession in the southern limits of Early Wisconsin glaciation as indicated by a pollen spectrum from Bacon Swamp, Marion County, Indiana. *Butler Univ. Bot. Studies* **4**:93–116, 1938.

Palmer, E. J.: Canyon flora of the Edwards Plateau of Texas. *Jour. Arnold Arb.* **1**:233–39, 1920.

———: The forest flora of the Ozark region. *Jour. Arnold Arb.* **2**:216–32, 1921.

———: The ligneous flora of Rich Mountain, Arkansas and Oklahoma. *Jour. Arnold Arb.* **5**:108–34, 1924.

Palmer, E. J., and J. A. Steyermark: An annotated catalogue of the flowering plants of Missouri. *Ann. Mo. Bot. Gard.* **22**:375–758, 1935.

Pammel, L. H.: A comparative study of the vegetation of swamp, clay, and sandstone areas in western Wisconsin, southeastern Minnesota, northeastern, central, and southeastern Iowa. *Proc. Davenport Acad. Sci.* **10**:32–126, 1905.

Parker, Dorothy: Affinities of the flora of Indiana. *Amer. Midl. Nat.* **17**:700–724, 1936.

Patrick, Ruth: Diatoms from the Patschke Bog, Texas. *Notulae Naturae,* No. **170**, Acad. Nat. Sci. Phila., 1946.

Peattie, Donald Culross: Flora of the Tryon region of North and South Carolina (Introduction). *Jour. Elisha Mitchell Sci. Soc.* **44**:95–113, 1928.

Penfound, Wm. T., and Julian R. Howard: A phytosociological study of an evergreen oak forest in the vicinity of New Orleans, Louisiana. *Amer. Midl. Nat.* **23**:165–74, 1940.

Penfound, Wm. T., and M. E. O'Neill: The vegetation of Cat Island, Mississippi. *Ecol.* **15**:1–16, 1934.

Penfound, Wm. T., and Allan G. Watkins: Phytosociological studies in the pinelands of southeastern Louisiana. *Amer. Midl. Nat.* **18**:661–82, 1937.

Pennell, Francis W.: Flora of the Conowingo Barrens of southeastern Pennsylvania. *Proc. Acad. Nat. Sci. Phila.* **62**:541–84, 1910.

———: Further notes on the flora of the Conowingo or Serpentine Barrens of southeastern Pennsylvania. *Proc. Acad. Nat. Sci. Phila.* **64**:520–39, 1912.

Pepoon, H. S.: The forest associations of northwestern Illinois. *Trans. Ill. State Acad. Sci.* **3**:143–56, 1910.

Perkins, G. H.: Tertiary lignite of Brandon, Vermont, and its fossils. *Geol. Soc. Amer. Bull.* **16**:499–516, 1905.

———: The lignite or brown coal of Brandon. *Rept. State Geol. Vt. 1905–06,* pp. 202–230.

Pessin, L. J.: Forest associations in the uplands of the Lower Gulf Coastal Plain (Longleaf pine belt). *Ecol.* **14**:1–14, 1933.

Pessin, L. J., and T. D. Burleigh: Notes on the forest biology of Horn Island, Mississippi. *Ecol.* **22**:70–78, 1941.

Pool, Raymond J.: A study of the vegetation of the sandhills of Nebraska. *Minn. Bot. Studies* **4**:189–312, 1914.

Pool, R. J., J. E. Weaver, and F. C. Jean: Further studies in the ecotone between prairie and woodland. *Univ. Nebr. Studies* **18**:1–47, 1918.

Potzger, J. E.: Topography and forest types in a central Indiana region. *Amer. Midl. Nat.* **16**:212–29, 1935.

———: The vegetation of Mackinac Island, Michigan: an ecological survey. *Amer. Midl. Nat.* **25**:298–323, 1941.

———: Pollen spectra from four bogs on the Gillen Nature Reserve along the Michigan-Wisconsin state line. *Amer. Midl. Nat.* **28**:501–11, 1942.

———: The vegetation of Round Island (Straits of Mackinac), Michigan. *Butler Univ. Bot. Studies* **6**:116–21, 1944.

———: The Pine Barrens of New Jersey, a refugium during Pleistocene times. *Butler Univ. Bot. Studies* **7**:182–96, 1945.

———: Phytosociology of the primeval forest in central-northern Wisconsin and Upper Michigan, and a brief post-glacial history of the Lake Forest formation. *Ecol. Monog.* **16**:211–50, 1946.

Potzger, J. E., and R. C. Friesner: Some comparisons between virgin forest and adjacent areas of secondary succession. *Butler Univ. Bot. Studies* **3**:85–98, 1934.

———, and ———: A phytosociological study of the herbaceous plants in two types of forests in central Indiana. *Butler Univ. Bot. Studies* **4**:163–80, 1940.

———, and ———: What is climax in central Indiana? A five-mile quadrat study. *Butler Univ. Bot. Studies* **4**:181–95, 1940a.

———, and ———: An ecological survey of Berkey woods, a remnant of forest primeval in Kosciusko County, Indiana. *Butler Univ. Bot. Studies* **6**:10–16, 1943.

Potzger, J. E., R. C. Friesner, and C. O. Keller: Phytosociology of the Cox Woods: a remnant of forest primeval in Orange County, Indiana. *Butler Univ. Bot. Studies* **5**:190–221, 1942.

Potzger, J. E., and James H. Otto: Post-glacial forest succession in northern New Jersey as shown by pollen records from five bogs. *Amer. Jour. Bot.* **30**:83–87, 1943.

Potzger, J. E., and B. C. Tharp: Pollen record of Canadian spruce and fir from Texas bog. *Science* **98**:584, 1943.

———, and ———: Pollen profile from a Texas bog. *Ecol.* **28**:274–80, 1947.

Pound, R., and F. E. Clements: "The Phytogeography of Nebraska." Jacob North & Company, Lincoln, 1900.

Quick, B. E.: A comparative study of the distribution of the climax association in southern Michigan. *Papers Mich. Acad. Sci., Arts, Letters* **3**:211–44, 1924.

Raisz, Erwin: "Map of the Landforms of the United States." Harvard University, Cambridge, Mass., 1939.

Raup, Hugh M.: Recent changes of climate and vegetation in southern New England and adjacent New York. *Jour. Arnold Arb.* **18**:79–117, 1937.

———: Botanical studies in the Black Rock Forest. *Black Rock For. Bull.* **7**, 1938.

Raup, Hugh M.: Old field forests of southeastern New England. *Jour. Arnold Arb.* **21**:266–73, 1940.
———: An old forest in Stonington, Connecticut. *Rhodora* **43**:67–71, 1941.
———: Botanical problems of Boreal America. *Bot. Rev.* **7**:147–248, 1941a.
———: Trends in the development of geographic botany. *Ann. Assoc. Amer. Geogr.* **32**:319–54, 1942.
Reed, Franklin W.: Report on an examination of a forest tract in western North Carolina. *U. S. Bureau of Forestry Bull.* **60**, 1905.
———: A working plan for forest land in central Alabama. *U. S. Dept. Agr. Forest Service Bull.* **68**, 1905a.
Reed, W. G.: Frost and the growing season. Pt. II, Section 1 of Atlas of American Agriculture, *U. S. Dept. Agr.*, 1920.
Reid, E. M., and M. E. J. Chandler: "The London Clay flora." *Brit. Mus. (Nat. Hist.)*. London, 1933.
Rich, John L.: Drainage changes and re-excavated valleys as measures of interstream degradation. *Geol. Soc. Amer. Proc. 1934:* pp. 102–3.
Richards, P. W.: The floristic composition of primary Tropical Rain Forest. *Biol. Rev.* **20**:1–13, 1945.
Romell, L. G., and S. Heiberg: Types of humus layer in the forests of northeastern United States. *Ecol.* **12**:567–608, 1931.
Rosendahl, C. O.: A contribution to the knowledge of the Pleistocene flora of Minnesota. *Ecol.* **29**:284–315, 1948.
Ruthven, A. G.: An ecological survey in the Porcupine Mountains and Isle Royale, Michigan; pp. 17–55 in "An ecological survey of northern Michigan." *State Board Geol. Surv. Rep. for 1905,* 1906.
Sampson, H. C.: Succession in the swamp forest formation in northern Ohio. *Ohio Jour. Sci.* **30**:340–56, 1930.
———: The mixed mesophytic forest community of northeastern Ohio. *Ohio Jour. Sci.* **30**:358–67, 1930a.
Sargent, C. S.: "Scientific Papers of Asa Gray." The Macmillan Company, 1889.
Schrock, Alta E.: A preliminary analysis of an unglaciated bog in Pennsylvania. *Univ. Pittsburgh Bull.* **41**: No. 4, 1945.
Schwarz, G. F.: The sprout forests of the Housatonic Valley of Connecticut. *Forestry Quat.* **5**:121–53, 1907.
Sears, P. B.: The natural vegetation of Ohio. *Ohio Jour. Sci.* **25**:139–49, 1925; **26**:128–46; 213–31, 1926.
———: Types of North American pollen profiles. *Ecol.* **16**:488–99, 1935.
———: Glacial and postglacial vegetation. *Bot. Rev.* **1**:37–51, 1935a.
———: Forest sequences in the North Central States. *Bot. Gaz.* **103**:751–61, 1942.
———: Postglacial migration of five forest genera. *Am. Jour. Bot.* **29:**684–91, 1942a.
———: Xerothermic theory. *Bot. Rev.* **8**:708–36, 1942b.
———: Forest sequence and climatic change in northeastern North America since Early Wisconsin time. *Ecol.* **29**:326–33, 1948.
Sears, P. B., and Glenn C. Couch: Microfossils in an Arkansas peat and their significance. *Ohio Jour. Sci.* **32**:63–68, 1932.
Shanks, R. E.: The original vegetation of a part of the Lake Plain of Northwestern Ohio: Wood and Henry counties. *Abstr. Doctor's Dissertations,* No. **29**, Ohio State Univ. Press, 1939.

———: The vegetation of Trumbull County, Ohio. *Ohio Jour. Sci.* **42**:220–36, 1942.

Shantz, H. L., and Raphael Zon: Natural Vegetation. Atlas of American Agriculture, Pt. I, Section E. *U. S. Dept. Agr.*, 1924.

Sharp, A. J.: A preliminary report on some phytogeographical studies in Mexico and Guatemala. *Amer. Jour. Bot.* **33**:844, 1946.

Shimek, B.: The prairie of the Mississippi River bluffs. *Ia. Acad. Sci.* **31**:205–12, 1924.

Shreve, Forrest: Ecological plant geography of Maryland, Mountain Zone; pp. 275–92 in "Plant Life of Maryland." The Johns Hopkins Press, 1910.

———: Ecological plant geography of Maryland, Midland Zone, Lower Midland district; pp. 199–220 in "Plant Life of Maryland." 1910a.

———: Ecological plant geography of Maryland, Coastal Zone, Eastern Shore district; pp. 101–48 in "Plant Life of Maryland." 1910b.

Shreve, F., M. A. Chrysler, F. H. Blodgett, and F. W. Besley: "The Plant Life of Maryland." *Md. Weather Serv. Spec. Publ.* **3**:1–533. The Johns Hopkins Press, Baltimore, 1910.

Shunk, I. V.: Oxygen requirements for germination of seeds of *Nyssa aquatica*, tupelo gum. *Science* **90**:565–66, 1939.

Society of American Foresters, Committee of the Southern Appalachian Section: A forest classification for the Southern Appalachian Mountains and the adjacent plateau and coastal region. *Jour. For.* **24**:673–84, 1926.

Society of American Foresters: Forest cover types of the eastern United States. *Jour. For.* **30**:451–98, 1932.

———: Forest cover types of the eastern United States. 3d ed, revised. *Soc. Amer. Foresters,* Washington, D. C., 1940.

———: "Forest Terminology: a Glossary of Technical Terms used in Forestry." *Soc. Amer. Foresters,* Washington, 1944.

Spalding, V. M., and B. E. Fernow: The white pine. *U. S. Dept. Agr. Div. For. Bull.* **22**, 1899.

Spring, S. N.: The natural replacement of white pine on old fields in New England. *U. S. Dept. Agr. Bur. For. Bull.* **63**, 1905.

Standley, Paul C.: Trees and shrubs of Mexico. *Contr. U. S. Nat. Herb.*, **23**, 1920–26.

Stallard, Harvey: Secondary succession in the climax forest formation of northern Minnesota. *Ecol.* **10**:476–547, 1929.

Sterrett, W. D.: Report on forest conditions in Delaware. *Del. Coll. Agr. Exp. Sta. Bull.* **82**, 1908.

Steyermark, Julian A.: Studies of the vegetation of Missouri. I. Natural plant associations and succession in the Ozarks of Missouri. *Field Mus. Nat. Hist. Bot. Ser.* **9**:349–475, 1940.

Stone, Witmer: The life-areas of southern New Jersey. *Proc. Acad. Nat. Sci. Phila.* **59**:452–59, 1908.

———: The plants of southern New Jersey with especial reference to the flora of the Pine Barrens and the geographic distribution of the species. *Ann. Rep. N. J. State Mus., 1910,* pp. 23–828. Trenton, 1911.

Stout, A. B.: The bur oak openings in southern Wisconsin. *Trans. Wis. Acad.* **36**:141–61, 1944. (Issued Jan. 1946.)

Taylor, Norman: The origin and present distribution of the Pine Barrens of New Jersey. *Torreya* **12**:229–42, 1912.

Taylor, Norman: A white-cedar swamp at Merrick, Long Island, and its significance. *N. Y. Bot. Gard. Mem.* **6**:79–88, 1916.

———: The vegetation of Long Island. I. The vegetation of Montauk: A study of grassland and forest. *Brooklyn Bot. Gard. Mem.* **2**, 1923.

Telford, C. J.: Third report on a forest survey of Illinois. *Ill. Nat. Hist. Surv. Bull.* **16**:1–102, 1926.

Tharp, B. C.: Structure of Texas vegetation east of the 98th meridian. *Univ. Tex. Bull.* **2606**, 1926.

Thomson, John W.: Relic prairie areas in central Wisconsin. *Ecol. Monog.* **10**:685–717, 1940.

Thornthwaite, C. W.: Atlas of climatic types in the United States, 1900–1939. *U. S. Dept. Agr. Soils Cons. Serv. Misc. Publ.* **421**, 1941.

Thwaites, Reuben Gold: "Early Western Travels." Arthur H. Clark Co., Cleveland, 1904.

Tight, W. G.: Drainage modifications in southeastern Ohio and adjacent parts of West Virginia and Kentucky. *U. S. Geol. Surv. Prof. Paper* **13**, 1903.

Toumey, J. W., and R. Kienholz: Trenched plots under forest canopies. *Yale Univ. School of Forestry Bull.* **30**, 1931.

Toumey, J. W., and C. F. Korstian: "Foundations of Silviculture upon an Ecological Basis." 2d ed. John Wiley & Sons, New York, 1937.

———, and ———: "Foundations of Silviculture upon an Ecological Basis." 2d ed. rev. John Wiley & Sons, New York, 1947.

Transeau, E. N.: The bogs and bog flora of the Huron River valley. *Bot. Gaz.* **40**:351–75; 418–48, 1905; **41**:17–42, 1906.

———: The relation of plant societies to evaporation. *Bot. Gaz.* **45**:217–31, 1908.

———: Successional relations of the vegetation about Yarmouth, Nova Scotia. *Plant World* **12**:271–81, 1909.

———: The prairie peninsula. *Ecol.* **16**:423–37, 1935.

———: Prehistoric factors in the development of the vegetation of Ohio. *Ohio Jour. Sci.* **41**:207–11, 1941.

Trask, Parker D., Fred B. Phleger, Jr., and Henry C. Stetson: Recent changes in sedimentation in the Gulf of Mexico. *Science* **106**:460–61, 1947.

Tryon, H. H.: The Black Rock Forest. *Black Rock For. Bull.* **1**, 1930.

Turner, Lewis M.: Notes on forest types of northwestern Arkansas. *Amer. Midl. Nat.* **16**:417–21, 1935.

———: Trees of Arkansas. *Coll. Agr. Univ. Ark. and U. S. Dept. Agr. Ext. Circ.* **180**, 1937.

Veatch, J. O.: The dry prairies of Michigan. *Papers Mich. Acad. Sci., Arts, Letters* **8**:269–78, 1927.

———: Reconstruction of forest cover based on soil maps. *Mich. Agr. Exp. Sta. Quat. Bull.* **10**:116–26, 1928.

———: Soil maps as a basis for mapping original forest cover. *Papers Mich. Acad. Sci., Arts, Letters* **15**:267–73, 1932.

Vermeule, C. C.: The forests of New Jersey. Report on forests, Part I. *Geol. Surv. N. J., Ann. Rep. State Geol. for 1899,* pp. 13–101, 1900.

Ver Steeg, Karl: The Teays River. *Ohio Jour. Sci.* **46**:297–307, 1946.

Vestal, A. G.: A preliminary vegetation map of Illinois. *Trans. Ill. State Acad. Sci.* **23**:204–17, 1931.

———: Bibliography of the ecology of Illinois. *Trans. Ill. State Acad. Sci.* **27**:163–261, 1934.

Viosca, Percy: Louisiana out-of-doors: a handbook and guide. New Orleans, La., 1933.

Voss, John: Pleistocene forests of central Illinois. *Bot. Gaz.* **94**:808–14, 1933.

Waksman, Selman A.: The peats of New Jersey and their utilization. *Dept. Conserv. & Devel., State of N. J., Geol. Ser. Bull.* **55**, 1943.

Waterman, W. G.: Ecology of Crystal Lake bar region. *Mich. Acad. Sci. Ann. Rep.* **19**:197–208, 1917.

———: Forests and dunes from Point Betsie to Sleeping Bear, Benzie and Leelanau counties, Michigan. 20 pp. *Northwestern Univ.*, Evanston, Ill., 1922.

———: Development of plant communities of a sand ridge region in Michigan. *Bot. Gaz.* **74**:1–31, 1922a.

———: Ecological problems from the Sphagnum bogs of Illinois. *Ecol.* **7**:255–72, 1926.

Weaver, J. E., and F. E. Clements: "Plant Ecology." 2d ed. McGraw-Hill Book Co., New York, 1938.

Weaver, J. E., H. C. Hanson, and J. M. Aikman: Transect method of studying woodland vegetation along streams. *Bot. Gaz.* **80**:168–87, 1925.

Weaver, J. E., and Albert F. Thiel: Ecological studies in the tension zone between prairie and woodland. *Univ. Nebr. Bot. Surv. n.s.* **1**, 1917.

Wells, B. W.: Plant communities of the coastal plain of North Carolina and their successional relations. *Ecol.* **9**:230–42, 1928.

———: Origin of the Southern Appalachian grass balds. *Science* **83**:283, 1936.

———: Andrews Bald: the problem of its origin. *Jour. So. App. Bot. Club* (*Castanea*) **1**:59–62, 1936.

———: Southern Appalachian grass balds. *Jour. Elisha Mitchell Sci. Soc.* **53**:1–26, 1937.

———: A new forest climax: the salt spray climax of Smith Island, N. C. *Bull. Torr. Bot. Club* **66**:629–34, 1939.

———: Ecological problems of the southeastern United States Coastal Plain. *Bot. Rev.* **8**:533–61, 1942.

Wells, B. W., and I. V. Shunk: A southern upland grass–sedge bog. *N. C. Agr. Exp. Sta. Tech. Bull.* **32**, 1928.

———, and ———: The vegetation and habitat factors of the coarser sands of the North Carolina Coastal Plain: an ecological study. *Ecol. Monog.* **1**:465–520, 1931.

———, and ———: Salt spray: an important factor in coastal ecology. *Bull. Torr. Bot. Club* **65**:485–92, 1938.

Wentworth, C. K.: The fossil swamp deposit at the Walker Hotel site, Connecticut Avenue and DeSales Street, Washington, D. C. *Jour. Wash. Acad. Sci.* **14**:1–11, 1924.

Westveld, M.: Suggestions for the management of spruce stands in the Northeast. *U. S. Dept. Agr. Circ.* **134**, 1930.

Westveld, R. H.: The relation of certain soil characteristics to forest growth and composition in the northern hardwood forest of northern Michigan. *Mich. State Coll. Agr. Exp. Sta. Tech. Bull.* **135**, 1933.

West Virginia State Planning Board: Original condition of West Virginia's forests. Morgantown, W. Va., 1943.

Wheeler, C. F.: A sketch of the original distribution of white pine in the Lower Peninsula of Michigan. *Mich. Agr. Exp. Sta. Bull.* **162**, 1898.

Wherry, E. T.: Ecological studies of serpentine-barren plants—I. Ash composition: *Proc. Pa. Acad. Sci.* **6**:32–38, 1932.

Wherry, E. T.: The box huckleberry as an illustration of the need for field work. *Bull. Torr. Bot. Club* **61**:81–84, 1934.

———: Distribution of the North American pitcherplants. Reprinted from "Illustrations of North American Pitcherplants" by Mary V. Walcott; with descriptions and notes on distribution, by E. T. Wherry. Smithsonian Inst., 1935.

White, I. C.: Origin of the high terrace deposits of the Monongahela River. *Amer. Geol.* **18**:368–79, 1896.

Whitford, H. N.: The genetic development of the forests of northern Michigan. *Bot. Gaz.* **31**:289–325, 1901.

Wilde, S. A.: The relation of soils and forest vegetation of the Lake States region. *Ecol.* **14**:94–105, 1933.

———: "Forest Soils and Forest Growth." Chronica Botanica Company, Waltham, Mass., 1946.

Williams, Arthur B.: The composition and dynamics of a beech–maple climax community. *Ecol. Monog.* **6**:317–408, 1936.

Williams, Ruby M., and H. J. Oosting: The vegetation of Pilot Mountain, North Carolina: a community analysis. *Bull. Torr. Bot. Club* **71**:23–45, 1944.

Wilson, L. R., and R. M. Kosanke: The microfossils in a pre-Kansan peat deposit near Belle Plaine, Iowa. *Torreya* **40**:1–5, 1940.

Wilson, Ira T., and J. E. Potzger: Pollen study of sediments from Douglas Lake, Cheboygan County and Middle Fish Lake, Montmorency County, Michigan. *Proc. Ind. Acad. Sci.* **52**:87–92, 1943.

Wing, Leonard W.: Evidences of ancient oak openings in southern Michigan. *Ecol.* **18**:170–71, 1937.

Wolfe, John N.: Species isolation and a proglacial lake in southern Ohio. *Ohio Jour. Sci.* **42**:2–12, 1942.

Wolfe, John N., R. T. Wareham, and H. T. Scofield: The microclimates of a small valley in central Ohio. *Trans. Amer. Geophysical Union*, pp. 154–66, 1943.

———, ———, and ———: Microclimates and macroclimate of Neotoma, a small valley in central Ohio. *Ohio Biol. Surv. Bull.* **41**, 1949.

Wood, Alphonso: "Class-book of Botany." Barnes & Co., New York, 1869.

Woodard, John: Factors influencing the distribution of tree vegetation in Champaign County, Illinois. *Ecol.* **6**:150–56, 1925.

Woollett, Marjorie L., and Dorothy Sigler: Revegetation of beech–maple areas in the Douglas Lake region. *Torreya* **28**:21–28, 1928.

Wright, A. H., and A. A. Wright: The habitats and composition of the vegetation of Okefenokee Swamp, Georgia. *Ecol. Monog.* **2**:109–232, 1932.

Wright, Jonathan W.: Genotypic variation in white ash. *Jour. For.* **42**:489–95, 1944.

Zon, Raphael: Chestnut in southern Maryland. *U. S. Dept. Agr. Bur. Forestry Bull.* **53**, 1904.

Zon, Raphael, and H. F. Scholz: How fast do northern hardwoods grow? *Wis. Agr. Exp. Sta. Res. Bull.* **88**, 1929.

Index of Scientific and Common Names

The following is an alphabetical list of scientific names of plants mentioned in the text. Common names of trees and shrubs, if used, are also included. Synonyms are given if the names used in the text cannot be located readily in the commonly used manuals, or if some other name occurs frequently in the literature of the region. For species frequently mentioned, the indexing is selective. Species listed in the Tables are not indexed unless the name does not appear in the text. Numbers refer to location in the text.

Subject Index

Page numbers set in **boldface** refer to definitions or definitive explanations; page numbers set in *italics* refer to figures or charts. See "Index of Scientific and Common Names," pp. 557–583, for references to species.